D. J. Martin.

MAGNETIC PROPERTIES
OF RARE EARTH METALS

MAGNETIC PROPERTIES
OF RARE EARTH METALS

Edited by

R. J. Elliott

University of Oxford
Department of Theoretical Physics
Parks Road
Oxford

PLENUM PRESS ● LONDON AND NEW YORK ● 1972

Library of Congress Catalog Card Number: 77–161302

SBN 306 30565 8

PRINTED IN GREAT BRITAIN BY
ADLARD AND SON LTD, BARTHOLOMEW PRESS
DORKING, SURREY

Preface

The rare earths have a unique place among the elements. Although very much alike chemically and in most physical properties they each have very different and striking magnetic properties. The reason, of course, lies in their $4f$ electrons which determine the magnetic properties but have little effect on other chemical and physical behaviour. Although they are not rare, some indeed are among the more common heavy elements in the earth's crust, the difficulty of separation has meant that their intricate magnetic properties have only recently been unravelled. Now, however, the general pattern of their magnetism is well charted and the underlying theory is well understood. Both are thoroughly summarised in this book. It provides an excellent example of the kind of extensive synthesis which is possible with modern solid state physics.

But it represents only a high plateau in the ascent to complete understanding. It will become clear to the reader that while the overall position is satisfactory there are many details still to be elucidated experimentally and much to be done theoretically before all the underlying forces are identified and estimated from *a priori* calculations.

It is hoped that the book will provide a useful stimulus in this direction. It should also be of use to those who are interested in related disciplines, for example the rare earth compounds, or the transition metals. In addition rare earths promise to be important technologically as alloy constituents. This vast field is only touched on here, but a thorough understanding of the pure metals is basic to an understanding of alloys.

The pattern of the book is as follows. After a general introduction in Chapter 1 the phenomenological theory of the magnetic ordering is summarised in Chapter 2. The experimental results on magnetic ordering and bulk magnetic properties are given in Chapters 3 and 4. Then follows a chapter on spin waves which looks back to the phenomenological theory and forward to an explanation of the exchange interaction. This is taken up in Chapter 6 where the band theory of the conduction electrons is discussed and related to the exchange. In Chapter 7 the electrical properties are reviewed —these depend on the conduction electrons and their interaction with the magnetic electrons. Finally Chapter 8 deals with the nuclear hyperfine interaction, an aspect of magnetism which has proved surprisingly informative in the solid state.

I have been fortunate in my collaborators. All are acknowledged experts in the field, and all have entered the project with enthusiasm. Indeed the rare

earths seem to generate a special mystique of their own, which infects everyone
who touches the subject.

I hope it will not be invidious to mention two schools which have, among
hundreds of scientific participants, provided the essential base on which this
pyramid of successful science is built. Nothing would have been possible
without pure separated elements and it is Professor F. H. Spedding and his
collaborators at Iowa State University, America, who made the important
breakthrough in this area and who have shown what is possible in the way
of materials. Of our authors, Professor Legvold has been associated with that
programme over many years and Professor Mackintosh for a briefer period.
Secondly the extraordinary complexity of the magnetic ordering could hardly
have been unravelled without slow neutron scattering techniques (the
classical magnets Fe, Co and Ni give no clue to rare earth behaviour).
More recently these techniques have given much detailed information through
the spin waves. The group at Oak Ridge National Laboratory have made
many important contributions in this field, and we are fortunate to have Dr.
Koehler as their representative here. But all modern science is a co-operative
effort, and many other groups have made notable contributions. We hope
that our co-operation here will be helpful to our many colleagues. We thank
them all for their help and guidance.

R. J. ELLIOTT

Oxford 1971

Contributors

B. Bleaney — University of Oxford, Clarendon Laboratory, Oxford, England.

B. R. Cooper — General Electric Research and Development Center, Schenectady, New York 12301, U.S.A.

R. J. Elliott — Department of Theoretical Physics, University of Oxford, Oxford, England.

A. J. Freeman — Department of Physics, Northwestern University, Evanston, Illinois 60201, U.S.A.

W. C. Koehler — Solid State Division, Oak Ridge National Laboratory, Oak Ridge, Tennessee, U.S.A.

S. Legvold — Professor of Physics, Senior Physicist, Ames Laboratory of the A.E.C. Iowa State University, Ames, Iowa 50010, U.S.A.

A. R. Mackintosh — H. C. Ørsted Institute, University of Copenhagen, Copenhagen, Denmark.

H. Bjerrum Møller — Atomic Energy Commission Research Establishment, Risø, Roskilde, Denmark.

J. J. Rhyne — U.S. Naval Ordnance Laboratory, White Oak, Silver Spring, Maryland 20910, U.S.A.

Contents

Chapter 1

Introduction

1.1. RARE EARTH IONS

1.1.1. Electronic Configurations

The most striking fact about rare earth chemistry is the extraordinary similarity in the properties of all the elements. They all appear to be group IIIA elements of the periodic table like yttrium (Y). This arises of course from the fact that the $4f$ shell of electrons is being filled at this point in the periodic table, while the number of outer valence electrons remains unchanged. The $4f$ electrons are closely bound inside the outer closed shells and therefore play a small role in chemical bonding. On the other hand the f electrons with their high angular momentum are responsible for the magnetic properties of rare earth materials and these vary strikingly from atom to atom since the interelectronic interactions on the atom give many states which change in detail as the number of f electrons changes. In addition they give many low-lying energy levels which may be investigated spectroscopically.

The similarity of the chemical properties has made the rare earths hard to separate with high purity. Since the war, largely due to the work of F. H. Spedding and his collaborators[1] at Ames, Iowa using ion exchange methods, this difficulty has been overcome and reasonable quantities of pure material are now available. Single crystals of all the stable elements have been made and studied, and these have proved invaluable in unravelling the intricate properties of the metals. However these materials are still not available in sufficient purity to allow Fermi surface studies of the type now common in other metals.

Most rare earths occur in compounds as the trivalent form and this is common in the metals as well. In this form the number of f-electrons runs from zero at La to fourteen at Lu with a half filled shell at Gd. Because of the stability of full and half-filled shells Yb and Eu appear in the divalent form while Ce may be quadrivalent. Y has very similar properties and is often considered as a rare earth although it appears in the preceding (fifth) period and is more like Lu in its properties than La. Pm has no stable isotopes but some of its properties have been determined.

The f-electrons behave, to a first approximation, like those in a free ion. The ground state is given by Hund's Rule, i.e. with maximum total spin S for the configuration and maximum L for that S. Spin–orbit coupling is relatively strong so that only the lowest J multiplet is normally populated

Table 1.1. Basic Properties of Free Ions. λ is Lande's Factor and Δ_0 the Spin Orbit Splitting to the Next J Level.

Config. $4f^n$	3^+ ion		L	S	J	λ	$(\lambda-1)^2 J(J+1)$	Δ_0 (cm^{-1})
0	La	Ce^{4+}	0	0	0	0	0	
1	Ce		3	$\frac{1}{2}$	$\frac{5}{2}$	$\frac{6}{7}$	0.18	2,200
2	Pr		5	1	4	$\frac{4}{5}$	0.80	2,150
3	Nd		6	$\frac{3}{2}$	$\frac{9}{2}$	$\frac{8}{11}$	1.84	1,900
4	Pm		6	2	4	$\frac{3}{5}$	3.20	1,600
5	Sm		5	$\frac{5}{2}$	$\frac{5}{2}$	$\frac{2}{7}$	4.46	1,000
6	Eu		3	3	0	—	0	350
7	Gd	Eu^{2+}	0	$\frac{7}{2}$	$\frac{7}{2}$	2	15.75	—
8	Tb		3	3	6	$\frac{3}{2}$	10.50	2,000
9	Dy		5	$\frac{5}{2}$	$\frac{15}{2}$	$\frac{4}{3}$	7.08	3,300
10	Ho		6	2	8	$\frac{5}{4}$	4.50	5,200
11	Er		6	$\frac{3}{2}$	$\frac{15}{2}$	$\frac{6}{5}$	2.55	6,500
12	Tm		5	1	6	$\frac{7}{6}$	1.17	8,300*
13	Yb		3	$\frac{1}{2}$	$\frac{7}{2}$	$\frac{8}{7}$	0.32	10,300
14	Lu	Yb^{2+}	0	0	0	0	0	

* $^3H_4$ is lower at 5,900 cm^{-1}.

at room temperature. This has the value $L \pm S$ according as the shell is more or less than half full. Values for L, S and J in the lowest multiplet are given in Table (1.1). For Russel-Saunders coupling which is appropriate here, the energy separation to the next excited multiplet is found to be $\langle \zeta \rangle (J+1)/2S$ if the shell is less than half full and $\langle \zeta \rangle J/2S$ if it is more. Here $\langle \zeta \rangle$ is the average of the single electron spin orbit–coupling

$$\sum_i \frac{\hbar^2}{2m^2c^2} \frac{1}{r_i} \frac{\partial V(r_i)}{\partial r_i} \mathbf{l}_i \cdot \mathbf{s}_i = \sum_i \zeta(r_i) \mathbf{l}_i \cdot \mathbf{s}_i$$

over the $4f$ wave functions. It increases across the series as the extra nuclear charge pulls the electrons into tighter orbits. Thus the splitting to the first multiplet has a small value at f^5 and f^6 where J is low. Actual values of this splitting Δ_0 obtained in the main from spectroscopic evidence[2] in rare earth salts as shown in Figure 1–1 are given in the Table (1.1).

1.1.2. Magnetic Properties of Free Ions

The rare earth atoms in a crystal interact with their environment through electrostatic forces. These take the form of a crystal field from the classical coulomb terms, and exchange terms. Both turn out, as will be seen in detail in Chapter 2, to give energies of interaction which are small compared to the spin–orbit coupling energy, and smaller than or comparable to kT at room temperature. Thus to a rough approximation we expect any rare earth crystal to have magnetic properties similar to those of an assembly of free ions, particularly above room temperature.

The magnetic properties of free ions are well known. The magnetic moment

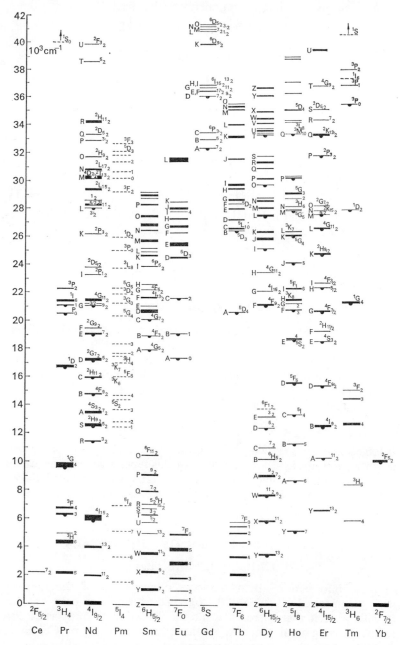

Figure 1–1. Position of *J*-multiplets in triply ionized rare earths on anhydrous LaCl₃ (after Dieke).[2]

of the electrons gives an interaction energy with a field H

$$\sum_i \beta(\mathbf{l}_i + 2\mathbf{s}_i).\mathbf{H} = \beta(\mathbf{L} + 2\mathbf{S}).\mathbf{H}. \qquad (1.1)$$

In the lowest J multiplet this is proportional to

$$\lambda\beta\mathbf{H}.\mathbf{J} \qquad (1.2)$$

where Lande's factor

$$\lambda = \tfrac{3}{2} + [S(S+1) - L(L+1)]/2J(J+1). \qquad (1.3)$$

Values of λ for this level are given in Table 1.1. We may note that in the lowest multiplet

$$\mathbf{L} \sim (2-\lambda)\mathbf{J} \qquad (1.4)$$

is always parallel to $\mathbf{J}$ but

$$\mathbf{S} \sim (\lambda-1)\mathbf{J} \qquad (1.5)$$

is antiparallel in the first half and parallel in the second half of the series.
  The saturation moment of the system will be

$$\mu = \lambda\beta J \qquad (1.6)$$

per atom although this may be reduced by the environmental forces. The susceptibility is given by Curies Law

$$\chi = N\lambda^2\beta^2 J(J+1)/3kT. \qquad (1.7)$$

In fact the susceptibility of the metals agrees fairly well with this at moderate T.
  In those configurations where the next excited multiplet is at low energy, there may be appreciable admixtures of this state by the crystal field and by the exchange. Even the free ion susceptibility is modified, as was first pointed out by Van Vleck.[3] At low T there is an extra constant term in the susceptibility. In the first half of the series it has the value

$$\chi' = 2\beta^2 N(L+1)S/3(J+1)\,\Delta_0 \qquad (1.8)$$

and is largest at f^5 and f^6. The latter configuration does not occur in the metals since Eu is divalent but the effects of this type, involving the exchange in particular, may prove important in Sm metal when a detailed theory is worked out. In the second half of the series where Δ_0 is large these corrections have proved to be negligible.

1.2. BASIC PROPERTIES OF THE METALS

1.2.1. Crystal Structures

  The basic information about the crystal structures of the rare earth metals is summarized in Table 1.2, together with similar data on Y and Sc. It will be seen that there is a simple gradation of properties across the series. Leaving until the next section any discussion of Ce, Eu and Yb; we see that all have, at room temperature, a close packed structure. Sc, Y and the heavy rare earths Gd–Lu have the usual hexagonal close packed structure with a c/a ratio somewhat smaller than the perfect value of

$$2\sqrt{2}/\sqrt{3} = 1.633.$$

Table 1.2. Basic Properties of Rare Earth Metals

	Density gm/cm³	Atom vol. cm³/mole	Atom rad. Å	Structure $T=25°C$	a	c	c/a†	b.c.c. transition $T_B°C$	a	Melting $T_M°C$
Sc	2.989	15.04	1.641	h.c.p.	3.309	5.268	1.592	1335		1539
Y	4.457	19.95	1.803	h.c.p.	3.650	5.741	1.573	1479	4.08	1523
La	6.166	22.53	1.877	hex	3.772	12.144	1.610	861	4.26	920
Ce	6.771	20.69	1.824	hex*	3.673	11.802	1.607	762	4.12	798
Pr	6.772	20.81	1.828	hex	3.672	11.833	1.611	795	4.13	931
Nd	7.003	20.60	1.822	hex	3.659	11.799	1.612	855	4.13	1018
Pm	7.26	—	—	hex ‡	3.65	11.65	1.60			
Sm	7.537	19.95	1.802	rhom (hex)	3.626	26.18	1.605	924	4.07	1072
Eu	5.253	28.93	1.983	b.c.c.	4.580			stable		822
Gd	7.898	19.91	1.801	h.c.p.	3.634	5.781	1.591	1260	4.05	1311
Tb	8.234	19.30	1.783	h.c.p.	3.604	5.698	1.581	1287	4.02	1360
Dy	8.540	19.03	1.775	h.c.p.	3.593	5.655	1.574	1384	3.98	1409
Ho	8.781	18.78	1.767	h.c.p.	3.578	5.626	1.572	1428	3.96	1470
Er	9.045	18.49	1.758	h.c.p.	3.560	5.595	1.572	?		1522
Tm	9.314	18.14	1.747	h.c.p.	3.537	5.558	1.571	?		1545
Yb	6.972	24.82	1.939	f.c.c.	5.483			792	4.44	824
Lu	9.835	17.79	1.735	h.c.p.	3.505	5.553	1.584	?		1656

Some of the data on the high temperature phases has been inferred.

* The stable phase of Ce at 25°C is f.c.c. with $a=5.16$ Å. At low T it is f.c.c. with $a=4.85$ Å. La also has a f.c.c. phase above 310°C with $a=5.30$ Å.

† In case of double hexagonal structures $c/2a$ is given.

‡ Results for Pm from P. J. Pallmer and T. D. Chikalle, *J. Less Comm. Met.* **24**, 233 (1971).

The atomic volume decreases by about 5% in going from Gd to Lu—the so called lanthanide contraction. We should remark here that this represents a small shrinkage of the outer closed electrons shells—the $4f$ electron orbits contract much more since they lie well inside the atom and feel the effect of the extra nuclear charge.

The light rare earths have a more complex double hexagonal structure in which the stacking sequence of the layers is ABACABAC rather than ABAB. The face centred cubic structure may also be written in this form by stacking (111) layers in the form ABCABC. Thus in the double structure the layers have alternately hexagonal and cubic environments. However since the c/a ratio is not exactly equal to the ideal value, being 1.611 in Pr and 1.612 in Nd, the near environment of the "cubic" sites only has trigonal symmetry.

The light elements La–Nd can also be made in the f.c.c. form with the same atomic volume. La is stable in this structure from 310°C–861°C but Pr and Nd are only metastable and the structure must be made by quenching. Sm has an even more complex structure which can be regarded as hexagonal with the stacking sequence ABABCBCAC of nine layers although this structure may be described with a smaller unit cell which is a rhombohedron. All those elements appear to become body centred cubic at high T although the results quoted for the heavy elements are deduced from studies on magnesium alloys.

Elementary excitations such as spin waves in Chapter 5 or band electrons in 6 are normally described in terms of their wave vector which lies in the Brillouin Zone. For the h.c.p. structure this is a hexagonal prism of side $4\pi/a\sqrt{3}$ and height $2\pi/c$ (Figure. 5–2). Since the structure contains two atoms per cell the number of **k** values in the zone is half the number of atoms. Thus for phonons and spin waves there will be two branches in the spectrum for each atomic degree of freedom. In the double hexagonal structure where there are four atoms per cell the Brillouin Zone is again a hexagonal prism but with half the height (Figure 6–16).

1.2.2. The Metals Ce, Eu and Yb

Eu and Yb metals are seen in Table 1.2 to have an atomic volume some 30% larger than that of the other rare earths. Magnetic measurements show Yb to have a very small moment characteristic of a full $4f$ shell while Eu shows a moment per atom similar to Gd and characteristic of a half filled shell. There seems no doubt that they are divalent, and behave rather as if they belonged to the Ca–Sr–Ba series.

In terms of band theory the question of valency is determined by the position of the $4f$ energy levels relative to the Fermi energy. Of course the $4f$ energies are not given in detail by band theory because of the large correlation arising from the interelectronic interaction on the atom. But the energy required to remove a $4f$ electron and place it in the conduction band at the Fermi level determines the position of the last occupied $4f$ level, and similarly the energy needed to trap a conduction electron in a $4f$ state gives the position of the next unoccupied $4f$ level. In most metals these are well below and well above the Fermi energy. But in Yb there is some evidence that the last full level is not far below,[5] the spin orbit splitting gives two levels about 1.3 and 2.6 eV below E_F. The position of the $4f$ levels as revealed by band structure calculations is discussed in detail in Chapter 6.

The resistivity of Yb increases with pressure and it is believed that an energy gap widens near E_F which makes it progressively a semi-metal and a semiconductor.[6] This behaviour also occurs in Ca and is not connected with the f-levels.[7]

Because of the different valence and structure the magnetic properties of Eu are not expected to be related in any simple way to those of the other rare earths. As an S-state ion it has small anisotropy. The different Fermi surface imposes quite different magnetic ordering though it still has spiral structures.

In cerium the energy difference between the $4f$ and $5d$ atomic level is very small. In the ground state of rare earth atoms the $5d$ shell begins to fill at the beginning of the series and is overtaken by the $4f$ in the vicinity of La and Ce. If the fourth electron in Ce outside the closed shells chooses the $5d$ state it will appear as an extra conduction electron. In fact the low T stable phase of Ce is quite different to the other rare earths,[8] being f.c.c. with a lattice constant of 4.85 Å. Below room temperature however it makes a transition to another f.c.c. phase with a lattice constant of 5.16 Å some 6% larger. At slightly higher temperatures the double hexagonal phase will also form. These close packed phases have a density comparable

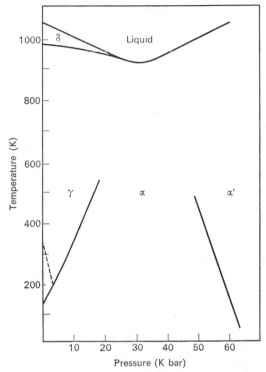

Figure 1–2. Phase diagram of Ce (after King *et al.*[8]).

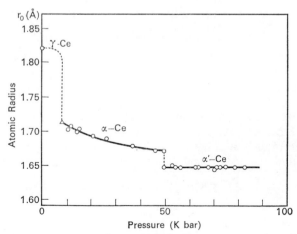

Figure 1–3. Atomic radius of Ce in Å at room temperature as a function of pressure (after Francheschi and Olcese[10]).

with La and Pr and represent the trivalent state with one f-electron. The dense low temperature phase is quadrivalent and has no f-electrons as is confirmed by magnetic experiments. The other phases have a magnetic susceptibility like that of Ce^{3+} ions and they can be obtained at low temperatures where they show ordered magnetism. The transition between the dense and light forms is rather sluggish with considerable hysteresis so that it occurs around 115 K on cooling at 180 K on heating. It can also be induced by a pressure of about 10 K bar at room temperature. A schematic phase diagram is shown in Figure 1–2 while the variation of the atomic radius with pressure is shown in Figure 1–3.

Band theory calculations and other considerations[9] suggest a fairly broad f-band in Ce with considerable hybridization with the conduction bands. In these circumstances the "valence" has a less precise meaning and the number of electrons per atom with f-character may be non-integral. With this definition the valency of Ce varies from 3.11 to 4 with increasing pressure.[10] It has also been suggested that La electrons have a small amount of f character, and that Yb has a small number 0.8% of f holes.[10]

1.3. CRYSTAL FIELDS

1.3.1. General Theory

The array of changes around any ion produces an electric field at the $4f$-electron, which has an appropriate symmetry. The potential energy then takes the form

$$V(\mathbf{r}_i) = \int \frac{e\rho(\mathbf{R})}{|\mathbf{r}_i - \mathbf{R}|} \, d\mathbf{R} \tag{1.9}$$

where $\rho(\mathbf{R})$ is the charge density. If $\rho(\mathbf{R})$ lies entirely outside the ion (1.9) is a solution of Laplace's equation and may be expanded in spherical harmonics in terms of the electron position $\mathbf{r}$ given in spherical polars (r, θ, ϕ)

$$V(\mathbf{r}) = \sum_{l, m} B_l^m r^l Y_l^m(\theta, \phi) \tag{1.10}$$

where

$$B_l^m = (-1)^m \frac{4\pi}{(2l+1)} \int \frac{\rho(\mathbf{R})}{R^{l+1}} Y_l^{-m}(\Theta, \Phi) \, dR \tag{1.11}$$

and $\mathbf{R} \rightarrow (R, \Theta, \Phi)$. Thus the B_l^m represents the electric field components of appropriate symmetry and the $r^l Y_l^m(\theta, \phi)$ various multipoles of the electron distribution. When the average of V is taken over the electronic states it is clear that all matrix elements for $l > 6$ will be zero for f electrons, because of the properties of spherical harmonics. f electrons with angular momentum 3 cannot have multipole distributions with $l > 6$.

The literature on this subject is somewhat confused by vagaries of notation. The definitions (1.10) and (1.11) use spherical harmonics normalized so that

$$\int |Y_l^m(\theta, \phi)|^2 \sin \theta \, d\theta \, d\phi = 1. \tag{1.12}$$

It is also convenient to use Tesseral harmonics

$$\begin{aligned} Z_l^{|m|} &= 2^{-\frac{1}{2}}[Y_l^m + (-1)^m \, Y_l^{-m}] \\ Z_l^{|m|'} &= 2^{-\frac{1}{2}}[Y_l^m - (-1)^m \, Y_l^{-m}] \end{aligned} \tag{1.13}$$

which are completely real. The potential energy can then be written

$$V(\mathbf{r}) = \sum_{l, |m|} r^l [B_l^{|m|} Z_l^{|m|} + B_l^{|m|'} Z_l^{|m|'}] \tag{1.14}$$

and

$$B_l^{|m|} = \frac{4\pi}{(2l+1)} \int \frac{\rho(\mathbf{R})}{R^{l+1}} Z_l^{|m|} \, dR. \tag{1.15}$$

Finally it has become usual to extract from $r^l Z_l^{|m|}$ the polynomial in x, y and z which has the lowest common multiple removed so that the coefficients are all integers. These are closely related to the Legendre polynomials when $m=0$. For example

$$\left.\begin{array}{l} r^2 P_2^0 = 3z^2 - r^2 \\[4pt] r^4 P_4^0 = 35z^4 - 30r^2 z^2 + 3r^4 \\[4pt] r^6 P_6^0 = 231z^6 - 315r^2 z^2 + 105r^4 z^2 - 5r^6 \\[4pt] r^6 P_6^{|6|} = x^6 - 15x^4 y^2 + 15x^2 y^4 - y^6 \end{array}\right\} \tag{1.16}$$

and so on. A full discussion of the relevant normalization factors connecting the Z's and the P's is to be found in review articles of Hutchings and Newman[11] or the book by Abragam and Bleaney.[12]

The effect of the crystal field terms on the energy levels of the system is normally adequately calculated in first order perturbation theory since the crystal field energy is small compared to the spin–orbit splitting Δ_0. Thus it is necessary to calculate the matrix elements of

$$\sum_i V(r_i)$$

inside the states of a given J manifold. Stevens[13] first pointed out that because of the Wigner–Eckhart theorem these matrix elements were proportional to those of an operator equivalent written in terms of the operators J (and also to Wigner coefficients). These operators $O_l^{|m|}$ are closely related to the $P_l^{|m|}$ of (1.16) and are obtained by substituting J_x, J_y, J_z and x, y, z on the right hand side and taking symmetric products (e.g. $2xy \rightarrow J_x J_y + J_y J_x$) to account for the non-commutation of the J operators. The constants of proportionality depend on l (but not on m) and also on the electronic states, i.e. on L, S, and J. In addition the operators r_i^l are averaged over the $4f$ wave functions. Hence we replace

$$\sum_i r_i^l P_l^m(i) \rightarrow \alpha_l \langle r^l \rangle O_l^{|m|}(\mathbf{J}). \tag{1.17}$$

Following Elliott and Stevens[14] we use α, β, γ for the constants of proportionality when $l = 2$, 4, 6 respectively. Values of these constants are in Table 1.3. Absorbing the other constants into the $B_l^{|m|}$ we define an effective operator equivalent for the crystal field

$$\begin{aligned} V(\mathbf{r}) &= \sum_{l, |m|} \alpha_l [A_l^{|m|} O_l^{|m|}(\mathbf{J}) + A_l^{|m|'} O_l^{|m|'}(\mathbf{J})] \langle r^l \rangle \\ &= \sum_{l, |m|} V_l^{|m|} O_l^{|m|}(\mathbf{J}) + V_l^{|m|'} O_l^{|m|'}(\mathbf{J}). \end{aligned} \tag{1.18}$$

Table 1.3. Properties of Electronic Multipoles. Steven's Factors α_l and Maximum Values of Multipole Moments $\alpha_l O_l(J)$.

3^+ ion	$\alpha \times 10^2$	$\beta \times 10^4$	$\gamma \times 10^6$	$\frac{1}{2}\alpha\, O_3{}^0(J)$	$\frac{1}{8}\beta\, O_4{}^0(J)$	$\frac{1}{16}\gamma\, O_6{}^0(J)$
Ce	-5.71	63.5	0	-0.286	0.0476	0
Pr	-2.10	-7.35	61.0	-0.294	-0.0771	0.0192
Nd	-0.643	-2.91	-38.0	-0.116	-0.0550	-0.0359
Pm	0.771	4.08	60.8	0.108	0.0428	0.0191
Sm	4.13	25.0	0	0.206	0.0188	0
Tb	-1.01	1.22	-1.12	-0.333	0.0909	-0.0116
Dy	-0.635	-0.592	1.03	-0.333	-0.1212	0.058
Ho	-0.222	-0.333	-1.30	-0.133	-0.0909	-0.116
Er	0.254	0.444	2.07	0.133	0.0909	0.116
Tm	1.01	1.63	-5.60	0.333	0.1212	-0.058
Yb	3.17	-17.3	148.0	0.333	0.0909	-0.0116

The algebraic form of $O_l{}^{|m|}$ and the numerical values of operators in various J manifolds are given in references (11) and (12). Values of $\langle r^l \rangle$ computed by Freeman and Watson are given in Table 6.2.

The use of operator equivalents is a computational convenience but it also has important physical significance. As was pointed out above the electronic operators represent the multipole moments of the $4f$ electrons charge clouds. By relating these to J operators it is easy to see how they affect the magnetism, since the magnetic moment is also proportional to J.

1.3.2. Crystal Field Estimates for the Rare Earth Metals

The crystal field energies depend therefore on the α_l and $\langle r^l \rangle$ which are properties of the free ions and vary from ion to ion. In addition they depend on the $B_l{}^m$ and $A_l{}^m$ which are properties of crystal lattice. They are restricted by the symmetry of $\rho(\mathbf{R})$. For sites with hexagonal symmetry only $A_2{}^0$, $A_4{}^0$, $A_6{}^0$ and $A_6{}^{|6|}$ are non zero. For the "cubic" sites in the double hexagonal and Sm structures which have strictly only trigonal symmetry $A_4{}^{|3|}$ and $A_6{}^{|3|}$ are also non zero. If the symmetry is really cubic $A_2{}^0 = 0$ and the $A_l{}^{|m|}$ with the same l have a simple relationship.

$$V^c = A_4{}^c \langle r^4 \rangle\, \beta[O_4{}^0 + 20\sqrt{2}\, O_4{}^3] + A_6{}^c \langle r^6 \rangle\, \gamma\left[O_6{}^0 - \frac{35}{2\sqrt{2}}\, O_6{}^3 + \frac{77}{8}\, O_6{}^6\right].$$
$$(1.19)$$

This expression is also given related to the four-fold axes in (2.122). The energy levels in cubic fields for all ratios of $A_4{}^c/A_6{}^c$ have been worked out by Lea, Leask and Wolf.[15]

The actual calculation of $A_l{}^m$ depends on the model of $\rho(\mathbf{R})$. In a metal with its conduction electrons this is extremely difficult, as is discussed in detail in §6.6.2. So far most calculations have been made for very simple models using point charges at the nearest neighbour sites. For the hexagonal metals preliminary results were obtained by Elliott,[16] and in more detail by Kasuya[17] who assumed only small deviations from a perfect close-packing ratio c/a. More recently Cooper[18] has given more complete expressions for this model. Bleaney[19] has made similar calculations for the f.c.c. case. For nearest neighbours A_2^0 is zero for perfect close packing, as it is in general for the cubic case. However, at the real c/a ratio it is quite large for both the hexagonal and "cubic" sites.

Even if reliable calculations of A_l^m could be performed there are additional effects which modify the results. The closed shells of $5s$ and $5p$ electrons can distort to screen the crystal field. They find it easier to make induced quadrupoles than to support higher multipoles so that this effect is largest for the $l=2$ terms. It may give a result comparable to the initial estimate.[20] Near the nucleus this quadrupolar shielding is very much larger (see §8.4). In addition hybridization and bonding[21] effects can give appreciable modifications which tend to be largest for $l=6$ terms which depend most strongly on the exterior of the atom. These effects can be approximately lumped into modifications of $\langle r^l \rangle$ to some effective value. As a result of all these uncertainties there are as yet no reliable *a priori* estimates of the V_l^m. Very surprisingly the simplest model of assuming point charges on nearest neighbour sites gives a remarkably good agreement with experimental results as far as sign and order of magnitude is concerned. Why this should be so remains a mystery and much work remains to be done to clarify the position (cf. §6.6).

It is usual to regard the V_l^m as parameters, satisfying symmetry requirements, which can be determined by experiment. In particular they are responsible for the magnetic anisotropy as is discussed in §4.3 where the experimental results are compared in detail with Kasuya's[17] point charge model. They have a marked effect on the spin wave results as shown in Chapter 5. Other effects of the crystal field are reviewed in §2.3.2.

The variation of magnetic anisotropy across the series is largely determined by the variation of $\alpha_l O_l^m(J)$. The factors $A_l^m \langle r^l \rangle$ vary only slowly from ion to ion. We therefore give in Table 1.3 values of $\alpha_l O_l^0(J_z = J)$. In particular it is the variation in sign of α_l which causes the main changes between ions. In the heavy ion rare earths for example, the axial anisotropy is dominated by $A_2^0 < 0$. Thus Tb, Dy and Ho where $\alpha < 0$ tend to align perpendicular to the hexagonal axis while Er and Tm where $\alpha > 0$, align parallel. The occurrence of cone structures in Ho and Er depends on the higher order harmonics The hexagonal anisotropy arises partly from $A_6^6 < 0$ and alternates in sign as γ does.

Although a specific set of A_l^m therefore give qualitative agreement with the properties, explicit evaluation of A_l^m from various data give variations between ions and even on the same ion from different experimental techniques. This presumably arises from shortcomings in the theory which relates the crystal field hamiltonian to the observed properties. For example, the V_2 defined by (5.10) and given in Table 5.2 are related to K_2 as defined in (4.3.1) by $K_2 = 2/3\, V_2 J J_1$ where K_2 in ergs/c.c. ~ 2 meV/atom. The agreement between the two values in Dy is excellent, but K_2 gives a value larger by some 30% in Tb. Other anisotropy constants are in less satisfactory agreement, for example the sign of K_4 in Tb given in §4.3.1 is different from that expected, although those of Dy and Ho agree.

1.3.3. Crystal Fields in Gd

An ion in a pure S-state has a completely spherical charge cloud and all its multipole moments are zero. Hence to this order there is no crystal field effect and all the $\alpha_l=0$. A breakdown of Russel-Saunders coupling by the spin–orbit coupling will modify this result, since then the state is simply one of $J=7/2$. Because it requires an extra order of spin–orbit coupling to increase the L value admixed by one, perturbation theory suggests that the effective $\alpha_2 > \alpha_4 > \alpha_6$. This is borne out by magnetic resonance measurements in the salts.[22] The effect of covalent bonding is also found to be important in determining the magnitude of the effects.[23] However whatever the origin, the salts can be explained phenomenologically in terms of a spin Hamiltonian like (1.18)

$$\Sigma\, V_l{}^m O_l{}^m(\mathbf{S}) \tag{1.20}$$

where now the V are much smaller than in the other ions. Typically $V_2{}^0 \sim 10^{-3}$ cm^{-1} and $V_6{}^0 \sim 10^{-7}$ cm^{-1}. We therefore expect a similar $\mathscr{H}$ to determine the small observed anisotropy in Gd.

The experimental results (§4.32 and §5.3.4) can be readily interpreted in terms of such an $\mathscr{H}$. The most striking feature in comparison to the salts is the relatively large values of $V_4{}^0$ and $V_6{}^0$ which are obtained. If this is confirmed—and the interpretation is not unambiguous because of the theory of anisotropy which must be employed—it means that the conduction electrons and bonding effects must play an important role in determining these quantities.

Eu also has an S state configuration and it will have small crystal field terms appropriate to cubic symmetry.

1.4. INTERIONIC INTERACTIONS

The crystal field is a single ion effect and as such cannot give rise to co-operative transitions. These—and in particular the magnetic ordering—must be caused by interaction between the ions. It is believed that interactions of an exchange type are normally responsible. In the simplest case this couples individual electronic spins on pairs of ions as

$$\mathscr{H}_{ex}=j\mathbf{s}_l.\mathbf{s}_{l'}. \tag{1.21}$$

If there are several electrons per atom this must be summed and projected on to the total spin. If j does not depend on the electronic orbital this becomes

$$\mathscr{H}_{ex}=j\mathbf{S}_l.\mathbf{S}_{l'}. \tag{1.22}$$

In the case where exchange is smaller than the spin orbit–splitting this can be further projected on to J using (1.5) so that

$$\mathscr{H}_{ex}=\mathscr{J}\mathbf{J}_l.\mathbf{J}_{l'} \tag{1.23}$$

where $\mathscr{J}=(\lambda-1)^2 j$. The exchange constant j depends on overlap and crystal structure and so is not expected to vary much along the series. Hence the exchange energy on this simple model is expected to be proportional to the de Gennes factor[24]

$$G = (\lambda - 1)^2 J(J+1) \qquad (1.24)$$

which is given in Table 1.1. In fact the transition temperatures are found to vary smoothly and roughly linearly with this factor across the heavy rare earths and a wide range of alloys (cf. Figure 4.18).

In addition the wave-like form of the magnetic ordering (cf. Ch. 3) indicates a $\mathscr{J}(\mathbf{R})$ which oscillates with interionic distance $\mathbf{R}$. Such a variation is expected if the exchange takes place through the conduction electrons via the Rudermann–Kittel interaction[25] when the oscillations reflect the properties of the Fermi surface. The properties of such systems are discussed in §2.2.1. The relationship of exchange to the band structures is discussed in detail in §6.3.5. The strong interaction between the local moments and the conduction electrons is confirmed by the striking way in which the electronic transport properties are affected by magnetic order, as is discussed in Ch. 7. Although the qualitative form of $\mathscr{J}(R)$ can be explained detailed calculations from *a priori* band structures are not yet able to predict the detailed forms which are measured from spin–wave spectra (cf. Ch. 5).

Although good qualitative agreement with experiment is obtained on the basis of this isotropic exchange, it is possible that other types of interaction may be important but have remained so far undetected.[26] If the exchange were due to direct overlap it would depend on the orientation of the charge clouds—this in turn can be related to the operator equivalents of the multipole moments to give a complex interaction in the form[27]

$$\sum_{\substack{n, m \\ n', m'}} \mathscr{J}_{nn'}{}^{mm'}(\mathbf{R})\, O_n{}^m(\mathbf{J}_l)\, O_{n'}{}^{m'}(\mathbf{J}_{l'})\, \mathbf{J}_l \cdot \mathbf{J}_{l'}. \qquad (1.25)$$

Some aspects of spin–wave experiments can rule out the existence of large terms whose symmetry is related to the direction of R (§5.3.2). So these overlap effects must be small. However terms with an anisotropy related to the crystal axes (and not to $\mathbf{R}$) may be present and would be difficult to distinguish from crystal field anisotropy. The simplest such term would take the form

$$\mathscr{K}(\mathbf{R})\, O_2{}^0(\mathbf{J}_l)\, O_2{}^0(\mathbf{J}_{l'})\, \mathbf{J}_l \cdot \mathbf{J}_{l'}. \qquad (1.26)$$

Such terms could arise via the Rudermann–Kittel interaction through the spin–orbit coupling in the d-bands.[28]

In addition electrostatic interactions between the electronic charge clouds on adjacent rare earth ions can give some contribution to the anisotropy. Quadrupole–quadrupole interaction is the most likely possibility.[29] It would take the form

$$C^{mm'}(\mathbf{R})\, O_2{}^m(\mathbf{J}_l)\, O_2{}^m(\mathbf{J}_{l'}) \qquad (1.27)$$

when transformed into operator equivalents.

To date there is no very direct evidence of the existence of anisotropic exchange or of multipole interactions in the heavy rare earth metals. Most of the experimental results can be adequately accounted for on the basis of isotropic exchange and the crystal field. However, there are relatively few experiments which can distinguish these effects in an unequivocal fashion and it may be that they are not negligible as is usually assumed. In particular they may be especially important in the light rare earths where the exchange is weaker and the electronic multipoles are larger since $\langle r^n \rangle$ is larger.

1.5 MAGNETOELASTIC COUPLING

Both the single ion crystal field and the interionic interactions have so far been discussed in terms of a static crystal lattice in its equilibrium configuration. Both are modified when the atomic positions are changed. Such changes can be dynamic, arising from the phonons, or due to static strains. The latter have the most spectacular effects in the metals, where they give rise to large magnetostrictive effects as discussed in detail in §4.4. They are also believed to be mainly responsible for the transitions between spiral structures and ferromagnetism which occurs in the heavy metals at low $T^{(30)}$ (cf. §2.2.2). In the presence of static strain new crystal field terms appear with the symmetry appropriate to the distorted lattice. These give rise to further magnetic anisotropy. For a strain of type $\epsilon(i)$ extra terms

$$\sum_{l, m} E_l^m(i) \; \epsilon(i) \sum_R O_l^m(\mathbf{J}) + \sum_{l, l'} D_{ll'}^{mm'}(i) \; \epsilon(i) \sum_{R, R'} O_l^m(\mathbf{J}) \, O_{l'}^{m'}(\mathbf{J}).$$

$$(1.28)$$

appear in the hamiltonian. Of these the former, the crystal field modification, is thought to be the more important. The latter term depends on $\mathbf{R}_l - \mathbf{R}_{l'}$ and is due to modification of the exchange. In addition the first term in second order, and similar terms which couple to the dynamics strains, can give an effective multipole–multipole interaction. Such effects occur in some concentrated rare earth salts[31] and give coupling of order of several meV, but there is no direct evidence for them in the metals.

1.6. DISCUSSION

The main properties of the rare earth metals can therefore be qualitatively understood on the basis of a phenomenological theory including crystal field, exchange, and magnetoelastic coupling. The basic electric and exchange fields due to the environment of an atom appear to vary only slightly across the series. The main difference in the magnetic properties arises from the way in which the magnetic moment and the electronic multipoles of the ions vary with the number of f-electrons, and can be understood when this variation is calculated.

As subsequent chapters will show, a great wealth of experimental data can be interpreted in this way. However the detailed form of the interactions is not always reflected in the data so that the interpretation is not completely unique. Moreover considerable theoretical calculation is necessary to go from the basic hamiltonian to a description of say spin waves at high T, or the T dependence of anisotropy energy. This also prevents a precise evaluation of some of the phenomenological constants and may account for some of the variation of parameters obtained by different techniques. In spite of these difficulties, however, it appears that the basic phenomenological hamiltonian describing the system is now known and understood.

A priori calculations of the phenomenological constants are still in a fairly primitive state. The present state of the theory is reviewed in detail in Chapter 6. It is clear that the future will bring considerable refinements of these calculations.

In addition we may expect further refinement of the determination of the phenomenological constants by experiments. In particular those on spin waves by neutron diffraction or magnetic resonance[32] still have much to reveal. The light rare earths, Ce, Pr, Nd and Sm, are now available in greater quantities, and a more concentrated attack on their magnetic properties may be expected.

Finally, it may be expected, now that the elements are understood in some detail that a more concentrated effort will be made on various rare earth alloys, some of which are likely to have useful practical applications. Although alloys among the rare earths are referred to from time to time in the subsequent chapters there is no serious attempt to correlate all their properties. Even less mention is made of alloys between rare earths and other elements. The subject is too vast to be conveniently treated here, and although some progress has been made to date, not many alloy systems are understood in the same kind of detail as the pure metals.

REFERENCES

1. Spedding, F. H., Beaudry, B. J., Croat, J. J. and Palmer, P. E. Les Elements des Terres Rares, I, 25 (Colloques Intern. du C.N.R.S. Nº 180, 1970).
   Spedding, F. H. and Daane, A. H. The Rare Earths (Wiley, 1961).
2. Dieke, G. H. Spectra and Energy Levels of Rare Earth Ions in Crystals (Wiley, 1968).
3. Van Vleck, J. H. Theory of Electric and Magnetic Susceptibilities, Ch. IX, p. 245 (O.U.P., 1932).
4. Gschneidner, K. A. The Rare Earths, Ch. 14 (Wiley, 1961).
5. Broden, G., Hagstrom, S. B. M. and Norris C. *Phys. Rev. Lett.* **24**, 1173 (1970).
   Hagstrom, S. B. M., Heden, P. O. and Lofgren, H. *Solid State Comm.* **8**, 121 (1970).
6. McWhan, D. B., Rice, T. M. and Schmidt, P. H. *Phys. Rev.* **177**, 1063 (1969).
   Jerome, D. and Rieux, M. *Solid State Comm.* **7**, 957 (1969).
7. Mackintosh, A. R. and Johansen, G. *Solid State Comm.* **8**, 121 (1970).
8. Coqblin, B. and Blandin, A. *Adv. in Phys.* **17**, 281 (1968). See this article for further references.
   King, E., Lee, J. A., Harris, I. R. and Smith, T. F. *Phys. Rev.* **B1**, 1380 (1970).
9. Waber, J. T. and Switendick, A. C. Proc. 5th Rare Earth Conf., II, p. 75, Ames (1965).
10. Coqblin, B. *J. de Phys.* **32**, C1–599 (1971). Les Elements des Terres Rares, II, p. 579. (Colloques Intern. du C.N.R.S. Nº 180, 1970.)
    Franceschi, E. and Olcese, G. L. *Phys. Rev. Lett.* **22**, 1299 (1969).
11. Hutchings, M. T. *Solid State Physics* **16**, 227 (Ed. Seitz and Turnbull. Academic Press, 1964).
    Newman, D. J. *Adv. in Phys.* **20**, 197 (1971).
12. Abragam, A. and Bleaney, B. Electron Paramagnetic Resonance of Transition Ions (O.U.P., 1970), particularly Ch. 16, 18 and Tables 15–20.
13. Stevens, K. W. H. *Proc. Phys. Soc.* A **65**, 209 (1952).
14. Elliott, R. J. and Stevens, K. W. H. *Proc. Roy. Soc.* **218**, 553 (1953).
15. Lea, K. R., Leask, M. J. M. and Wolf, W. P. *J. Phys. Chem. Solids* **23**, 1381 (1962).
16. Elliott, R. J. *Phys. Rev.* **124**, 346 (1961).
17. Kasuya, T. Magnetism IIB, p. 215 (Ed. Rado and Suhl. Academic Press, 1966).

18. Cooper, B. R. G.E. Tech. Rep. 70–C–372 (1970).
19. Bleaney, B. *Proc. Roy. Soc.* **A276**, 19 (1963).
20. Burns, G. *Phys. Rev.* **128**, 2121 (1962).
    Lenander, C. J. and Wong, E. Y. *J. Phys. Chem.* **47**, 1986 (1963).
    Ghatikar, M. N., Raychandhuri, A. K. and Ray, D. K. *Proc. Phys. Soc.* **84**, 297 (1965).
    Sternheimer, R. M., Blume, M. and Peierls, R. F. *Phys. Rev.* **173**, 376 (1968).
21. Ellis, M. and Newman, R. *J. Chem. Phys.* **47**, 1986 (1967).
22. See for example Ref. 12, p. 339.
23. Wybourne, B. G. *Phys. Rev.* **148**, 317 (1966).
    Gabriel, J. R., Johnston, D. F. and Powell, M. J. D. *Proc. Roy. Soc.* **264**, 503 (1964).
24. De Gennes, P. G. *C.R. Acad. Sci.* **247**, 1836 (1966).
25. Rudermann, M. A. and Kittel, C. *Phys. Rev.* **96**, 99 (1954).
    Kasuya, T. *Prog. Theor. Phys.* **16**, 45 and 58 (1956); **22**, 227 (1959).
26. Blandin, A. Les Elements des Terres Rares, II, p. 505 (Colloques Intern. du C.N.R.S. Nº 180, 1970).
27. Wolf, W. P. *J. de Phys.* **32**, C1–26 (1970).
28. Kaplan, T. A. and Lyons, D. H. *Phys. Rev.* **128**, 2072 (1962).
    Specht, F. *Phys. Rev.* **162**, 389 (1967).
29. Birgeneau, R. J., Hutchings, M. J. and Rogers, R. M. *Phys. Rev.* **175**, 1116 (1968).
    Finkelstein, R. and Mencher, A. *J. chem. Phys.* **21**, 472 (1953).
    Bleaney, B. *Proc. Phys. Soc.* **77**, 113 (1961).
30. Cooper, B. R. *Phys. Rev. Letters* **19**, 900 (1967).
31. Orbach, R. and Tachiki, M. *Phys. Rev.* **158**, 524 (1967).
    Sugihara, K. *J. Phys. Soc. Japan* **14**, 1231 (1959).
    Elliott, R. J. Proc. 2nd Int. Conf. on Light Scattering (Flamarion Paris, 1971).
    Baker, J. M. *Rep. Prog. Phys.* **34**, 109 (1971).
32. Vigren, D. T. and Liu, S. H. *Phys. Rev* (to be published).
    Brookes, M. S. S. *Phys. Rev.* **1B**, 3748 (1970).

Chapter 2

Phenomenological Theory of Magnetic Ordering: Importance of Interactions with the Crystal Lattice

Bernard R. Cooper

General Electric Research and Development Center, Schenectady, New York 12301, U.S.A.

2.1. INTRODUCTION

The diverse, and sometimes exotic, magnetic behavior of the rare earth elements and their alloys as observed in the past fifteen or so years is basically understood in terms of a very simple physical picture. The key element of this picture is that one makes a sharp distinction between localized, magnetic, $4f$ electrons and outer-shell conduction electrons; and one takes the magnetic system for these metals as a lattice of localized tripositive rare earth ions (divalent for Eu) with moments corresponding to the unfilled $4f$ shells. (The ionic moment is quite well given by the application of Hund's rules so that in general, J, the total spin plus orbital angular momentum, is treated as a good quantum number for the magnetic system of tripositive ions.) This lattice of localized ions, with their corresponding localized moments, is then immersed in a sea of conduction electrons to which each rare-earth atom contributes its three outer electrons. This picture is excellent for the heavy rare earth metals[1, 2] (gadolinium through thulium), and is reasonably good for most of the light rare earths. (The most complex behavior in the rare earth series, requiring concepts beyond those of this simple picture, is found for the end members, cerium and ytterbium.)

Thus, so far as the externally observed magnetic properties of the rare earth metals (excepting cerium and ytterbium) are concerned, one can think in terms of the behavior of a system of localized $4f$ shell moments with various effective forces acting on and between these moments. The theory that describes the magnetic behavior of this localized moment system, acting under the influence of the effective forces, is the phenomenological theory. In this theory, the conduction electrons disappear from view. As is always the object of a phenomenological theory, here it provides a reasonable set of parameters describing the experimentally observed properties; and, of course, also has the objective of suggesting new experiments clarifying the type and form of the important effective interactions. It is then the object of "funda-mental", i.e. microscopic, theory to provide a first-principles justification for the behavior of the effective interactions. (The detailed nature of the conduc-tion electron behavior plays a key role in understanding the way in which the effective forces acting on the localized $4f$ moments arise.) For understanding the equilibrium magnetic ordering in the rare earth metals, the phenomenolo-gical theory has been very successful, even though the microscopic theory presents formidable, at present in some respects insurmountable, difficulties. The phenomenological theory is equally successful in dealing with dynamic properties, and such questions are treated in Chapter 5 on the spin wave behavior. As will be apparent, symmetry considerations play a very important part in development of the phenomenological theory.

The first part of the present chapter (Section 2.2) then deals with the phenomenological theory of magnetic ordering, i.e. understanding of the equilibrium magnetic arrangements and transitions between them. The discussion of magnetic ordering will show the importance of the crystal lattice, acting through crystal-field and elastic effects, in conjunction or competition with effective exchange forces in determining the nature of the magnetic ordering in the rare earth metals.

The second part of this chapter (Section 2.3) will discuss the role played by crystal-electric-field effects in rare earth magnetism, and will discuss the experimental measurements of the size of crystal-field effects. So far as the heavy rare earth metals are concerned, lattice effects play a very important role in determining the detailed nature of the magnetic ordering; however, this role is still secondary in so far as concerns the question of whether the metals order magnetically at all. On the other hand, for the light rare earths the f shells are markedly larger, and the effects of the crystal-field are so large as to affect the fundamental question of the existence of magnetic ordering. The second part of this chapter (Section 2.3) will begin by treating the behavior of induced moment systems where the relative strengths of crystal-field and exchange interactions determine whether any magnetic ordering at all occurs. Emphasis will be laid on the induced moment effect as it exists in fcc Pr and Pr–Th alloys and in cubic rare earth intermetallic compounds, especially, the rare earth-group V compounds of NaCl structure. The relevance of induced moment effects to understanding the magnetic behavior of Pr in the ordinary double hexagonal crystal structure will be discussed. As in Section 2.2, the crystal-field effects will be treated in a phenomenological way, emphasizing symmetry considerations. Finally, we shall discuss experimental measurements showing the size of crystal-field effects, both for rare earth elements and intermetallic compounds, and for dilute rare earths in noble metals. Also we discuss how the results of such measurements relate to the fundamental question of the origin and nature of the crystal-electric-field in metals.

2.2. PHENOMENOLOGICAL THEORY OF MAGNETIC ORDERING

2.2.1. Effective Forces and Magnetic Structures

The great variety of magnetic structures for the rare earth metals described in Chapter 3 can be understood as the consequence of two types of interaction for the localized rare earth ion moments.

$$\mathcal{H} = \mathcal{H}_{\text{iso ex}} + \mathcal{H}_{\text{orb}} \qquad (2.1)$$

The first contribution in (2.1) arises from a long-range oscillatory exchange interaction of the Ruderman–Kittel type via polarization of the conduction electrons. So long as one does not explicitly take into account the way in which the presence of an orbital contribution to the ionic moment modifies this interaction (i.e. in practice the limit that the product of the effective radius of the $4f$ orbital wave function and the Fermi radius for the conduction electrons (in k-space) is negligible), the Ruderman–Kittel interaction depends only on the scalar product of the total spins of the two interacting ions.

$$\mathcal{H}_{\text{iso ex}} = - \sum_{i \neq j} j(\mathbf{R}_i - \mathbf{R}_j)\, \mathbf{S}_i \cdot \mathbf{S}_j \qquad (2.2)$$

One can then map this interaction onto the manifold of states of the total angular momentum as an isotropic interaction

$$\mathcal{H}_{\text{iso ex}} = -(\lambda - 1)^2 \sum_{i \neq j} j(\mathbf{R}_i - \mathbf{R}_j)\, \mathbf{J}_i \cdot \mathbf{J}_j = - \sum_{i \neq j} \mathcal{J}(\mathbf{R}_i - \mathbf{R}_j)\, \mathbf{J}_i \cdot \mathbf{J}_j \qquad (2.3)$$

The second contribution in (2.1) consists of those interactions whose presence depends on the orbital contribution to the ionic moment. These interactions are characteristically anisotropic with respect to the crystal axes and/or depend on the elastic strains.

$$\mathcal{H}_{\text{orb}} = \mathcal{H}_{\text{an ex}} + \mathcal{H}_{\text{cf}} + \mathcal{H}_{\text{ms}} \qquad (2.4)$$

The first term on the right of (2.4) is the anisotropic exchange[3, 4] resulting from taking account of the nonsphericity of a $4f$ wave function of finite radius. The detailed theory of such interactions is quite complex and depends greatly on the particular exchange mechanism, viz., direct, via polarization of conduction electrons, superexchange. For exchange via polarization of a free electron sea the form of the anisotropic exchange, with the simplest approximations,[3, 4] is

$$\mathcal{H}_{\text{an ex}} \simeq \sum_{i \neq j} K(\mathbf{R}_i - \mathbf{R}_j)\, \mathbf{S}_i \cdot \mathbf{S}_j \{[\mathbf{L}_i \cdot (\mathbf{R}_i - \mathbf{R}_j)]^2 + [\mathbf{L}_j \cdot (\mathbf{R}_i - \mathbf{R}_j)]^2\} \qquad (2.5)$$

The second contribution in (2.4) is the anisotropy energy of the unstrained lattice resulting from interaction with the crystalline electric field caused by each rare earth ion seeing the other charged rare earth ions. The crystal-field exhibits the symmetry of the ionic lattice. For the hcp lattice pertinent to the heavy rare earths, the crystal-field interaction consists of a large axial and smaller planar anisotropy.

$$\mathcal{H}_{cf} = \sum_i \{V_2{}^0 Y_2{}^0(\mathbf{J}_i) + V_4{}^0 Y_4{}^0(\mathbf{J}_i) + V_6{}^0 Y_6{}^0(\mathbf{J}_i)$$
$$+ V_6{}^6[Y_6{}^6(\mathbf{J}_i) + Y_6{}^{-6}(\mathbf{J}_i)]\} \quad (2.6)$$

The $Y_l{}^m(\mathbf{J}_i)$ are operator equivalents of spherical harmonics as discussed in Chapter 1.

The final contribution to $\mathcal{H}_{orb}$ comes from magnetostriction effects. There are both single-ion and two-ion contributions to the magnetostriction effects which arise from modulation by the strain of the crystal-field and anisotropic exchange interactions, respectively.

$$\mathcal{H}_{ms} = \mathcal{H}_e + \mathcal{H}_m \quad (2.7)$$

Here $\mathcal{H}_e$ is the elastic energy associated with the homogeneous strain components, and $\mathcal{H}_m$ is the magnetoelastic interaction, coupling the spin system to the strains.

We will now show how the Hamiltonian of (2.1) can lead to the various types of magnetic structures found in the rare earth metals. As is usual in the phenomenological theory,[5-7] we treat the interactions involved in the magnetic ordering within the molecular field approximation. We shall restrict our discussion to the hcp structure characterizing the heavy rare earth metals. (Any discussion of the light rare earths involves treating the peculiarities of each element—to the extent they are understood—individually.)

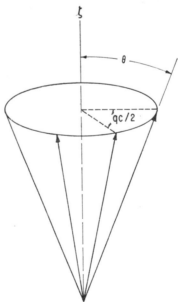

Figure 2–1. Cone spin arrangement. Arrows indicate moments of successive planes perpendicular to the c-axis.

In all of the various moment arrangements found in the heavy rare earths the moments of the ions lying in a given hexagonal layer are parallel. In this chapter, for illustrative purposes, in the absence of applied field effects, in addition to the ferromagnet we shall consider only the simplest magnetic structures showing periodic effects. These are the arrangements where the moments of successive planes looking down the c-axis lie on the surface of a cone or form a planar spiral. As shown in Figure 2–1, one can describe the moment components for cone, planar spiral, and ferromagnet in terms of a wave vector $\mathbf{q}$ parallel to the c-(hexagonal)-axis giving the periodicity, and a polar angle θ giving the moment orientation with respect to the c-axis.

$$J_{i\zeta}=\bar{J}\cos\theta, \quad J_{i\xi}=\bar{J}\sin\theta\cos(\mathbf{q}\cdot\mathbf{R}_i), \quad J_{i\eta}=\bar{J}\sin\theta\sin(\mathbf{q}\cdot\mathbf{R}_i) \quad (2.8)$$

Here ζ is along the c-axis; ξ points toward a nearest neighbor in the hexagonal plane, and η is orthogonal to ξ in the hexagonal plane. $\bar{J}$ denotes the equilibrium value of total angular momentum per ion at a given temperature. For general θ and $\mathbf{q}$, (2.8) describes a conical moment arrangement as in Er or Ho at low temperatures. If $\theta=\pi/2$ and $q\neq0$, the moment arrangement is a planar spiral in the hexagonal plane as in Dy, Tb, and Ho at high T. While if $\mathbf{q}=0$, the arrangement is ferromagnetic, and θ gives the angle of the ferromagnetic axis with respect to the c-crystal-axis. For $\theta=\pi/2$ and $q=0$, one has ferromagnetic alignment as in Tb and Dy at low T.

Of the interactions in (2.1) the exchange interactions are the only ones that can have the tendency to give a periodic moment arrangement. While the contributions to the free energy of all the types of interactions mentioned will change as $\mathbf{q}$ in (2.8) varies between different nonzero values, for the crystal-field and magneto-elastic terms this variation is quite weak compared to that for the exchange terms. (On the other hand, the tendency of these terms to favor ferromagnetism over a periodic arrangement is strong; and, as discussed in Section 2.2.2, this tendency plays a key role in the transition from spiral to ferromagnetic arrangements.)

To consider the dependence of the energy on $\mathbf{q}$, we shall consider only isotropic exchange. Indeed in our further discussion we shall not explicitly include anisotropic exchange effects. There are two related reasons for this omission. First is that in the phenomenological theory we use the molecular field approximation. Then any anisotropic exchange effects acting on a given ion will have the overall symmetry of the hcp lattice. Therefore their effects on the magnetic structure will be difficult to separate from crystal-field and magnetoelastic interactions having the same symmetry. Partially because of this there is no clear experimental evidence for the presence of significant anisotropic exchange effects in the heavy rare earth metals. Also, the form of anisotropy in the exchange arising from the simplest theory would manifest itself by removing certain degeneracies in the spin wave spectrum.[8] This splitting is found to be small in neutron inelastic scattering experiments[9] (see Figure 5–4). This absence of large observed effects is our second reason for omitting anisotropic exchange from our molecular field treatment of the magnetic structure behavior. (This is discussed somewhat further at the end of this section.)

To consider the dependence of the energy on $\mathbf{q}$, it is convenient to define the Fourier transform of the exchange energy, $\mathscr{J}(\mathbf{q})$:

$$\mathscr{J}(\mathbf{q}) = \sum_j \mathscr{J}(\mathbf{R}_i - \mathbf{R}_j) \exp\left[i\mathbf{q} \cdot (\mathbf{R}_i - \mathbf{R}_j)\right] \tag{2.9}$$

In deciding which of two ordered magnetic arrangements holds at a given temperature, if the ordered moment per ion in each arrangement is the same, then entropy considerations can be neglected, and one is justified in comparing the internal energies. Then the exchange energy per ion is

$$E_{ex}/N = -\mathscr{J}(\mathbf{q}) J^2 \sin^2 \theta - \mathscr{J}(0) J^2 \cos^2 \theta \tag{2.10}$$

and the spiral pitch is determined by taking $\mathbf{q} = \mathbf{k}_0$, the value maximizing $\mathscr{J}(\mathbf{q})$ and thereby minimizing the energy.

Thus the exchange energy determines the periodicity of the magnetic structure. If the maximum of $\mathscr{J}(\mathbf{q})$ comes at $\mathbf{q} = 0$, then exchange favors a ferromagnetic arrangement. If the maximum comes at the edge of the Brillouin Zone, then exchange favors a conventional antiferromagnet; while if k_0 is some wave vector intermediate between the origin and the zone boundary, exchange favors a spiral arrangement. In general this latter case holds for the heavy rare earth metals.

One of the major objectives of the microscopic theory of the magnetic properties of the heavy rare earth metals, is to explain the occurrence of the maximum of $\mathscr{J}(\mathbf{q})$ at some intermediate $\mathbf{q}$ in the Brillouin Zone. As discussed in Chapter 6, the microscopic origin of the $\mathscr{J}(\mathbf{q})$ behavior and of the corresponding effective exchange interactions in real space, is generally thought to lie in the polarization of the conduction electrons by the localized $4f$ spin at a given site, and the response to that polarization by the localized spin at a different site. Thus, in a sense the $\mathbf{q}$ dependent susceptibility $\chi(\mathbf{q})$ giving the response of the electron gas to the effective field associated with the ionic moment, is the fundamental quantity calculated in the first-principles theory. Detailed calculations of $\chi(\mathbf{q})$ for realistic band structures are quite involved. A further complication is that to calculate $\mathscr{J}(\mathbf{q})$ even with a known $\chi(\mathbf{q})$ involves matrix elements of the s–f exchange that themselves have quite complicated wave vector dependence Thus, while calculations are available[10] qualitatively confirming the concepts involved in the conduction-electron-polarization mechanism giving $\mathscr{J}(\mathbf{q})$, the currently available calculations are still too crude to give a complete and accurate first-principles calculation of $\mathscr{J}(\mathbf{q})$. For this reason a simple semiempirical phenomenological scheme for evaluating $\mathscr{J}(\mathbf{q})$ is particularly useful. This phenomenological $\mathscr{J}(\mathbf{q})$ provides both a meeting place at which one hopes to match experiment and first-principles theory, and also serves as a device for conveying information obtained about $\mathscr{J}(\mathbf{q})$ in one set of experiments in a way useful for analyzing other experiments.

The fact that the ionic moments in a given plane perpendicular to the c-axis are all parallel leads to a simple semiempirical scheme for evaluating $\mathscr{J}(\mathbf{q})$. This is the planar interaction model. The existence of a spiral structure depends on the presence of competing exchange interactions.[11–16] For example, a positive interaction between nearest neighbor planes, and a negative interaction between second-nearest neighbor planes can give a spiral.

Thus, the simplest model giving a $\mathcal{J}(\mathbf{q})$ showing the experimental behavior found in the heavy rare earth metals is the three plane interaction model originally used by Enz,[17] and adopted by Elliott,[5] Cooper,[18] and other workers. This gives a good picture of the behavior of $\mathcal{J}(\mathbf{q})$ for reasonably low energies, say up to a quarter or so of the maximum value. One can get a more complete picture by taking interactions out to further neighbor planes.[19]

For $\mathbf{q}$ along the c-axis in the three plane interaction model,

$$\mathcal{J}(\mathbf{q})\,\bar{J}^2 = \sigma^2(B_0 + 2B_1 \cos \tfrac{1}{2}qc + 2B_2 \cos qc) \tag{2.11}$$

Here the B_n are the effective exchange interactions between planes, and $\sigma = \bar{J}/J$ is the relative magnetization. Then the wave vector for the equilibrium magnetic structure, $\mathbf{k}_0$, is that value of $\mathbf{q}$ maximizing $\mathcal{J}(\mathbf{q})$. So that

$$\cos \tfrac{1}{2}k_0 c = -B_1/4B_2 \tag{2.12}$$

Thus the spiral wave number is simply related to the phenomenological exchange parameters. One can use other experimental information to complete the evaluation of the B_n as called for in any particular instance. To calculate thermodynamic quantities one often wishes to have values of $\mathcal{J}(\mathbf{q})$ throughout the Brillouin Zone. The interplanar coupling scheme again offers a simple way to do this. One makes some reasonable identification between the interplanar effective exchange constants and those between individual ions. For example, as illustrated in Figure 2–2, the present author[18] identi-

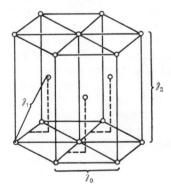

Figure 2–2. Exchange interactions for hcp structure of heavy rare earths including exchange with only nearest neighbors in the same plane and in the first and second nearest planes (after Cooper[18]).

fied the interplanar exchange with the exchange of a given ion with its nearest neighbors in the same plane and the first and second nearest planes. So that,

$$B_0 = 6\mathcal{J}_0 J^2, \qquad 2B_1 = 6\mathcal{J}_1 J^2, \qquad 2B_2 = 2\mathcal{J}_2 J^2 \tag{2.13}$$

The values of $\mathcal{J}_0$, $\mathcal{J}_1$ and $\mathcal{J}_2$ can be used in (2.9) to give a complete evaluation of $\mathcal{J}(\mathbf{q})$. Goodings[19] used a larger number of inter-ionic exchange constants to fit the $\mathcal{J}(\mathbf{q})$ for $\mathbf{q}$ along the c-axis (as obtained by analysis of neutron spin wave scattering experiments[9]) and then in turn used the inter-ionic exchange constants to evaluate $\mathcal{J}(\mathbf{q})$ at all $\mathbf{q}$ in the Brillouin Zone.

2

We should point out that it is a simple matter to allow for temperature dependence of the spiral angle in the phenomenological theory. This is done by letting one or more exchange constants vary with temperature through dependence on the magnetization. For example Elliott[5] let

$$B_2 \rightarrow B_2 - C\sigma^2 \tag{2.14}$$

This replacement is helpful in using the phenomenological theory in its role as a link between different experiments or between experiment and first-principles theory. However, the phenomenological theory provides no insight into the origin of the temperature dependence of the periodicity. This is one property that seems to be particularly dependent on the detailed behavior of the conduction electrons, so one must turn to the microscopic theory. There are a number of candidates for the principal mechanism; however, the major effects as studied by Miwa,[20] following the work of Elliott and Wedgwood[21] and of deGennes and Saint James,[22] is the combined effect on $\chi(\mathbf{q})$ due to redistribution of the conduction electrons across the energy gaps at super-zone boundaries introduced by periodic magnetic ordering and to $4f$ spin disorder scattering. Magnetoelastic effects may also play a significant role.[10]

Thus, the basic question of the magnetic structure periodicity is determined by the form of $\mathcal{J}(\mathbf{q})$, where (as discussed below) the competition with magnetostriction, planar anisotropy, and applied fields may give transitions from periodic to ferromagnetic structures. Given that a periodic structure occurs, the anisotropic terms in the Hamiltonian, reflecting the symmetry of the hcp lattice determine whether the magnetic moment arrangement is a planar spiral or a cone. The form of these terms in the effective Hamiltonian is shown in Eqn (2.6). While these terms are identified in (2.6) as the crystal-field terms for the unstrained lattice, magneto-elastic effects can also give such contributions.[23, 24] (Within the molecular field approximation, anisotropic exchange effects for a ferromagnetic arrangement can also be put into this form. For a periodic arrangement, the anisotropic exchange can give somewhat different effects, but none qualitatively—or probably even quantitatively—significantly different in determining the magnetic structure.) The values of the various axial terms in (2.6) determine θ, the angle the moments make with the c-axis. To deal with the question of determining θ, and for other uses below, it is convenient to regroup the terms of the crystal-field Hamiltonian of (2.6) into sums over the various powers of $J_{i\zeta}$ present. Then the overall $\mathcal{H}$ for the system (omitting anisotropic exchange) becomes

$$\mathcal{H} = -\sum_{i \neq j} \mathcal{J}(\mathbf{R}_i - \mathbf{R}_j)\, \mathbf{J}_i \cdot \mathbf{J}_j + \sum_i \{V_2 J_{i\zeta}^2 + V_4 J_{i\zeta}^4 + V_6 J_{i\zeta}^6$$
$$+ V_6{}^6[Y_6{}^6(\mathbf{J}_i) + Y_6{}^{-6}(\mathbf{J}_i)]\} + \mathcal{H}_{ms} \tag{2.15}$$

For the moment arrangement of (2.8) (with $\mathbf{q} = \mathbf{k}_0$), the energy per ion in the molecular field approximation for this Hamiltonian is

$$E/N = -\mathcal{J}(\mathbf{k}_0)\, \bar{J}^2 \sin^2\theta - [\mathcal{J}(0) - V_2]\, \bar{J}^2 \cos^2\theta + V_4 \bar{J}^4 \cos^4\theta + V_6 \bar{J}^6 \cos^6\theta \tag{2.16}$$

(The temperature dependence indicated for the crystal-field terms is correct for temperatures in the high temperature part of the ordered regime. At lower temperatures, the behavior is significantly different.[25] Our present purpose is simply to point out that for axial terms of competing sign one can obtain values of θ anywhere between 0 and $\pi/2$). The cone angle θ is given by $\partial E/\partial\theta = 0$. This has three possible solutions:

$$\sin \theta = 0 \qquad \text{axial ferromagnet} \qquad (2.17a)$$

$$\cos \theta = 0 \qquad \text{planar spiral} \qquad (2.17b)$$

$$3V_6 \bar{J}^4 \cos^4 \theta + 2V_4 \bar{J}^2 \cos^2 \theta + [\mathscr{J}(k_0) - \mathscr{J}(0) + V_2] = 0$$
$$\text{general cone} \qquad (2.17c)$$

The conditions on the parameters for the various arrangements to be stable are discussed by Cooper et al.[26]

In discussing the origin of the anisotropic terms in the effective Hamiltonian we have indicated the difficulty of experimentally distinguishing between the various sources of anisotropy, viz., the crystal-field of the undistorted lattice, magneto-elastic effects, and anisotropic exchange. (Indeed the fact that magneto-elastic effects can arise from modulation by the strains of either crystal-field or anisotropic exchange interactions further complicates the situation.) Actually there are experiments that, at first sight, one might think indicate the importance of anisotropic exchange effects. These are neutron diffraction studies on the magnetic structure of mixed rare earth alloys.[27, 28] For example in Ho–Er alloys,[27] both components form a uniform magnetic structure with respect to the cone angle. In a comparison between the expected behavior for crystal-field anisotropy as opposed to anisotropic exchange, one might think that this observed behavior would indicate anisotropic exchange. This is because, if crystal-field effects were the source of anisotropy, one would expect each component to have its own individual cone angle as in the pure element. However, the interpretation of such experiments does not appear to be so simple. This is because isotropic exchange interaction within the hexagonal plane tends to "tie together" large numbers of rare earth ions and thus give uniform behavior for the magnetic structure even if the anisotropy is of a single-ion origin. Also magnetoelastic interactions will have the effect of rather long-range interactions between rare earth ions even if they originate in modulation by the strains of the single ion crystal-field anisotropy. This is because the elastic interactions also "tie together" large numbers of rare earth ions.

As yet there has been no attempt to analyze theoretically the question of the relative contributions of anisotropic exchange and magnetoelastic interactions to the magnetic structure in the heavy rare earth alloys. This is understandable since the major part of the experiments[28] are quite recent. Also it is a safe guess that detailed theoretical understanding will depend on the availability of elastic constant,[29, 30] magnetostriction,[31, 32] and anisotropy constant[33] measurements of the sort that have been performed for the pure elements. While the questions involved in analysis of the alloy behavior are clearly very formidable, success in such an effort would complete our basic physical picture of the dominant interactions determining the magnetic behavior in the heavy rare earth metals.

To summarize the main points of this Section: A periodic moment arrangement occurs only if this is favored by exchange. Whether the periodic arrangement is a planar spiral or a cone is determined by the competition between the three axial anisotropy terms. Even when exchange favors a periodic moment arrangement, planar anisotropy terms and magnetostriction effects can overcome this tendency and give ferromagnetism. These driving mechanisms for ferromagnetism are discussed in the next Section.

2.2.2. Transitions Between Magnetic Structures

2.2.2.1. Applied Magnetic Field Effects

When a magnetic field is applied to a period magnetic structure, for sufficiently large fields one ultimately arrives at a ferromagnetic arrangement. However, in general the transition from spiral or cone to ferromagnet is complex in nature going through various intermediate stages. (The experimental behavior[34] in Ho is an example of the complexity possible in real situations.) Theoretical studies of the magnetization process of a helical spin structure were first made by Herpin and Meriel[14, 35] and by Enz.[17, 36] The most complete studies of the magnetization process for periodic magnetic structures can be found in the papers by Nagamiya and coworkers.[16, 37-39] These papers include effects in three-dimensional structures such as cones as well as the simpler case where the moments are confined to a plane for all fields. The change in magnetic structure with applied field is of interest both in itself, as a means of showing and measuring the competition between the different types of interaction for the localized moment system, and also for discussing the magnetic resonance and inelastic neutron scattering experiments used to investigate spin wave excitations in the heavy rare earth metals. Here we will confine ourselves to the planar case, where the transition is from a planar spiral, and the final arrangement is a ferromagnet lying in the plane. (We assume very large anisotropy holding the moments in the plane.) This case shows the general character of the transition effects and serves as a basis for understanding the behavior in Tb and Dy.

For the present we will omit the effects of planar anisotropy or magnetostriction. Such effects will be discussed below. We then discuss a magnetic system described by the Hamiltonian,

$$\mathscr{H} = - \sum_{i \neq j} \mathscr{J}(\mathbf{R}_i - \mathbf{R}_j)\, \mathbf{J}_i \cdot \mathbf{J}_j - \sum_i J_{i\zeta}^2 - \lambda\beta H \sum_i J_{i\zeta} \qquad (2.18)$$

Here the axial anisotropy V_2 is taken as very large and negative, so the moments are always confined to the hexagonal plane. The applied field is along the ξ-axis in the hexagonal plane. We consider a situation where the equilibrium magnetic structure in the absence of an applied magnetic field is a planar spiral. As we shall now discuss, quite moderate magnetic fields applied in the spiral plane can have very marked effects on the magnetic structure.

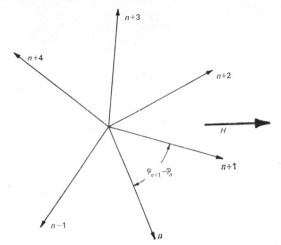

Figure 2–3. Distortion of spiral moment arrangement by small applied field. Arrows indicate moments of successive planes perpendicular to the c-axis.

(a) Low Fields

For small H, the situation is as illustrated in Figure 2–3. For $H=0$, the spiral angle $\phi_n = nk_0c/2$, so that $\phi_{n+1} - \phi_n = k_0c/2$. For small fields there is a slight distortion of the spiral with a small net moment along H.

The variation of ϕ_n with H can be found from the condition that $\partial E_0 / \partial \phi_n = 0$ where E_0 is the equilibrium energy in the molecular field approximation.

$$\frac{\partial E_0}{\partial \phi_n} = 0 = 2 \sum_p \mathscr{J}(\mathbf{R}_n - \mathbf{R}_p) J^2 \sin(\phi_n - \phi_p) + \mu H \sin \phi_n \qquad (2.19)$$

where

$$\mu \equiv \lambda \beta J \qquad (2.20)$$

is the ionic moment (μ_0 will sometimes be used to denote the value at $T=0$). If ϕ_n is of the form

$$\phi_n = \phi_n^{(0)} + \sum_{i=1}^{N} \delta \phi_n^{(i)} \qquad (2.21)$$

where $\phi_n^{(0)} = nk_0 c/2$ is the value for the unperturbed spiral and $\delta \phi_n^{(i)}$ is the correction to ϕ_n of order H^i, then (2.21) can be solved[26, 40] as an iterative equation to any desired order in H.

To order H^3,

$$\phi_n = nk_0c/2 + [XH + AH^3] \sin(nk_0c/2)$$
$$+ YH^2 \sin(nk_0c) + BH^3 \sin^3(nk_0c/2) \qquad (2.22)$$

with

$$X = \mu/(a' + b') \qquad (2.23a)$$

$$Y = \mu^2(4b' - c')/4(a' + b')^2 \, c' \qquad (2.23b)$$

$$A = [\mu^3/2(a' + b')^4 \, (b' + d') \, c'] \{(b' + d') \, (-a'c' + b'c')$$
$$+ 4a'b'd' - 12b'^3 - 8a'b'^2\} \qquad (2.23c)$$

$$B = [\mu^3/3(a' + b')^3 \, (b' + d') \, c'] \{c'(b' + d') + 6b'(3b' - d')\} \qquad (2.23d)$$

and

$$a' \equiv \bar{J}^2[\mathscr{J}(\mathbf{k}_0) - \mathscr{J}(0)] \qquad (2.24a)$$

$$b' \equiv \bar{J}^2[\mathscr{J}(\mathbf{k}_0) - \mathscr{J}(2\mathbf{k}_0)] \qquad (2.24b)$$

$$c' \equiv \bar{J}^2[\mathscr{J}(\mathbf{k}_0) - \mathscr{J}(3\mathbf{k}_0)] \qquad (2.24c)$$

$$d' \equiv \bar{J}^2[\mathscr{J}(\mathbf{k}_0) - \mathscr{J}(4\mathbf{k}_0)] \qquad (2.24d)$$

(b) High Fields

We now examine the behavior when the field is sufficiently large so ϕ_n is small. Nagamiya et al.[37] studied the behavior in this regime at $T=0$ by expanding the energy as a power series in

$$\sin\frac{\phi_n}{2} \equiv x_n \ll 1 \qquad (2.25)$$

If one Fourier expands x_n

$$x_n = \sum_q \xi_q e^{inq} \qquad (2.26)$$

then to the fourth order in ξ the energy per ion is,

$$E = -\mathscr{J}(0)J^2 - \mu H + 2\sum_q [2\mathscr{J}(0)J^2 - 2\mathscr{J}(q)J^2 + \mu H]|\xi_q|^2$$
$$- 2J^2 \sum_{q,q',q''} [\mathscr{J}(q) - \mathscr{J}(q+q') - \mathscr{J}(q+q'') + \mathscr{J}(q+q'+q'')]$$
$$\times \xi_q \xi_{q'} \xi_{q''} \xi_{-q-q'-q''} \qquad (2.27)$$

[The expansion is needed to fourth order to recognize both the threshold value of H where all $\xi_q = 0$ (and hence $\phi_n = 0$, i.e. ferromagnetism occurs), and the amplitude of ξ_q below the threshold. As is usual in such instability problems, the second order effects allow one to recognize the threshold value of H, while the fourth order terms supply the "restoring" effects giving the amplitude.]

For the energy to be minimized, if the coefficient of $|\xi_q|^2$ is positive for all q, ξ_q must vanish for all q. So that one has ferromagnetism when H attains a value such that the $|\xi_q|^2$ coefficient is positive for all q. Since the maximum of $\mathscr{J}(q)$ occurs at k_0, this means that for

$$H > H_f = 2J^2[\mathscr{J}(k_0) - \mathscr{J}(0)]/\mu \qquad (2.28)$$

one has complete ferromagnetic alignment along the field ($\phi_n = 0$).

For $H < H_f$ there is a range of q near k_0 for which the $|\xi_q|^2$ coefficients are negative. Nagamiya et al.[37] were able to prove that the only nonvanishing ξ_q is for $q = k_0$. Then E given by (2.27) reduces to,

$$E = -\mathscr{J}(0)J^2 - \mu H - 4[2J^2\mathscr{J}(k_0) - 2J^2\mathscr{J}(0) - \mu H]|\xi_{k_0}|^2$$
$$+ 8[3J^2\mathscr{J}(k_0) - J^2\mathscr{J}(2k_0) - 2J^2\mathscr{J}(0)]|\xi_{k_0}|^4 \quad (2.29)$$

The amplitude of $|\xi_{k_0}|$ is the value minimizing E, so that

$$|\xi_{k_0}|^2 = \frac{\mu[H_f - H]}{4J^2[3\mathscr{J}(k_0) - \mathscr{J}(2k_0) - 2\mathscr{J}(0)]} \qquad (2.30)$$

and one obtains a fan arrangement[15, 36, 37] of the moments,

$$\sin \frac{\phi_n}{2} = 2\,|\xi_{k_0}|\,\sin\left(\tfrac{1}{2}nk_0c\right) \tag{2.31}$$

(c) Intermediate Fields; Sharpness of Transition; Planar Anisotropy and Magnetostriction Effects

As we have seen above, at small fields the spiral is slightly distorted as the moments rotate by a small amount toward H; while at fields approaching H_f the moment arrangement is a fan deviating only slightly from complete ferromagnetic alignment along H. The question then arises as to the behavior at intermediate fields. Kitano and Nagamiya[39] were able to show that, for the type of $\mathscr{J}(q)$ behavior generally expected in the spiral case, at T not too far below T_N the transition from spiral to fan with increasing field is continuous. However, for lower temperature the transition becomes discontinuous.

It is easy to estimate the field of the discontinuity at $T=0$. This can be done by equating the energies for the low field and high field cases as described in a and b above. This gives

$$-\mathscr{J}(k_0)J^2 - \tfrac{1}{4}\,\frac{\mu^2 H_c^2}{J^2[2\mathscr{J}(k_0)-\mathscr{J}(2k_0)-\mathscr{J}(0)]}$$

$$= -\mathscr{J}(0)J^2 - \mu H_c - \frac{\mu^2(H_f-H_c)^2}{2J^2[3\mathscr{J}(k_0)-\mathscr{J}(2k_0)-2\mathscr{J}(0)]} \tag{2.32}$$

So that the transition field from spiral to fan, H_c, is given by

$$H_c/H_f = [(1+\alpha)(2+\alpha)]^{\frac{1}{2}} - (1+\alpha) \tag{2.33}$$

with

$$\alpha \equiv \frac{2J^2[\mathscr{J}(k_0)-\mathscr{J}(2k_0)]}{\mu H_f} \tag{2.34}$$

For the $\mathscr{J}(q)$ behavior generally found in the rare earths, $\alpha \gg 1$ and

$$H_c/H_f \simeq \tfrac{1}{2} \tag{2.35}$$

Actually, depending on the value of k_0, there can be other stable structures occurring between the spiral and the ferromagnet besides the fan.[37, 38]

To study the question of when the transition between distorted spiral and fan is first order (discontinuous) and when second order (continuous) Kitano and Nagamiya[39] used a molecular field theory and an expansion of the molecular field about the value for $H=0$ which is valid not too far below the Neel temperature, T_N. (One anticipates that the transition will be first order up to some T not too far from T_N. Actually the conditions as to the temperature range to which the theory is restricted are not very strong.)

The treatment of Kitano and Nagamiya involves solving the self-consistent equation resulting from equating the sum of the exchange and external fields to the molecular field, H_m, which is proportional to the inverse Brillouin function of the magnetization

$$\mu H_m \frac{\sigma_i}{\sigma_i} = \mu \mathbf{H} + 2\sum_j \mathscr{J}(\mathbf{R}_i-\mathbf{R}_j)J^2\frac{\sigma_j}{\sigma} \tag{2.36}$$

Here σ is the relative magnetization for $H=0$ (identical for all sites), while σ_i is the vector relative magnetization in the presence of nonzero H (so $|\sigma_i| \neq \sigma$ in general).

Then one expands $\mu H_m / \sigma_i$ about the value at $H=0$.

$$\frac{\mu H_m}{\sigma_i} = \frac{\mu H_m}{\sigma} + \mu \frac{\partial}{\partial \sigma^2} \frac{H_m}{\sigma} (\sigma_i^2 - \sigma^2) + \ldots \qquad (2.37)$$

One substitutes (2.37), keeping terms only up to the second order, into (2.36) and solves for $\sigma_{i\xi}$ (component parallel to H) and $\sigma_{i\eta}$ (component perpendicular to H).

To carry out the solution one Fourier expands $\sigma_{i\xi}$ and $\sigma_{i\eta}$,

$$\sigma_{i\xi} = \sigma \left\{ \psi_0 + \psi_1 \cos \left(\tfrac{1}{2} nk_0 c \right) + \psi_2 \cos (nk_0 c) + \ldots \right\} \qquad (2.38a)$$

$$\sigma_{i\eta} = \sigma \left\{ \nu_1 \sin \left(\tfrac{1}{2} nk_0 c \right) + \nu_2 \sin (nk_0 c) + \ldots \right\} \qquad (2.38b)$$

Here as $H \to 0$, $\psi_0 = 0$, $\psi_1 = \nu_1 = 1$, all other ψ_n, $\nu_n = 0$; and as $H \to H_f$, $\psi_0 = 1$, ψ_1, ν_1, all other ψ_n, $\nu_n = 0$. Equating the coefficients of 1, $\cos \left(\tfrac{1}{2} nk_0 c \right)$, $\cos (nk_0 c)$, ..., $\sin \left(\tfrac{1}{2} nk_0 c \right)$, $\sin (nk_0 c)$ ..., determines ψ_0, ψ_1, $\psi_2, \ldots$, ν_1, $\nu_2, \ldots$.

If we retain only the Fourier coefficients ψ_0, ψ_1, and ν_1 then there are three regimes of behavior for varying H,

(1) $\psi_0 \leq \tfrac{1}{2}$

$$(1 + \gamma/2) \psi_0 - \tfrac{5}{2} \gamma \psi_0^3 = H/H_f \qquad (2.39a)$$

$$\psi_1^2 = 1 - 4\psi_0^2, \qquad \nu_1 = 1 \qquad (2.39b)$$

(2) $\tfrac{1}{2} \leq \psi_0 \leq 1$

$$(1 - \gamma/6) \psi_0 + \frac{\gamma}{6} \psi_0^3 = H/H_f \qquad (2.40a)$$

$$\psi_1 = 0, \qquad \nu_1^2 = \tfrac{4}{3}(1 - \psi_0^2) \qquad (2.40b)$$

(3) $\psi_0 > 1$

$$\left(1 - \frac{\gamma}{2} \right) \psi_0 + \frac{\gamma}{2} \psi_0^3 = H/H_f \qquad (2.41a)$$

$$\psi_1 = \nu_1 = 0 \qquad (2.41b)$$

Here the quantity γ is defined as

$$\gamma = \frac{\mu \sigma^2}{J^2 [\mathscr{J}(k_0) - \mathscr{J}(0)]} \frac{\partial}{\partial \sigma^2} (H_m / \sigma) \qquad (2.42)$$

and γ is a decreasing function of temperature which vanishes at T_N. Near T_N

$$\gamma \simeq \frac{2 \mathscr{J}(k_0)}{\mathscr{J}(k_0) - \mathscr{J}(0)} \frac{T_N - T}{T_N} \qquad (2.43)$$

Equations (2.39a), (2.40a), and (2.41a) determine the net magnetization (per ion), $M/N = \mu \sigma \psi_0$, as a function of H/H_f. In Figure 2–4a the behavior

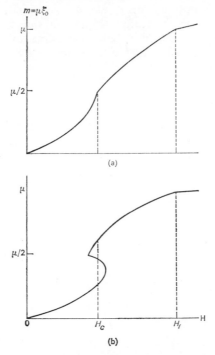

Figure 2–4. Net magnetization (per ion) for spiral structure with increasing field at temperature not too far below T_N. In (a) for $T_0 < T < T_N$, magnetization varies continuously; while in (b) for $T < T_0$, there is a discontinuity at H_c (after Kitano and Nagamiya[39]).

is shown for small γ (i.e. high T). The magnetization varies continuously with H/H_f, and has a cusp at $M/N = \frac{1}{2}\mu\sigma$, i.e. at a field

$$\text{high } T, \qquad H_c' = \left(\frac{1}{2} - \frac{\gamma}{16}\right) H_f \qquad (2.44)$$

As γ increases (i.e. T decreases), the slope of the magnetization just below H_c becomes steeper and becomes infinite at a critical temperature T_0 for which $\gamma = 8/11$, so that

$$\frac{T_N - T_0}{T_N} \simeq \frac{8}{11} \frac{\mathscr{J}(k_0) - \mathscr{J}(0)}{2\mathscr{J}(k_0)} \qquad (2.45)$$

For $\mathscr{J}(k_0) - \mathscr{J}(0)$ small compared to $\mathscr{J}(k_0)$, as is typical, T_0 is close to T_N. The magnetization behavior for $T < T_0$ is shown in Figure 2–4b. A discontinuous (i.e. first order) transition from distorted spiral to fan occurs at a field H_c separating equal areas enclosed by the double valued magnetization curve.

Actually, Kitano and Nagamiya show that the approximations involved in their calculations are quite good at all temperatures so long as the quantities defined in Eqns (2.24), b', c', d', and similar higher harmonic differences between $\mathscr{J}(nk_0)$ and $\mathscr{J}(k_0)$ satisfy the condition, b', c', d', etc. $\gg a'$. Taking into account the temperature dependence of H_c implicitly expressed through γ and H_f, they found the value of H_c/H_f falls in the range between 0.5 and 0.414 for all values of γ and b'/a'. Thus the effects of an applied magnetic field on a spiral structure are as indicated in Figure 2–5. For small fields there is a slight distortion of the spiral with a small net moment along H. For temperatures until T approaches T_N, at a critical field H_c, there is an abrupt transition to a fanlike structure with a large net moment along H. As the field increases still further, the angular amplitude of the fan decreases continuously until complete ferromagnetic alignment is achieved in a second order transition at $H=H_f$ which is approximately twice H_c. For parameters typical of the heavy rare earths, H_c is of the order of ten thousand Oersted or less, so that all these structures can occur in conventional magnetization experiments.

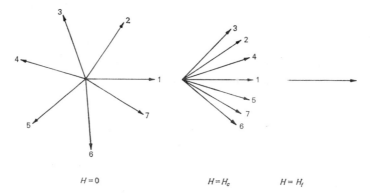

$$H=0 \qquad\qquad H=H_c \qquad\qquad H=H_f$$

Figure 2–5. Effect on magnetic structure of applied magnetic field in plane of spiral.

Planar anisotropy effects arising from both the undistorted crystal-field and from magnetostriction can have marked effects on the applied field dependence of the magnetic structure. As shown by Cooper and Elliott[40] such effects can be seen most elegantly by use of spin wave theory. (See Chapter 5 for a discussion of spin wave theory.) The transition from fan to ferromagnet occurs when, approaching from the ferromagnetic regime, the spin wave of wave vector $\mathbf{k}_0$ becomes unstable, i.e. $\omega(\mathbf{k}_0)=0$.

When planar anisotropy effects are included, the spin wave energy $\omega(\mathbf{k}_0)$ for the ferromagnet is given by[40]

$$\hbar\omega(\mathbf{k}_0)=\{[-2V_2J-2a'/J+\lambda\beta(6H_h+H)]$$
$$\times[-2a'/J+\lambda\beta(36H_h+H)]\}^{\frac{1}{2}} \quad (2.46)$$

where

$$H_h=-V_6{}^6J^5/\lambda\beta \quad (2.47)$$

at $T=0$ (and the temperature dependence can be appropriately included as shown in Ref. 23). So $\omega(\mathbf{k}_0)=0$, and the transition from fan to ferromagnet in the presence of planar anisotropy, is given by

$$H_f'=H_f-36H_h \quad (2.48)$$

On the other hand, from energetic considerations[5] the transition from fan to spiral occurs at

$$H_c' = H_c - H_h \qquad (2.49)$$

The difference in a factor of 36 for the effect of planar anisotropy on shifting the two critical fields can be easily understood. In determining H_c one compares the anisotropy energy to the Zeeman and exchange energies. On the other hand, at H_f the pertinent comparison is between torques or second derivatives of energy with angle. For a hexagonal anisotropy this introduces a factor of 36.

As the planar anisotropy gets larger, there are striking qualitative changes from the behavior shown in Figure 2–5. Because the effect of the planar anisotropy is greater on H_f than on H_c, as H increases the fan phase is eliminated; and the transition at H_c is directly to the ferromagnet. Presumably magnetostriction can have the same effect. For sufficiently large planar anisotropy or magnetostriction effects, one has a ferromagnet rather than a spiral even without any applied field. As discussed in Section 2.2.2.2, which follows, such effects account for the spontaneous transition at the Curie temperature from spiral to ferromagnet with lowering temperature in Tb and Dy, where it has been shown[41] that the magnetostriction effects provide the dominant driving mechanism for the transition.

2.2.2.2. Thermal First Order Transition from Spiral to Ferromagnetic Arrangement

For Tb and Dy a spiral magnetic structure has been found in the range of temperature between the Neel temperature, T_N, and the Curie temperature, T_C. For example, in Dy this holds between[42] 178.5 and 85 K. As shown in Figure 2–6, the spiral turn angle between the moments of successive hexagonal layers varies with temperature.[43] For Dy the spiral angle decreases

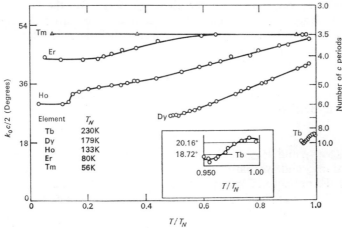

Figure 2–6. Temperature variation of the interlayer phase change $k_0c/2$ (e.g. interlayer turn angle for spiral structure) giving the periodicity of the magnetic structure for the heavy rare earths. Inset gives data for Tb on an expanded scale (after Koehler[43]).

from about $43°$ at T_N to about $26°$ at T_c. For Tb the spiral arrangement holds for only a narrow temperature range between 230 K and about 221 K. The spiral angle is much smaller than in Dy and varies from about $20°$ to $17°$. For both Dy and Tb there is a spontaneous transformation at T_c to a ferromagnetic configuration. It should be emphasized that the wave vector giving the spiral angle immediately above T_c has a substantial nonzero value in both cases. For Dy a thermal hysteresis[44] of 5 or 6 K has been observed for this transition. For both Dy and Tb there is appreciable lattice distortion[31, 45] ($\delta l/l \sim 10^{-2} - 10^{-3}$) at the transition.

In this section we will discuss the driving mechanism for the thermal first order transition from spiral to ferromagnet with decreasing temperature. The transition occurs because there is a competition between the exchange interaction for the undistorted hcp lattice favoring a spiral arrangement on the one hand, and the magnetostriction or the hexagonal anisotropy effects (of the undistorted lattice) favoring ferromagnetic alignment on the other hand. The reason that the transition occurs is that the temperature dependence of these different contributions to the free energy differ sharply. The magnetostriction and planar anisotropy terms are larger at low temperature but fall off more sharply with increasing T, so that at the high temperature end of the ordered regime, exchange effects for the unstrained lattice dominate. In particular, the present author[23, 41] has shown that the dominant driving force for the spiral-to-ferromagnetic transition in Dy and Tb is the energy of cylindrical symmetry associated with the lowest order magnetostriction effects. Basically, the spiral arrangement serves to restrain ("clamp") each successive plane along the c-axis from developing the strain that would minimize the combined elastic and magnetoelastic energy. Transition to a ferromagnet allows such energetically favorable strains to develop. This cylindrically symmetric magnetostrictive energy equally favors ferromagnetism pointing in any direction in the hexagonal plane, and therefore does not enter into the experimentally measured hexagonal anisotropy energy. There is a second, much smaller, contribution to the driving energy for the transition, the hexagonal planar anisotropy energy of the undistorted lattice. (Actually the hexagonal anisotropy could also arise from magnetoelastic effects.[23]) This second contribution does determine the easy direction once the transition to ferromagnetism occurs. We will now present the detailed treatment of these ideas and the experimental evidence for the conclusions just stated.

The Hamiltonian for the system of interacting $4f$ ionic moments is that discussed in Section 2.2.1.

$$\mathcal{H} = \mathcal{H}_{ex} + \mathcal{H}_{cf} + \mathcal{H}_{ms} + \mathcal{H}_{Zeeman} \qquad (2.50)$$

Here $\mathcal{H}_{ex}$, $\mathcal{H}_{cf}$, and $\mathcal{H}_{ms}$ are the exchange, crystal-field, and magnetostriction contributions; and we have added a fourth term, the Zeeman energy, to allow for the presence of an applied magnetic field. (For the reasons stated in Section 2.2.1 we omit anisotropic exchange effects for the undistorted crystal. Anisotropic exchange interactions would contribute to the two-ion magnetostriction effects. While most of our discussion will be devoted to showing how the spiral to ferromagnetic transition can come about through single-ion contributions to the magnetostriction, we will briefly mention the contribution of two-ion effects). The magnetostriction contribution is given by Eqn (2.7) as the sum of $\mathscr{H}_e$, the elastic energy, and $\mathscr{H}_m$, the magnetoelastic interaction coupling the spin system to the strains.

We begin by considering the change in the magnetostriction contribution to the free energy of an hcp crystal when the magnetic structure changes from spiral to ferromagnet. The total strain-dependent energy density E_{ms} leading to the magnetostriction for the ferromagnetic arrangement is

$$E_{ms} = E_e + E_m \tag{2.51}$$

where E_e is the elastic contribution and E_m is the magnetoelastic contribution.

To consider the spiral to ferromagnetic transition we consider only those terms contributing to the strains $\epsilon_1{}^\gamma$ and $\epsilon_2{}^\gamma$ and the associated magnetostriction coefficient λ^γ corresponding to the distortion of the circular symmetry of the basal plane by the rotation of the magnetization within the hexagonal plane. (A complete discussion of magnetostriction effects is given in Chapter 4.) Then the relevant elastic contribution to the free-energy density is of the form

$$E^\gamma = \tfrac{1}{2}c^\gamma[(\epsilon_1{}^\gamma)^2 + (\epsilon_2{}^\gamma)^2] \tag{2.52}$$

and the magnetoelastic contribution is

$$-E_m{}^\gamma = \cdot B^{\gamma,\,2}\epsilon_1{}^\gamma\tfrac{1}{2}(\alpha^2{}_\xi - \alpha_\eta{}^2) + B^{\gamma,\,2}\epsilon_2{}^\gamma\alpha_\xi\alpha_\eta \tag{2.53}$$

where α_ξ, α_η are the direction cosines of the magnetization relative to the crystal axes. $E_m{}^\gamma$ in (2.53) comes from the lowest order magnetoelastic effects, i.e. those arising from terms in the Hamiltonian having quadratic dependence on spin components.

Here $\epsilon_1{}^\gamma$ and $\epsilon_2{}^\gamma$ are irreducible strains with the symmetry of the hcp lattice.[46] They are related to the usual strains defined with respect to Cartesian axes as follows.

$$\epsilon_1{}^\gamma = \tfrac{1}{2}(\epsilon_{\xi\xi} - \epsilon_{\eta\eta}) \tag{2.54a}$$

$$\epsilon_2{}^\gamma = \epsilon_{\eta\xi} \tag{2.54b}$$

The elastic constant c^γ is related to the Cartesian stiffness constants by

$$c^\gamma = 2(c_{11} - c_{12}) \tag{2.55}$$

The equilibrium values of the strain obtained by minimizing E_{ms} with respect to $\epsilon_1{}^\gamma$ and $\epsilon_2{}^\gamma$ are

$$\bar{\epsilon}_1{}^\gamma = \frac{1}{2c^\gamma}\,B^{\gamma,\,2}(\alpha_\xi{}^2 - \alpha_\eta{}^2) \tag{2.56a}$$

$$\bar{\epsilon}_2{}^\gamma = \frac{1}{c^\gamma}\,B^{\gamma,\,2}\alpha_\xi\alpha_\eta \tag{2.56b}$$

Then the decrease in the total strain-dependent energy density E_{ms} on transition from spiral to ferromagnet aligned along the ξ-axis as in Dy is

$$\text{Dy,}\quad \bar{E}_{ms}{}^\gamma = -\tfrac{1}{2}c^\gamma(\bar{\epsilon}_1{}^\gamma)^2 \tag{2.57}$$

The equilibrium strains are given in terms of the magnetostriction coefficient by

$$\bar{\epsilon}_1{}^\gamma = \tfrac{1}{2}\lambda^\gamma(\alpha_\xi{}^2 - \alpha_\eta{}^2) \tag{2.58a}$$

$$\bar{\epsilon}^\gamma = \lambda^\gamma\alpha_\xi\alpha_\eta \tag{2.58b}$$

so that

$$\bar{E}_{ms}{}^\gamma = -\tfrac{1}{8}c^\gamma(\lambda^\gamma)^2 \tag{2.59}$$

(For Tb, where the ferromagnetic alignment is at $30°$ to the $\xi-$axis, (2.57), is replaced by

$$\text{Tb,} \qquad E_{ms}{}^{\gamma} = -\frac{c^{\gamma}}{2}\,[(\bar{\epsilon}_1{}^{\gamma})^2 + (\bar{\epsilon}_2{}^{\gamma})^2] \qquad (2.60)$$

However, the final result (2.59) is unchanged, as it must be since $\bar{E}_{ms}{}^{\gamma}$ has cylindrical symmetry with respect to the direction of magnetization.)

Our picture[2, 23, 41] for the spiral to ferromagnetic transition is that $\bar{E}_{ms}{}^{\gamma}$ gives an energy favoring ferromagnetism. Throughout the spiral temperature regime it competes with the exchange forces favoring a spiral moment structure. However, with decreasing temperature $\bar{E}_{ms}{}^{\gamma}$ increases in magnitude more rapidly than the exchange energy favoring a spiral over a ferromagnet; and the transition occurs when $\bar{E}_{ms}{}^{\gamma}$ becomes equal to that exchange energy.

It is simple to estimate the temperature dependence of $\bar{E}_{ms}{}^{\gamma}$. Throughout the spiral regime c^{γ} has little temperature dependence, so that the temperature dependence of $E_{ms}{}^{\gamma}$ is mostly due to the variation of $(\lambda^{\gamma})^2$. (There is a fairly substantial anomaly in c^{γ} right at T_c where the lattice distorts.[30]) The temperature dependence of λ^{γ} is given by the theory of Callen and Callen[46] for single ion magnetostriction as

$$\lambda^{\gamma} = \lambda^{\gamma}(0)\,\hat{I}_{5/2}[\mathscr{L}^{-1}(\sigma)] \qquad (2.61)$$

Here $\hat{I}_{5/2}[\mathscr{L}^{-1}(\sigma)]$ is the ratio of the hyperbolic Bessel function of order $5/2$ to that of order $1/2$, where the argument of the Bessel functions is the inverse Langevin function of the reduced magnetization. This temperature dependence is in excellent agreement with the experimental behavior[31, 32] in both Dy and Tb. This indicates that the single ion contribution to the magnetostriction dominates, or that the one-ion and the anisotropic two-ion temperature dependence are very similar. For our purposes, which of these is true does not really matter. The expression (2.61) in terms of hyperbolic Bessel functions is an approximation valid for large J, i.e. a classical approximation for the single-ion anisotropy. More generally, one can show that terms in the magnetoelastic Hamiltonian which are lth order in J give contributions to

$$\lambda^{\gamma} \sim \sigma\frac{l(l+1)}{2} = \sigma^3 \qquad \text{for } l = 2 \text{ at low } T$$

and $\sim \sigma^l = \sigma^2$ for $l = 2$ as T approaches T_c. This low and high temperature limiting behavior is the same for both single-ion and two-ion anisotropy. For two-ion anisotropy the transition between these limits is governed by the range of the two-ion interaction. Callen and Callen[46] have performed calculations indicating fairly generally that the two-ion behavior goes over to the σ^2 limit rather quickly with increasing temperature. This indicates that of the two possibilities for understanding the experimental temperature dependence of the magnetostriction in Dy and Tb—dominance of the single-ion effects or similarity of single-ion and two-ion effects—the former is the explanation.

Until the temperature is quite close to T_N, $\hat{I}_{5/2} \sim \sigma^3$, so that $E_{ms}{}^{\gamma} \sim \sigma^6$ throughout the temperature range of interest. Thus the energy favoring ferromagnetism increases sharply with lowering temperature. There is also a second, much smaller contribution to the driving energy, the hexagonal planar anisotropy energy of the undistorted lattice. This is given by $V_6{}^6 J^6 \hat{I}_{13/2}[\mathscr{L}^{-1}(\sigma)]$. (This gives the temperature dependence if the hexagonal anisotropy is that of the undistorted crystal-field. If it comes from magneto-elastic effects,[23] $\hat{I}_{13/2}$ is replaced by $\hat{I}_{5/2}\hat{I}_{9/2}$. The somewhat different temperature dependence is not significant for the present purposes.) This follows from the temperature dependence of the thermal average $\langle Y_6{}^6(\mathbf{J}) \rangle$ in the theory of Callen and Callen.[25] In that theory, the anisotropic contribution to the free energy is treated as a first-order perturbation on the cylindrically symmetric, about the magnetization direction, energy consisting of the Zeeman and isotropic exchange terms; and thus all the $\langle Y_1{}^m \rangle$ have the same temperature dependence as $\langle Y_1{}^0 \rangle$, viz. $\sim \hat{I}_{13/2}[\mathscr{L}^{-1}(\sigma)]$.

So the total driving energy, E_d, favoring the transition from spiral to ferromagnet is given by

$$E_d = -\tfrac{1}{8}c^\gamma(\lambda^\gamma[T=0])^2(\hat{I}_{5/2}[\mathscr{L}^{-1}(\sigma)])^2 + V_6{}^6 J^6 \hat{I}_{13/2}[\mathscr{L}^{-1}(\sigma)] \qquad (2.62)$$

The $V_6{}^6$ term has a temperature dependence $\sim \sigma^{21}$, so that contribution drops off much more rapidly with temperature than that from $\bar{E}_{ms}{}^\gamma$. Thus although for Dy the two contributions are comparable for $T=0$, by T_c and above σ has dropped off sufficiently so that the first term in (2.62), $\bar{E}_{ms}{}^\gamma$, dominates. Using experimental values of the elastic, magnetoelastic,[31] and hexagonal anisotropy[47] coefficients for Dy at $T=0$, the present author[41] calculated E_d. At $T=0$,

$$\text{Dy,} \qquad E_{ms}{}^\gamma(T=0) = -\tfrac{1}{8}c^\gamma(\lambda^\gamma)^2 = -2.0 \text{ K/atom} \qquad (2.63)$$

$$V_6{}^6 J^6 = -2.4 \text{ K/atom}$$

and the temperature dependence was calculated using the hyperbolic Bessel function as in Eqn (2.62); where the reduced magnetization for Dy was obtained from the data of Behrendt et al.[42]

For Dy, an analysis by Elliott gives the experimental values for the term in the free energy which tries to drive the system ferromagnetic in the spiral temperature range between 85 and 179 K. Elliott used the fact that in the spiral regime, a magnetic field of a certain critical size is sufficient to give a discontinuous transition from a spiral to a ferromagnet.[42] This field then supplies the difference in energy between the spiral and ferromagnetic arrangements, and goes to zero at T_c.

$$E_{ex\ spiral} = -\mu_0 H\sigma + E_d + E_{ex\ ferro} \qquad (2.64)$$

Elliott used the three-layer interaction model discussed in Section 2.2.1 to evaluate the exchange energy from experimental information. Thus using (2.11), (2.64) becomes

$$-[B_0 + 2B_1 \cos \frac{k_0 c}{2} + 2(B_2 - C\sigma^2) \cos k_0 c] \sigma^2$$

$$= -\mu_0 H_c\sigma + E_d - [B_0 + 2B_1 + 2(B_2 - C\sigma^2)] \sigma^2 \qquad (2.65)$$

where we use the replacement of (2.14) to allow for the temperature dependence of the spiral angle. Equation (2.65) can be simplified using (2.12) for the spiral angle to give

$$E_d/\sigma^2 = \mu_0 H_c/\sigma - B_1 \left(1 - \cos \frac{k_0 c}{2}\right)^2 \Big/ \cos \frac{k_0 c}{2} \qquad (2.66)$$

All the quantities on the right side of (2.66) are known experimentally. (B_1 was determined[5] from the values of k_0 and H_c as $T \to T_N$ where E_d is negligible.

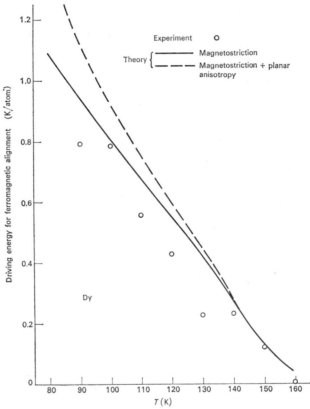

Figure 2–7. Temperature dependence of driving energy for ferromagnetic alignment in Dy (after Cooper[41]).

The experimentally determined E_d for Dy is shown in Figure 2–7 together with the theoretical result from (2.62). As can be seen from the figure, the values of the theoretical driving energy give excellent agreement with the experimental values. As shown in the figure, for temperatures approaching and exceeding T_c, the magnetostriction term in the driving energy dominates. Clearly, whichever of the two contributions is dominant in assisting the applied field in bringing about the transition for T just above T_c, is also dominant in driving the spontaneous transition at T_c. Thus we conclude that the driving force for the spontaneous spiral to ferromagnet transition in

Dy is the energy of cylindrical symmetry associated with the lowest order magnetostriction effects.

One can also[23, 41] obtain experimental values of $\bar{E}_{ms}$ ($T=0$) and $V_6{}^6J^6$ for Tb from the work of Rhyne and Legvold,[32] and of Rhyne and Clark,[33] respectively. This gives for Tb,

$$\text{Tb,} \qquad \bar{E}_{ms}{}^\gamma(T=0) = -\tfrac{1}{8}c^\gamma(\lambda^\gamma)^2 = 1.97 \text{ K/atom}$$

$$V_6{}^6J^6 = -0.57 \text{ K/atom} \qquad\qquad (2.67)$$

Hence the dominance of the magnetostriction contribution to the driving energy is even more pronounced for Tb than for Dy.

Evenson and Liu[10] have given a more detailed theory of the way in which magnetoelastic effects serve as the principal driving force stabilizing the ferromagnetic state. Their treatment gives a microscopic demonstration of the lattice clamping effects and allows for refinements on our picture of that effect. They allow for inhomogeneous strain effects and show that when the magnetic ordering is periodic, there is a thin layer—for typical parameters only a few lattice constants—at the surface of the crystal where the lattice is distorted with the same periodicity. The bulk of the crystal remains unstrained. In agreement with the assumption of the present author,[23] they found the lattice to be completely clamped with respect to the ϵ^γ strains. However, they included clamping effects on the strains $\epsilon^{\alpha,1}$ and $\epsilon^{\alpha,2}$ coming from two-ion magnetoelastic effects; and they found these strains to be only partially clamped. Including such effects and the change in elastic constants at T_c, for Dy Evenson and Liu found a difference in exchange energy between the spiral and ferromagnetic states (i.e. the energy to be supplied by the driving force) of 2.2 K/atom. This compares to the earlier value of about 1.2 K/atom (see Figure 2–7) found by the present author[41] and an experimental value estimated at about 0.9 K/atom (extrapolating experimental points in Figure 2–7 to 85 K). Rosen and Klimker[30] have recently measured the elastic constants of Dy between 4.2 K and 300 K. Using their results and the theory of Evenson and Liu, they estimated the change in magnetoelastic energy at T_c to be about 3.2 K/atom. These differences are probably not too significant considering the problems involved in a realistic microscopic picture for the lattice clamping effect and that different contributions are included in the various calculations. The real significance is that the microscopic theory basically supports the original simple physical idea of lattice clamping in the spiral regime.

The quantitative arguments presented above show that the dominant driving force for the spiral-to-ferromagnetic transition in Dy and Tb is the energy of cylindrical symmetry associated with the lowest order magnetostriction effects. This is in contrast with other possibilities that may be considered, in particular the planar crystal-field anisotropy of the unstrained crystal, or explicit temperature dependence of the exchange mechanism. We have already discussed the former of these. So far as the latter is concerned, one would envision the ferromagnetic transition occurring because of a qualitative change in the nature of the RKKY interaction caused by the difference in exchange splitting of the conduction bands between the ferromagnetic and spiral states (i.e. superzone energy gaps have to be considered in the spiral state). The possibility of a first order transition from the spiral to the ferromagnet exists if, for some range of temperature, $\mathscr{J}(\mathbf{q})$ for a spiral structure has a maximum at $\mathbf{k}_0$, and $\mathscr{J}(\mathbf{q})$ for a ferromagnetic structure has a maximum at $\mathbf{q}=0$. Then if the difference $\mathscr{J}_{\text{spiral}}(\mathbf{k}_0)-\mathscr{J}_{\text{ferro}}(0)$ (for finite $\mathbf{k}_0$) change from positive to negative as temperature is lowered, a first order transition occurs.[47a] (We note, however, that for a simple one-dimensional parabolic band (Evenson and Liu[10] have calculated the total energy at a given temperature for a continuous variation of $\mathbf{k}_0$ from 0 through all finite values [see Figure 8 of Evenson and Liu and the related discussion]; and they find that the exchange band splitting effect gives a second order transition, i.e. $\mathbf{k}_0$ goes to zero continuously on approaching T_c). Our discussion above shows that for Dy the magnetoelastic effects, rather than the temperature dependence of the exchange mechanism, provide the dominant driving force for the ferromagnetic transition. For Tb, and dilute alloys of Ho with Tb, Mackintosh[47a] suggests that a decrease in size of the peak in $\mathscr{J}(\mathbf{q})$ as T is lowered is a consequence of the temperature dependence of the superzone energy gaps, i.e. explciit temperature dependence of the exchange mechanism; but that the transition to ferromagnetism itself occurs because of magneto-elastic forces. These ideas and our study then indicate that the situation observed in the neutron scattering experiments,[48] viz., the presence of a clearly defined maximum of $\mathscr{J}(\mathbf{q})$ in the spiral regime in a Tb–Ho alloy for a $\mathbf{q}$ corresponding to the spiral period and the considerable decrease or possible absence of such a maximum in the ferromagnetic regime, is the result of the transition (i.e. of the change in lattice and magnetic structure), and is not the principal driving mechanism for the transition. Conceptually, $\mathscr{J}(\mathbf{q})$ could retain its maximum at some incommensurate $\mathbf{q}$ in the ferromagnetic regime as is observed in[48a] Ho–10% Tb and[48b] Dy.

Finally, we point out that the same terms in the effective spin Hamiltonian for the localized $4f$ ions that give the lowest order magnetostriction energy (of cylindrical symmetry) have profound effects for the spin wave behavior as discussed in Chapter 5. This is because for long wavelength spin waves, the appropriate picture for the magnetoelastic effects is the frozen lattice approximation.[23, 41, 49] In that picture the uniform mode spin wave frequency is determined by keeping the strain frozen at its equilibrium position. Thus in the excited spin wave state the relative orientation of moment and strain changes, and there is a net increase of energy relative to the equilibrium state even though the equilibrium energy associated with the magneto-striction has cylindrical symmetry. The frozen lattice effect has been experimentally observed in the last year or two in neutron scattering,[50] far infrared reflectivity,[51] and high frequency microwave resonance[52] experiments.

2.3. THE CRYSTAL-FIELD AND RARE EARTH MAGNETISM

In this Section we will focus our attention on the part played by crystal-electric-field effects in rare earth magnetism. In Section 2.3.1 we will discuss situations, relevant to the light rare earths, where crystal-field effects are comparable to exchange, and fundamentally affect the question of whether the material orders magnetically at all. In Section 2.3.2 we will discuss the experimental measurement of the size of crystal-field effects and how the results of such measurements relate to the fundamental question of the origin and nature of the crystal-electric-field in metals.

2.3.1. Induced Moment Systems

As we have discussed in Section 2.2, for the heavy rare earth metals crystal-field effects play a very important role in determining the detailed nature of the magnetic ordering; however, this role is still secondary with regard to the question of whether the metals order magnetically at all. For the light rare earths, crystal-field effects are larger relative to exchange, and this is no longer true, especially for Pr.

For a rare earth material at temperatures of the order of the crystal-field splitting, typically tens of K, the crystal-field can diminish or destroy the orbital contribution to the ionic moment. However, because the spin-orbit coupling is very strong, if the crystal-field succeeds in restraining the orbital moment, the orbital moment also holds back the spin moment from following the effect of an applied magnetic field; and thus the crystal-field diminishes or destroys the total ionic moment.

In this Section we will discuss the situation when crystal-field effects, which tend to destroy the ionic moment and hence tend to destroy magnetic ordering, are comparable to, or dominant over, the exchange effects which tend to have exactly the opposite effect. Such a situation can lead to a number of interesting magnetic properties. The most striking situation occurs when the crystal-field-only ground state of the rare earth ion is a singlet. (This can occur only for non-Kramers ions, i.e. those rare earth ions with an even number of f electrons and thus an integral J for the ground state multiplet.) Then as exchange increases, magnetic ordering at zero temperature occurs not through the usual process of alignment of permanent moments, but rather through a polarization instability of the crystal-field-only singlet ground state wave function.

For such induced moment systems, there is a threshold value for the ratio of exchange to crystal-field interaction necessary for magnetic ordering even at zero temperature.[53-59] We shall discuss the behavior as this threshold is approached and exceeded.

Bleaney[60] first recognized the relevance of singlet crystal-field ground state and induced moment behavior to magnetism in Pr metal. His discussion was with regard to the magnetic behavior of the ordinarily stable phase, the double hexagonal (dhcp) crystal structure where there are two types of site, cubic and hexagonal. Very recent experiments[61, 62] on single crystals of dhcp Pr, discussed below, show the relevance of these ideas. However, the ultimate fulfillment of expectations growing out of Bleaney's ideas was the discovery by Bucher *et al.*[63] of a new elemental ferromagnet fcc Pr, where the ferromagnetism arises from the induced moment effect on a singlet crystal-field ground state.

We will begin this Section in Part 2.3.1.1 with a discussion of the theoretically expected behavior for singlet ground state systems. The way in which magnetic ordering occurs will be demonstrated with molecular field theory and also by considering a "soft mode" instability for the collective excitations.

In Part 2.3.1.2 of Section 2.3.1 we shall discuss the experimental results pertaining to induced moment behavior. The most complete quantitative studies of singlet ground state magnetism performed up to the present time are in the rare-earth–group V compounds of NaCl structure;[64–69] and we shall discuss the observation and quantitative understanding of the ordering threshold in [68]$Tb_\zeta Y_{1-\zeta}Sb$. We shall discuss the behavior for dhcp Pr, single-crystal[61, 62] and polycrystalline,[70] for fcc Pr–Th alloys[71] and for fcc Pr, as yet studied[63] only in polycrystalline samples. Also, we shall mention interesting effects in[72] PrB_6 and[73–75] $TmAl_3$. In addition to singlet ground state materials, we shall briefly mention materials with rare earth ions having Kramers doublet ground states, where the crystal-field and induced moment effects strongly affect the magnetic behavior. In this regard we will discuss elemental[62, 76] Nd metal and the Ce–group V compounds of NaCl structure.[77–80]

Because of the large orbital contribution to the moment, for systems in which crystal-field effects are comparable to exchange, the effective exchange tends to be higher degree than quadratic in J and/or to be aniso-tropic.[3, 4, 81–85] In discussing the experimental situation, we shall mention several cases[62, 68, 86, 87] where such effects occur.

2.3.1.2. Theory of Singlet Ground State Magnetism

Figure 2–8 shows the crystal-field splitting of the ground state multiplet for Pr^{3+} and Tm^{3+} placed in sites of hexagonal (shown only for Pr^{3+}) or cubic symmetry. The crystal-field levels are labeled with their degeneracies and symmetry types. The ions shown have integral values of J, the total angular momentum, in their ground state multiplets; and the figure shows the common case where the crystal-field ground state is a singlet. In a hexagonal crystal-field the first excited state is also a singlet, while in a cubic

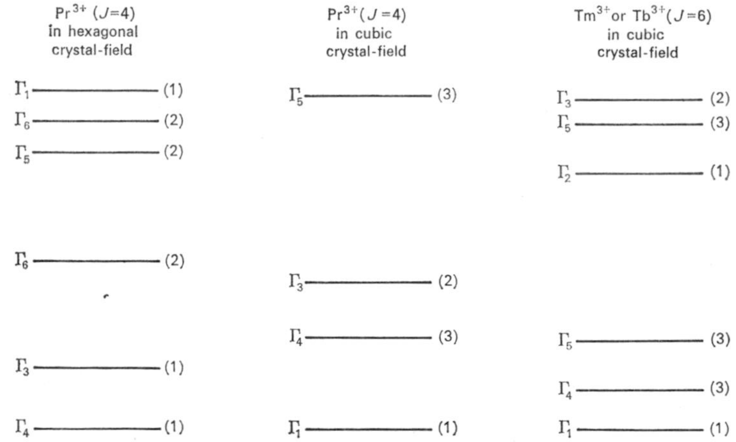

Figure 2–8. Typical crystal-field splitting of the ground state multiplet for Pr^{3+} in sites of hexagonal and cubic symmetry, and for Tm^{3+} in cubic site, when crystal-field parameters are such that ground state is singlet. The crystal-field levels are labeled with their degeneracies and symmetry types.

crystal-field the first excited state is a triplet. Typically the splitting from the ground state to the first excited state is between 10 and 100 K; while the overall splitting is several hundred K. (It is useful to remember that 1 meV = 11.606 K.)

The singlet ground state means that the ionic moment vanishes. In the absence of any exchange effects, we get a Van Vleck susceptibility at low temperature because an applied field admixes some excited state wave function into the ground state thereby inducing a magnetic moment.

We can ask what happens to the magnetization at low temperature as exchange increases. In particular, how does magnetic ordering first come about as exchange increases from zero for such a singlet ground state system?

This question has attracted much attention during the past ten years. We begin our discussion by showing that one can arrive at many of the gross features of induced magnetic behavior with a rather simple molecular field theory.

(a) Molecular Field Theory

We can qualitatively understand how magnetic ordering occurs in systems with crystal-field level schemes such as those shown in Figure 2–8 on the basis of a simple molecular field, as originally treated by Trammell[53, 54] and by Bleaney.[55] For simplicity, rather than treating the full crystal-field level scheme, we follow Bleaney[55] and consider the two-level model shown in Figure 2–9, where the excited state is also a singlet. Besides simplicity, this

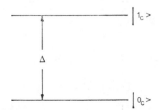

Figure 2–9. Crystal-field levels in two singlet level model.

two singlet state model has the advantage of showing purely induced moment effects. (Physically, the two singlet state model would apply, say for Pr^{3+} in a hexagonal environment when the other excited states were at much higher energies. We shall discuss this correspondence further below.)

We consider the Hamiltonian,

$$\mathcal{H} = \sum_i V_{ci} - \sum_{i \neq j} \mathcal{J}(\mathbf{R}_i - \mathbf{R}_j)\, \mathbf{J}_i \cdot \mathbf{J}_j - \lambda \beta H \sum_i J_{iz} \qquad (2.68)$$

This Hamiltonian has a crystal-field term giving for each ion the states $|0_c\rangle$ and $|1_c\rangle$ with energy splitting Δ, an exchange term, and a Zeeman term. The exchange is taken to be isotropic. Actually in real systems with large orbital contribution to the moment, there may be substantial higher degree and anisotropic exchange interactions. The way in which these have to be taken into account depends on their particular form. The isotropic exchange taken here will illustrate the physical effects of interest. Below, we shall mention the extent to which anisotropic and/or higher degree exchange is called for in understanding several systems of experimental interest.

In the absence of exchange the applied field admixes the two wave functions, and thereby induces the Van Vleck susceptibility. By conventional first-order perturbation theory, the susceptibility per ion is

$$1/\chi_{CF} = (\Delta/2\lambda^2\beta^2\alpha^2) \, [1/\tanh (\Delta/2kT)] \qquad (2.69)$$

where

$$\alpha \equiv \langle 0_c | J_z | 1_c \rangle \qquad (2.70)$$

(The z-axis has been chosen so that only the z component of angular momentum, J_z, has a nonzero matrix element between the crystal-field only singlet states. The off-diagonal matrix elements of J_x and J_y vanish. Of course, all diagonal matrix elements of angular momentum vanish for a singlet state. Such a choice of axes is always possible for an even-electron two-singlet-level system.[88] For a hexagonal system, the z-axis is the axis of hexagonal symmetry).

Now in a similar way a molecular field induces a magnetic moment in the singlet ground state (and the opposite moment in the excited state). So if one includes exchange in the molecular field approximation the magnetization (per ion) is

$$M = \lambda\beta\bar{J} = \chi_{CF} \left(H + \frac{2\mathcal{J}(0)}{\lambda\beta} \bar{J} \right) \qquad (2.71)$$

or

$$\frac{1}{\chi} = \frac{1}{\chi_{CF}} - \frac{2\mathcal{J}(0)}{\lambda^2\beta^2} \qquad (2.72)$$

and defining

$$A \equiv \frac{4\mathcal{J}(0) \, \alpha^2}{\Delta}, \qquad (2.73)$$

the only change in $1/\chi$ from (2.69) is the replacement

$$1/\tanh (\Delta/2kT) \rightarrow [1/\tanh (\Delta/2kT)] - A \qquad (2.74)$$

Thus the effect of exchange on $1/\chi$ in the molecular field theory,[55] as illustrated in Figure 2–10a, is simply to shift the curve of $1/\chi$ versus T rigidly downward (ferromagnetic exchange) or upward (antiferromagnetic exchange). The quantity A giving the shift is just the ratio of exchange interaction to crystal-field splitting. The value of A for which the susceptibility diverges at $T=0$ gives the threshold value of exchange for ferromagnetic ordering to occur at $T=0$. This is $A=1$.

For A exceeding the threshold value, it is a simple procedure to find the magnetization self-consistently as a function of temperature. The molecular field Hamiltonian is

$$\mathcal{H}_0 = \sum_i V_{ci} - 2 \, \mathcal{J}(0) \, \bar{J} \sum_i J_{iz} \qquad (2.75)$$

and the molecular field eigenstates are

$$|0\rangle = \cos \theta \, |0_c\rangle + \sin \theta \, |1_c\rangle \qquad (2.76a)$$

$$|1\rangle = -\sin \theta \, |0_c\rangle + \cos \theta \, |1_c\rangle \qquad (2.76b)$$

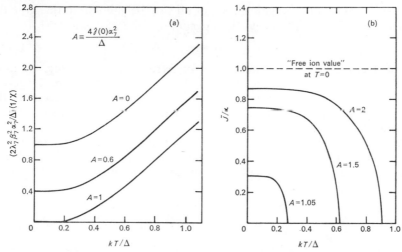

Figure 2-10. (a) Inverse susceptibility and (b) magnetization (in dimensionless form) versus temperature for the two singlet level model for varying ferromagnetic exchange. Here $A = 4\mathscr{J}(0)\alpha^2/\Delta$ (after Cooper[50]).

The energy eigenvalues are

$$E_0 = -E_1 = -\frac{\Delta}{2}\left(\cos 2\theta + \frac{A\bar{J}}{\alpha}\sin 2\theta\right) \qquad (2.77)$$

and the rotation angle which diagonalizes $\mathscr{H}_0$ is given by

$$\tan 2\theta = A(\bar{J}/\alpha) \qquad (2.78)$$

Then for A exceeding the threshold value, it is a simple matter to find the magnetization self-consistently as a function of temperature. This is given by

$$\bar{J}/\alpha = \sin 2\theta \tanh\{(\Delta/T)\,[\tfrac{1}{2}\cos 2\theta + \tfrac{1}{2}A(\bar{J}/\alpha)\sin 2\theta]\} \qquad (2.79)$$

This expression reflects the fact that in the absence of an applied field, the existence of a molecular field presupposes the existence of an ordered moment. This leads to a threshold value of A at $T=0$ for a finite $\bar{J}$ (i.e. ferromagnetism) to exist. This value,

$$A_f = 1 \qquad (2.80)$$

is the critical value for the polarization instability of the ground state wave function giving ferromagnetism, and corresponds to the divergence of the susceptibility.

One can go through the corresponding theory for an antiferromagnet, where the sublattice magnetization takes the place of the magnetization. For antiferromagnetism, the threshold value of A depends on the particular type of antiferromagnetic ordering. The simplest situation is that for which the moment of any ion is antiparallel to the moments of all neighbors with which it has exchange interaction (i.e. most simply when there is exchange with only one type of neighbor; that exchange is antiferromagnetic; and the moment alignment is antiparallel that of all such neighbors). In that case the critical value for antiferromagnetic ordering is

$$A_{a1} = -1 \qquad (2.81)$$

For A exceeding the threshold value for ferromagnetic ordering, the most notable features in the magnetization behavior shown in Figure 2–10b are the reduction of the Curie temperature for small A, and the reduction of the normalized magnetization (i.e. $\bar{J}/\alpha$) at zero temperature from its maximum "free ion" value of unity.

As shown in Figure 2–11, the rise in both the magnetization per ion at $T=0$ and the Curie temperature as A increases beyond the critical value of unity is quite sharp. Eventually, of course, the ordered moment at $T=0$

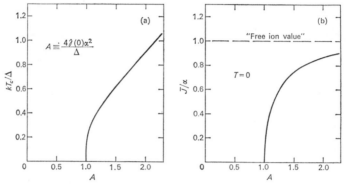

Figure 2–11. (a) Spontaneous magnetization (in dimensionless form) versus $A \equiv 4\mathscr{J}(0)\alpha^2/\Delta$ for the two singlet level model at $T=0$. (b) T_c/Δ versus A for the two singlet level model.

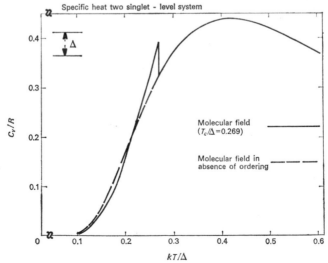

Figure 2–12. Specific heat versus temperature for the two singletl evel model in the molecular field approximation.

saturates at the free ion moment, and the Curie temperature varies linearly with A (corresponding to the Curie–Weiss Law regime in the susceptibility behavior).

Figure 2–12 illustrates another interesting characteristic of singlet ground state systems, the specific heat behavior, which can be found from the same sort of self-consistent theory used to find the magnetization. The broad peak is the Schottky anomaly associated with the crystal-field splitting from the ground state to the first excited level (i.e. the contribution to the specific heat caused by thermally populating a crystal-field level at an energy Δ above the ground state.) Location of this peak unambiguously identifies the crystal-field splitting. A second, lower temperature and sharper, peak is shown. This is the anomaly associated with magnetic ordering. This is shown in Figure 2–12 for a value of A not too far above threshold, so that the ordering temperature is below the temperature of the Schottky anomaly. Now this specific heat anomaly is much less prominent than that usual for magnetic ordering. This decrease in the specific heat associated with magnetic ordering basically comes about because much of the entropy of the system has already been removed by ions dropping into the singlet crystal-field state before magnetic ordering occurs.[55]

As we shall see in Section 2.3.1.2, the behavior shown in Figures 2–10, 2–11 and 2–12 corresponds to some of the qualitative features that are most striking in the experimental behavior of induced moment systems.

(b) Collective Excitation Theory

In Part (a) above, we have seen that a rather simple molecular field theory yields a picture of induced magnetic behavior showing a number of striking characteristics. As we shall see, these provide a good basis for understanding the experimental behavior. On the other hand, we would also like to discuss the behavior of the collective excitations for singlet ground state systems. This behavior is of interest for several reasons. First, one obtains what is probably the best physical picture of the threshold behavior for magnetic ordering in terms of the instability of one of the collective modes. This "soft" mode instability is analogous to the phonon instability in some ferroelectrics. Second, the dispersion behavior for the energy of the elementary excitations is of interest since understanding it provides the basis for calculating the various thermodynamic properties. Third, and of great potential importance, is that one can hope to observe directly the excitation spectrum by neutron scattering, and thereby directly determine the form of the exchange interaction. The final reason for studying the collective excitations is that these are a type of exciton that exists both in the paramagnetic and magnetically ordered regimes, and that are quite different from ordinary spin waves. Indeed, as we shall see, there are no spin waves for a two singlet level system. As discussed below for the physical example of Pr^{3+} in a hexagonal crystal-field, this is because there are no "spin-flip" transitions between two singlet states. For Pr^{3+} such transitions giving rise to spin waves, occur to higher lying crystal-field states. *In this respect, the present discussion is worth keeping in mind when considering the spin behavior for the heavy rare earth metals as treated in Chapter 5. Clearly the Holstein–Primakoff transformation used in such treatments is meaningless, for large enough crystal-field. (Indeed the question of the applicability, or inapplicability, of the Bogoluibov-type approximation discussed in (i) which follows is really a generalization of the question of the applicability of the Holstein–Primakoff transformation.) As exchange and crystal-field effects become comparable, the true nature of the excited states lies somewhere between conventional spin waves, modified for large anisotropy effects, and the type of exciton to be discussed in this section.*

There have been no experiments performed as yet attempting to observe directly (either by inelastic neutron scattering or by the absorption of electromagnetic radiation) the excitons we shall discuss for singlet crystal-field ground state systems. Thus the discussion which follows is in part intended to serve as a guide in planning and understanding such experiments.

Note added in proof: Initial reports of such experiments on dhcp Pr [Rainford, B. R. and Gylden Houmann *Phys. Rev. Lett.* **26**, 1254 (1971)] and Tb 5b [Holden, T. M. *et al. Bull. Amer. Phys. Soc.* **16**, 325 (1971) Paper BC 7] have now appeared while Birgeneau and Als-Nielsen (private communication) have observed singlet ground state excitons in fcc Pr and Pr₃Tl.

(*i*) *Bogoluibov-type Bosons*. The behavior of the collective excitations for singlet crystal-field ground state systems was first studied by Van Vleck[89] and by Trammell[53] using a Bogoluibov-type approximation.[90] Subsequently this technique was also employed by Grover[91] and by the present author.[56]

We continue to consider the two singlet level system and the Hamiltonian of (2.68). The Bogoluibov-type theory proceeds by assigning fermion operators to each molecular-field energy level at each ion site. The molecular field Hamiltonian becomes,

$$\mathcal{H}_0 = \sum_i (E_0 d^*_{i0} d_{i0} + E_1 d^*_{i1} d_{i1}) + N \mathcal{J}(0) \bar{J}^2 \tag{2.82}$$

and the correction, $\mathcal{H}_1$ giving the difference between $\mathcal{H}$ and $\mathcal{H}_0$ is

$$\mathcal{H}_1 = -\sum_{i \neq j} \sum_{n,m,n',m'} \mathcal{J}(\mathbf{R}_i - \mathbf{R}_j)\langle n|j_i|m\rangle \cdot \langle n'|j_j|m'\rangle d^*_{in} d_{im} d^*_{jn'} d_{jm'} \tag{2.83}$$

with

$$\mathbf{j}_i \equiv \mathbf{J}_i - \bar{J}\epsilon_z \tag{2.84}$$

(ϵ_z denotes a unit vector along the z-axis).

It should be noted that the Hamiltonian in this fermion representation is not exact since $\mathcal{H}_0$ allows the unphysical states where both molecular field levels at one site are unoccupied or occupied simultaneously.

The approximation proceeds further by defining operators a and a^*,

$$a_{i1} = d^*_{i0} d_{i1}, \qquad a^*_{i1} = d^*_{i1} d_{i0} \tag{2.85}$$

These can be approximated as boson operators at low temperature by neglecting occupation of the excited states on considering the commutators, and on treating the ground state occupation number operator as a c-number

$$[a_{i1}, a_{j1}] = [a^*_{i1}, a^*_{j1}] = 0 \tag{2.86}$$

$$[a_{i1}, a^*_{j1}] = \delta_{ij} n_{i0} - \delta_{ij} n_{i1} \simeq \delta_{ij} \tag{2.87}$$

These further approximations, together with the original fermion approximation, restrict the treatment to low temperatures and neglect correlation effects. A further source of inaccuracy even at zero temperature is the assumption of complete occupation of the molecular field ground state. This is not strictly correct since the true ground state differs from the molecular field ground state. However, once these approximations are made, the Hamiltonian can be written as a quadratic form in boson operators and Fourier transformed to give,

$$\mathcal{H} = \mathcal{H}_0 + \mathcal{H}_1 \tag{2.88}$$

$$\mathcal{H}_0 = NE_0 + N\mathcal{J}(0)\,\bar{J}^2 + \sum_q (E_1 - E_0)\,a^*_{q1}a_{q1} \tag{2.89}$$

$$\mathcal{H}_1 = -\sum_q \mathcal{J}(q)\langle 1\,|J_z|\,0\rangle^2\,[2a^*_{q1}a_{q1} + a_{q1}a_{-q1} + a^*_{q1}a^*_{-q1}] \tag{2.90}$$

$\mathcal{H}$ can then be diagonalized in a conventional manner (e.g. see Chapter 5) to give the excitation spectrum. In the paramagnetic regime with H and $\bar{J} = 0$ this is,

$$E_q = \Delta(1 - A\gamma_q)^{\frac{1}{2}} \tag{2.91}$$

with

$$\gamma_q \equiv \mathcal{J}(q)/\mathcal{J}(0) \tag{2.91b}$$

The Bogoluibov-type boson excitation spectrum found as A varies at $T = 0$ is shown in Figure 2–13. For zero exchange one has the sharp crystal-field-only level. As ferromagnetic exchange increases, one has greater dispersion about the crystal-field-only level. As the exchange increases, the $\mathbf{q} = 0$

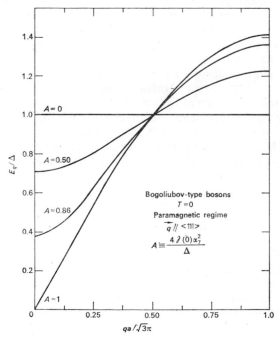

Figure 2–13. Dispersion curve at $T = 0$ of elementary excitations in the two singlet level model for a simple cubic lattice with nearest-neighbor ferromagnetic exchange. This is shown calculated in the Bogoliubov type approximation for several values of A in the paramagnet regime.

mode energy drops more rapidly, until at $A=1$ the $\mathbf{q}=0$ mode becomes unstable signaling a transition to ferromagnetism. The criterion for ferromagnetism is exactly the same as in the molecular field theory. Indeed, this is a built-in feature because of the various approximations.

This theory has a number of deficiencies: the restriction to low temperature, the assumption that the molecular field ground state is exact, the built-in neglect of correlation effects between the excitations. In (ii) which follows, we discuss a theoretical treatment that overcomes these difficulties.

(ii) *Pseudospin Formalism.* The most appropriate technique for overcoming the difficulties of the effective boson formulation and for finding the collective excitation spectrum in the two singlet level model is the pseudospin formalism used by Wang and Cooper.[57, 58] Pink[92] has also used a very similar technique. To develop this theoretical technique and to understand the nature of the excitonic states, we use the formal correspondence to the magnetic behavior, and the formal technique used to study that behavior, in ordinary magnetic materials. These analogies pertain to both the nonmagnetic and the magnetic regimes in singlet ground state systems.

In the pseudospin treatment for determining the behavior of the collective excitations, one develops a theory where the difference in occupation number of crystal-field-only states in the paramagnetic regime, or of molecular-field states in the ordered regime, plays a role analogous to that of magnetic moment, or of the expectation value of the spin component along the direction of magnetization, for ordinary ferromagnetic systems. Another way of saying this is that the quantities analogous to the components of spin, the equations of motion of which are studied, are operators projecting the true wave functions of the system onto the molecular field wave functions.

The pseudospin formalism for the two singlet level problem is a variation of the well-known effective spin Hamiltonian technique,[93] and is based on the fact that the Hamiltonian for any two level system can be written in terms of four spin $\frac{1}{2}$ operators which correspond to 2×2 matrices with unity as one element, and zero as the other elements. (To emphasize that the pseudospin $\mathbf{S}$ is a spin in an abstract vector space, we label the pseudospin axes ζ, ξ, and η. These axes are, of course, not the cartesian axes in real space z, x, and y.) Then we assign $S_\zeta = -\frac{1}{2}$ to the molecular field ground state and $S_\zeta = +\frac{1}{2}$ to the molecular field excited state.

$$|1\rangle = -\sin\theta\,|0_c\rangle + \cos\theta\,|1_c\rangle \to |+\rangle = \begin{pmatrix} 1 \\ 0 \end{pmatrix} \qquad (2.92\text{a})$$

$$|0\rangle = \cos\theta\,|0_c\rangle + \sin\theta\,|1_c\rangle \to |-\rangle = \begin{pmatrix} 0 \\ 1 \end{pmatrix} \qquad (2.92\text{b})$$

Then we can project any operator, Ω_i, onto the manifold of molecular field states, and write it in terms of pseudospin $\frac{1}{2}$ operators.

$$\hat{\Omega}_i = \langle 0|\hat{\Omega}_i|0\rangle\,|0\rangle\langle 0| + \langle 1|\hat{\Omega}_i|1\rangle\,|1\rangle\langle 1| + \langle 1|\hat{\Omega}_i|0\rangle\,|1\rangle\langle 0|$$
$$+ \langle 0|\hat{\Omega}_i|1\rangle\,|0\rangle\langle 1| \quad (2.93)$$

where we then use the equivalences,

$$|0\rangle\langle 0| = \binom{0}{1}\ (01) = \binom{00}{01} = S_i^- S_i^+ \tag{2.94a}$$

$$|1\rangle\langle 1| = \binom{1}{0}\ (10) = \binom{10}{00} = S_i^+ S_i^- \tag{2.94b}$$

$$|1\rangle\langle 0| = \binom{1}{0}\ (01) = \binom{01}{00} = S_i^+ \tag{2.94c}$$

$$|0\rangle\langle 1| = \binom{0}{1}\ (10) = \binom{00}{10} = S_i^- \tag{2.94d}$$

to write $\hat{\Omega}_i$ in terms of the components of the pseudospin $\mathbf{S}$.

This allows us exactly to transform the Hamiltonian to the following form in terms of pseudospin $\frac{1}{2}$ operators.

$$\mathscr{H} = -2\sum_i E_0 S_{i\zeta} - \sum_{i\neq j} \mathscr{J}(\mathbf{R}_i - \mathbf{R}_j)\,\{\alpha_{01}S_i^- + \alpha_{10}S_i^+ + \alpha_{00}S_i^- S_i^+$$
$$+\alpha_{11}S_i^+ S_i^-\}\,\{\alpha_{01}S_j^- + \alpha_{10}S_j^+ + \alpha_{00}S_j^- S_j^+ + \alpha_{11}S_j^+ S_j^-\} + N\mathscr{J}(0)\bar{J}^2 \tag{2.95}$$

The α_{ij} are matrix elements of angular momentum between the molecular field states. With $\mathbf{j}$ defined as in (2.84),

$$\alpha_{10} = \alpha_{01} = \langle 1|j_z|0\rangle = \alpha\cos 2\theta \tag{2.96a}$$

$$\alpha_{00} = \langle 0|j_z|0\rangle = \alpha\sin 2\theta - \bar{J} \tag{2.96b}$$

$$\alpha_{11} = \langle 1|j_z|1\rangle = -\alpha\sin 2\theta - \bar{J} \tag{2.96c}$$

With α given by (2.70) and θ given by (2.78).

Actually, the two singlet level problem is sufficiently simple so that one can perform the transformation of (2.94) to the pseudospin operators by inspection. However, the explicit representation of the projection operators in terms of spinors is useful in more involved cases such as when the first excited state is a triplet.

In the transformed Hamiltonian (2.95), E_0 is the molecular field ground state energy; and the first term, having the form of a simple Zeeman term, is the molecular field Hamiltonian. The remaining expression gives the correction to molecular field theory. *The virtue of this pseudospin representation is that by means of an exact transformation we have put the Hamiltonian into a form where one can use the powerful techniques developed for treating conventional spin systems.*

In the paramagnetic regime, the angle θ in the molecular field wave functions becomes zero, and the molecular field wave functions become identical to the crystal-field-only wave functions. Then the pseudospin Hamiltonian simplifies, and has the form of an Ising interaction in a transverse magnetic field.

$$\text{paramagnetic,}\quad \mathscr{H} = \sum_i \Delta S_{i\zeta} - 4\sum_{i\neq j} \mathscr{J}(\mathbf{R}_i - \mathbf{R}_j)\,\alpha^2 S_{i\xi}S_{j\xi} \tag{2.97}$$

In the paramagnetic phase, one can find the collective excitation dispersion behavior at zero temperature by considering the equations of motion of the pseudospin operators, $|1\rangle\langle0|=S_i^+$ and $|0\rangle\langle1|=S_i^-$, that generate excited states.

$$i\hbar\dot{S}_g^{\pm}=[S_g^{\pm},\mathscr{H}]=\mp\Delta S_g^{\pm}\mp4\alpha^2\sum_f\mathscr{J}(\mathbf{R}_f-\mathbf{R}_g)(S_{g\zeta}S_f^++S_{g\zeta}S_f^-)\quad(2.98)$$

To limit the equations of motion to a finite set of coupled equations, in the usual way we have to make approximations to terminate the chain of equations. The simplest approximation allowing us to decouple and linearize the equations of motion is the Random Phase Approximation. This involves replacing S_ζ in the equations of motion by its expectation value at zero temperature.

$$\text{RPA,}\quad S_{g\zeta}S_f^{\pm}\to\bar{S}_\zeta S_f^{\pm}\quad(2.99)$$

Once this is done, one can in the usual way perform a spatial Fourier transformation of the equations of motion, and find the dispersion relationship.

$$\text{RPA paramagnetic,}\quad E_q=\Delta(1+2\bar{S}_\zeta A\gamma_q)^{\frac{1}{2}}\quad(2.100)$$

with

$$\gamma_q\equiv\mathscr{J}(\mathbf{q})/\mathscr{J}(0)\quad(2.101)$$

The generating operators for the excitation modes are

$$\mathscr{S}_q^+=\cosh\beta_q S_q^+-\sinh\beta_q S_{-q}^-\quad(2.102)$$

with

$$\tanh2\beta_q=-\bar{S}_\zeta A\gamma_q/(1+\bar{S}_\zeta A\gamma_q)\quad(2.103)$$

From (2.100) we seen that a knowledge of $\bar{S}_\zeta$, the expectation value of the operator $S_{i\zeta}$, is necessary to determine the excitation spectrum. For this purpose $\bar{S}_\zeta$ must be determined self-consistently.[57, 58] At $T=0$ $\bar{S}_\zeta$ is an expectation value over the true ground state, $|\phi_0\rangle$ which satisfies the condition,

$$\mathscr{S}_q^-|\phi_0\rangle=0\quad(2.104)$$

for all $\mathbf{q}$. To evaluate $\bar{S}_\zeta$, we make use of the relationship,

$$S_{i\zeta}=-\tfrac{1}{2}+S_i^+S_i^-,\quad(2.105)$$

replace S_i^+ and S_i^- by their Fourier expansions, and take the ground state expectation value. Then use of (2.102), (2.103), and the property given by (2.104) yields,

$$\bar{S}_\zeta=-\tfrac{1}{2}\left(1+\frac{2}{N}\sum_q\sinh^2\beta_q\right)^{-1}\quad(2.106)$$

Then $\bar{S}_\zeta$ is determined by the self-consistent solution of Eqns (2.103) and (2.106).

If the exact ground state wave function were identical to the molecular field ground state wave function, $\bar{S}_\zeta$ at zero temperature would be $-\tfrac{1}{2}$. Instead, for a simple cubic lattice with nearest neighbor ferromagnetic exchange, as A approaches the critical value for ferromagnetism at zero temperature, $\bar{S}_\zeta$ differs by about 4% from $-\tfrac{1}{2}$. This deviation is a measure of the extent to which the true ground state wave function contains an admixture of the molecular field excited state.

One can extend the RPA theory to finite temperature by defining the retarded time Green's functions corresponding to the operators generating excited states,

$$G^-(g, l) \equiv -i\langle[S_g^-(t), S_l^+(0)]\rangle\theta(t) = \langle\!\langle S_g^-(t) ; S_l(0)\rangle\!\rangle, \qquad (2.107)$$

and by considering their equations of motion,

$$EG_E^-(g, l) = \frac{1}{2\pi}\langle[S_g^-, S_l^+]\rangle + \langle\!\langle[S_g^-(t), \mathcal{H}] ; S_l(0)\rangle\!\rangle_E \qquad (2.108)$$

To solve the equations of motion, one adopts approximations,

$$\langle\!\langle S_{g\zeta}S_f^- ; S_l^+\rangle\!\rangle \xrightarrow[g \neq f]{} \bar{S}_\zeta G_E^-(f, l) \qquad (2.109)$$

analogous to those involved in the decoupling at zero temperature. One finds the same dispersion relationship given by (2.100), where $\bar{S}_\zeta$ now denotes the self-consistently determined thermal average.

Figure 2–14 shows the excitation spectrum at zero temperature as exchange increases from zero in the paramagnetic regime. This is for a simple cubic crystal with nearest neighbor ferromagnetic exchange. At zero exchange

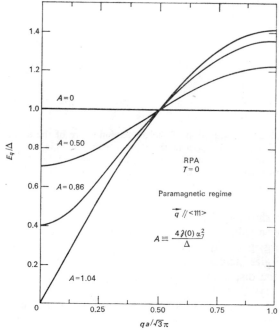

Figure 2–14. Dispersion curve at $T=0$ of elementary excitations in the two singlet level model for a simple cubic lattice with nearest-neighbor ferromagnetic exchange. This is shown calculated in the RPA for several values of A in the paramagnetic regime.

one has the sharp crystal-field-only level. For finite exchange, one has dispersion about the crystal-field only level, which increases with increasing exchange. As the exchange continues to increase, the $\mathbf{q}=0$ mode becomes unstable signaling a transition to ferromagnetism. This critical value of A for ferromagnetic ordering is 4% greater than the molecular field critical value. This just corresponds to the difference between the true ground state and the molecular field ground state wave function.

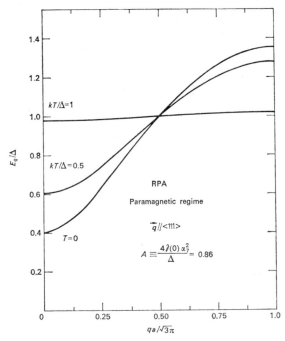

Figure 2–15. Temperature variation of the dispersion curve of the elementary excitations for the two singlet level model in the RPA for exchange less than the critical value necessary for ferromagnetism at $T=0$. Calculations are for a simple cubic lattice with nearest-neighbor ferromagnetic exchange.

Figure 2–15 shows typical theoretical dispersion behavior with increasing temperature for a value of A less than the critical value. At zero temperature, there is an energy gap at $\mathbf{q}=0$. As temperature increases, the long wavelength soft mode energies increase, at first abruptly and then more slowly. At high temperatures the dispersion tends to disappear, and the excitation energies approach the crystal-field-only values.

The RPA treatment does not include self-consistent correlation effects between excitations on different sites. To include such effects, Wang and Cooper[57, 58] generalized the RPA to include two site correlation effects. At zero temperature this involves replacing S_ζ in the equations of motion by expression (2.105), and then after Fourier transforming these equations, replacing products of three spin operators by the approximation,

$$\frac{1}{N}\sum_{q_1, q_2} \mathcal{J}(q-q_1+q_2) S^+_{q-q_1+q_2}S_{q_1}^+S_{q_2}^- \to \frac{1}{N}\sum_{q_1} \mathcal{J}(q)\langle S_{q_1}^+S_{q_1}^-\rangle S_q^+$$

$$+\frac{1}{N}\sum_{q_1} \mathcal{J}(q_1)\langle S_{q_1}^+S_{q_1}^-\rangle S_q^+ + \frac{1}{N}\sum_{q_1} \mathcal{J}(q_1)\langle S_{q_1}^+S_{-q_1}^+\rangle S_{-q}^-$$

(2.110)

where the pointed brackets denote taking the expectation value. One can then find the excitation spectrum,

$$E_q = \Delta[1+2\bar{S}_\zeta A(\gamma_q-2\epsilon)]^{\frac{1}{2}} \tag{2.111}$$

self-consistently including the function

$$\epsilon = -2\bar{S}_\zeta \frac{1}{N}\sum_q \gamma_q \left[\langle S_q^+S_q^-\rangle + \langle S_q^+S_{-q}^+\rangle\right] \tag{2.112}$$

giving the correlation between excitations on differing sites.

One can give the corresponding treatment at finite temperature by using the symmetric decoupling approximation of Callen[94] on the Green's function equations of motion.

$$\langle\langle S_{g\zeta}S_f^+; S_l^+\rangle\rangle \xrightarrow[g\neq f]{} \bar{S}_\zeta \{G_E^+(f,l) - 2\langle S_g^-S_f^+\rangle G_E^+(g,l)$$

$$-2\langle S_g^+S_f^+\rangle G_E^-(g,l)\} \tag{2.113}$$

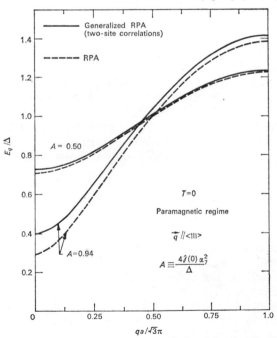

Figure 2–16. Comparison of dispersion curve calculated in the Two-Site Correlation Approximation with that calculated in the RPA for the two singlet level model. Calculations are for a simple cubic lattice with nearest-neighbor ferromagnetic exchange at $T=0$.

3

The difference at zero temperature between the RPA and the Two-Site Correlation Approximation behavior is shown in Figure 2–16. There is little difference for small exchange, but as A approaches the critical value, the soft mode drops more rapidly for the RPA. The critical value of A for ferromagnetism at $T=0$ in the simple cubic lattice with nearest neighbor exchange is about 4% larger in the Two-Site Correlation Approximation. (This is complicated by the question of a discontinuity in the magnetization as discussed below.)

We next consider the pseudospin Hamiltonian in the ferromagnetic regime (or in the paramagnetic regime in the presence of an applied field so there is a moment). It is useful to rewrite the Hamiltonian of (2.95) in such a way as to make apparent one fundamental difficulty[58] with the calculations.

$$\mathcal{H} = \sum_i C_\zeta S_{i\zeta} + 2 \sum_i C_\xi S_{i\xi}$$
$$+ \sum_{i \neq j} \mathcal{J}(\mathbf{R}_i - \mathbf{R}_j) [C_{\zeta\zeta} S_{i\zeta} S_{j\zeta} + 4 C_{\xi\xi} S_{i\xi} S_{j\xi} + 4 C_{\xi\zeta} S_{i\xi} S_{j\zeta}] \quad (2.114)$$

with

$$C_\zeta = \Delta \cos 2\theta + 2\lambda\beta H\alpha \sin 2\theta \qquad (2.115a)$$

$$C_{\zeta\zeta} = -4\alpha^2 \sin^2 2\theta \qquad (2.115b)$$

$$C_{\xi\xi} = -\alpha^2 \cos^2 2\theta \qquad (2.115c)$$

$$C_\xi = 2\mathcal{J}(0) \alpha \bar{J} \cos 2\theta \qquad (2.115d)$$

$$C_{\xi\zeta} = 2\alpha^2 \sin 2\theta \cos 2\theta \qquad (2.115e)$$

We are interested in small deviations from the equilibrium direction in pseudospin space. For the paramagnetic regime, the equilibrium direction in the pseudospin space is the ζ-axis; and as we have seen, the analogies to the magnetic behavior in ordinary ferromagnetic systems are clear. On the other hand, in general when the system has a moment, the ζ-axis, defined by projecting onto the manifold of molecular field states, is no longer the equilibrium direction; and one must perform a transformaiton in the pseudospin space such that the ζ'-axis is the equilibrium direction.

$$S_{\zeta'} = S_\zeta \cos \phi + S_\xi \sin \phi \qquad (2.116a)$$

$$S_{\xi'} = -S_\zeta \sin \phi + S_\xi \cos \phi \qquad (2.116b)$$

$$S_{\eta'} = S_\eta \qquad (2.116c)$$

This transformation is required so that one can calculate the elementary excitations which correspond to small deviations of the pseudospin from the equilibrium direction. The difficulty arises because of the presence of the C_ξ and $C_{\xi\zeta}$ terms in $\mathcal{H}$. Because $S_{j\zeta}$ has a nonvanishing time average, the C_ξ and $C_{\xi\zeta}$ terms act like static fields transverse to the assumed equilibrium direction. This means that unless the effect of the C_ξ and $C_{\xi\zeta}$ terms vanishes, the assumed equilibrium direction is incorrect, and the equilibrium direction of the pseudospin changes so that there is no effective static transverse field.

The condition determining the vanishing of transverse field effects, and hence the determining the angular rotation of the equilibrium direction from the ζ-axis is

$$C_{\xi'} + 2C_{\xi'\zeta'}\mathcal{J}(0)\,\bar{S}_{\zeta'}(1-2\epsilon) = 0 \qquad (2.117)$$

Now for RPA theory, it is easy to show[58] that this condition is satisfied with $\phi = 0$. This means that the molecular field states offer a good basis for representing the pseudospin behavior in a way directly analogous to the motion of magnetization in an ordinary ferromagnetic system. One can formally treat the transformation to a new equilibrium direction in improved treatments such as the Two-Site Correlation Approximation, but a detailed self-consistent treatment of the resulting excitation and magnetization behavior has been too formidable to carry out in practice.

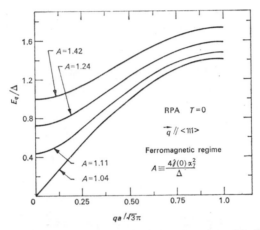

Figure 2–17. Dispersion curve at $T=0$ of elementary excitations in the two singlet level model for a simple cubic lattice with nearest-neighbor ferromagnetic exchange. This is shown calculated in the RPA for several values of A in the ferromagnetic regime.

On the other hand, in the Random Phase Approximation, where no rotation of the pseudospin axis is necessary, the calculations proceed in a way very similar to that in the paramagnetic regime. Figure 2–17 shows the dispersion curve for the collective excitations in the ferromagnetic regime at zero temperature in the RPA. This is for a simple cubic lattice with nearest neighbor ferromagnetic exchange. As indicated earlier (see Figure 2–14), in the paramagnetic regime the energy gap at $\mathbf{q}=0$ decreases as the exchange increases, and vanishes when A attains the critical value of 1.04. At this value of A, the $\mathbf{q}=0$ soft mode becomes unstable, and there is a second-order phase transition to ferromagnetic ordering. For nearest neighbor antiferromagnetic exchange the critical value of A has the same magnitude, but the instability is at the Brillouin Zone boundary. For A exceeding the critical value for ferromagnetic ordering, the energy gap at $\mathbf{q}=0$ increases with increasing A; and the dispersion tends to disappear for large exchange.

Figure 2–18 shows the temperature dependence of the dispersion curve for exchange only slightly larger than the critical value for ferromagnetism at zero temperature. This is for a Curie temperature equal to one tenth the

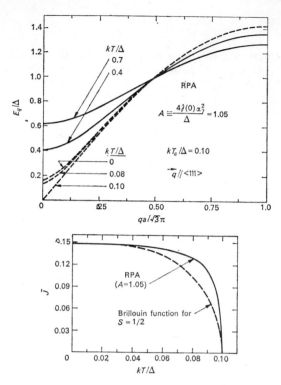

Figure 2–18. (upper) Change of dispersion curve as temperature is raised through T_c for $A = 1.05$, $T_c/\Delta = 0.1$. Calculations are with two singlet level model for a simple cubic lattice with nearest neighbor ferromagnetic exchange. (lower) Corresponding thermal variation of magnetization (after Wang and Cooper[58]).

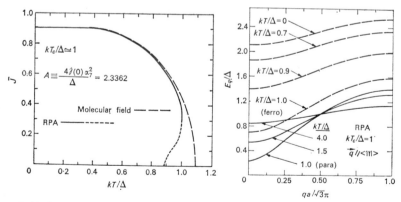

Figure 2–19. (left) Comparison of magnetization in RPA and molecular-field approximation for $A = 2.3362$ (after Wang and Cooper[58]); (right) Thermal variation of excitation spectrum in RPA for $A = 2.3362$. Calculations are with two singlet level model for a simple cubic lattice with nearest neighbor ferromagnetic exchange.

crystal-field splitting. As T increases from zero, the $\mathbf{q}=0$ mode energy drops toward zero. The rate of decrease is, however, slow until T is close to the Curie temperature, whereupon the $\mathbf{q}=0$ mode frequency drops to zero precipitously. For T greater than T_c, the $\mathbf{q}=0$ mode frequency increases again. For high temperatures the collective excitation energies approach Δ, and the dispersion tends to disappear.

From the self-consistent determination of the excitation behavior, one also finds the temperature dependence of the magnetization. The most striking feature is the sharp drop in magnetization close to the Curie temperature. This corresponds to the sharp drop in the soft mode energy at the same temperatures.

As shown in Figure 2–19, for increasing exchange and Curie temperature, one obtains a double valued magnetization behavior near the Curie temperature. This indicates that the magnetization is discontinuous at T_c. This is in distinction to the behavior in molecular field theory where the transition is always second order for the two singlet level system. The discontinuity in magnetization is most significant for values of T_c/Δ near unity, where it is quite substantial. Also, as shown in Figure 2–19 the excitation spectrum is considerably narrower than for lower T_c.

Basically, this first-order transition occurs because the long wavelength, "soft" modes dropping in energy offer a catastrophically effective channel for depopulation of the ground state. Actually, in the Two-Site Correlation Approximation, calculations[58] indicate that at zero temperature, for increasing exchange the transition to the ferromagnetic state is discontinuous. A question that arises is whether this first order transition would occur in an exact treatment of the two singlet level problem. While the answer is not known, there are some complementary studies[95, 96] currently being carried out that may help in answering this question at least at zero temperature. These studies develop series expansions for the energy in the ratio of exchange to crystal-field interaction in a way similar to the expansion in exchange over temperature used to treat critical phenomena in conventional magnetic systems.

As T_c/Δ increases still further in the RPA, while the discontinuity in magnetization persists, as shown in Figure 2–20 for $T_c/\Delta \simeq 5$, the magnetization approaches the molecular field behavior, and the excitation spectrum approaches the molecular field levels.

The approach to molecular field behavior in the RPA for large exchange is easily understood.[58] First, we note that there are no spin waves for the two singlet level system even in the limit $A \to \infty$. This is because the two states of a single ion go to $|\pm m\rangle$ with $m > \frac{1}{2}$ in our model as $A \to \infty$. [Here m denotes the J_z quantum number. For example, for Pr^{3+} in a hexagonal crystal-field, the two lowest singlet states are $1/\sqrt{2}(|+3\rangle - |-3\rangle)$ and $1/\sqrt{2}(|+3\rangle + |-3\rangle)$. As exchange increases from zero, these states become $|+3\rangle$ and $|-3\rangle$.] The spin waves involve transitions to $|m \pm 1\rangle$, and these are higher lying states not taken into account for the two singlet level model system. Second, as $A \to \infty$, there are also no excitations of the type we have considered, since the coupling vanishes between the two single-ion states, i.e. $\langle m|J_z|-m\rangle = 0$.

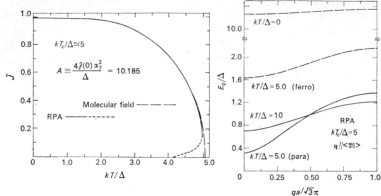

Figure 2–20. (left) Comparison of magnetization in RPA and molecular-field approximation for $A = 10.185$ (after Wang and Cooper[58]); (right) Thermal variation of excitation spectrum in RPA for $A = 10.185$. Calculations are with two singlet level model for a simple cubic lattice with nearest-neighbor ferromagnetic exchange.

Of course, in a real physical system there are spin waves. For Pr^{3+} in a hexagonal crystal-field these would involve transitions to higher lying crystal-field levels. Then in a real physical singlet ground state system, with increasing exchange one would expect to go from a regime where the excited states are basically the sort of excitons we have been discussing to a regime where the excited states are spin waves.

Our attention is then drawn to the singlet-triplet system, shown in Figure 2–21, where one has the possibility of both excitonic and spin wave type states. This case is also of practical interest since it describes the behavior in cubic systems such as fcc Pr metal or the rare-earth–Group V compounds of NaCl

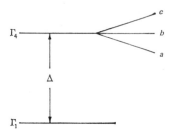

Figure 2–21. Crystal-field levels for singlet-triplet case.

structure; and these are the class of singlet ground state materials whose macroscopic magnetic properties have been most studied to date. (The question of whether the magnetic ordering transition is first or second order is more complex for a singlet-triplet system. Either type of transition can occur even in molecular field theory depending on the system parameters for a particular instance.)

Pink[92] has suggested the idea of projecting the operators describing the behavior of this singlet-triplet system onto the space spanned by the direct product of two spin $\frac{1}{2}$ systems. Each single-ion molecular field theory state then corresponds to a two spin state.

$$|0\rangle = \cos\theta\,|\Gamma_1\rangle + \sin\theta\,|\Gamma_{4b}\rangle \rightarrow (|+-\rangle - |-+\rangle) \tag{2.118a}$$

$$|1\rangle = -\sin\theta\,|\Gamma_1\rangle + \cos\theta\,|\Gamma_{4b}\rangle \rightarrow (|+-\rangle + |-+\rangle) \tag{2.118b}$$

$$|\Gamma_{4a}\rangle \rightarrow |++\rangle \tag{2.118c}$$

$$|\Gamma_{4c}\rangle \rightarrow |--\rangle \tag{2.118d}$$

If we denote the two spin $\frac{1}{2}$ operators at each site as $\mathbf{L}$ and $\mathbf{S}$, one can write each term in the Hamiltonian in terms of the components of $\mathbf{L}$ and $\mathbf{S}$. For example, for the crystal-field term we obtain,

$$V_{ci} = \frac{\Delta}{4} + \Delta S_{i\zeta}L_{i\zeta} + \frac{\Delta}{2}\cos 2\theta(S_i{}^+L_i{}^- + S_i{}^-L_i{}^+) + \frac{\Delta}{2}\sin 2\theta(S_{i\zeta} - L_{i\zeta}) \tag{2.119}$$

Treating this singlet-triplet case is considerably more involved than the two singlet level problem, and writing the various projection operators explicitly in the spinor representation is quite useful in performing the transformation to the pseudospin representation in terms of $\mathbf{L}$ and $\mathbf{S}$.

This $\mathbf{L}$, $\mathbf{S}$ representation lends itself naturally to a singlet-triplet splitting, and also helps to reveal some of the symmetry of the problem. For example, in the paramagnetic regime the crystal-field term becomes,

$$V_{ci} = \frac{\Delta}{4} + \Delta \mathbf{S}\cdot\mathbf{L} \tag{2.120}$$

However, as shown below, the generating operators for excited states are much more complicated than the simple pseudospin raising operator, S^+, that pertained to the two singlet level case:

$$\left.\begin{aligned}
|1\rangle\langle 0| &= \tfrac{1}{2}(S_\zeta - L_\zeta) + \tfrac{1}{2}(-S^+L^- + S^-L^+) \\[4pt]
|a\rangle\langle 0| &= \frac{1}{2\sqrt{2}}(-S^+ + L^+) + \tfrac{1}{2}(S_\zeta L^+ - S^+L_\zeta) \\[4pt]
|c\rangle\langle 0| &= \frac{1}{2\sqrt{2}}(S^- - L^-) + \tfrac{1}{2}(-S^-L_\zeta + S_\zeta L^-)
\end{aligned}\right\} \begin{array}{l} \text{generating-operators} \\ \text{singlet-triplet} \\ \text{case} \end{array} \tag{2.121}$$

Here the analogies to ordinary magnetic systems are difficult to draw, and the problem is a complicated one. Pink[92] treated only two extreme limiting cases that are not sufficient to illustrate the behavior involved in the coexistence of the excitonic states and spin waves. However, more detailed calculations[97] are in progress that should hopefully yield such results. These calculations should also relate to future experiments studying the excitations in cubic systems.

2.3.1.2. Experimental Situation for Induced Moment Behavior

(a) Rare-Earth–Group V Compounds of NaCl Structure

The most complete quantitative studies done on singlet ground state systems have been for the rare-earth–Group V compounds of NaCl structure (i.e. R.E.N., R.E.P, R.E.As, R.E.Sb, R.E.Bi). These compounds are metallic (except possibly for the nitrides which in the pure state might be narrow band gap semiconductors), so our expectation is that exchange forces may be rather long range.

An important contribution at an early stage was the neutron diffraction investigation of the magnetic properties of such compounds by the Oak Ridge group.[66] There were two key qualitative features found in those experiments. First, for any anion, say the nitrides or antimonides, the ordering temperature steadily decreased on going from the center of the rare earth row toward the end. Also, the ratio of the ordered moment at the lowest temperature to the free ion moment tended to be less than unity. This trend was most strikingly shown by the behavior of the Tm compounds which did not order magnetically at all. Much more data has been accumulated since that time,[67] and one sees the same trends for the complete family of these compounds. In particular, the Pr compounds, at the light end of the rare earth series, show the same behavior as the Tm compounds at the heavy end, and do not order magnetically at all.

This behavior led Trammell[53, 54] to recognize the sort of competition, discussed in Section 2.3.1.1 above, between the crystal-field and the exchange effects in determining the magnetic ordering in these compounds. As illustrated in Figure 2–10, the characteristic trends noted above, particularly the absence of ordering for Tm and Pr compounds, can be understood with a simple molecular field theory. Thus the overall trend of behavior found becomes clear when one recognizes the increase of the orbital contribution relative to the spin contribution to the total moment on going from the center of the rare earth series toward either end. On going from Gd toward Tm the crystal-field effects, which depend on the orbital contribution to the moment, become more important relative to exchange which depends on the spin contribution.

Experiments have been done on rare-earth–Group V compounds that directly and quantitatively study the magnetic ordering process, particularly near the threshold value of exchange for magnetic ordering. To do this Cooper and Vogt[68] studied the effect of substituting Y for Tb in TbSb. (TbSb orders antiferromagnetically at[68] 15.1 K.) Yttrium is essentially identical to the heavy rare earths in its valence electron behavior, but has an empty $4f$ shell. Thus substituting Y for Tb essentially does not change the crystal-field, but does reduce the effective exchange field acting on a given Tb ion. (This picture was confirmed experimentally[66] by measuring the susceptibility and high-field magnetization in $Tm_{0.53}Y_{0.47}Sb$ where one expects no exchange effects. The behavior, normalized per Tm, was the same as for pure TmSb.) In this way one could reduce the exchange below the critical value necessary for magnetic ordering. Mixed crystals were made by Vogt across the whole composition range. The lattice constant had only a very small variation indicating a correspondingly very small variation in the crystal-field. (This was also confirmed by specific heat experiments.[69])

The results found for the variation of susceptibility with temperature for differing Tb concentration are shown in Figure 2–22. At low Tb concentrations, one has Van Vleck susceptibility behavior. The exchange is antiferro-

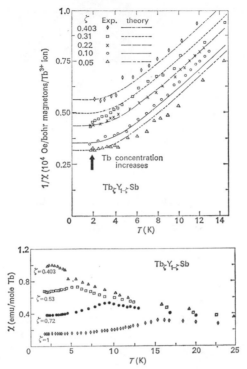

Figure 2–22. (upper) Inverse susceptibility per Tb ion versus temperature for $Tb_\zeta Y_{1-\zeta} Sb$ in paramagnetic regime ($\zeta \leqslant 0.403$). (lower) Susceptibility versus temperature for $Tb_\zeta Y_{1-\zeta} Sb$ in antiferromagnetic regime ($\zeta \geqslant 0.403$) (after Cooper and Vogt[68]).

magnetic, and increasing Tb concentration corresponds to increasing the antiferromagnetic exchange. Thus for increasing Tb concentration the inverse susceptibility curve shifts upward until at concentrations above 40% Tb, the exchange has increased sufficiently to give the peak in susceptibility characteristic of antiferromagnetic ordering. As the Tb concentration increases to 100%, the Neel temperature increases to a little over 15 K.

In the paramagnetic regime, one can analyze the inverse susceptibility behavior as a function of Tb concentration on the basis of molecular field theory. To do this one first recognizes that the crystal-field Hamiltonian for a rare earth ion in an octahedral crystal-field has the form,

$$\mathscr{H}_{CF\ cubic} = V_4^0(0_4^0 + 5 \times 0_4^4) + V_6^0(0_6^0 - 21 \times 0_6^4) \qquad (2.122)$$

Here, 0_4^0, 0_4^4, 0_6^0, and 0_6^4 are specified operators for a given J ($J=6$ for Tm^{3+} and Tb^{3+}), and the axis of quantization has been chosen parallel to a crystal axis. The operators 0_4^0 and 0_4^4 are fourth-order in the components of J, while 0_6^0 and 0_6^4 are sixth-order in J. Thus the crystal-field Hamiltonian is completely determined by symmetry considerations except for the constants V_4^0 and V_6^0. Rather than deal with V_4^0, and V_6^0, it is often more convenient to treat two other parameters,[98] x and W. The ratio of fourth-to-sixth-order anisotropy is given by x; while W gives the absolute scaling of the crystal-field energy levels.

$$\frac{B_4}{B_6}=\frac{x}{1-|x|}\frac{F(6)}{F(4)} \tag{2.123}$$

$$B_4 F(4)=Wx \tag{2.124}$$

Here $F(4)$ and $F(6)$ are numerical factors known for a given J.

For Tb in TbSb (and for Tm in TmSb) one can show[66, 68] convincingly that the crystal-field is predominantly fourth-order. Then one attempts[68] to fit the data for the paramagnetic regime shown in Figure 2–22 with the molecular field theory expression for $1/\chi$ versus T.

$$1/\chi=1/\chi_{\mathrm{CF}}-C_{\mathrm{ex}}\zeta \tag{2.125}$$

This involves only two parameters for all values of the concentration. These are W giving the crystal-field splittings and hence $1/\chi_{\mathrm{CF}}$, and C_{ex} giving the exchange (assuming exchange linearly dependent on concentration).

For $\mathrm{Tb}_\zeta \mathrm{Y}_{1-\zeta}\mathrm{Sb}$,

$$\mathrm{Tb}_\zeta \mathrm{Y}_{1-\zeta}\mathrm{Sb}, \quad W=-0.396\ \mathrm{K}, \quad C_{\mathrm{ex}}=-0.697\times 10^4\ \mathrm{Oe}/\beta \tag{2.126}$$

and one obtains the theoretical curves shown in Figure 2–22. The overall agreement of theory and experiment is quite good, considering the large amount of data to be fitted. (Since the data is normalized per Tb ion, the experimental points at low concentration are quite sensitive to the correct determination of the Tb concentration.) The value of W for Tb $\mathrm{Y}_{1-\zeta}\mathrm{Sb}_\zeta$ corresponds to a Γ_1 to Γ_4-state splitting of 11.9 K. This is close to the value found in specific heat measurements,[69] and is less than half the value for [66] TmSb.

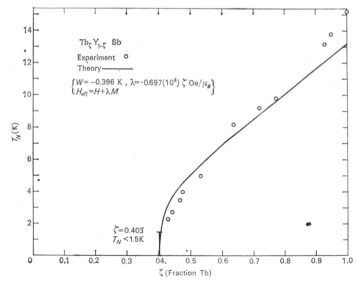

Figure 2–23. Neel temperature versus terbium concentration in $\mathrm{Tb}_\zeta \mathrm{Y}_{1-\zeta}\mathrm{Sb}$ for $\zeta \geqslant 0.403$ (after Cooper and Vogt[68]).

As shown in Figure 2–23 one can also calculate[68] the variation of the Neel temperature with Tb concentration on the same theoretical basis, and again the agreement is good. Thus one has good agreement for the threshold behavior for the magnetic ordering and for the behavior below the exchange threshold at low fields where only a small magnetization is developed.

However, at high pulsed fields in the paramagnetic regime the nonlinear magnetization is not explained[68] by the simple theory using the value of crystal-field and exchange parameters that describe the low field paramagnetic susceptibility and the Neel temperature behavior. (This involves anisotropic magnetization measurements at high fields in the nonlinear magnetization regime. The anisotropic magnetization, described in Section 2.3.2 below, is one of the striking characteristics[66, 68, 99, 100] of singlet ground state systems in the paramagnetic regime.) This led one to suspect[68] the presence of higher degree and/or anisotropic exchange effects as might be expected for a system with a large orbital contribution to the moment. Recently, d.c. measurements[87] up to 100 kOe for differing concentrations have been performed which allow one to confirm the presence of such higher degree exchange terms and to measure them. [The term in effective field linear in M is quite close to that given by (2.126), so the analysis of the small moment behavior as described by Figures 2–22 and 2–23 is basically unchanged.]

(b) Praseodymium

(ii) Double Hexagonal Phase. Ordinarily the stable crystallographic phase of Pr metal has a double hexagonal structure. The layer packing is ABAC. As discussed by Bleaney,[60] ions in A layers are subjected to a crystal-field approximating that of an fcc lattice; while those in B and C layers have crystal-fields approximating to hcp. Bleaney's analysis indicated a singlet crystal-field ground state for both types of site, but the splitting to the first excited state is much smaller for the hexagonal sites.

Until very recently all studies of the magnetic properties of Pr metal were on polycrystalline samples. These present somewhat contradictory results indicating the presence and amount of magnetic ordering to be sample dependent. Both specific heat measurements[101] between 2 and 180 K and susceptibility measurements between 1.5 and 300 K show no evidence for magnetic ordering. As shown in Figure 2–24, the magnetic contribution to the specific heat shows the Schottky anomaly expected for a singlet ground state (note that this involves separating out lattice, electronic, and nuclear contributions); while the susceptibility, shown in Figure 2–25, shows Van Vleck-type behavior also characteristic of a singlet ground state.

Bleaney[60] was able to analyze the specific heat behavior to yield the ratio of fourth-to-sixth-order anisotropy terms, and to give the splitting in each type of site. For the hexagonal sites he found the first excited state (singlet) to be 23 K, and the second excited state (doublet) to be 63 K, above the singlet ground state. For the cubic sites, he found the first excited state (triplet) to lie 87 K above the ground state. On this basis, as shown in Figure 2–24 Bleaney found excellent agreement with the specific heat Schottky anomaly. He also calculated the susceptibility as shown in Figure 2–25.

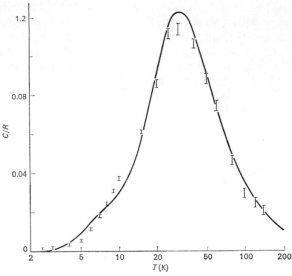

Figure 2–24. Specific heat of praseodymium metal (C/R per gram atom). The curve shows the calculated values of Bleaney,[60] and the points are the experimental values of Parkinson *et al.*[101] (after Bleaney[60]).

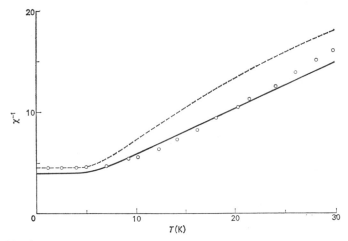

Figure 2–25. $1/\chi$ versus T for polycrystalline praseodymium. Experimental points are from Lock[102] and theoretical curves are shown for two different exchange fields calculated by Bleaney[60] (after Bleaney[60]).

The crystal-field-only inverse susceptibility lay well above the experimental values. However, as discussed above in connection with Figure 2–10, Bleaney[55, 60] recognized that a net ferromagnetic exchange would shift the curve of $1/\chi$ downward. As shown in Figure 2–25, he performed two calculations, including different range of exchange, which gave reasonably good agreement with experiment. The value of exchange found in these calculations indicated a value close to the critical value necessary for induced ferromagnetic ordering on the hexagonal sites.

At that point in time, neutron diffraction measurements[70] were performed on polycrystalline Pr at 1.4 K which indicated a rather complex antiferromagnetic structure with a moment of 0.7 per ion if all sites ordered (1.0 per ion if only hexagonal sites ordered) compared to the free ion moment (λJ) of 3.2β. The Neel temperature deduced from the neutron diffraction data was 25 K.

Very recently, Rainford and Wedgwood[61] reexamined the neutron diffraction behavior of polycrystalline Pr (on a 2 cm³ cast specimen) and observed the same pattern earlier associated with magnetic ordering by Cable et al.[70] However, both they[61] and Johansson et al.[62] found no evidence for magnetic ordering in single-crystal Pr.

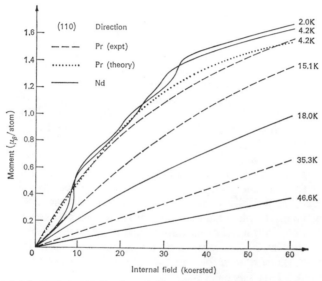

Figure 2–26. $\langle 110 \rangle$ magnetization of dhcp Pr and Nd (after Johansson et al.[62]).

As shown in Figure 2–26, the application of a magnetic field along a $\langle 110 \rangle$ direction in single-crystal Pr produces a large induced moment.[62] By observing the neutron diffraction intensities at different reciprocal lattice points, Johansson et al.[62] were able to separate the contributions to the induced magnetization from the cubic and hexagonal sites. They obtained a fairly good fit to their results using Bleaney's[60] molecular field model. However, they found a splitting to the second excited crystal-field state (i.e the first excited doublet state) of about 20 K, considerably smaller than Bleaney's value of 63 K. Their value for the molecular field constant is positive and close to the critical value necessary for the occurrence of ferromagnetism. Also they find a very marked anisotropy in the susceptibility; that along an $\langle 001 \rangle$ direction is an order of magnitude smaller than that in the basal plane. A large susceptibility anisotropy was predicted by Bleaney,[60] whose calculations were based on isotropic exchange, but in the opposite sense to that observed. This suggests[62] the presence of very large anisotropy in the exchange interaction.

Johansson et al.[62] suggest that the apparent presence of magnetic ordering in polycrystalline Pr and the absence in single-crystal Pr indicates that the exchange is close to the threshold value for ordering. Then a small modification of the crystal-field splittings, due perhaps to strains, can lead to magnetic ordering. A somewhat different, but perhaps related idea is the suggestion of the possible presence of small clusters of magnetically ordered material in the polycrystalline sample, so that only a small fraction of the entire sample is magnetically ordered.

This suggestion arose as part of Bleaney's[60] discussion of the nuclear hyperfine specific heat[103, 104] of Pr metal. Most of the rare earth metals have huge nuclear specific heats, C_N, associated with the interaction between the nuclear magnetic moment and the very large effective hyperfine field associated with the ordered moment of the rare earth $4f$ electrons. (See Chapter 8 for a discussion of hyperfine effects.) For Pr, because of the crystal-field quenching of the electronic moments, one expects a much smaller C_N than in other of the rare earth metals. C_N is indeed much smaller than for the other rare earths, but it is still larger than would be expected if the ionic moment is completely quenched. Among possible explanations for this, Bleaney suggested that there might be small clusters, involving only a few per cent of the entire sample, of magnetically ordered Pr. Subsequent to Bleaney's analysis, Lounasmaa[105] remeasured C_N. While he found a somewhat different value than the earlier investigations,[103, 104] the basic discrepancy already noted was still present. Lounasmaa found that the discrepancy for the nuclear specific heat would be removed if 2% of the sample were magnetically ordered. He also found evidence for a small anomaly in the electronic magnetic specific heat, C_M, between 3.0 and 3.5 K. He separated C_M into a Schottky contribution (corresponding to an excited crystal-field level at 28 K—compared to Bleaney's[60] 23 K from analyzing the data of Parkinson et al.[101] and a smeared-out cooperative peak with a maximum at 3.2 K. The entropy under the cooperative peak was very close to the value expected if 2% of the Pr ions were magnetically ordered.

The possibility that some perturbation in the polycrystalline phase causes magnetic ordering for some fraction of the Pr ions implies both that the ratio of exchange to crystal-field splitting of the unperturbed crystal is quite close to the threshold for ordering, and that the crystal-field and/or exchange interactions are quite sensitive to small local distortions of the lattice. Since the crystal-field has a very strong dependence (a^5 for fourth-order, a^7 for sixth-order, where a is the lattice parameter), this is quite conceivable. If the rise in T_N and M per ion above the threshold is sharp enough (see Figure 2–11 and Figure 2–23), one might even be able to explain a T_N as large as 25 K and an M of $\sim 1\beta$ per ion. Such effects are amenable to a fairly detailed study, and hopefully in the near future the studies of single-crystal Pr will provide more information that is valuable in this regard.

There is a possibility that any clusters of ordered Pr actually are contamination by a second phase. The possibility that this might be fcc Pr provided

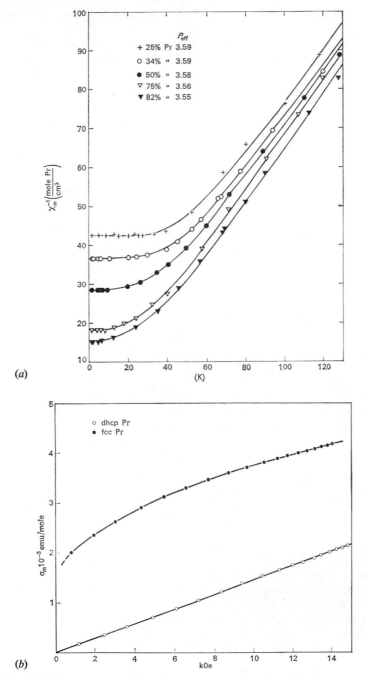

Figure 2–27. (a) Inverse susceptibility (normalized to Pr concentration) versus temperature for fcc Pr–Th alloys (after Bucher *et al.*[71]); (b) magnetization versus field at 1.43 K for fcc Pr (after Bucher *et al.*[63]).

part of the motivation for the studies by Bucher et al.[63] on the magnetic properties of fcc Pr.

(ii) fcc Pr. As shown in Figure 2–27(a), Bucher and co-workers[71] studied the susceptibility behavior of the system Th–Pr in which the fcc phase persists up to 90 at.% Pr. The alloys showed Van Vleck susceptibility behavior; and the inverse susceptibility, normalized to Pr concentration, shifted steadily downward with increasing Pr concentration indicating that the behavior for pure fcc Pr would be quite close to the onset of ferromagnetism. (Both the exchange and crystal-field interactions appear to vary strongly with varying Pr concentration, so an analysis of the sort discussed above for $Tb_\zeta Y_{1-\zeta} Sb$ is impossible.)

This behavior led Bucher et al.[63] to prepare and study fcc Pr. From the susceptibility behavior, they recognized the onset of ferromagnetism at 8.7 K. Both the low temperature magnetization, shown in Figure 2–27(b), and the hyperfine contribution to the specific heat indicate that the ordered moment at zero temperature is only about 20% of the free ion moment.

At the ferromagnetic transition in fcc Pr, there is[63] no detectable specific heat anomaly. This is understandable if we consider the specific heat behavior shown in Figure 2–12 and the earlier discussion accompanying that Figure. From analysis of the Schottky contribution to the specific heat, Bucher et al.[63] found the crystal-field splitting from the singlet ground state to the first excited (triplet) level to be about 69 K. Then one would expect the magnetic ordering anomaly to be of much less relative importance even than that shown in Figure 2–12.

Their results for fcc Pr led Bucher et al.[63] to suggest that the presence of contamination by fcc Pr led to various anomalies in the behavior of polycrystalline dhcp Pr as discussed above. This may be the case for much of the behavior, e.g. the specific heat anomaly at about 3.2 K found by Lounasmaa.[105] However, this does not seem to explain the neutron diffraction results,[70] especially an apparent T_N at approximately 25 K.

(b) Other Materials Showing Induced Moment Effects

Here we briefly mention some other materials showing induced moment effects, with special characteristics deserving further study.

$TmAl_3$ is one[73] of the large class of Cu_3Au-structure (a cubic structure) intermetallic compounds in which rare earths have nonmagnetic ground states. The special interest in $TmAl_3$ is the possibility[74] that the ground state may be a nonmagnetic doublet (Γ_3) rather than a singlet state. Recent nuclear magnetic resonance studies[75] support this possibility. A Γ_3 ground state would lead to dramatic high field magnetization effects in single crystals (not presently available). Also if some related compound (in the sense that TmSb and TbSb are related NaCl-structure compounds) is found to order through induced moment effects involving a Γ_3 ground state, this would be quite interesting.[54]

The rare earth hexaborides are a family of materials of great current interest.[106] Lee et al.[72] have recently studied the magnetic ordering in PrB_6 through specific heat and electrical resistivity measurements. They find an extremely sharp specific heat anomaly, indicating that PrB_6 may undergo a first-order transition.

There is another experiment of great possible interest, although quite speculative in nature, for another of the rare earth hexaborides. This is SmB_6, which is a semiconductor at low temperatures and which shows no magnetic ordering[107] down to 0.35 K. Here the samarium is thought to be divalent.[107] If the divalent Sm has a $4f^6$ configuration, the ground state multiplet would have $J=0$, and this would explain the absence of magnetic ordering. (There is an alternative electronic behavior suggested by some of the experimental behavior.[108]) It would be very interesting to dope SmB_6 to increase the number of conduction electrons. Perhaps this could be done by substituting La or Y for Sm, unfortunately diluting the potentially magnetic species. This could provide a coupling mechanism to give effective exchange interaction between Sm^{2+} ions. If one can get enough conduction electrons and hence coupling, one could presumably get magnetic ordering by the induced moment mechanism. Such a system should also have very striking transport properties.

Finally, we mention two systems where the rare earth is a Kramers ion, and therefore cannot have a singlet state; but where there are important crystal-field effects on the ordered magnetic moment. First is Nd metal. As with Pr metal, the usual stable phase is dhcp. (There has been a very limited study[63] showing the existence of an fcc phase which orders ferromagnetically at 29 K.)

Moon et al.[76] and Johansson et al.[62] have both performed neutron diffraction investigations of the magnetic structure of dhcp Nd, and are in good agreement with each other. The moments on hexagonal sites order first, at 19 K in a rather complex structure, while the moments on cubic sites order at 7.5 K. For both kinds of sites, the maximum ordered moment is reduced well below the free ion value of 3.27 by crystal-field effects. The amplitude of the modulated moments is[70] 1.8 for cubic sites, and 2.3 for hexagonal sites. As shown[62] in Figure 2–26, the moment per ion can be substantially increased by an applied magnetic field. It is suggested[62] that abrupt changes in the magnetization may be caused by the crossing of crystal-field levels.

The behavior of the Ce–Group V compounds of NaCl structure serves to illustrate[80, 86] the interesting behavior that can come about by competition between exchange and crystal-field effects giving induced moment effects for a Kramers ion. For such a system in the antiferromagnetic phase, in contrast to the behavior of conventional antiferromagnets the inverse susceptibility of a polycrystalline sample need not increase monotonically as temperature decreases below T_N, and can have a maximum and an additional minimum. The anomalous behavior arises because[80] there is a range of exchange field for which Γ_7 ground state doublet splitting decreases with increasing field until the doublet levels cross. Such energy level behavior leads to the expectation of approximately normal susceptibility behavior at low and high values of the ratio of exchange to crystal-field strength. Anomalous behavior is expected in the intermediate regime where the splitting decreases with increasing exchange field. Corresponding to this variation of the ratio of exchange to crystal-field interaction, the susceptibility behavior is normal for the light compounds (CeP and CeAs), most anomalous for CeSb, and returns toward normal for CeBi.

There is also dramatic evidence[86] for substantial anisotropic exchange in CeSb. Contrary to the prediction of crystal-field theory, at high fields for CeSb the easy magnetization direction at 1.5 K is $\langle 100 \rangle$. However, with Y or La dilution there is a striking change to a $\langle 111 \rangle$ easy direction as predicted by crystal-field theory. At a Ce concentration of 30% in $Ce_xY_{1-x}Sb$ or 60% in $Ce_xLa_{1-x}Sb$, the high field (to 100 kOe) magnetization is almost isotropic. Above this Ce concentration there is large anisotropy with a $\langle 100 \rangle$ easy direction; below this Ce concentration, large anisotropy with a $\langle 111 \rangle$ easy direction. This can be understood on the basis of diluting aniso-tropic exchange, but leaving crystal-field strength largely unchanged on replacing Ce with Y or La.

2.3.2. Measurement of Crystal-Field Splitting

As is discussed in a number of places in this volume, parameters reflecting crystal-field effects can be deduced via various theoretical interpretations from a number of magnetic properties, e.g. anisotropy constants, spin wave energies, magnetic structure behavior. However, in this Section we are interested in experiments directly measuring the crystal-field splitting of rare earth ions. In Section 2.3.2.1, we will discuss the results of such experi-ments in concentrated rare earth systems; while in Section 2.3.2.2 we will discuss experiments for very dilute concentrations of rare earths (1 at.% or less) in noble metals and in yttrium or lanthanum. These experiments give valuable information on the origin and nature of crystal-field effects in metals.

2.3.2.1. Concentrated Systems

There is rather little in the way of direct measurements of the crystal-field splitting of rare earth ions in the pure elemental rare earth metals. This is presumably a consequence of the fact that for the heavy rare earth metals the magnetic ordering temperatures are typically larger than the crystal-field splittings. The only direct measurements of crystal-field splittings for the heavy rare earth metals are for Tm, which is the material with the lowest ordering temperature, and which has relatively large crystal-field splittings. Pr and Nd are the two light rare earth metals, the behaviors of which are sufficiently simple to offer promising situations for direct observation of crystal-field splittings. (This is particularly true for dhcp Pr for which single crystals do not order magnetically.) For Pr, as described above, the crystal-field splitting has been measured directly in both the dhcp and fcc phases. One can be optimistic in the case of Nd, although no direct measurements have as yet been made. Here we briefly mention the techniques available for direct measurement of crystal-field splittings and any results obtained.

(a) Specific Heat Measurements

In Section 2.3.1 we have discussed (Figure 2–12) how observation of the Schottky anomaly in the specific heat gives the crystal-field splitting to the first excited state. For dhcp Pr, as discussed in Part (b) of Section 2.3.1.2 in connection with Figure 2–24, this gives a splitting[60] of 23 K; while for fcc Pr the splitting is[63] 69 K.

(b) Low Field Susceptibility and High Field Nonlinear Magnetization

The low field Van Vleck-type susceptibility as $T \to 0$ can be used to determine the crystal-field splitting if one has some way of judging the way in which the exchange shifts the curve of $1/\chi$ versus T from the crystal-field-only behavior. For example this procedure was quite successful for[68] $Tb_\zeta Y_{1-\zeta} Sb$ as discussed in connection with Figure 2–22. For Pr the reverse procedure has been used,[60] and as discussed in connection with Figure 2–24, the susceptibility behavior has been used to deduce the exchange constants for known crystal-field.

As discussed in connection with Figure 2–26 for dhcp Pr, the high field nonlinear magnetization behavior can be analyzed to yield the crystal-field splittings. This is because the ordered moment is well below the saturation value for a free ion. Then an applied field induces additional moment by admixing higher lying crystal-field states into the ground state. The size of the induced moment depends directly on the crystal-field splitting, and serves to measure these splittings.

In the high field nonlinear regime, the induced magnetization is expected to be highly anisotropic.[66, 68, 99, 100] This is illustrated in Figure 2–28 for[66] TmSb which does not order and which shows evidence of no significant exchange. The form of the anisotropic magnetization depends much more

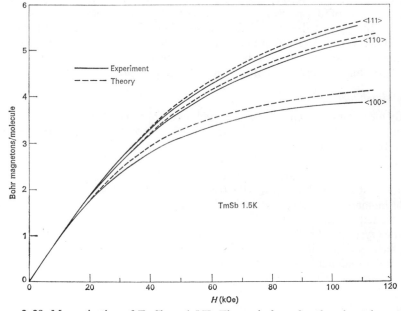

Figure 2–28. Magnetization of TmSb at 1.5 K. Theory is for a fourth-order-only crystal-field with a splitting of 26.6 K from the singlet ground state to the triplet first excited state. There is no exchange interaction (after Cooper and Vogt[66]).

strongly on the detailed nature of the crystal-field splittings than does the low field isotropic Van Vleck susceptibility. (Basically, for TmSb with NaCl crystal structure the difference at low temperature between the magnetization with field in the $\langle 111 \rangle$ direction and in the $\langle 100 \rangle$ direction occurs because for the fields shown, in the $\langle 100 \rangle$ direction, the only significant admixture into the Γ_1 singlet ground state is from the Γ_4 first excited triplet state. On the other hand, for H along $\langle 111 \rangle$, in the nonlinear regime there is significant admixture from the second excited triplet state, of Γ_5 symmetry. Such anisotropic magnetization measurements allow one to separate[66] the fourth and six-order contributions to the crystal-field for cubic materials, and also allow one to distinguish between[68, 86, 87] crystal-field anisotropy and anisotropic exchange.

For dhcp Pr such anisotropic magnetization experiments would be more involved to interpret than for TmSb because of the more complicated crystal structure, but should still yield much valuable information. Such measurements would be particularly useful in separating out anisotropic exchange from crystal-field effects.

(c) Neutron Inelastic Scattering

If exchange is small enough so that one may think of reasonably sharp crystal-field energy levels, then one can measure the crystal-field splittings directly by neutron inelastic scattering. Such an experiment has been very successful for[109] PrBi. Recently such experiments have been successful in other Pr compounds[109a] and in TmSb.[109b] However, for the rare earth metals such experiments on terbium[110] and on holmium and erbium[111] do not reveal the presence of discrete energy levels, although a moment analysis[110–112] of the energy distribution of the scattered neutrons can yield information about the crystal-field and exchange effects. Inelastic neutron scattering experiments would seem to have a better chance of directly observing crystal-field levels in thulium and neodymium, and especially praseodymium.

(d) Mossbauer Effect

By measuring the temperature dependence (from 59 to 156 K) of the nuclear quadrupole splitting as seen in the 8.42-keV Mossbauer transition, Uhrich and Barnes[113] determined the crystal-field splitting for Tm metal. They found an overall splitting of the ground state multiplet equal to 109.5 K. This experiment measures the interaction of the electric quadrupole moment of the first excited level of Tm^{169} (the nuclear ground state has no quadrupole moment) with the electric field gradient caused by the $4f$ electron cloud as affected by the crystal-field. The low ordering temperature (56 K) relative to the crystal-field splitting makes Tm a particularly favorable case for such a Mossbauer determination of the crystal-field splitting involving measurements in the paramagnetic regime. The measurements of Uhrich and Barnes indicate a predominantly second-order crystal field.

An interesting possibility for other rare earths would be to dilute them with Y sufficiently to significantly reduce the ordering temperature relative to the crystal-field splitting, so that one could use the Mossbauer technique to determine the crystal-field splitting. One would expect the crystal-field splittings of rare earths diluted by Y to be quite similar to those of the pure rare earth metals. (In the next section, we shall mention the results of a very limited number of experiments determining, by susceptibility measurements, the crystal-field splitting of very dilute rare earths in Y.)

2.3.2.2. Dilute Systems

(a) Dilute Rare Earths in Noble Metals

There is little in the way of first-principles theory and calculations of the crystal-field effects in the rare earth metals. However, one tends to think in terms of a picture where point charges, and the changes induced in the localized charge distribution associated with the rare earth ion at each lattice site, basically determine the crystal-electric-field. The conduction electrons enter in only by providing a shielding (or antishielding) factor reducing (or increasing) this electric field. The crystal-field effects found for dilute rare earths in noble metals are instructive for indicating how crystal-field effects may arise in metals when the nature and localization of charge giving rise to such a field is not so apparent as it seems to be in the concentrated rare earth metals. Indeed, as we shall see, the results of such experiments suggest the possibility that a picture differing from that sketched above may provide a better description for the origin of the crystal-field in the concentrated elemental rare earth metals.

Williams and Hirst[114, 115] have measured the susceptibility of samples containing several tenths of an atomic % of each of the heavy rare earths in Ag and similar samples for Er, Tm, and Yb in Au. They have used these measurements to deduce the fourth and sixth-order crystal-field parameters given in Eqn (2.122). The parameters V_4^0 and V_6^0 in (2.122) can each be taken as the product of two multiplicative factors ($V_4^0 = C_4\beta$, $V_6^0 = C_6\gamma$; i.e. $C_4 \equiv A_4^0\langle r^4 \rangle$, $C_6 \equiv A_6^0\langle r^6 \rangle$). Now β and γ are specified constants for a given rare earth ion (cf. Table 1.3). On the other hand, C_4 and C_6 are determined by the crystal potential giving the crystal-electric-field. If the host (Ag or Au) provided a crystal potential which did not vary with the rare earth impurity, then C_4 and C_6 would be constant independent of the rare earth impurity (i.e. neglecting the variation of $\langle r^4 \rangle$ and $\langle r^6 \rangle$). Williams and Hirst find that in fact this is not the case; however, they fit their data for the various rare earths unambiguously by putting some rather mild restrictions on the allowed variation of C_4 and C_6. These are that C_4 retains the same sign and similar magnitudes for varying rare earths within the same host. The same is true for C_6. Once these restrictions are adopted, Williams and Hirst can evaluate C_4 and C_6 individually. (This procedure is aided by the identification of the symmetry-type of the ground state in paramagnetic resonance experiments.[114, 116]) They find that C_4 is negative and C_6 positive for both Ag and Au host materials. They find that for each host: C_4 and C_6 each vary by almost a factor of two for the series of rare earth alloys; the overall crystal-field splittings are typically about 100 K for the Au alloys and about 200 K for the Ag alloys; and that in general the fourth-order terms are only marginally more important than the sixth-order terms in determining the overall splitting.

These results can be compared to those expected for an effective point charge model. (The conduction electron charge density is, of course, nonuniform. The question is whether the charge density variations are purely radial or not. If the variation is purely radial, then simple screening ideas apply, and one takes the conduction electrons into account by simply changing the effective point charge value associated with each lattice site.) For a reasonable effective charge (one electron charge) the magnitudes of both C_4 and C_6 are too small by up to a factor of five to agree with experiment. Moreover, such a model gives the wrong sign for C_4. *The relative signs and magnitudes of C_4 and C_6 are determined from purely symmetry conditions for an effective point charge model. Thus the failure to predict the correct signs is a very strong failure of the effective charge model.*

As an alternative explanation for the origin of the crystal-field, Williams and Hirst quote a private communication from B. R. Coles and R. Orbach, suggesting the occurrence of a $5d$ nonmagnetic virtual bound state on the rare earth impurity site. This nonmagnetic $5d$ virtual bound state is acted on by the crystal-field of the lattice ions, and itself in turn interacts with the $4f$ core to provide the splitting of the $4f$ states which is directly observed in the magnetic properties. The hypothesis of this nonmagnetic $5d$ virtual bound state appears to be capable of explaining the experimental values of C_4. However, there is still no obvious explanation as to why C_6 has values as large as those measured. In any case, the ideas suggested in this work bear further study, and could have bearing on the crystal-field behavior in the pure rare earth metals.

It is worth pointing out that if one made single-crystal samples of the type studied by Williams and Hirst, one could do anisotropic magnetization measurements of the type shown in Figure 2–28 for TmSb. Such measurements would be especially valuable[66] in confirming the relative sizes of C_4 and C_6.

We also point out that the apparent failure of the point charge model for dilute rare earths in silver and gold contrasts with the behavior[66] for TmSb and[109] PrBi, which are metals and for which the point charge model works quite well.

(b) Dilute Rare Earths in Yttrium, Lutetium, or Lanthanum

It would be interesting to know the crystal-field splittings for dilute rare earths in yttrium, lutetium, or lanthanum. (One would expect Y or Lu to be the better hosts with regard to matching ionic size—and hence lattice structure—and the conduction electron behavior for the heavy rare earth metals, and La to be a better host for Pr and Nd.) There is little such data available. Sugawara and Soga[117] have briefly reported susceptibility measurements for about 0.5 at.% Er, Nd, and Tb in Y. They say their measurements indicate a splitting between the ground state and first excited state of about 15 K for Er and 30 to 45 K for Nd. Their data suggests a splitting for Tb small compared to liquid helium temperatures.

Nagasawa and Sugawara[118] measured the susceptibility of 2 at.% Pr in La. Using a crystal-field Hamiltonian of the form suggested by Bleaney[60] (i.e. of the form given by general symmetry considerations for the cubic sites, and of the form given by the point charge model with ideal c/a ratio for the hexagonal sites),

$$\mathscr{H}_{\text{CF cubic}} = V_4{}^0(O_4{}^0 + 20\sqrt{2}\, O_4{}^3) + V_6{}^0 \left(O_6{}^0 - \frac{35}{\sqrt{8}}\, O_6{}^3 + \frac{77}{8}\, O_6{}^6\right) \quad (2.126a)$$

$$\mathscr{H}_{\text{CF hex}} = V_4{}^0 O_4{}^0 + V_6{}^0 \left(O_6{}^0 + \frac{77}{8}\, O_6{}^6\right), \quad\quad\quad (2.126b)$$

they found the crystal-field splittings by fitting the temperature dependence of the susceptibility. [The form of $\mathscr{H}_{\text{CF cubic}}$ in (2.126a) differs from that in (2.122) because a $\langle 111 \rangle$ axis is the axis of quantization in (2.126a) and a $\langle 100 \rangle$ axis in (2.122).] They found a splitting from the singlet ground state to the first excited state of 11 K for the hexagonal sites and of 25 K for the cubic sites. [There seems to be somewhat of an inconsistency in the numbers quoted.[118] The fourth-order contribution would have to be somewhat larger relative to the sixth-order term than the values quoted in order to have a singlet ground state for the cubic site.]

It would be valuable to attempt the same sort of analysis for dilute heavy rare earths in Y or Lu as Williams and Hirst[115] performed for Ag and Au as hosts. (As already noted, La would be an appropriate host for the light rare earths.) One would think that the effective crystal-field felt by a heavy rare earth in Y or Lu would correspond closely to that felt in a concentrated rare earth metal.

The analysis would be somewhat more complicated than for Ag or Au as hosts. The crystal-field Hamiltonian contains four independent terms rather than two as in the cubic expression given by (2.122) or (2.126a).

$$\mathscr{H}_{\text{CF hex}} = V_2{}^0 O_2{}^0 + V_4{}^0 O_4{}^0 + V_6{}^0 O_6{}^0 + V_6{}^6 O_6{}^6 \quad (2.127)$$

(The $V_2{}^0$ term would vanish in the nearest-neighbor point charge model for a c/a ratio equal to the ideal value for the hcp lattice. In practice the departure from the ideal c/a ratio is such that the $V_2{}^0$ term is significant.) Even for an ideal hcp lattice, on symmetry grounds alone the $V_6{}^0$ and $V_6{}^6$ coefficients remain independent. Only if one chooses the point charge model is the ratio $V_6{}^6/V_6{}^0 = 77/8$ fixed. (This contrasts to the cubic case where that ratio is fixed wholly on symmetry grounds. For the cubic point groups there is only one linear combination of $Y_6{}^m$'s giving a Γ_1 representation; while for the point group D_{3h} corresponding to the site symmetry in the hcp lattice, there are two linear combinations of $Y_6{}^m$'s giving Γ_1 representations).

On the other hand, there are several experimental techniques available, in addition to susceptibility measurements, to aid in the analysis. Because of the induced moment effect, one can anticipate that, especially for Er and Tm, valuable information about the crystal-field can be obtained from nonlinear high field magnetization measurements. Mossbauer measurements and neutron inelastic scattering experiments, provided they can be done on samples of the desired dilution, could also supply valuable information about the crystal-field.

REFERENCES

1. Elliott, R. J., in Magnetism (G. T. Rado and H. Suhl, eds). Academic Press, New York (1965), Vol. IIA, p. 385.

2. Cooper, B. R., in Solid State Physics (F. Seitz, D. Turnbull and H. Ehrenrecih, eds). Academic Press, New York (1968), Vol. 21, p. 393.
3. Kaplan, T. A. and Lyons, D. H. *Phys. Rev.*, **129**, 2072 (1963).
4. Specht, F. *Phys. Rev.*, **162**, 389 (1967).
5. Elliott, R. J. *Phys. Rev.*, **124**, 346 (1961).
6. Yosida, K. and Miwa, H. *J. Appl. Phys.*, **32**, 8S (1961); Miwa, H. and Yosida, K. *Progr. Theoret. Phys.*, (Kyoto) **26**, 693 (1961).
7. Kaplan, T. A. *Phys. Rev.*, **124**, 329 (1961).
8. Brinkman, W. F. and Elliott, R. J. *Proc. Roy. Soc.*, **294A**, 343 (1966); Brinkman, W. F. *J. Appl. Phys.* **38**, 939 (1967).
9. Bjerrum Møller, H. and Gylden Houmann, J. C. *Phys. Rev. Letters*, **16**, 737 (1966).
10. Evenson, W. E. and Liu, S. H. *Phys. Rev.*, **178**, 783 (1969).
11. Yoshimori, A. *J. Phys. Soc. Japan*, **14**, 807 (1959).
12. Villain, J. *J. Phys. Chem. Solids*, **11**, 303 (1959).
13. Herpin, A., Meriel, P. and Villain, J. *Compt. Rend.*, **249**, 1334 (1959).
14. Herpin, A. and Meriel, P. *Compt. Rend.*, **251**, 1450 (1960).
15. Kaplan, T. A. *Phys. Rev,*, **116**, 888 (1959).
16. Nagamiya, T., in Solid State Physics (F. Sitez, D. Turnbull, and H. Ehrenreich, eds). Academic Press, New York (1967), Vol. 20, p. 306.
17. Enz, U. *J. Appl. Phys.*, **32**, 22S (1961).
18. Cooper, B. R. *Proc. Phys. Soc.* (London), **80**, 1225 (1962).
19. Goodings, D. A. *J. Appl. Phys.*, **39**, 887 (1968).
20. Miwa, H. *Proc. Phys. Soc.* (London), **85**, 1197 (1965).
21. Elliott, R. J. and Wedgwood, F. A. *Proc. Phys. Soc.* (London), **84**, 63 (1964).
22. de Gennes, P. G. and Saint James, D. *Solid State Comm.*, **1**, 62 (1963).
23. Cooper, B. R. *Phys. Rev.*, **169**, 281 (1968).
24. Feron, J. L. Thesis, University of Grenoble, 1969 (unpublished).
25. Callen, H. B. and Callen, E. *J. Phys. Chem. Solids*, **27**, 1271 (1966).
26. Cooper, B. R., Elliott, R. J., Nettel, S. J. and Suhl, H. *Phys. Rev.*, **127**, 57 (1962).
27. Shirane, G. and Pickart, S. J. *J. Appl. Phys.*, **37**, 1032 (1966).
28. Millhouse, A. H. Ph.D. Thesis, Virginia Polytechnic Institute, Sept. 1969 (unpublished).
29. Fisher, E. S. and Dever, D. *Trans. AIME*, **239**, 48 (1967).
30. Rosen, M. and Klimker, H. *Phys. Rev.*, **B1**, 3748 (1970).
31. Clark, A. E., DeSavage, B. F. and Bozorth, R. *Phys. Rev.*, **138**, A216 (1965).
32. Rhyne, J. J. and Legvold, S. *Phys. Rev.*, **138**, A507 (1965).
33. Rhyne, J. J. and Clark, A. E. *J. Appl. Phys.*, **38**, 1379 (1968).
34. Koehler, W. C., Cable, J. W., Child, H. R., Wilkinson, M. K. and Wollan, W. O. *Phys. Rev.*, **158**, 450 (1967).
35. Herpin, A. and Meriel, P. *J. Phys. Radium*, **22**, 337 (1961).
36. Enz, U. *Physica*, **26**, 698 (1960).
37. Nagamiya, T., Nagata, K. and Kitano, Y. *Progr. Theoret. Phys.* (Kyoto), **27**, 1253 (1962).
38. Nagamiya, T. *J. Appl. Phys.*, **33**, 1029 (1962).
39. Kitano, Y. and Nagamiya, T. *Prlgr. Theoret. Phys.* (Kyoto), **31**, 1 (1964).
40. Cooper, B. R. and Elliott, R. J. *Phys. Rev.*, **131**, 1043. (1963). Erratum: *Phys. Rev.*, **153**, 654 (1967).
41. Cooper, B. R. *Phys. Rev. Letters*, **19**, 900 (1967).
42. These values of T_N and T_C are taken from the magnetization measurements of Behrendt *et al.* (D. R. Behrendt, S. Legvold and F. H. Spedding, *Phys. Rev.*, **109**, 1544 (1958) and differ slightly from the values found by electrical resistivity measurements in Chapter 7.
43. Koehler, W. C. *J. Appl. Phys.*, **37**, 1076 (1965).
44. Wilkinson, M. K., Koehler, W. C., Wollan, E. O. and Cable, J. W. *J. Appl. Phys.*, **32**, 485 (1961).
45. Darnell, F. J. *Phys. Rev.*, **132**, 128 (1963).
46. Callen, E. and Callen, H. B. *Phys. Rev.*, **139**, A455 (1965).
47. Liu, S. H., Behrendt, D. R., Legvold, S. and Good, Jr, R. H. *Phys. Rev.*, **116**, 1464 (1959).

47a. Mackintosh, A. R. 1970 NATO Advanced Summer Institute on Magnetism of Rare Earth Materials, Alta Lake, British Columbia, August, 1970.

48. Bjerrum Møller, H., Gylden Houmann, J. C. and Mackintosh, A. R. *Phys. Rev. Letters*, **19**, 312 (1967).

48a. Bjerrum Møller, H., Nielsen, M. and Mackintosh, A. R. *Les Eléments des Terres Rares*. (Proceedings of the Grenoble Conference May, 1969) Colloques Intern. du CNRS, N° 180, Tome II, page 277.

48b. Nicklow, R. M. *J. Appl. Phys.*, **42**, 1672 (1971).

49. Turov, E. A. and Shavrov, V. G. *Fiz. Tverd. Tela*, **7**, 217 (1965). [Translation: *Soviet Phys. Solid State*, **7**, 166 (1965)].

50. Nielsen, M., Bjerrum Møller, H. and Mackintosh, A. R. *J. Appl. Phys.*, **41**, 1174 (1970).

51. Sievers, A. J. *J. Appl. Phys.*, **41**, 980 (1970).

52. Wagner, T. K. and Stanford, J. L. *Phys. Rev.*, **184**, 505 (1969).

53. Trammell, G. T. *J. Appl. Phys.*, **31**, 362S (1960).

54. Trammell, G. T. *Phys. Rev.*, **131**, 932 (1963).

55. Bleaney, B. *Proc. Roy. Soc.* (London), **A276**, 19 (1963).

56. Cooper, B. R. *Phys. Rev.*, **163**, 144 (1967).

57. Wang, Y.-L. and Cooper, B. R. *Phys. Rev.*, **172**, 539 (1968).

58. Wang, Y.-L. and Cooper, B. R. *Phys. Rev.*, **185**, 696 (1969).

59. Cooper, B. R. *J. Appl. Phys.*, **40**, 1344 (1969).

60. Bleaney, B. *Proc. Roy. Soc.* (London), **A276**, 39 (1963).

61. Rainford, B. D. and Wedgwood, F. A. 1969 (unpublished).

62. Johansson, T., Lebech, B., Nielsen, M., Bjerrum Møller, H. and Mackintosh, A. R., *Phys. Rev. Letters*, **25**, 524 (1970).

63. Bucher, E., Chu, C. W., Maita, J. P., Andres, K., Cooper, A. S., Buehler, E. and Nassau, K. *Phys. Rev. Letters*, **22**, 1260 (1969).

64. Child, H. R., Wilkinson, M. K., Cable, J. W., Koehler, W. C. and Wollan, E. O. *Phys. Rev.*, **131**, 922 (1963).

65. Busch, G. *J. Appl. Phys.*, **38**, 1386 (1967).

66. Vogt, O. and Cooper, B. R. *J. Appl. Phys.*, **39**, 1202 (1968); Cooper, B. R. and Vogt, O. *Phys. Rev.*, **B1**, 1211 (1970).

67. Junod, P., Menth, A. and Vogt, O. *Phys. Kondens. Materie*, **8**, 323 (1969).

68. Cooper, B. R. and Vogt, O. *Phys. Rev.*, **B1**, 1218 (1970).

69. Stutius, W. *Phys. Kondens. Materie*, **9**, 341 (1969).

70. Cable, J. W., Moon, R. M., Koehler, W. C. and Wollan, E. O. *Phys. Rev. Letters*, **12**, 553 (1964).

71. Bucher, E., Andres, K., Maita, J. P. and Hull, Jr., G. W. *Helv. Phys. Acta*, **41**, 723 (1968).

72. Lee, K. N., Bachmann, R., Geballe, T. H. and Maita, J. P., *Phys. Rev.*, **B2**, 4580 (1970).

73. Buschow, K. H. J. and Fast, J. F. *Z. Physik. Chem.*, Neue Folge, **50**, 1 (1966).

74. Cooper, B. R. *Helv. Physica Acta*, **41**, 750 (1968).

75. de Wijn, H. W., van Diepen, A. M. and Buschow, K. H. J., *Phys., Rev.* **B1**, 4203 (1970).

76. Moon, R. M., Cable, J. W. and Koehler, W. C. *J. Appl. Phys.*, **35**, 1041 (1964).

77. Tsuchida, T. and Wallace, W. E. *J. Chem. Phys.*, **43**, 2087 (1965).

78. Tsuchida, T. and Wallace, W. E. *J. Chem. Phys.*, **43**, 2885 (1965).

79. Busch, G. and Vogt, O. *Phys. Letters*, **20**, 152 (1966).

80. Wang, Y. L. and Cooper, B. R., *Phys. Rev.*, **B2**, 2607 (1970).

81. Stevens, K. H. W. *Rev. Mod. Phys.*, **25**, 166 (1953).

82. Van Vleck, J. H. *Rev. Mat. Fis. Teor.* (Tueuman, Argentina), **14**, 189 (1962).

83. Levy, P. M. *Phys. Rev. Letters*, **20**, 1366 (1968); *Phys. Rev.*, **177**, 509 (1969).

84. Elliott, R. J. and Thorpe, M. F. *J. Appl. Phys.*, **39**, 802 (1968).

85. Birgeneau, R. J., Hutchings, M. T., Baker, J. M. and Riley, J. D. *J. Appl. Phys.*, **40**, 1070 (1969).

86. Cooper, B. R. and Vogt, O., *J. de Physique*, **32**, C1–1026 (1971).

87. Cooper, B. R., Jacobs, I. S., Graham, C. D. and Vogt, O., *J. de Physique*, **32**, C1–359 (1971).
88. Griffith, J. S. *Phys. Rev.*, **132**, 316 (1963).
89. Bozorth, R. M. and Van Vleck, J. H. *Phys. Rev.*, **118**, 1493 (1960).
90. Bogoliubov, N. N. and Tiablikov, S. Ia. *Izv. Akad. Nauk. SSSR Ser Fiz.*, **21**, 849 (1957).
91. Grover, B. *Phys. Rev.*, **140**, A1944 (1965).
92. Pink, D. A. *J. Phys.*, **C1**, 1246 (1968).
93. Stevens, K. W. H., in Magnetism (G. T. Rado and H. Suhl, eds). Academic Press, New York (1963), Vol. I, p. 1.
94. Callen, H. B. *Phys. Rev.*, **130**, 890 (1963).
95. Elliott, R. J. and Pfeuty, P., private communication (1969); also see Elliott, R. J., Pfeuty, P. and Wood, C. *Phys. Rev. Letters*, **25**, 443 (1970).
96. Kitano, Y., abstract (in Japanese) for talk presented at the joint annual meeting of the Physical and Applied Physical Societies of Japan, 1969. Translation provided by the Japanese Liaison Office of the General Electric Research and Development Center.
97. Hsieh, Y. Y. and Blume, M., private communication, 1970.
98. Lea, K. R., Leask, M J. M. and Wolf, W. P. *J. Phys. Chem. Solids*, **23**, 1381 (1962).
99. Cooper, B. R., Jacobs, I. S., Fedder, R. C., Kouvel, J. S. and Schumacher, D. P. *J. Appl. Phys.*, **37**, 1384 (1966).
100. Cooper, B. R. *Phys. Letters*, **22**, 24 (1966).
101. Parkinson, D. H., Simon, F. E. and Spedding, F. H. *Proc. Roy. Soc.* (London), **207A**, 137 (1951).
102. Lock, J. M. *Proc. Phys. Soc.* (London), **B70**, 566 (1957).
103. Dreyfus, B., Goodman, B. B., Lacaze, A. and Trolliet, G. *Compt. Rend.*, **253**, 1764 (1961).
104. Dempsey, C. W., Gordon, J. E. and Soller, T. *Bull. Am. Phys. Soc.*, **7**, 309 (1962).
105. Lounasmaa, O. V. *Phys. Rev.*, **133**, A211 (1964).
106. Matthias, B. T., Geballe, T. H., Andres, K., Corengwit, E., Hull, G. W., and Maita, J. P. *Science*, **159**, 530 (1968); Geballe, T. H., Matthias, B. T., Andres, K., Maita, J. P., Cooper, A. S. and Corengwit, E. *Science*, **160**, 1443 (1968).
107. Menth, A., Buehler, E. and Geballe, T. H. *Phys. Rev. Letters*, **22**, 295 (1969).
108. Geballe, T. H., Menth, A., Buehler, E. and Hull, G. W. *J. Appl. Phys.*, **41**, 904 (1970).
109. Birgeneau, R. J., Bucher, E., Passell, L., Price, D. L. and Turberfield, K. C. *J. Appl. Phys.*, **41**, 900 (1970).
109a. Turberfield, K. C., Passell, L., Birgeneau, R. J. and Bucher, E. *J. Appl. Phys.*, **42**, 1746 (1971).
109b. Birgeneau, R. J., Bucher, E., Passell, L. and Turberfield, K. C. *Phys. Rev.*, **B4**, 718 (1971).
110. Cable, J. W., Collins, M. F. and Woods, A. D. B. Proc. 6th Rare Earth Research Conference, Gatlinburg, Tenn. (May 3–5, 1967), page 297.
111. Holden, T. M., Powell, B. M., and Woods, A. D. B. *J. Appl. Phys.*, **39**, 457 (1968).
112. Woods, A. D. B. *Canadian J. of Physics*, **46**, 1499 (1968).
113. Uhrich, D. L. and Barnes, R. G. *Phys. Rev.*, **164**, 428 (1967).
114. Hirst, L. L., Williams, G., Griffiths, D. and Coles, B. R. *J. Appl. Phys.*, **39**, 844 (1968).
115. Williams, G. and Hirst, L. L. *Phys. Rev.*, **185**, 407 (1970).
116. Griffiths, D. and Coles, B. R. *Phys. Rev. Letters*, **16**, 1093 (1966).
117. Sugawara, T. and Soga, R. *J. Phys. Soc. Japan*, **18**, 1102 (1963).
118. Nagasawa, H. and Sugawara, T. *J. Phys. Soc. Japan*, **23**, 701 (1967).

Chapter 3

Magnetic Structures of Rare Earth Metals and Alloys†

W. C. Koehler

Solid State Division, Oak Ridge National Laboratory, Oak Ridge, Tennessee

3.1. INTRODUCTION

In this chapter we review the magnetic structures of the rare earth metals and of certain of their alloys. Much of this information has been obtained in the last ten years from neutron diffraction experiments. Complementary

† Research sponsored by the U.S. Atomic Energy Commission under contract with the Union Carbide Corporation.

structural data have been provided by the results of classical magnetizaton experiments. A detailed survey of such data is given in Chapter 4. Ideally, both techniques should be, and indeed have been, used to obtain the maximum structural information.

Some indication of the complexity of the magnetic structures in erbium was revealed by some early neutron diffraction studies with polycrystalline samples[1] and polycrystalline specimens of holmium gave diffraction patterns characteristic of a helical spin structure. Only after single crystals of the rare earths were studied, at Ames by classical magnetization methods and at Oak Ridge by neutron diffraction techniques,[2] did the full panoply of exotic spin configurations in these metals begin to be established.

Subsequently, a number of alloy systems were investigated, at first with polycrystalline specimens, later with single crystal samples, until there is now a considerable literature on the physical properties of rare earth alloys and intermetallic compounds.[3]

We shall, in this review, restrict ourselves primarily to the pure metals, and to alloys of the heavy rare earths with each other and with the diamagnetic diluents Y and Th and we shall preferentially utilize results obtained from single crystals when they exist. Since the situation prior to 1965 regarding rare earth metal and alloy structures has been summarized,[4] we shall concentrate on results obtained within the past five years, although for completeness we give a brief summary of earlier work as well.

Among the new developments is the accurate measurement of the form factors of a few of the rare earth metals by polarized beam methods. This represents a natural evolution in magnetic structure studies for once the equilibrium spin configuration has been determined with the aid of an approximate form factor, attention can be given to the form factor itself and to the details of the distribution of magnetically active electrons.

3.2. THE HEAVY METALS

3.2.1. Physical Properties

It will be convenient to consider first the metals of the second half of the rare earth series since all of them with the exception of Yb, exhibit at room temperature the simple hexagonal close-packed structure. The c/a ratio ranges from 1.57 to 1.59, a value significantly lower than that for ideal closest packing 1.633. Ytterbium crystallizes in the cubic close-packed structure. It is of little interest here since it has a full complement of fourteen 4-f electrons and is only weakly paramagnetic. The endpoint of the rare earth series, Lu, likewise has a full 4-f shell and it too is weakly paramagnetic.

Each of the remaining heavy metals exhibits an effective magnetic moment in the paramagnetic state which is nearly identical to that of the corresponding tripositive ion. In the metals, the valence electrons, $5d^16s^2$, are lost to conduction bands; there is left a highly localized moment due to unpaired electrons in the 4f shell. The moment is quite well given by application of Hund's Rules.[5]

As the temperature is reduced below a certain critical temperature there is a transition from the paramagnetic state to an ordered state in which the moments adopt one or another ordered magnetic configuration. The temperature at which the order-disorder transitions occur are evidenced by anomalies or discontinuities in the measured electrical, thermal, or magnetic properties of the metals. For all except Gd, there is evidence, at lower temperatures, for order-order transformations to other magnetic configurations.

Some pertinent physical properties of these heavy rare earth metals are summarized in Table 3.1.[6] The entries under T_N and T_C refer, respectively, to the Néel temperature, and the Curie point as determined from resistivity measurements; θ_p is the paramagnetic Curie point as obtained from susceptibility data. The magnetization, even in the paramagnetic temperature region, is generally anisotropic and this is indicated by the difference in θ_p as measured parallel ($\theta_{\parallel}$) and perpendicular ($\theta_{\perp}$) to the crystallographic c-axis. The column headed μ_f gives results obtained for saturation magnetization which have been measured parallel to the easy axes.

Table 3.1. Some Physical Properties of the Heavy Rare Earth Metals

Metal	$T_N(K)$[a]	$T_C(K)$[a]	$\theta_p K$ [b]	$\mu_f(\beta)$ [b]	$\lambda J(\beta)$	σ_{capt} (barn)
Gd		292.7	$\theta_{\parallel}=\theta_{\perp}=317$	7.55	7.0	46,000
Tb	230.2	219.6	$\theta_{\parallel}=195, \ \theta_{\perp}=239$	9.34	9.0	44
Dy	176	88.3	$\theta_{\parallel}=121, \ \theta_{\perp}=169$	10.20	10.0	1,100
Ho	130	19	$\theta_{\parallel}=73, \ \theta_{\perp}=88$	10.34	10.0	64
Er	85	19.5	$\theta_{\parallel}=61.7, \ \theta_{\perp}=32.5$	8.0	9.0	166
Tm	57.2	32	$\theta_{\parallel}=51, \ \theta_{\perp}=-17$	1.0	7.0	118

[a] Transition temperatures quoted here are from electrical resistivity measurements. See Chapter 7 for references.
[b] References to original investigation are given in Chapter 4.

In Gd, Tb, Dy, and Ho, the measured saturation moment exceeds slightly the value predicted from the Hund's Rules' ground state λJ. The excess is attributed to polarization of the conduction electrons and this has been sought, as we discuss later, in the polarized neutron experiments. The low values listed for Er and Tm are due to the magnitude of the anisotropy energy in these materials. In the former case, as will be seen below, the configuration of moments is a ferromagnetic spiral and the anisotropy energy keeps the basal plane component from turning into the c-axis. Magnetization measurements on a single crystal of thulium in low fields applied parallel to the c-axis gave the saturation moment per atom as $1.001 \pm 0.005 \beta$. This is consistent with the ferrimagnetic antiphase domain type configuration for thulium found in the neutron diffraction experiments. For fields greater than 28 kOe applied parallel to the c-axis thulium becomes ferromagnetic with a saturation moment of $7.14 \pm 0.02 \beta$ per atom.

In the last column of the table are listed quantities of interest for neutron diffraction experiments, namely, the thermal neutron absorption cross sections in barns/atom. For many experiments the cross section of dysprosium, 1100 b, is already uncomfortably large and gadolinium is extremely difficult to study except in certain specialized experiments. Fortunately isotopes of

Gd, Dy, and Er exist with much more favorable absorption cross sections and these have been, or are being, utilized.

3.2.2. Neutron Diffraction Results

Procedure. The experimental procedures which have been used in single crystal studies of the heavy rare earth metals have been described in a number of publications and need be mentioned only briefly here. We refer specifically to those which have been adopted at the Oak Ridge National Laboratory.

In a typical case a disk-shaped single crystal (sometimes a pillar) was cut so that the c-axis was normal to the plane of the disk and this was mounted in one of the low-temperature goniometers with the c-axis parallel to the horizontal rotation axis of the instrument. Systematic scans of important regions of reciprocal space were made at various temperatures above and below the critical temperature. The magnetic intensities were measured in the usual way by comparison with the nuclear intensities. This procedure has been used as well in the single crystal alloy studies to be described later.

To illustrate the diffraction effects observed and to prepare an introduction for the later discussion of experiments on erbium based alloys we describe briefly results obtained from a study of erbium.[7] These are shown, schematically, in Figure 3–1, where is represented a section of the reciprocal lattice for four temperature regions.

Above 80 K, Er is paramagnetic and the coherent reflections indicated in Figure 3–1(a) by the dark spots at reciprocal lattice points are due only to

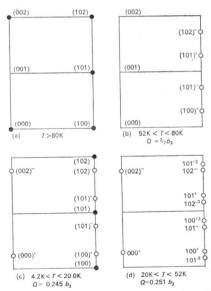

Figure 3–1. Schematic representation in the reciprocal lattice of diffraction effects from single crystal erbium. (a) Paramagnetic scattering region—nuclear scattering only is shown. (b) CAM configuration. (c) Conical configuration. (d) Squared elliptical configuration. In b, c, d only the magnetic scattering is shown.

nuclear scattering. (This same contribution will be present below the Néel temperature but for clarity it is omitted in Figure 3–1(b), (c), and (d) in which the magnetic scattering only is shown.) There is no contribution to (001) because of destructive interference of the nuclear scattering from the two atoms of the unit cell. The intensities in the diffraction maxima are proportional to the quantities $|F|^2_{hkl} = b^2_{Er}|G|^2_{hkl}$ where b_{Er} is the coherent scattering amplitude for Er, and G_{hkl} is the geometric structure amplitude for the hcp structure.

In Figure 3–1(b) is shown the magnetic scattering observed in the temperature range 52 K $< T <$ 85 K. Coherent magnetic reflections appear as satellites of the allowed nuclear reflections. A single pair of satellites, indicated by $(hkl)^\pm$, is associated with each reciprocal lattice point except for the points $(00l)$, and these are found on lattice rows parallel to the reciprocal c-axis, b_3, with spacings which are generally not commensurable with chemical cell dimensions. The existence of a single set of satellites, or harmonics, implies a structure in which the scattering amplitudes of the atoms are modulated sinusoidally. (This may be likened to a grating with a strictly sinusoidal ruling error.) From the location of the satellites on lines parallel to b_3 it follows that the wave vector of the modulation is parallel to that direction. The absence of magnetic satellites of the $(00l)$ reflections implies a uniaxial configuration in which the magnetic moments are parallel or antiparallel to the crystallographic c-axis. This follows because the neutron senses only the component of the moment which is perpendicular to the scattering vector. Thus one concludes that the magnetic structure of erbium in the high temperature ordered phase is a c-axis modulated structure in which the moments in a given plane normal to the c-axis all have the same thermally averaged value, but that this value varies sinusoidally with distance along the c-direction, a_3. Analytically this c-axis modulated structure, CAM, may be represented by

$$u_{nx} = u_{ny} = 0, \quad u_{nz} = A_1 \sin(n\omega + \alpha_1) \tag{3.1}$$

where n denotes the layer normal to the c-axis referred to an arbitrary reference, ω is the interlayer angle determined by $Q \cdot a_3/2$ where Q is the wave vector of the modulation, and α_1 is an arbitrary phase angle. As given in (1) A_1 is the thermal average of the component of the moment parallel to the c-axis. The wave vector Q, or the interlayer angle ω, is found experimentally from the observed separation of the satellites from the reciprocal lattice points with which they are associated. In the temperature range 52 K $< T <$ 80 K the wave vector Q is commensurable with the lattice and has a value equal to 2/7 b_3.

The intensities of the magnetic reflections for such a structure are proportional to the squares of the magnetic structure factors

$$|F|^2_{hkl}{}^\pm = \tfrac{1}{4}(p_0 f)^2 \sin^2\theta \, A_1{}^2 |G|^2_{hkl} \tag{3.2}$$

where θ is the angle between the scattering vector and a_3, p_0 is a constant appearing in the magnetic scattering amplitude equal to 0.269×10^{-12} cm and f is the magnetic form factor. The quantity A_1 is the amplitude of the modulation. In erbium the amplitude saturates at a value of 7.6 β/Er atom at a temperature near 52 K.

In the intermediate temperature range $20 \text{ K} < T < 52 \text{ K}$ the diffraction effects illustrated in Figure 3–1(d) are found. There are now first-order satellites of all reflections, including the origin, as well as third-order satellites of the reflections (hkl). In addition the wave vector varies with temperature and shows a thermal hysteresis effect.

The development of the satellites of reflections $(00l)$ implies an ordering in the basal plane components of the moments. At the same time the magnitude of the c-axis component exceeds its maximum value consistent with the anisotropy in the crystal and the configuration of z-components changes, necessarily, and tends toward an antiphase domain type configuration. The introduction of higher harmonics in the series representing the moment configuration results in the introduction of corresponding overtones of the fundamental satellites in the diffraction patterns.

Two structural models can be envisaged which would produce these diffraction effects. Both are modified spirals[8] in which the moments when projected to a common origin lie in a plane. The observed intensities show that the z-component is larger than the x- or y-component so that the hodograph of the moments is an ellipse. Presumably this is due to crystal field anisotropy effects (the axial anisotropy constant K_2^0 is negative and hence favors alignment along the c-axis). In one model the major axis is parallel to the c-axis so that x- and z-components oscillate while the y-component is zero. Miwa and Yosida[9] have considered this to be the nature of the configuration near the temperature of ordering of the planar component on the basis of a molecular field calculation. The x-axis would be determined, presumably, by the basal plane anisotropy, and the six-fold symmetry of that anisotropy would undoubtedly lead to domains in which the xz planes of spin rotation would assume the preferred orientations. In the second model the major axis of the ellipse is tilted away from the c-axis so that both the x- and y-components oscillate to execute a regular spiral. The presence of the third order satellites shows that neither model is an adequate description and that there must be some distortion of the ellipse. This model too has been extracted by Miwa and Yosida from their calculations and is expected to develop at a lower temperature than the first model considered. Unfortunately, with multidomain crystals both models lead to the same diffraction effects, and they cannot be distinguished. However, the tilted ellipse seems more realistic from the following arguments. Below 20 K the configuration, as we shall consider presently, is a conical or ferromagnetic spiral structure with an interlayer angle of about 44° per layer. The hexagonal anisotropy is evidently not too important since the basal plane spiral is undistorted. It is expected to be even less important at higher temperatures. Thus there is no reason to select one or another direction in the basal plane as a preferred x-axis. Accordingly, the data have been analyzed on the tilted axis model.

The structure in this intermediate temperature range may be represented by

$$\begin{aligned}
\mu_{nx} &= \mu_\perp \cos (n\omega + \beta_1) \\
\mu_{ny} &= \mu_\perp \sin (n\omega + \beta_1) \\
\mu_{nz} &= A_1 \sin (n\omega + \alpha_1) + A_3 \sin (3n\omega + \alpha_3)
\end{aligned} \tag{3.3}$$

Expressions for the structure factors of the several types of satellites contain the coefficients of these expressions in such a way that they can be determined experimentally; thus

$$|F|^2_{00l\pm} = \tfrac{1}{4}(p_0 f)^2 (1 + \cos^2 \theta)\, \mu_\perp{}^2 |G|^2_{00l}$$

$$|F|^2_{hkl\pm} = \tfrac{1}{4}(p_0 f)^2 |G|^2_{hkl}[(1 + \cos^2 \theta)\, \mu_\perp{}^2 + A_1{}^2 \sin^2 \theta] \qquad (3.4)$$

$$|F|^2_{hkl\pm^3} = \tfrac{1}{4}(p_0 f)^2 |G|^2_{hkl}\, A_3{}^2 \sin^2 \theta$$

Just above 20 K, on warming, the interlayer angle is equal to 45°. The values obtained for the coefficients at that temperature are $\mu_\perp = 3.6\,\beta$, $A_1 = 9.72\,\beta$ and $A_3 = 2.97\,\beta$. The configuration is determined if the phase angles α_1 and α_3 can be given. Since the maximum total moment on any atom is $9.0\,\beta$, the z-component of the moment cannot exceed $8.2\,\beta$. This suggests that the two terms in μ_{nz} of Eqn (3.3) be as nearly out of phase as possible. The most symmetrical phasing which satisfies the $8.2\,\beta$ condition is that which puts $\alpha_1 = 5\pi/8$ and $\alpha_3 = 15\pi/8$. This gives a sequence of moment components μ_{nz} as follows:

$$\mu_{0z} = -\mu_{3z} = -\mu_{4z} = \mu_{7z} = 7.84\,\beta$$

$$\mu_{1z} = -\mu_{2z} = -\mu_{5z} = \mu_{6z} = 6.46\,\beta$$

which is approaching an antiphase domain configuration with four moments up followed by four moments down. The phase angle β_1 is indeterminate. Below 20 K, Figure 3–1(c), the conical configuration exists and this is described by

$$\mu_{nx} = \mu_\perp \cos (n\omega + \beta_1)$$

$$\mu_{ny} = \mu_\perp \sin (n\omega + \beta_1) \qquad (3.5)$$

$$\mu_{nz} = \mu_\parallel$$

In this configuration there is a constant component along the c-axis so that there are magnetic contributions to the reciprocal lattice points. Since this part of the structure is uniaxial along the c-axis there are no magnetic contributions to the reflections $(00l)$. The satellites due to the helical arrangement of the x and y components are still observed.

The intensities produced by a conical configuration are proportional to

$$|F|^2_{hkl} = (p_0 f)^2 \sin^2 \theta\, \mu_\parallel{}^2 |G|^2_{hkl}$$

$$|F|^2_{hkl\pm} = \tfrac{1}{4}(p_0 f)^2 (1 + \cos^2 \theta)\, \mu_\perp{}^2 |G|^2_{hkl} \qquad (3.6)$$

At 4.2 K $\mu_\parallel = 7.6\,\beta$, and $\mu_\perp = 4.3\,\beta$ yielding a total moment of $9.0\,\beta$. The semi-cone angle, that is, the angle made by the moment with the c-axis is 29°.

Diffraction data found for other structures characteristic of the heavy rare earths may be illustrated by the same figure. The pure helix, for instance, gives rise to the same satellite structure as Figure 3–1(c); in that case there is no magnetic contribution to the normal lattice sites ($\mu_\parallel = 0$). A basal plane ferromagnet would contribute magnetic intensity at all allowed nuclear positions; it would be represented by Figure 3–1(a) where nuclear and magnetic intensities would be superimposed.

4

The Magnetic Structures. From detailed analyses of single crystal data, similar to that just described for erbium, magnetic configurations for the heavy rare earth metals were deduced. These are represented schematically in Figure 3–2. In each of these structures the period of the modulation varies with temperature in the manner shown in Figure 3–3.[10] The wave vector Q, on the left, and the period, on the right are plotted as functions of the reduced temperature T/T_N for the five metals Tb, Dy, Ho, Er, and Tm. In inserts to this figure are shown on expanded scales, the temperature dependence of the wave vector for Tb, and, near the transition temperature, for Tm. We discuss these figures now in a little more detail.

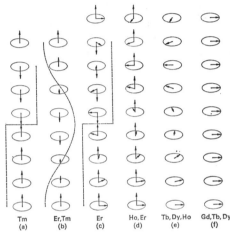

Figure 3–2. Schematic representation of the magnetic structures of heavy rare earth metals. The moments are supposed to be parallel in a given hexagonal layer. The ranges of stability of the various structures are given in the text.

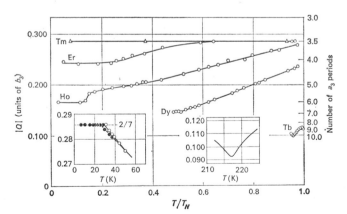

Figure 3–3. Variation of the wave vector Q with temperature. The magnitude of the wave vector, on the left, and the period on the right, is given as a function of reduced temperature for the heavy rare earths. Data for Tb (Ref. 21) and Tm (Ref. 33) are given in the right and left hand inserts, respectively.

(a) Gadolinium

The elements from Tb to Tm have been found to possess antiferromagnetic oscillatory spin structures in some temperature region, whereas gadolinium has been reported to be ferromagnetic with a Curie temperature of 293 K. Belov and Pedko[11] observed some very low field magnetization anomalies in polycrystalline Gd above 210 K which they attributed to a spiral spin structure. Very small fields were sufficient, according to their report, to transform the spiral into a ferromagnet. Graham has repeated their measurements on a single crystal specimen but found no such anomalies.[12] Three neutron diffraction investigations have been made to search for direct evidence for the existence of oscillatory spin structures in gadolinium.

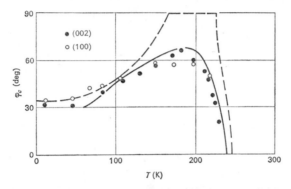

Figure 3–4. Temperature dependence of ϕ_c, the angle between the moment direction and the c-axis in gadolinium. The dotted curve represents torque measurements of Graham (Ref. 16) and the dashed curve those of Corner, Roe, and Taylor (Ref. 15).

Will, Nathans, and Alperin[13] studied a single crystal of gadolinium of naturally occurring isotopic abundance. The extremely high cross section for absorption of thermal neutrons implies a very short penetration depth and their investigation at 1.055 Å was restricted to a study of reflections with scattering vectors nearly normal to the face of the crystal. Nevertheless, they were able to conclude, within relatively modest limits, that there were no observable satellites and that Gd behaves as a normal ferromagnet. Cable and Wollan[14] used a beam of wavelength 0.353 Å at which wavelength the resonance absorption cross section is down by two orders of magnitude, and a more complete neutron study was possible. They too found no evidence for satellite structure. All the magnetic scattering appeared at normal lattice positions. Furthermore, they observed no broadening of the rocking curves below T_C whereas a spiral turn angle $> 2°$ would have produced a noticeable broadening. There was no indication of a field dependence of the intensities in the range 0.1 to 5 Oe. The evidence is strong that Gd is spontaneously ferromagnetic throughout the entire ordered region.

It is possible that the anomalous behavior of Belov and Pedko is associated with the curious crystalline anisotropy of Gd. The neutron results of Cable and Wollan relative to this point are shown in Figure 3–4 along with curves representing the torque measurements of Corner, Roe, and Taylor[15] and Graham.[16] The moment direction is parallel to the c-axis from T_C to 232 K, moves away from that axis to a maximum deviation 65° near 180 K, and then back to within 32° of the c-axis at low temperatures.

In a recent investigation by Kuchin et al.[17] a single crystal of Gd enriched in the low capturing isotope ^{160}Gd was investigated at temperatures ranging from 78 K to room temperature. They, too, conclude that Gd is a normal ferromagnet below T_C, and they too find that the direction of the moment depends strongly on the temperature. We shall return to this study later on under the discussion of the form factor of gadolinium.

The magnetic structure of gadolinium, then, will be represented by Figure 3–2(f) if allowance is made for the fact that the moments have the complicated directional dependence with temperature described above.

(b) Terbium

As seen in Table 3.1, the antiferromagnetic temperature range for Tb is quite narrow. Nevertheless it has been established that the structure in this temperature range is an oscillatory one.[18–19] It is not possible to say with certainty that it is a helical structure because the moments are far from their saturation values when the structure transforms spontaneously to a planar ferromagnet. Arguments can be given,[20] that when the first order aniso- tropy constant is positive, as it is in Tb, the initial ordering which is stable is the helical structure. Moreover, the addition of Y to Tb[18] is known to widen the range of stability of the antiferromagnetic configuration and at saturation, intensity arguments can be made in favor of the helical structure.

Near 225 K the transformation, helix-planar ferromagnet begins and is complete near 217 K, according to the neutron measurements. There is some temperature hysteresis and there is a narrow temperature region where the helical and ferromagnetic configurations coexist. Below 217 K, the sub- stance is ferromangetic with the moments in the basal plane. From neutron diffraction measurements on a system with random domain population it is possible to determine the direction of the moments in the layer. Magnetization data, however, show that at low temperatures the easy direction is a b-direc- tion, normal to (10$\bar{1}$0) planes. From the neutron data at 4.2 K a moment of 9.0 β was obtained which is in agreement with that expected for the Hund's Rules' ground state. The magnetic structures of terbium then are illustrated by Figure 3–2(e) and 3–2(f).

The turn angle of Tb varies somewhat with temperature over the antiferro- magnetic temperature range. As first measured[19] the results appeared as shown in the unexpanded part of Figure 3–3. The interlayer angle has a value of about 20 deg/layer at the Néel point, falls to a minimum near 18 deg/ layer and starts to rise again so that the temperature-turn angle curve is somewhat s-shaped. Recently Dietrich and Als-Nielsen,[21] and Umebayashi et al.[22] have, in the course of other experiments, confirmed some of the general features of the magnetic structure of Tb. In addition they were able to measure the turn angle to somewhat lower temperatures and found that it continued to rise, reaching nearly 19° per layer. A smoothed averaged set of data for Tb is given in the right-hand insert to Figure 3–3.[23]

(c) Dysprosium

Dysprosium has the simple helical structure from its Néel point to about 90 K at which temperature the transformation helix-ferromagnet begins.[24] This transformation takes place over a narrow range of temperatures with the midpoint occurring at 87 K. Thermal hysteresis was observed so that on warming the corresponding temperature occurs at 92 K. At low temperatures the moments are in or nearly in the basal plane and directed along easy *a*-directions according to the magnetization data. The structures of Dy, as are those of Tb, are represented by Figure 3–2(e) and 3–2(f).

The turn angle at the Néel temperature is 43.2° and it decreases to 26.5° at the temperature at which the ferromagnetic state starts to occur.

The total amount of coherent scattering was found to be continuous across the transition which implies that the moments are merely reorienting themselves at the transition. In both the ferromagnetic and antiferromagnetic structures the moment derived from the data was 9.5 β, a value slightly smaller than the expected value of 10.0 β It is not yet certain that the discrepancy is significant.[25]

(d) Holmium

The properties of holmium depend, in detail, on the particular crystal studied but the general features which are exhibited by all the specimens so far investigated are the following.[26-27] Below the Néel point a simple spiral develops and this is retained with some distortion below 45 K down to a Curie point near 20 K at which the moments tilt out of the basal plane to form a conical structure with a net moment, parallel to *c*, of 1.7 β. The basal plane component, $\mu_\perp = 9.5 \beta$, is ordered in a distorted spiral. Only when a magnetic field is applied parallel to the *b* direction is a ferromagnet produced. At 4.2 K the full moment of 10.0 β is induced along the easy direction by an applied field.

The approximate low temperature structure of holmium is the conical configuration, Figure 3–2(d), in which the semi-cone angle is about 80°. The approximate structure above T_C is the simple helix Figure 3–2(e).

In Figure 3–3 the wave-vector-temperature variation is that which appears to be representative of most samples. This is reproduced in Figure 3–5 where it is labeled Ho(B) along with data for another crystal, Ho(A). Both crystals have an initial value of the turn angle near 50° per layer. In both cases the turn angle decreases with decreasing temperature. In Ho(A) it becomes constant at a value of 36.7° per layer at a temperature near 35 K. For Ho(B) the turn angle reaches a value of precisely 30° per layer below about 19 K. Some significant parameters for the two crystals are summarized in tabular form in the figure. The Néel points differ by 11 K, but, within experimental observation the Curie points are identical. With a minimum value of $\omega = 30.0°$ the magnetic cell for Ho(B) is commensurable with the chemical cell. The temperature measured on warming at which ω departs from its minimum value, T_ω is about 35 K for Ho(A) and 18.7 K for Ho(B). Two other specimens investigated at Oak Ridge, Ho(C) and Ho(D), showed a final turn angle of 30.0°. A specimen investigated at Chalk River exhibited a final turn angle

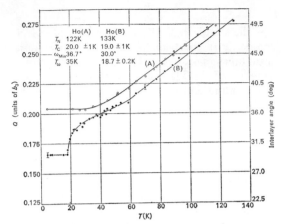

Figure 3–5. Temperature dependence of the wave vector of holmium crystals. The scale on the left is the wave vector in units of b_3; that on the right the corresponding interlayer angle. Pertinent data characteristic of the two crystals are collected in the insert.

of 32.5°.[28] The reason for the variety of final turn angles is undoubtedly related to a delicate balance between exchange, anisotropy, and magneto-elastic energies in this metal.

The diffraction patterns of both specimens exhibited weak but measurable 5th and 7th harmonics of the primary satellites of (00l) reflections which indicate that the configuration of moments in the basal plane is distorted from the simple spiral. For Ho(B), with a turn angle of 30°, moments would alternately be found near hard and easy axes. To minimize the energy the moments tend to "bunch" around the easy directions. This "bunching" persists to temperatures of the order of 45 K for both specimens, and this reflects the influence of the hexagonal anisotropy energy in holmium. That T_ω and T_C are equal for Ho(B) is evidently coincidental.

In the original investigation of Ho(B) other "harmonics" were observed which could, except for a peak occurring at the position of the second harmonic of the origin, $(000)^{+2}$, be explained as spurious double reflections. Evidence has recently been obtained from a polarization analysis investigation[29] of the second harmonic that it has its origin in double scattering as well.

(e) Erbium

The magnetic structures of erbium have been described in the section on procedures above. They are represented by Figure 3–2(b) for the high temperature phase, Figure 3–2(c) for the intermediate temperature range. The turn angle for Er is constant at 51.4° from T_C to 53 K. Below 20 K it is constant again at a value of 44.0°. It is to be noted that the "squared up" configuration of Figure 3–2(c) is never quite achieved. The structure transforms to the conical configuration before the squaring process is complete.[30]

(f) Thulium

Thulium is another case apparently where the structural properties obtained from investigations of different specimens differ in detail. This is probably due to the fact that the more recent specimens are of significantly higher purity. The general features of the magnetic structures of thulium are, however, confirmed, and these are illustrated in Figure 3–2(b) for the higher temperature region and Figure 3–2(a) at very low temperatures.[31] At T_N thulium orders in the CAM structure which it retains until the amplitude of the modulation saturates. When the amplitude begins to exceed 7.0 β, the structure must change, and it does. Higher harmonics of the fundamentals are observed beginning at about 40 K. This "squaring up" process continues, until at 4.2 K an antiphase domain type structure with 4 moments up and 3 moments down is the stable configuration. The net moment per atom is 7.0 β and, in the 3–4 configuration, this gives rise to a net moment $\mu_\| = 1.0\ \beta$ in excellent agreement with the low field magnetization data.

The existence of a ferromagnetic component corresponds to a zero'th order harmonic. It was not possible to determine precisely the temperature at which this harmonic appeared because it requires the measurement of a small ferromagnetic contribution superimposed on a relatively large nuclear peak. Nevertheless it was clearly well above the value of 22 K then reported for the Curie point.[32] In this earlier study it was concluded that the wave vector remained constant at 2/7 b_3 over the entire temperature range of magnetic order.

Recently Brun et al.[33] have reexamined the temperature dependence of the wave vector for thulium because of evidence from Mössbauer data that the wave vector is not precisely commensurable with the lattice above about 32 K.[34] Their results, shown in the left-hand insert to Figure 3–3, clearly show this incommensurability.[35] Brun et al.[36] have also exploited the sensitivity of the polarized beam technique to study the temperature dependence of the ferromagnetic component. The wave vector increases with decreasing temperature and locks into the seven layer periodicity at 32 K. The sine wave of the high temperature phase squares up symmetrically from 40 K to 32 K at which point the zero'th order harmonic sets in. The conclusion that a strictly antiferromagnetic configuration could exist only in a narrow temperature range below 40 K was reached in the first investigation as well.

3.3. HEAVY METAL ALLOYS

The heavy metals Gd-Lu (Yb excepted) form continuous solid solutions among themselves and with nonmagnetic yttrium. Yttrium has an outer electron configuration similar to that of the rare earths and its lattice constants are nearly identical to those of Gd at room temperature. It serves as a nearly ideal diluent for the rare earths.

Neutron diffraction studies have now been made on a number of these heavy metal alloy systems.

Results of investigations of polycrystalline specimens of alloys of the metals Tb-Tm with Y[37] have been included in an earlier review paper and will be summarized only briefly here. This study has recently been extended to include compositions in the Gd-Y system by means of metal enriched in the low capturing isotope ^{160}Gd.[38] In that system alloys of high Gd content are ferromagnetic, those with high Y content are helical. In all the other alloys the high temperature ordered phases are the same as in the pure metals.

In all the systems studied the initial transition temperatures were found to fall on a universal curve when plotted against the effective spin parameter $x = c(\lambda - 1)^2 J(J+1)$ where c is the concentration of the rare earths with angular momentum J and λ is the Landé factor. In addition the initial turn angles, ω_i, form, approximately, a universal curve in x such that ω_i approaches a temperature independent value of about 50° per layer for small x regardless of the magnetic atom in the alloy. For x at about 11.5, ω_i extrapolates to zero. These observations are summarized in Figure 3–6. In the high-tempera-

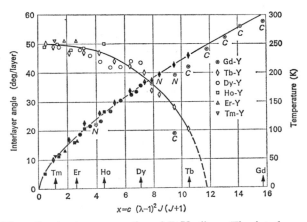

Figure 3–6. Magnetic structure properties of R-Y alloys. The interlayer angles at the ordering temperature, on the left, and the ordering temperature, on the right, are universal functions of the average squared projection of S on J. The interlayer angle curve projects to zero for $x = 11.5$ approximately.

ture form of the alloys the moments have the same direction as in the parent metal (Gd-Y excepted) indicating little change on dilution in the axial anisotropy constants of lowest order. In the ferromagnetic $Gd_{0.80}$–$Y_{0.20}$ alloy there is an easy cone with angular opening of about 70° at all temperatures below T_C. There are no universal curves at low temperatures; each alloy system exhibits its own characteristic behavior.

As little as 10% Y in Er, and 5% Y in Dy, reduces the Curie point to below 4.2 K. The temperature of ordering of the planar component in Er decreases with yttrium concentration at about the same rate as T_N. In the Tb-Y system the Curie point is depressed below 4.2 K at 25% Y.

Gray and Spedding[39] made magnetization measurements on single crystals of alloys with the approximate compositions $Y_{0.25}Er_{0.75}$, $Y_{0.75}Er_{0.25}$, $Lu_{0.25}Er_{0.75}$, $Lu_{0.50}Er_{0.50}$, and $Lu_{0.75}Er_{0.25}$. None of these alloys was ferromagnetic in zero field. Anomalies in the c-axis magnetization data were observed for some compositions which are presumably associated with the ordering of the planar component. In others the origin of the anomaly is unknown. In the same experiment they detected an anomaly in the c-axis magnetization of pure Er which extrapolates to a zero field temperature of 28 K. The origin of this is not understood either.

Magnetization measurements have been carried out by Bozorth and Gambino[40] on a large number of polycrystalline specimens of binary rare earth alloys and Child[41] has examined a number of polycrystalline specimens by neutron scattering. In these systems, also, the upper transition temperatures follow the universal curve but the Curie points are characteristic of each system. Recently a number of systems have been studied in single crystal form by magnetization and by neutron diffraction methods. It is to these investigations that we turn now. Unless otherwise stated, all results quoted are from single crystal data.

3.3.1. The Er-Dy System

Results of an extensive neutron diffraction study of specimens in the Dy-Er[42] system are summarized in Table 3.2 and presented in another way in a structure–constitution diagram shown in Figure 3–7. In these alloys, and in fact, all the alloy systems to be discussed here, the structures are

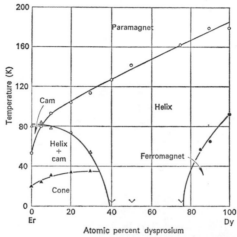

Figure 3–7. Magnetic structure–composition diagram for the Er-Dy system.

uniform in the sense that the moments take a common direction. This is interesting because the anisotropy of Er is such as to force alignment along the c-axis (at least initially, and for a large component at all temperatures) while that of Dy favors alignment normal to the c-axis.

Curie and Néel points for the system had been measured by Bozorth and Gambino[40] and they observed an abrupt discontinuity in the Curie point near the composition $Er_{0.60}Dy_{0.40}$. This is shown in Figure 3–8. They attributed this to a change from Er-like anisotropy to Dy-like anisotropy. The detailed structure studies confirm this interpretation. As seen in the phase diagram, however, there is an extensive region from 40% to about 75% Dy where the alloy is not ferromagnetic at all under the zero field conditions of the diffraction experiments but is in fact a simple helix at all temperatures below the Néel point. In the dysprosium rich region the structures characteristic of pure Dy are observed. The structure in the ferromagnetic region

Table 3.2. Properties of Er-Dy Alloys

Alloy	$\langle \lambda J \rangle_{ave}$	$\langle (\lambda-1)^2 J(J+1) \rangle_{ave}$	T_N(K) c-axis	T_N(K) basal plane	Structure	T_C(K)	Structure θ = semi-cone angle	Ref.
Er	9	2.55	80	53	CAM $\omega_i=51°$ CAM+HELIX and gradual squaring of CAM	20	CONE, $\theta=29°$	(a)
95Er-5Dy	9.05	2.78	85	80	CAM $\omega_i=51°$ CAM+HELIX and gradual squaring of CAM	24	CONE, $\theta=31°$	(b)
90Er-10Dy	9.10	3.00	78	93	HELIX $\omega_i=51°$ HELIX+CAM and gradual squaring of CAM	31	CONE, $\theta=34°$	
80Er-20Dy	9.20	3.46	77	104	HELIX $\omega_i=50°$ HELIX+CAM	35	CONE, $\theta=39°$	
70Er-30Dy	9.30	3.91	55	114	HELIX $\omega_i=49°$ HELIX+CAM	38	CONE, $\theta=43°$	
60Er-40Dy	9.40	4.36		127	HELIX $\omega_i=49°$ $\omega_f=38°$			
50Er-50Dy	9.50	4.82		140	HELIX $\omega_i=49°$ $\omega_f=37°$			
25Er-75Dy	9.75	5.95		158	HELIX $\omega_i=47°$ $\omega_f=31°$			
15Er-85Dy	9.85	6.40		160	HELIX $\omega_f=30°$	58	Ferromagnetic $\omega_f=0°$	
10Er-90Dy	9.90	6.63		179	HELIX $\omega_i=47°$ $\omega_f=28°$	65	Ferromagnetic $\omega_f=0°$	
Dy	10	7.08		179	HELIX $\omega_i=43°$ $\omega_f=26.5°$	87	Ferromagnetic $\omega_f=0°$ μ along a axis	(c)

(a) Cable, J. W., Wollan, E. O., Koehler, W. C. and Wilkinson, M. K., *Phys., Rev.* **140**, 1896 (1965).
(b) Millhouse, A. H. and Koehler, W. C., *Proc. Int. Conf. on Rare Earth Metals*, Paris-Grenoble, 1969, published by C.N.R.S., Paris, Vol. II, 214 (1970).
(c) Wilkinson, M. K., Koehler, W. C., Wollan, E. O. and Cable, J. W. *J. Appl. Phys.*, **32**, 48S (1961).

is a planar ferromagnet. To our knowledge no magnetization data exist, as yet, which yield the easy direction in these alloys. One expects, from the hexagonal anisotropy of the pure metals that the moments at low temperatures will be in the a-direction.

Bozorth[43] has reported magnetization data for single crystal $Er_{0.5}Dy_{0.5}$. At 4.2 K the virgin state helix is transformed to a ferromagnet after possibly going through some intermediate configurations. The easy axis in low fields is the a-direction but this apparently changes to a b-direction for fields in excess of 15 kOe.

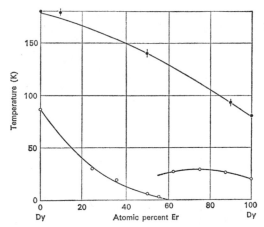

Figure 3–8. Curie temperature for Er-Dy alloys as determined from magnetization measurements (Ref. 40) (open circles). Néel points as determined from neutron scattering data are shown as closed circles (Ref. 42).

The most interesting region is the Er-rich part of the constitution diagram. In this composition range the structures resemble those found for pure Er but with a remarkable inversion of the upper two ordering temperatures.

In the $Er_{0.95}Dy_{0.05}$ alloy the CAM structure develops first at a Néel temperature of 85 K, a value slightly greater than that of the pure metal. The ordering temperature for the perpendicular component now occurs at 80 K as compared to 52 K for pure Er. With a further reduction in temperature third harmonics develop. The transition to the cone structure occurs at 24 K. Just above this temperature the structure is similar to that of Er just above its Curie point. The coefficient A_3 was measured to be 2.97 β. At 4.2 K the moments measured in the conical region were $\mu_\perp = 4.8\ \beta$ and $\mu_\parallel = 7.8\ \beta$. Thermal hysteresis in the moment values and of the turn angle were observed near T_C.

With 10% Dy in Er the Néel point increases to 93 K, but it is the perpendicular component, now, which orders first. From 93 K to 78 K the moments are ordered in a helix in the basal plane.[44]

Below 78 K the c-axis component develops, at first in the simple CAM structure and at a temperature near 43 K exhibits very weak third harmonics. This is also the temperature at which the moment saturates. The conical transformation occurs at 31 K. Hysteresis is again observed at the transition. At 4.2 K the parameters of the cone are $\mu_\perp = 5.1\ \beta$, $\mu_\parallel = 7.6\ \beta$.

The 20% Dy alloy exhibits the planar helical structure from the Néel point to 77 K. Below 77 K a simple CAM configuration sets in. In this alloy, the planar component of the moment is sufficiently great that the maximum allowable moment on any site is never exceeded. No third harmonics are needed, and none were observed. Just above the Curie point of 36 K the structure is evidently the undistorted tilted elliptical oscillation described above in the section on pure Er. The moments in this configuration and at 4.2 K are $\mu_\perp = 6.0\ \beta$, $\mu_\parallel = 7.1\ \beta$.

The properties of the 30% Dy alloy are similar to those just described. Helical ordering exists between 114 K and 55 K, and the tilted ellipse is found from 55 K to the conical transition at 38 K. The moments at low temperatures have the values $\mu_\perp = 6.8\ \beta$, $\mu_\parallel = 6.3\ \beta$.

In the 40% alloy the conical structure is gone. A value for $\mu_\perp = 8.6\ \beta$ obtained at 4.2 K indicates that a small sinusoidal c-axis component may still be present at this temperature.

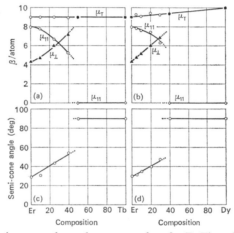

Figure 3–9. Semi-cone angles and moment values for Er-Tb and Er-Dy systems.

In summary, the c-axis and planar modulated structures exist in the Er-rich end of the composition diagram. Squaring of the c-axis component diminishes rapidly with increasing Dy. The conical transition is abruptly suppressed between 30% and 40% Dy.

The variation with composition of the ordered moment components is illustrated for the Er-Dy system in Figure 3–9(b); that for the semi-cone angle in Figure 3–9(d). The cone angle increases linearly to a composition around 40% Dy at which it changes abruptly to 90°. Bozorth et al.[45] and Pickart[46] have examined a single crystal of DyEr₃ at 4.2 K by magnetization and neutron diffraction, respectively, with results in good agreement with the parameters of the conical structures reported by Millhouse and Koehler. Similar parameters are given in Figure 3–9 for the system Er-Tb which system we now describe.

3.3.2. The Er-Tb System

Single crystal data for the Er-Tb system exist only for the Er-rich region ;[42] polycrystalline samples have been investigated in the Tb-rich range.[41] Together, the two sets of data suffice to define the structure–constitution diagram shown in Figure 3–10. Some parameters of these alloys are listed in Table 3.3. As in the Er-Dy system the addition of small amounts of Tb to Er produces an increase in the ordering temperature of the planar component.

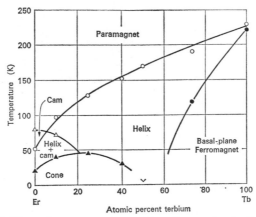

Figure 3–10. Magnetic structure–composition diagram for the Er-Tb system.

The 10% Tb alloy has structures similar to those found in the 10% Dy sample. Between $T_N = 97$ K and 72 K only the helical structure is found. Below 72 K, the parallel component A_1 grows in, and at 50 K, on cooling, third harmonics indicating a nonvanishing value of A_3 are observed. Just above 24 K, the Curie point measured on cooling, the coefficients describing the configuration are $\mu_\perp = 4.6\ \beta$, $A_1 = 8.82\ \beta$, and $A_3 = 0.81\ \beta$. As in pure Er, the phases are adjusted to give an antiphase domain-like configuration. In the conical structure, at 4.2 K, the moment components are $\mu_\perp = 4.7\ \beta$, $\mu_\parallel = 7.8\ \beta$ to give a total moment of 9.0 β. Temperature hysteresis in the turn angle, and moment values are observed. This is illustrated in Figure 3–11 where the intensities of some representative reflections are plotted as a function of temperature. The Curie point on warming is about 40 K.

For the 25% Tb alloy, the moments order initially in the basal plane, and the helical structure is retained to a temperature of 45 K. Below this temperature, the moments lift out of the basal plane, as in holmium, to form a conical arrangement. In this alloy the CAM structure does not develop. The components of the moments at 4.2 K are $\mu_\perp = 6.0\ \beta$, $\mu_\parallel = 6.7\ \beta$.

Similar structures are found in the 40% Tb alloy. The Curie point in the alloy is near 30 K.

In the magnetic phase diagram of Figure 3–10 it is apparent that for Tb concentrations less than 5% the sequence of magnetic structures which exists when the sample is cooled to 4.2 K is the same as that for pure Er. For Tb concentrations greater than 5% the moments order in the basal plane first

Table 3.3. Properties of Er-Tb Alloys

Alloy	$\langle \lambda J \rangle_{ave}$	$\langle (\lambda-1)^2 J(J+1) \rangle_{ave}$	T_N(K) c-axis	T_N(K) basal plane	Structure	T_C(K)	Structure θ=semi-cone angle	Ref.
Er	9	2.55	80	53	CAM	20	CONE, $\theta=29°$	(a)
90Er-10Tb	9	3.35	72	97	CAM+HELIX and gradual squaring of CAM; HELIX	40	CONE, $\theta=31°$	(b)
75Er-25Tb	9	4.54		127	CAM with gradual squaring; HELIX	45	CONE, $\theta=42°$	(b)
60Er-40Tb	9	5.72		150	HELIX	40	CONE, $\theta=53°$	(c)
50Er-50Tb*	9	6.52		170	HELIX		Planar Ferromagnet	(d)
26.3Er-73.7Tb*	9	8.41		192	HELIX	32	Planar Ferromagnet	(d)
Tb	9	10.5		228	HELIX	220	Planar Ferromagnet	(e)

* Polycrystalline sample.

(a) Cable, J. W., Wollan, E. O., Koehler, W. C. and Wilkinson, M. K., *Phys. Rev.*, **140**, 1896 (1965).
(b) Millhouse, A. H. and Koehler, W. C., *Proc. Int. Conf. on Rare Earth Metals*, Paris-Grenoble, 1969, published by C.N.R.S., Paris, Vol. II, 214 (1970).
(c) Khan, Q., Millhouse, A. H. and Koehler, W. C., unpublished.
(d) Child, H. R., *O.R.N.L. Report* T.M. 1063 (1965).
(e) Koehler, W. C., Child, H. R., Wollan, E. O. and Cable, J. W., *J. Appl. Phys. Suppl.*, **34**, 1335 (1963).

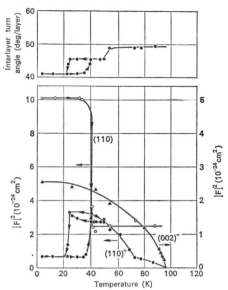

Figure 3–11. Temperature variation of the interlayer turn angle and the structure factors of several reflections for a 90 Er–10 Tb alloy single crystal. Arrows indicate cooling or warming through regions having thermal hysteresis.

with the CAM appearing at a lower temperature. Above 20% Tb the c-axis modulation is suppressed altogether with the simple helix being the stable antiferromagnetic structure in the remainder of the alloy system. The semicone angle in the ferromagnetic spiral region increases slowly with increasing Tb content, as shown in Figure 3–9(a) and 3–9(c). Between 40% and 50% Tb, the conical structure disappears. With increasing Tb content the structures characteristic of Tb are found, the sequence of transitions being helix to planar ferromagnet.

Magnetization measurements have not, apparently, been made on single crystal specimens in this system.

3.3.3. The Er-Ho System

Bozorth and Gambino found that the Curie points of specimens in this system rose to a maximum of 35 K near the 50–50 composition. Shirane and Pickart[47] investigated a single crystal of the alloy $Er_{0.5}$-$Ho_{0.5}$ and determined that the magnetic structure of the alloy was similar to that of pure holmium. A transition to the simple helical structure was observed at 104 K. No evidence for a CAM structure was observed. At the Curie point of 33 K the moments tilt out of the basal plane to form a conical configuration.

Millhouse and Koehler[42] investigated one alloy in this system, 10% Ho, and found the initial ordering at 88 K to the CAM structure. The planar component ordered at 77 K. At lower temperatures a transformation to the conical configuration took place.

Shirane and Pickart studied two polycrystalline specimens $Er_{0.3}Ho_{0.7}$ and $Er_{0.6}Ho_{0.4}$ at 4.2 K and found, again, helical structures at high temperatures

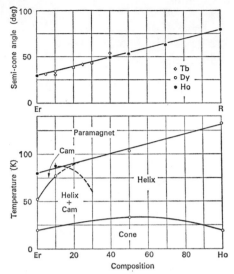

Figure 3–12. Magnetic structure–composition diagram for the Er-Ho system (lower diagram) semi-cone angles for rare earth-erbium alloys (upper diagram).

and a conical configuration at low temperatures. A rudimentary phase diagram for this system is given in Figure 3–12 (lower). In Figure 3–12 (upper) are shown as full circles the semi-cone angles which they deduced. The three points fall quite well on a straight line drawn through the semi-cone angle of the pure metals, 29° for Er, 80° for Ho. The semi-cone angles measured at 4.2 K for the compositions in the Er-Tb and Er-Dy systems which have the cone structures likewise fall on the same straight line. Magnetization measurements have now been made on single crystal Ho_3Er, and HoEr at 4.2 K to fields up to 140 kOe.[45] The low field results are in good agreement with the zero field neutron diffraction data.

3.3.4. The Er-Gd System

A single specimen with 7.2% Gd (enriched in ^{160}Gd) has thus far been only partially studied.[42] Results obtained thus far are as follows: between 104 K, the Néel point, and 91 K the CAM structure alone is found. Below 91 K the basal plane component of the moment begins to order and third harmonics in the parallel components are observed. A conical configuration is found at low temperatures, the Curie point of which, on warming, is 59 K.

3.3.5. The Ho-Tb System

Properties of selected specimens in the Ho-Tb system have been investigated by Bente Lebech,[48] and by Bjerrum-Møller et al.[49] Recently, Spedding, Jordan and Williams[50] have made magnetization measurements on three single crystal specimens containing 10%, 50%, and 90% Ho. Neutron diffractions studies were made on the same specimens by Spedding, Ito, and Jordan,[51] and on specimens rich in Ho by Spedding, Ito, Jordan, and Croat.[52]

Table 3.4. Properties of Ho-Tb Alloys

Alloy	$\langle\lambda J\rangle_{ave}$	$\langle(\lambda-1)J(J+1)\rangle_{ave}$	T_N(K)	Structure	T_C(K)	Structure	Ref.
Ho	10	4.50	133	HELIX—Distorted below about 45 K, $T_\omega=18.7$ K, $\omega_i=50.0$, $\omega_f=30.0$	19	Cone with distorted helix, $\mu=1.7\ \beta$	(a)
98Ho-2Tb	9.98	4.62		HELIX—Distorted below about 40 K $T_\omega=20$ K, $\omega_i=50.0$, $\omega_f=30.0$	12	Cone? Distorted helix + small net planar moment	(b)
95Ho-5Tb	9.95	4.80	138	HELIX—Distorted below about 40 K, $T_\omega=20$ K, $\omega_i=50.0$, $\omega_f=30.0$	16	Tilted ferromagnet?	(b)
93Ho-7Tb	9.93	4.92	140	HELIX—Distortion not sought	18	Tilted ferromagnet?	(b)
90Ho-10Tb	9.9	5.10	144	HELIX—Undistorted $T_\omega=30$ K, $\omega_i=50.0$, $\omega_f=30.0$	22	Tilted ferromagnet $(0.9\ \beta$ parallel to $c)$	(c) (d)
50Ho-50Tb	9.5	7.50	185	HELIX, $\omega_i=42.0$, $\omega_f=23.5$	80	Planar ferromagnet	(c) (d)
20Ho-80Tb	9.2	9.3	213	HELIX	171	Planar ferromagnet	(e)
10Ho-90Tb	9.1	9.9	221	HELIX, $\omega_i=27.8$, $\omega_f=20.3$	196	Planar ferromagnet	(c) (d)
Tb	9	10.5		HELIX		Planar ferromagnet	(f)

(a) Koehler, W. C., Cable, J. W., Wilkinson, M. K. and Wollan, E. O., *Phys. Rev*, **151**, 414 (1966).
(b) Spedding, F. H., Ito, Y., Jordan, R. G. and Croat, J., *J. Chem. Phys.*, **54**, 1995 (1971).
(c) Spedding, F. H., Jordan, R. G. and Williams, R. W., *J. Chem. Phys.*, **51**, 509 (1969).
(d) Spedding, F. H., Ito, Y. and Jordan, R. G., *J. Chem. Phys.*, **53**, 1455 (1970).
(e) Lebech, Bente, *Solid State Comm.*, **6**, 791 (1968).
(f) Koehler, W. C., Child, H. R., Wollan, E. O. and Cable, J. W., *J. Appl. Phys. Suppl.*, **34**, 1335 (1963).

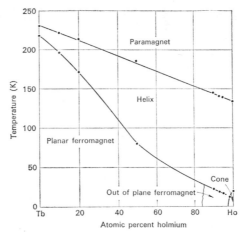

Figure 3–13. Magnetic structure–composition diagram for the Tb-Ho system. The wavy line indicates lack of knowledge of the composition at which the planar ferromagnetic configuration sets in.

Properties of this alloy system are collected in Table 3.4. A phase diagram for the system is shown in Figure 3–13. With the addition of Ho to Tb the Néel temperature for ordering in the simple helical configuration falls, almost linearly, with composition. The Curie temperature for transformation to the planar ferromagnet falls more rapidly. This transition in the alloys, as in pure Tb, shows pronounced thermal hysteresis.

The sequence HELIX to essentially b-axis ferromagnet is followed by all the alloys up to and including the 90% Ho specimen.[53] That the easy direction is a b-direction is not unexpected in view of the planar anisotropies of the constituents.

The 93% and 95% alloys were found to have ordered phases similar to that of the 90% alloy except that the distortion of the simple helix observed for pure Ho was observed as well in the 95% Ho alloy. The distortion manifests itself by the presence of fifth and seventh harmonics in the diffraction pattern. These harmonics persisted to about 40 K on warming.

The 98% alloy, according to Spedding et al. corresponds to a structure near the boundary between the pure holmium-like structure (conical, distorted helix in the basal plane) and the 90% Ho structure (slightly out of plane ferromagnet). There is a definite indication at low temperatures of a small net moment in the basal plane as evidenced by magnetic contributions to an (002) reflection. The distorted helical structure, however, accounts for most of the diffraction data. It is not clear from the data whether a c-axis component exists, but it probably does since it is present in the 90% Ho alloy. The out-of-plane ferromagnet goes to the planar ferromagnet between 50% and 90% Ho. The uncertainty of the exact composition is indicated by the wavy line at the bottom right-hand corner of the diagram.

Magnetization measurements on the 98% and 93% alloys could help to clarify this situation.

Pure Ho, Ho(B), that is, has a final turn angle of 30°/layer. This "locked-in" value is found as well for the Ho rich alloys. The temperature, T_ω, at which the turn angle breaks away from the commensurate value is, coincidentally, at the Curie point for Ho, but it occurs at a much higher value in the 90% Ho alloy. Moreover, the same value of 30°/layer is found for the undistorted and for the distorted helix configuration. Conversely, in Ho(A) distortion exists in a specimen whose final turn angle is not commensurable. T_ω for Ho(A) is greater than the Curie temperature.

3.3.6. The Ho-Dy System

The structural parameters of specimens in the Ho-Dy system have been studied by neutron diffraction.[42] Magnetization measurements have not apparently been made for single crystals in this system.

Some properties of Ho-Dy alloys are listed in Table 3.5, and also in the partially determined phase diagram of Figure 3–14. As in the previous case of the Ho-Tb system the Dy-rich end of the series presents no surprise. The

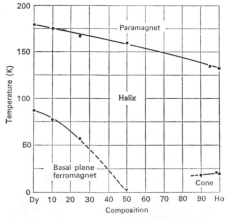

Figure 3–14. Magnetic structure–composition diagram for the Dy-Ho system.

addition of Ho to Dy decreases the ordering temperature to the helical configuration uniformly. The Curie point for the transformation to the planar ferromagnet[54] decreases rapidly until at a composition between 25% and 50% Ho it is below 4.2 K.

In the Ho-rich region the c-axis component characteristic of pure Ho is present in the 98% Ho alloy but was not measurable in the usual experiment in the 95% and 90% Ho alloys. Depolarization of a polarized beam was, however, observed for the 90% specimen. This may be indicative of a small c-axis moment. The limit of detection in the ordinary experiment is of the order of 0.4 β.

The final turn angle of 30°/layer was observed for alloys as concentrated in Dy as the 50–50 alloy. Fifth and seventh harmonics, indicative of a distorted helix, were observed in the diffraction patterns of the Ho-rich specimens including the 50–50 composition. The abrupt "locking-in" at 30° typical of Ho and the Ho rich Ho-Tb alloys was not observed in this system.

Table 3.5. Properties of Ho-Dy Alloys

Alloy	$\langle \lambda J \rangle_{ave}$	$\langle (\lambda-1)^2 J(J+1) \rangle_{ave}$	$T_N(K)$	Structure	$T_C(K)$	Structure	Ref.
Ho	10	4.50	133	HELIX $\omega_i=50°$, $\omega_f'=30°$	20	Distorted spiral in base plane. Ferromagnetic comp. along c-axis.	(a)
98Ho-2Dy	10	4.55	133	HELIX	23	Distorted spiral in base plane. Ferromagnetic comp. along c-axis.	(b)
95Ho-5Dy	10	4.63	135	HELIX, $\omega_i=50°$		Distorted spiral in base plane at 4.2 K, $\omega_f=30°$. Ferromagnetic comp. along c-axis not detected.	
90Ho-10Dy	10	4.76	135	HELIX, $\omega_i=50°$		Ferromagnetic comp. along c-axis not detected. $\omega_f=30°$.	
50Ho-50Dy	10	5.79	160	HELIX, $\omega_i=49°$		Ferromagnetic comp. along c-axis not detected. $\omega_f=29°$.	
25Ho-75Dy	10	6.44	167	HELIX, $\omega_i=45°$	50	Ferromagnetic $\omega_f'=27°$, $\omega_f=0°$.	
10Ho-90Dy	10	6.82	175	HELIX, $\omega_i=45°$	78	Ferromagnetic $\omega_f'=28°$, $\omega_f=0°$.	
Dy	10	7.08	179	HELIX, $\omega_i=43°$	87	Ferromagnetic $\omega_f=0°$ μ along a-axis.	(c)

(a) Koehler, W. C., Cable, J. W., Wilkinson, M. K. and Wollan, E. O., *Phys. Rev.*, **151**, 414 (1966).
(b) Millhouse, A. H. and Koehler, W. C., *Proc. Int. Conf. on Rare Earth Metals*, Paris-Grenoble, 1969, published by C.N.R.S., Paris, Vol. II, 214 (1970).
(c) Wilkinson, M. K., Koehler, W. C., Wollan, E. O. and Cable, J. W., *J. Appl. Phys.*, **32**, 48S (1961).

3.4. SUMMARY AND DISCUSSION

Before going on to discuss other types of experiments on heavy rare earth metals and alloys, it may be appropriate to point out some general conclusions which can be drawn from the data obtained thus far.

As we have seen above, the initial ordering temperatures of the pure metals, and of their alloys with yttrium, fall on a universal curve when plotted against the de Gennes factor $x=\langle(\lambda-1)^2 J(J+1)\rangle$. This relationship has been verified for all the intra-rare earth alloys discussed so far. One may conclude from the universal behavior among binary alloys and dilution alloys that both types of moments in an A-B alloy order at the same initial ordering temperature. (If for example, only one set of moments, A, in the alloy ordered, the ordering temperature should be close to that of the corresponding A-Y alloy, contrary to observation.) The same conclusion follows from arguments based on intensities observed slightly below the Néel points.

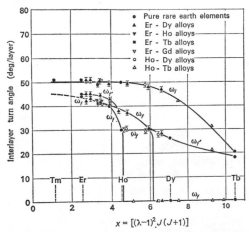

Figure 3–15. Summary of interlayer turn angle data for the heavy rare earth metals and binary alloys as determined from single crystal neutron diffraction experiments. The turn angles are plotted against the de Gennes factor for several important temperatures ω_i, just below T_N, ω_f', just above T_C and ω_f at 4.2 K. Specific references are given in the text and tables.

Universal behavior is observed as well for the value of the wave vector at the ordering temperature. This is illustrated in Figure 3–15 where the interlayer angles for a number of single crystal specimens are plotted against the de Gennes factor. At lower temperatures each alloy system behaves differently. In any case there is for any given alloy at any given temperature a single wave vector. In particular, at the initial ordering temperature the wave vector is characteristic of the alloy, not of the individual constituents.

The magnetic structures in the alloys are uniform structures in the sense that the moment directions of the elements in the alloys are the same, and in general different from their directions in the pure metals.

Many of the properties of the pure metals can be understood in terms of a

Hamiltonian consisting of three terms[55]

$$H = H_{ex} + H_{cf} + H_{me} \tag{3.7}$$

The first term is a long range oscillatory type exchange interaction of the Ruderman-Kittel-Kasuya-Yosida type which operates by polarization of the conduction electron system.

$$H_{ex} = - \sum_{i \neq j} \mathbf{S}_i \cdot \mathbf{S}_j \mathscr{J}(\mathbf{R}_i - \mathbf{R}_j) \tag{3.8}$$

When the exchange energy is dominant the magnetic structure which is stable is determined by a maximum in the Fourier transformed exchange energy

$$\mathscr{J}(\mathbf{q}) = \sum_i \mathscr{J}(\mathbf{R}_i - \mathbf{R}_j) \exp i\mathbf{q} \cdot (\mathbf{R}_i - \mathbf{R}_j) \tag{3.9}$$

If the maximum $\mathscr{J}(\mathbf{Q})$ occurs for a value of $\mathbf{Q} = 0$, a ferromagnetic configuration, as in Gd results. If the maximum occurs at some general point in the Brillouin zone the exchange favors an oscillatory arrangement with interlayer angle determined by $\mathbf{Q}$. Whether $\mathbf{Q}$ is a general vector or not is strongly dependent on the geometry of the Fermi surface.[56-57]

The second term is a single ion anisotropy energy resulting from crystalline field interactions on the distribution of 4-f electrons in a given ion. This may be represented classically by [19-20]

$$E_{cf} = K_2{}^0 Y_2{}^0(\theta) + K_4{}^0 Y_4{}^0(\theta) + K_6{}^0 Y_6{}^0(\theta) + K_6{}^6 \sin^6 \theta \cos 6\phi \tag{3.10}$$

where $Y_n{}^0$ is the nth order Legendre polynomial, and θ and ϕ are polar and azimuthal angles of the vector $\mathbf{J}_i$ measured from the c- and a-axes. When the crystal field anisotropy is the dominant one the directions of the moments are determined by minima in the expression for E_{cf}.

The temperature dependence of the various terms in the crystal field anisotropy energy is such that near the Néel temperature only the leading term $K_2{}^0$ is significant. For Tm and Er, $K_2{}^0$ is negative and the crystal field anisotropy favors an initial alignment parallel to the c-axis. In Ho, Dy, and Tb, $K_2{}^0$ is positive and the moments are constrained to the plane normal to c. At lower temperatures the higher order axial terms may become important. Thus in Ho, and in Er, the conical structures at low temperatures are thought to reflect the influence of these terms. The hexagonal anisotropy $K_6{}^6$ if sufficiently large may, in competition with the exchange energy, drive the system from a spiral to a planar ferromagnetic state.

The third term in the Hamiltonian is the magnetostriction contribution which is now thought to be at least as important as the planar anisotropy in driving the helix-ferromagnet transition in Tb and Dy.[55, 57] Since the magnetoelastic energy varies with temperature as a higher power of the relative magnetization, it is relatively unimportant at the initial ordering temperature but it can become competitive with the exchange energy at lower temperatures.

The observations which we have cited for the alloys (universal behavior, uniform structures) imply that these systems are to a first approximation exchange dominated, with the crystal field anisotropy a small perturbation for a temperature near the ordering temperature. If this be the case the concept of an average exchange energy and an average anisotropy energy is approximately valid and the theoretical considerations which have been made for the pure metals should be applicable to the alloys. Inspection of the data shows that such a procedure is qualitatively correct. Quantitatively, there are serious discrepancies.

The magnetic structure data which have been obtained near the initial ordering temperature can, for the most part, be grouped into three regions according to the nature of the initial ordering of the alloy. This grouping is illustrated in Figure 3–16 and is based on the single-crystal data discussed above and on data obtained from polycrystalline samples in the R-Y series, data points for which are not explicitly shown.

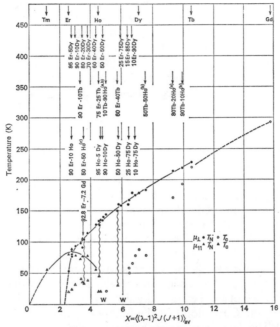

Figure 3–16. Summary of transition point data for the heavy rare earths and binary alloys as determined from single crystal neutron diffraction experiments. Wavy lines show direction transitions from helix to cone. (a) Ref. 48. (b) Ref. 51. (c) Ref. 47. All others, Ref. 42.

In pure erbium and pure Tm the initial ordering is to a CAM structure. With the addition of small amounts of Ho, Dy, or Tb, to Er there is a rapid rise in the temperature of ordering of the planar components, $T_{x, y}$, and a decrease in the ordering temperature of the parallel component T_z, until in the alloys containing 10% Tb and 10% Dy these two temperatures are inverted with respect to those for pure Er. Holmium is less effective than Tb or Dy in producing the change; the 10% Ho alloy still has $T_z > T_{x, y}$. The gadolinium alloy is remarkable in that it exhibits the same structures in the same sequence as pure erbium but at higher temperatures. In Er-Y alloys the same structures in the same sequence are observed at lower temperatures. In these two cases the anisotropy should be relatively unchanged (the anisotropy constants of Gd are an order of magnitude smaller than in the other rare earths) but the exchange energy is enhanced in the one case and reduced in the other.

The change in initial ordering from CAM to helix with the addition of Tb Dy, or Ho to Er, the efficiency of Tb and Dy relative to Ho in bringing about the change, and the observation that the dilute Gd alloy has an initial CAM ordering are to be expected in view of the signs and magnitudes of the crystal field anisotropy constants.

One may argue from the molecular field theoretical calculations of Miwa and Yosida[9] and Nagamiya[58] that the temperature at which T_z becomes equal to $T_{x, y}$ in the Er-based alloys is also the temperature at which the average anisotropy constant $\langle K_2^0 \rangle$ changes sign. With typical experimental and theoretical estimates of these constants[59-61] $\langle K_2^0 \rangle$ changes sign at Er concentrations of about 0.70, 0.70, and 0.50 for the Tb, Dy and Ho systems, respectively, whereas the observed inversion takes place at a much higher erbium concentration. It seems necessary to invoke some other anisotropic interaction, to account for the very rapid change in initial ordering properties of the Er-rich alloys. For x between about 2.8 and 11 (Gd-Er excepted) the initial ordering is to a helical configuration. Except for the Er-based systems just discussed these are all alloys of materials whose initial ordering is to the helical configuration. It is not surprising that their alloys take the same initial structure. For x greater than about 11 (Gd, Gd-Y) the initial ordering is to a ferromagnetic configuration.

At lower temperatures the properties of the alloys are, as we have indicated above, generally characteristic of the individual alloy systems. Even here, however, there are some general observations which can be made.

The elements Dy and Tb undergo transitions at 87 K and 220 K, respectively, from the helix to the planar ferromagnetic configuration. It has been suggested, recently,[55, 57] that at the relatively high temperatures of the Curie points in the metals, the planar anisotropy energy is small and that the magnetoelastic effect is the driving force which stabilizes the ferromagnetic state. (The influence of the hexagonal anisotropy manifests itself at temperatures well above the Curie point in the magnetization data for single crystal Dy, however.) The helix-planar ferromagnet transition is observed as well in some compositions of the systems Tb-Ho, Dy-Ho, Tb-Er, and Dy-Er. The Curie points fall more or less well on a line in the temperature x-diagram which line is the one followed more or less well by the Curie points of the Tb-Y and Gd-Y alloys.

In the Ho rich Tb-Ho and Dy-Ho alloys the conical configuration of Ho exists only for small Tb and Dy additions. In both cases as little as 2% Tb or Dy modifies the conical structure of Ho profoundly, However, the structure to which the alloy is tending with, say, 5% of the minor constituent is the ferromagnet for addition of Tb, the distorted helix with the addition of Dy.

The rapid disappearance of the cone structure of Ho with small additions of Tb and Dy can be understood on the basis of average anisotropy constants. Keeping only the K_2^0 and K_4^0 terms for the sake of simplicity, the axial anisotropy energy becomes

$$E_{\text{cf}} = \xi^2 \frac{35 \langle K_4^0 \rangle}{8} + \left[\frac{3}{2} \langle K_2^0 \rangle - \frac{30 \langle K_4^0 \rangle}{8} \right] \xi + \frac{3}{8} \langle K_4^0 \rangle - \frac{\langle K_2^0 \rangle}{2} \quad (3.11)$$

where $\xi = \cos^2 \theta$, with θ the semi-cone angle, and has a minimum value for

$$\xi_{min} = \frac{3}{7} - \frac{6}{65} \frac{\langle K_2{}^0 \rangle}{\langle K_4{}^0 \rangle} \qquad (3.12)$$

In pure Ho, the cone angle of 80° corresponds to $\xi_{min} = 0.030$ and a ratio of $K_2{}^0/K_4{}^0 = 2.32$. The experimentally determined ratio at 4.2 K is 2.34. This limiting value occurs, with the experimentally determined anisotropy constants for 5.7% Tb, and 6.1% Dy in fair agreement with the observations.

The conical configuration exists in the Er-rich region for Er-Gd, Er-Tb, and Er-Dy, and over the entire range of composition for Er-Ho. The Curie point for the transformation to the conical structure increases from 20 K for pure Er to a maximum of 33 K in the Ho-Er system, of 38 K in the Dy-Er system, of 45 K in the Tb-Er system, and to at least 59 K in the Gd-Er system. According to theory the competition between the conical and helical structure depends upon the relative values of the axial anisotropy energy and on the exchange energy. This seems to be borne out by the experimental observations.

The variation of the wave vectors of the pure metals with temperature and with the de Gennes factor is attributed to (1) the variation of the Fermi surface in the rare earth series,[56, 57, 62] (2) spin disorder scattering at the Néel point,[63, 64] (3) the band-gap affect due to ordering of the rare earth moments,[65-66] (4) the higher order anisotropy[20] and (5) the magneto-elastic effect.[57] The turn angle variation in the alloys can, presumably, be attributed to the same mechanisms. Detailed calculations for the Tb-Y system, for example, are in quite good agreement with the observations.[65]

3.5 OTHER ALLOYS AND OTHER EXPERIMENTS

The alloys which have been discussed up to now are simple solid solution type alloys which have the same crystal structure, similar outer electronic configurations and comparable interatomic spacings. A number of other systems has been investigated, the results of which studies are relevant to the nature of the exchange interaction in the rare earths. Of these, we consider two. Neutron diffraction experiments in which parameters such as pressure and applied field are varied have been carried out recently also. We consider some of these briefly.

3.5.1. hcp Rare Earth Thorium Alloys

An important parameter with regard to the coupling mechanism in the heavy rare earths is the effective number of conduction electrons per atom. Thorium has an outer electron configuration $(6d^2 7s^2)$ which in the metal forms the conduction band. Moreover, Th enters the hcp structure of the rare earths over an appreciable concentration range, of the order of 20 atomic percent. It thus serves as a tetravalent diluent.

Previous studies of polycrystalline specimens in the rare earth-thorium series[67] showed that the addition of Th to Er and Ho stabilizes a ferro-magnetic, at the expense of the oscillatory, configuration in contrast to the predictions of the simple RKKY theory. Details of the structures of the alloys were obscured by overlapping reflections in the powder patterns, and, as a result, the study has now been repeated with single crystal specimens.[68] Transition temperatures for the powders are somewhat different than in the annealed single crystals.

The Er-Th System. Three alloys in this system have been studied. In the 95% Er alloy the three structure types observed in pure Er are found, except that the transition temperatures for the planar component, and for the *c*-axis ferromagnetic configuration are inverted. The initial ordering is to a CAM configuration at 68 K (needless to say none of the rare earth-thorium alloys follows the universal curves of the previous section) and at 46 K this transforms to a *c*-axis ferromagnet. At 26 K, satellites characteristic of the planar component develop. At 4.2 K the configuration is a conical one with apparently some disorder since the maximum moment of 9.0 β per erbium is not achieved; $\mu_\perp \simeq 1.7\ \beta$, $\mu_\parallel \simeq 7.9\ \beta$ if the Er form factor is used to analyze the data In the 90% and 85% Er alloys the only configuration found is the *c*-axis ferromagnet having moments $\mu_\parallel = 8.0\ \beta/\text{Er}$ and 7.8 β/Er, respectively. The normal component remains disordered, evidently. The transition temperatures to the ferromagnetic state are 52 K, and 46 K in the two cases.

The Ho-Th System. Three specimens have been studied in the Ho-Th system. The structure–composition diagram is quite complicated and more specimens will have to be studied before it can be determined accurately. In the 95% alloy the initial ordering is to a helix at 107 K. At 40 K there is a transition to an as yet undetermined structure which is characterized by a large ferromagnetic component normal to the *c*-axis as well as very weak satellites characteristic of an oscillatory structure. The 90% alloy exhibits the same structures but with $T_N = 96$ K, $T_C = 30$ K. The weak satellites in this case occur with a different wave vector than in the 95% alloy. The 84% alloy goes to a helical configuration at 72 K with an initial turn angle of 19°/layer. At 53 K it transforms to a planar ferromagnet. The extra satellites are not observed in this material. Just above the Curie point $\omega = 13.5°$.

3.5.2. Rare Earth-Scandium Alloys

Scandium has an outer electron configuration similar to that of the heavy rare earths ($3d^14s^2$ as compared to $5d^16s^2$) and it has the same hcp structure. In spite of the very much smaller lattice constants of Sc as compared to the rare earths it forms continuous solid solutions with them.

Up to now rare earth scandium alloys have been studied only in polycrystalline form. Neutron diffraction measurements have been carried out for alloys with Gd,[38] Tb, Ho, and Er.[69]

The initially ordered structures in the Sc alloys are, to the resolution of the powder patterns, the same as in the pure rare earth metal. Thus Tb-Sc, and Ho-Sc alloys have spiral structures, Er-Sc alloys the CAM structure. In the Gd-Sc system the concentrated alloys are ferromagnetic, more dilute alloys are helical at the initial ordering temperature.

The rare-earth-scandium alloys and the rare-earth-yttrium alloys have generally similar properties, but there are a number of significant differences between them. Dilution of the rare-earth metal with scandium reduces the Néel temperature to a greater degree than by dilution (to the same extent) with yttrium. The transition temperatures for the scandium alloys seem more nearly linear at larger values of x than in the corresponding yttrium alloys. The turn angles in the two systems are not dissimilar, a limiting value of about 50°/layer being indicated. Perhaps the most striking difference between the Sc-based alloys and the Y-based alloys is that magnetically ordered structures are found in the Y-based systems for very dilute alloys; 5% Tb in Y for example. In the Sc-based system, alloys containing 25% Tb, 18% Ho, 39% Er, and 15% Gd showed no evidence for magnetic ordering to 4.2 K.

The existence in the alloys of the helical and modulated moment structures and their similarity to those of the pure metals indicates that the magnetic interaction in the alloys is not greatly changed in character—namely, it is still of relatively long range and oscillatory. On the other hand, the absence of magnetic order in what are still relatively concentrated alloys shows that the magnetic interaction is greatly modified when scandium is put into the system. The overall effect apparently is abruptly to weaken the interaction at a critical concentration in the range of 70–80 at. % scandium.

The most obvious difference between Sc and Y as a nonmagnetic diluent for the rare earths is the much smaller atomic volume of Sc, about 25% less than that of Y. In the simple Ruderman–Kittel–Kasuya–Yosida theory, no dependence of exchange energy on volume is predicted although there is a very slight variation of exchange energy with the c/a ratio. That volume change is an important parameter has been discussed by Wollan[70] who has compared, quantitatively, the effects of such change with the corresponding changes brought about in the pure metals by the application of high pressures.

3.5.3. Neutron Diffraction Under High Pressures

In the past few years the effects of high pressure on the magnetic properties of the heavy rare earths has been studied extensively.[71] Below certain critical pressures $\simeq 25$ kbar where crystallographic transitions take place the magnetic properties change smoothly as a function of pressure. Both the Néel temperature and ferromagnetic transition temperature decrease linearly with an increase of pressure. Data exist as well for pressure effects on other physical properties; saturation magnetization, resistivity, etc. Efforts have been made to interpret the high pressure data in terms of an indirect exchange interaction involving the conduction electrons.

Recently Umebayashi et al.[22] have made measurements on single crystals of Tb and Ho under high pressures in which they paid special attention to the behavior of the interlayer turn angle, a quantity which is uniquely observable in a neutron diffraction experiment. The results which they obtained for Tb are shown in Figure 3–17. The curves for the lowest pressures are in general agreement with those obtained previously. The curves appear complicated at the lower temperatures because $T_N - T_C$ varies with pressure. It is found, apart from such complications near the transition temperature, that ω increases with pressure. A similar increase was found for Ho, but with

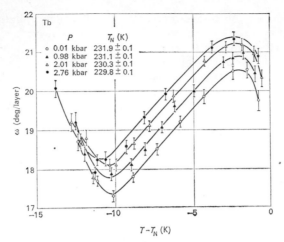

Figure 3–17. Turn angle variation of Tb as a function of $T - T_N$ at various pressures. After Umebayashi *et al*. (Ref. 22).

much smaller magnitude. For $T - T_N = -4.0$ K, $(1/\omega)\,(d\omega/dp) \times 10^3$ (kbar^{-1}) = 20 ± 2 for Tb and 1.2 ± 0.2 for Ho. They interpreted their measurements in terms of a theory due to Miwa[66] which incorporates the de Gennes–St. James effect and the band gap effect and concluded that the longer the wavelength of the periodic modulation the more sensitive it is to the small change of the conduction band characteristics, or of the $s-f$ interaction, produced by the pressure. This is evidently related to the observation (Figure 3–6, and Figure 3–15) that ω is a rapidly decreasing function of x when ω is small.

3.5.4. Neutron Diffraction with External Fields

Holmium. An extensive study of holmium in an applied field was undertaken to investigate the origin of the many anomalies in the single crystal magnetization data of Strandburg, Legvold, and Spedding[72] and to understand the difference between the magnetization studies which indicate a transition to a ferromagnetic state at low temperatures and the zero field neutron diffraction studies. Neutron diffraction measurements were made at temperatures ranging from 4.2 K to 120 K in applied fields up to 22.3 kOe in order to study the magnetization process.

Theoretical investigations of the magnetization process in a helical spin structure were made first by Herpin and Mériel[73] and by Enz.[74] The most complete investigations have been made by Nagamiya and his associates[75, 58] who studied the changes in helical and other oscillatory spin structures due to the application of a magnetic field at finite temperatures, as well as absolute zero, and for a number of cases of different anisotropy energy.

The results of the neutron diffraction experiments may be summarized as follows.[27] The application of a magnetic field to Ho at 4.2 K induces a transformation from the conical configuration to a ferromagnetic structure in which the moments are tipped out of the plane in low fields and become parallel to the basal plane in higher fields. The easy direction is the *b*-direction and the moments align themselves parallel to the easy axes nearest the field direction. The transformation, helix to ferromagnet, takes place without going through an intermediate oscillatory configuration, except possibly, through a slightly deformed helix.

The theory of Nagamiya *et al.* is not strictly applicable to Ho. A rigorous discussion does exist for the case that the moments are strictly restricted by anisotropy to the basal plane and with hexagonal anisotropy energy, the magnitude of which is treated as a parameter but is not so large as violently to deform the helix. A treatment has been given as well for the case that the zero field configuration is a cone, with hexagonal anisotropy neglected. In either case in low fields a slightly deformed configuration of the basal plane components is to be expected in which the moments tip slightly in the direction of the field.

This distorted helix should give rise to a second harmonic and to a zero'th harmonic (the net magnetization) in the diffraction pattern but in Ho, their intensities would have been too small to detect.

At higher fields a sine oscillation, or fan configuration, in which the magnetization vectors of the layers oscillate about the field direction, may develop. With sufficiently great hexagonal anisotropy the transition from slightly distorted helix to ferromagnet may occur directly. This is apparently the case for Ho at 4.2 K.

At higher temperatures, 40–70 K, the effect of a magnetic field applied parallel to a *b*-direction is to transform the system to a *b*-axis ferromagnet after causing it to pass through one or two, depending upon the temperature, intermediate fan-like oscillatory phases each of which is characterized by wave vectors of significantly different magnitude. More than one fan-phase and the change of wave vector are not predicted by the theory.

When the field was applied parallel to the *a*-direction, a net moment along the field direction was induced in oscillatory phases but the complete collapse of the system to an *a*-axis ferromagnet was never achieved with the fields available. A characterization of the structures in terms of the Fourier coefficients of the series describing the configurations is given in the original paper to which the reader is referred for details. The "humps" in the magnetization curves are shown to correspond to the onset of different intermediate configurations. An *H-T* phase diagram deduced from the measurements is shown in Figure 4–6 of Chapter 4.

Dysprosium. Magnetization data for Dy[76] show that the helix-ferromagnet transition can be raised to higher temperatures by application of an applied field and that anisotropy exists in the layers below about 110 K.

A few neutron diffraction measurements were carried out with the crystal in an external field.[24] With the field applied parallel to the *c*-axis, its magnetic properties were the same as those observed in zero field. When the field was applied normal to the *c*-axis the satellites disappeared with no perceptible change in turn angle, and intensity characteristic of the ferromagnetic structure appeared at normal nuclear positions.

3.6. THE LIGHT METALS

3.6.1. Physical Properties

Although the rare earth metals of the first half of the series have not been studied as thoroughly as those of the second half, there is nevertheless a considerable literature based on measurements with polycrystalline specimens. Recently single crystals of some of these metals have been produced, and physical measurements including neutron diffraction, undertaken with them.

The light rare earth metals are more complicated than those of the second half of the series in at least two respects; they have different and more complex crystal structures and the crystal field splittings of the magnetic energy levels of the more extended $4f$ electron distribution are comparable with the exchange energies.

In Pr, Nd, and a common allotropic form of La and Ce the crystal structure which is found is the double hexagonal close-packed structure (dhcp). This is formed by the stacking of hexagonal layers in the sequence ABAC. . . . In these metals the effective c/a ratio is 1.61. The stacking arrangement in the simple hexagonal structure (hcp) is ABAB . . . and in the face-centered cubic structure (fcc) it is ABCABC. . . . Thus, in the dhcp structure atoms on the A layers have near neighbor distributions typical of the cubic structures whereas those on B and C layers have hexagonal nearest neighbor environments. The existence of two kinds of sites with different local symmetry, and different crystalline fields, is a complication which is not found in the heavy metals.

Cerium undergoes a complex series of allotropic transformations when cooled from room temperature where both the fcc (γ-Ce) and dhcp (β-Ce) phases are found. Below about 100 K, transformation to a new collapsed fcc phase (α-Ce) with a volume 16.5% smaller is found. The amounts of the several phases, at low temperature, can be varied by different mechanical and thermal cycling procedures.

Following Nd in the series is Pm ($4f^4$) which has no stable isotope. The metal has now been produced in gram lots but as yet its physical properties have not been extensively studied. It is intensely radioactive and heat producing.

Samarium has an even more complicated crystal structure than that of Pr; it may be considered to be a nine layer sequence ABCBCACAB. . . . Two thirds of the sites in the structure have local hexagonal symmetry, one third, cubic.

Europium has the body centered cubic structure, the only one of the rare earth metals to be found in any but the compact forms at room temperature.

Anomalies in the specific heat and maxima in the susceptibility suggestive of cooperative phenomena have been observed in all the metals except Pr. Two such transitions have been observed in Nd. We turn now to a description of magnetic structure information for these metals.

3.6.2. Magnetic Structure Studies

Samarium. Recent polycrystalline susceptibility data show characteristic antiferromagnetic behavior at 14 K with an extremely broad minimum at about 200 K. For single crystal Sm, the susceptibility data are characteristic of a uniaxial spin alignment along the c-axis below 14 K.[77] The origin of the minimum at 200 K is not established. Mössbauer techniques have revealed no ordered magnetic structure at 77 K.[78] Preliminary neutron diffraction data obtained at the Oak Ridge National Laboratory from a single crystal of ^{154}Sm show antiferromagnetic ordering transitions near 100 K and 14 K which involve the ordering of moments on the hexagonal and cubic sites respectively.

Europium. Some early observations of the magnetic properties of polycrystalline Eu required a model of Eu^{+2} at low temperatures changing to Eu^{+3} at high temperatures.[79] Recent single crystal data are consistent with antiferromagnetic order at 4.2 K with Eu^{+2} in an $^8S_{7/2}$ state.[77] The transformation at the Néel point, near 90 K, has been shown to be a first-order transition.[80]

Polycrystalline neutron diffraction data have been obtained which show that the metal is indeed antiferromagnetic below 91 K.[81] A model which has been deduced from the data consists of a helical spin arrangement in which the moments are parallel to, and the screw axis normal to a cube face. The interlayer angle was found to vary only slightly with temperature: it varied from 51.4° per layer at T_N to 50.0° per layer at 4.2 K. A magnetic moment of $5.9 \pm 0.4 \beta$ was deduced from the data on the assumption of the helical model. This is rather less than the expected value of 7.0β and may reflect the influence of impurities which are hard to eliminate in the polycrystalline foils. Neutron diffraction studies of single crystal Eu are currently in progress at Risö.[82]

Cerium. An antiferromagnetic transition at 12.5 K in the dhcp structure of cerium has been observed in a study of polycrystalline cerium.[83] A possible structure is shown in Figure 3–18. It is formed by a stacking of ferrimagnetic layers of the type shown in Figure 3–18(a). To conform with the intensities of the magnetic reflections the moment must be parallel to the c-axis. If this model is correct the magnitude of the ordered moment is 0.62β.

Neodymium. The magnetic structure of neodymium has been investigated by single-crystal neutron diffraction in the temperature range from 1.6 to 20.0 K.[84] The intensity distribution in the diffraction pattern is quite complex; nevertheless, many of the features of the diffraction data are explicable in terms of a relatively simple model.

In the temperature range between 7.5 and 19 K there are found six magnetic satellites around each reciprocal lattice point and they are separated from the reciprocal lattice points by vectors $\pm Q_h$, where Q_h is parallel to one of the three equivalent b_1 directions. The magnitude of Q_h varies with temperature from $0.13 b_1$ at 18 K to $0.11 b_1$ at 8 K. The magnetic structure in this temperature range may be described as one having antiferromagnetic coupling between neighboring hexagonal site layers (B and C layers) and with a sinusoidal modulation within each layer. The direction of the moments and of the propagation vector (longitudinal oscillation) is in the basal plane parallel to a b_1 direction. The six satellites around each reciprocal lattice point are supposed to result from three equally populated domain orientations. The magnitude of the moments on the hexagonal sites is $\mu_h = 2.3 \pm 0.2 \beta$. In this temperature range there is no ordered moment on the cubic site layers (A layers). A schematic representation of the structure is shown in Figure 3–19(a).

At 7.5 K, new satellites appear around certain reciprocal lattice points, and these are interpreted as arising from ordering on the cubic A layers. In the model the moments on the A layers, A_1 and A_2 say, are given by $\mu_{A_1} = -\mu_{A_2} = \mu_c \, \hat{a} \cos (2Q_c \cdot x)$, where $\mu_- = 1.8 \pm 0.2 \beta$. The unit vector $\hat{a}$ is in the basal plane and rotated 30° from the vector b_1. The propagation vector Q_c is parallel to b_1, but has a magnitude of $Q_c = 0.15 b_1$ at 1.6 K. This structure, illustrated in Figure 3–19(b) is not dissimilar to that taken by the

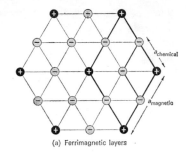

(a) Ferrimagnetic layers

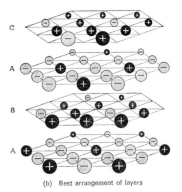

(b) Best arrangement of layers

Figure 3–18. Possible magnetic ordering in hexagonal cerium.

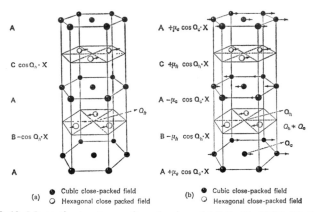

Figure 3–19. Magnetic structures of neodymium. (a) $7.5 < T < 19$ K. (b) $T < 7.5$ K.

moments on the hexagonal sites, but it has different moment directions, and a different magnitude for the propagation vector.

The models described above are adequate to explain most of the observations and must represent to a very good first approximation the magnetic structures of neodymium. A number of details in the diffraction data remain unaccounted for, however, and it seems probable that explanations for these must be sought in a weak interaction between the moments on the cubic and hexagonal sites.

Johansson *et al.* have recently investigated the effect of a magnetic field in the b_2 direction on these structures by a combination of magnetic susceptibility and neutron diffraction experiments.[85] At 4.2 K, a single domain crystal is formed at 7.0 kilogauss with the propagation vectors normal to the field. In low fields the moments on the hexagonal sites turn parallel to those on the cubic site, and this process is complete at 12 kilogauss. At the same time a net ferromagnetic moment is developed in the cubic and hexagonal sites in the ratio of about 5:2. With higher fields, presumably at a crossing of crystal field levels in the field, the ferromagnetic moment on the hexagonal sites grows at the expense of the oscillatory component. The periodic moment on the cubic sites decreases steadily with field and disappears at about 31 kilogauss. The crystal field levels in Nd restrict the magnitude of the ordered moment in zero fields. The total moment is substantially increased by the application of a magnetic field.

Praseodymium. The metal Pr exhibits no thermal or magnetic properties indicative of cooperative effects. A broad specific heat anomaly, centered near 30 K was observed by Parkinson, Simon, and Spedding[86] and was attributed by them to crystal field splittings of the electronic levels, and Lock[87] found that the magnetic susceptibility became temperature independent below about 4 K, indicating that the lowest level is a nonmagnetic singlet state. More recently the nuclear-hyperfine coupling contribution to the specific heat has been obtained from measurements in the range 0.4 to 4.0 K.[88] This contribution is very much smaller than that calculated for Pr on the assumption that the moments have their maximum allowed values, and this also has been taken as an indication of the absence of magnetic order in Pr.

Neutron diffraction data obtained from polycrystalline samples of Pr at 1.4 K revealed,[89] however, a series of magnetic reflections very similar to that of Nd; the positions of the reflections yield a modulation vector of 0.13 b_1. The magnitudes of the intensities, on the Nd-type structure model, lead to a value for the ordered moment of 0.7 β in the event that all the moments are ordered and to 1.0 β if only one set of sites has ordered moments. The Néel temperature deduced from the diffraction data was 25 K.

A single crystal specimen of Pr has been investigated by Johansson *et al.*[85] and they find no trace of magnetic order at 4.2 K. Antiferromagnetism in polycrystalline Pr, and its absence in a single crystal has now also been confirmed by Rainford and Wedgwood.[90]

Application of a magnetic field along the b_2 direction does, however, produce a large induced moment on both sets of sites $\mu_h = 1.8$ β/atom, $\mu_c = 0.9$ β/atom at 4.2 K and 46 kilogauss. The results are in reasonable agreement with a molecular field model and the level scheme deduced by Bleany[91] except that the best value for the energy separation between the ground state and the first excited doublet on the hexagonal sites is 20 K instead of Bleany's value of 63 K.

3.7. THE MAGNETIC FORM FACTORS

3.7.1. Introduction

An accurate measurement of the form factor of a magnetic element may provide information about the spatial distribution of magnetic moment

5

densities on a microscopic scale. Form factors for some of the 3-d transition group elements have been determined with great accuracy by exploiting the polarized beam technique, but only very recently have such measurements been made on rare earths.

Of course, every investigation of the magnetic structure of a rare earth metal or alloy with the usual unpolarized beam methods includes, to a greater or lesser degree, a form factor measurement; results of moderate accuracy have been reported for Ho,[26] Er,[7] Gd,[17] and for some of the Ho-Tb[51] alloys.

One of the interesting features of the magnetic scattering of neutrons by rare earth ions is the large contribution made by the orbital moment, a contribution which is comparatively very small in the 3-d elements. Moreover, since the rare earth metals of the second half of the series, particularly, appear to be well described by the Hund's Rule ground state, the theory is relatively straightforward, although the algebra is somewhat cumbersome.

The theory of paramagnetic scattering by rare earth ions was first given by Trammell.[92] His treatment was later extended to other ions, and to aligned moment systems by Odiot and St. James.[93] Koehler and Wollan,[94] and Koehler, Wollan, and Wilkinson[95] reported observations on the paramagnetic scattering by rare earth oxides. For comparison with the theory, screened hydrogenic radial wavefunctions were used with the screening constant as an adjustable parameter. Subsequently, Freeman and Watson[96] calculated free ion Hartree-Fock wavefunctions with which Blume, Freeman, and Watson[97] derived the radial integrals required in Trammell's formulation to calculate the form factors. Agreement between the theory and experiment was remarkably good. More recently Child *et al.*[98] and Koehler *et al.*[99] have measured the paramagnetic scattering cross section of specimens of Gd_2O_3 enriched in the isotope ^{160}Gd. Since Gd^{+3} is an S state ion, the complications of orbital moment scattering are not present. Over the range in which the cross section can be measured meaningfully, the observations are in excellent agreement with the free ion calculations except at very low scattering angles.

The form factors for ordered systems, except for Gd, are more complicated but careful measurements of these in suitably designed experiments can lead to information about the anisotropy of the magnetic moment distribution.

Trammell's formulation can be utilized to write down expressions for the x, y and z components of the Fourier transform of the magnetic moment density which involves certain spherical harmonics and the radial integrals referred to above. A more elegant formulation in terms of tensor operators has recently been given by Johnston,[100] Johnston and Rimmer,[101] and Lovesey and Rimmer.[102] In their approach the number of radial integrals (because of certain identities among them) is halved. Moreover, their formulation is ideally suited to machine programming. Lander and Brun[103] have prepared tables of coefficients, based on the tensor operator approach, for application to rare earth and possibly also to actinide form factors.

3.7.2. Polarized Beam Methods

We give here a very brief discussion of the polarized beam method for the measurement of form factors. The theoretical foundation was described many years ago by Halpern and Johnson.[104] They showed that the differential scattering cross section for a magnetic ion is given by

$$\frac{d\sigma}{d\Omega} = (b^2 + p^2 q^2 + 2bp\ \mathbf{q}\cdot\boldsymbol{\lambda}) \tag{3.13}$$

where b and p are the nuclear and magnetic scattering amplitudes described previously, $\boldsymbol{\lambda}$ is the unit vector describing the neutron polarization and $\mathbf{q} = \hat{\mathbf{K}}(\hat{\mathbf{K}}\cdot\hat{\mathbf{m}}) - \hat{\mathbf{m}}$ with $\hat{\mathbf{K}}$ the unit scattering vector and $\hat{\mathbf{m}}$ the unit vector parallel to the moment of the magnetic ion. If the experiment is arranged so that $\hat{\mathbf{m}}$ is perpendicular to the scattering vector $\hat{\mathbf{K}}$, then $|\mathbf{q}|$ becomes unity and the scattering cross sections for the two neutron spin states ($\mathbf{q}\cdot\boldsymbol{\lambda} = \pm 1$) become

$$\frac{d\sigma_+}{d\Omega} = (b + p)^2 \qquad \mathbf{q}\cdot\boldsymbol{\lambda} = +1 \tag{3.14}$$

$$\frac{d\sigma_-}{d\Omega} = (b - p)^2 \qquad \mathbf{q}\cdot\boldsymbol{\lambda} = -1.$$

It follows that a ferromagnetic crystal will preferentially scatter neutrons of one spin state of the equally populated states of the initially unpolarized beam. This will result in a polarized diffracted beam where the polarization P is defined as

$$P = \frac{I_+ - I_-}{I_+ + I_-} \tag{3.15}$$

and $I_{\pm}$ are the intensities, proportional to the cross sections, for the two spin states. This may be put into the form

$$P = \frac{2pb}{b^2 + p^2} \tag{3.16}$$

which shows that the polarizing efficiency will depend on the ratio p/b. If for a given Bragg reflection $p = b$ then the polarization will be complete since only one spin state will be scattered. Polarized beams with a high degree of polarization have been produced from the (220) reflection of Fe_3O_4, and the (200) reflection of the crystal $Co_{92}Fe_8$.

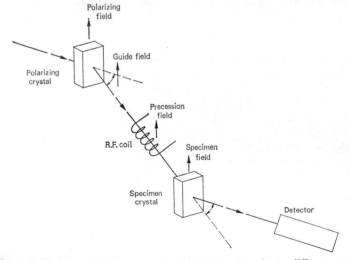

Figure 3–20. Schematic diagram of a polarized neutron beam diffractometer.

Figure 3–20 shows a schematic diagram of a typical polarized beam instrument. A polarizing crystal, Co-Fe say is magnetized vertically to satisfy the conditions of Eqn (3.14). After polarization the beam passes through a guide field which maintains the polarization until it strikes the sample, also magnetized vertically. Between the monochromator-polarizer and the sample is placed a radio-frequency coil which, when excited with the proper frequency reverses the direction of spin polarization. The experiment consists in measuring the peak intensity in a Bragg peak with the neutron spin parallel and antiparallel to the magnetization of the specimen. The so-called flipping ratio $R = I_+/I_-$ is related to the ratio of magnetic and nuclear scattering amplitudes, as seen from Eqn (3.14) in the following way.

$$R = \left[\frac{1+p/b}{1-p/b}\right]^2. \tag{3.17}$$

The magnetic scattering amplitude p is proportional to the magnetic form factor. In principle one has only to measure the flipping ratio for a large number of reflections. From a knowledge of the nuclear scattering amplitude b, one is then able to determine p as a function of scattering angle.

In practice there are a number of corrections to the data which have to be made. In any real installation the beam which reaches the sample is not perfectly polarized, nor in general is the r.f. system perfectly efficient in reversing the neutron spin. The degrees of imperfection are instrumental parameters which can be determined experimentally. They have the effect of reducing the flipping ratio R, for $p = b$, from infinity to, in a good installation, a large value (100–200).

When $p > b$ as it is for the ferromagnetic rare earths at small scattering angles the sensitivity of the polarized beam technique is low but not prohibitively so. Near the transition region where $p \simeq b$ instrumental corrections, and crystal effects such as depolarization, and extinction, all have to be accurately known. At large scattering angles where $p \ll b$, the technique has its greatest power for here the sensitivity of R to p is high, and the experimental corrections mentioned above are less significant.

3.7.3. Results

We summarize here the form factors which have been determined with the aid of polarized neutrons. Two features of rare earth form factors, as compared to those of the 3-d elements should be emphasized. Firstly, in the rare earths the spin-orbit coupling is much larger than crystalline field effects. Consequently when a magnetic field of sufficient strength to align the moment is applied along a given crystallographic direction (saturation is required in the polarized beam experiment to prevent serious depolarization of the beam) the field direction becomes the unique axis of the moment distribution. That is to say the direction of the spin follows the field, and, because of the strong spin-orbit coupling its spatial distribution is also influenced by the field.

The usual polarized beam experiment of measuring different reflections with the same $\sin\theta/\lambda$ values to assess the degree of asymmetry of the moment distribution will not work because the moment distribution always follows the field. It is in principle possible to detect asphericity in the form factor by measuring reflections off the zero-layer.

Secondly it should be noted that the form factor which is measured in a polarized beam experiment is not necessarily identical to that measured in the ordered state with unpolarized neutrons (sample unmagnetized) nor to that obtained with the sample in the paramagnetic state. The differences are usually, though not always, small at small scattering angles.

Gadolinium. The experimental form factor obtained for ^{160}Gd with both polarized and unpolarized neutrons[99] is shown in Figure 3-21. These are data obtained at 96 K from three different crystals at four neutron wavelengths. Other experiments were made at 312.5 K in which a moment was

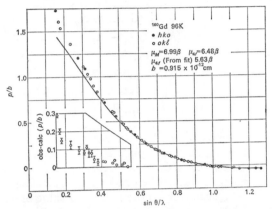

Figure 3–21. The magnetic form factor of gadolinium. The solid curve is a theoretical fit with free-ion wave functions to the high angle data. Extrapolation to $\sin \theta/\lambda = 0$ gives a 4-f moment of 5.63 β. The insert is the difference $(p/b)_{obs} - (p/b)_{calc}$ (Ref. 99).

induced in the paramagnetic state, and the distribution of induced moment density studied. To the precision of the observations the two form factors are identical.

The solid line is a theoretical curve calculated with the Freeman and Watson free ion radial wave functions. Since it is expected that the conduction electron contribution, if any, will contribute to reflections at small angles (the conduction electrons are 5d-like) only the data further out than $\sin \theta/\lambda = 0.5$ were used in the analysis. It may be seen that for those higher angle reflections the free-ion wave functions do give a remarkably good fit to the data, but the 4-f moment obtained on extrapolation to $\sin \theta/\lambda = 0$ is 5.63 β instead of the value of 6.48 β estimated from the temperature dependence of the saturation magnetization or of 6.99 β, the value obtained from the magnetization at 96 K and which includes a contribution from the conduction electrons.

In the insert are plotted values for the difference between observed and calculated values of p/b. This difference does have the character of a d-like distribution and it would be tempting to ascribe it to the conduction electrons. However, the deficiency in 4-f moment is too great to be accepted. One is forced to conclude that the free-ion wave functions do not adequately represent the situation in the metal.[105] As we shall see presently, the same conclusion can be reached from an analysis of data obtained for Tb andTm. In contrast, the Freeman–Watson wave functions do give a good fit to the Gd^{+3} form factor measured in the oxide.[98–99]

Note added in proof: Further analysis of the magnetic form factor data for ^{160}Gd has shown that the total magnetic moment density can be separated in a logical manner into a local part and a diffuse part. The local part corresponds to the 4f polarization and the diffuse part to the conduction electron polarization. Contour maps show the diffuse part to be long range and oscillatory. (R. M. Moon and W. C. Koehler, *Phys. Rev. Letters*, **27**, 407 (1971); R. M. Moon, W. C. Koehler, J. W. Cable and H. R. Child, *Phys. Rev.*, to be published.)

Terbium. The form factor for Tb has been measured by Steinsvoll *et al.*[106] at 4.2 K with both polarized and unpolarized neutrons. They have also given expressions for the theoretical form factors which are appropriate to the two experiments. For the polarized beam case

$$f(K) = f_s(K) - \Delta_p \qquad (3.18)$$

where $f_s(K) = \frac{1}{3}\langle g_0 \rangle + \frac{2}{3}\langle j_0 \rangle$ is the spherical part of the form factor[107] and

$$\Delta_p = -0.093 \langle j_2 \rangle + 0.034 \langle j_4 \rangle + 0.278 \langle g_2 \rangle - 0.102 \langle g_4 \rangle \qquad (3.19)$$

is the aspherical part.[108] The integrals $\langle g_n \rangle$ and $\langle j_n \rangle$ are tabulated in Blume, Freeman, and Watson.[97] A similar expression for the unpolarized beam experiment has been given in terms of a Δ_u the numerical values of which are listed by Steinsvoll *et al.* Over the range of scattering angles in which unpolarized beam data were collected Δ_u (and Δ_p) is very small.

We have taken their published data and expressed it as the series of points shown in the plot, Figure 3–22, of the product μf against $\sin \theta / \lambda$. The upper two curves correspond respectively to the measured value of 9.34 β and to the full 4-f moment of 9.0 β. It may be seen that both are too high. When only the high angle data are fitted to the theoretical curve, the extrapolated value of the moment is 8.18 β so that in this case again the wave functions appear to be incorrect.

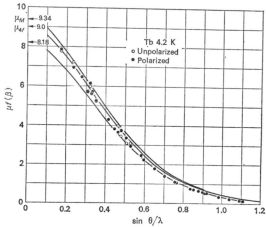

Figure 3–22. The magnetic form factor of terbium at 4.2 K. The upper two lines are theoretical curves normalized to 9.34 β and 9.0 β, respectively. The lower one, extrapolation to 8.18 β, is the result of a fit to high angle data only (Ref. 106).

A similar set of data is presented in Figure 3–23 which is derived from measurements of Brun.[109] Fitting at high angles results in a scale factor of 0.86. The theoretical form factor for Tb in the paramagnetic region may be written[109]

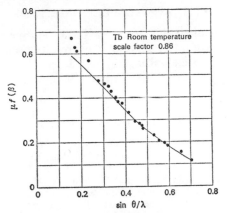

Figure 3–23. The magnetic form factor of terbium at room temperature. Fitting at high angles leads to a scale factor of 0.86 (Ref. 109).

$$f(K) = \langle j_0 \rangle - 0.3704 \langle j_2 \rangle \qquad (3.20)$$

Thulium. The form factor for thulium was determined at 4.2 K, and, as first reported, it was interpreted in terms of a low 4-*f* moment and a *d*-like conduction electron term.[36] The results, shown in Figure 3–24, have now been reinterpreted[109] to imply, as in the previous cases, that the 4-*f* radial wave functions are inadequate. For the basal plane reflections the theoretical form factor is given by[109]

$$f(K) = \langle j_0 \rangle + 0.5952 \langle j_2 \rangle - 0.0828 \langle j_4 \rangle + 0.0095 \langle j_6 \rangle. \qquad (3.21)$$

For the out-of-plane reflections different expressions must be used. The out-of-plane points shown in the figure reflect the asphericity of the moment density of thulium.

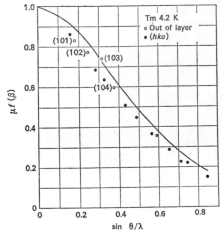

Figure 3–24. The magnetic form factor of thulium. The solid line is a theoretical curve normalized to 1.0 β. Out-of-plane reflections show the asphericity of the moment density distribution (Ref. 36, 109).

3.8. ACKNOWLEDGMENTS

It is a pleasure for the author to thank Dr. F. H. Spedding and Dr. T. O. Brun, for permission to quote some of their results prior to publication. He expresses his thanks to his colleagues, Dr. J. W. Cable, Dr. H. R. Child, Dr. R. M. Moon, and Dr. E. O. Wollan, for many discussions and much assistance in preparing this chapter.

REFERENCES

1. 1. Koehler, W. C. and Wollan, E. O., *Phys. Rev.*, **97**, 1177 (1955).
2. In many cases the same crystals were investigated at the two laboratories due to the courtesy of Dr. F. H. Spedding.
3. See, for example, Bozorth, R. M. and Graham, Jr., C. D., in Rare Earth Metals, Compounds, and Solid Solutions, Handbook of Magnetic Materials and Their Properties (P. Albert and F. Luborski, eds.). Reinhold Press (1970); and Tebble, R. S. and Craik, D. J., in Magnetic Materials. Wiley-Interscience (1969). See especially Chapter 6.
4. Koehler, W. C., *J. Appl. Phys.*, **36**, 1078 (1965); Koehler, W. C., in AIME Symposium on Magnetic and Inelastic Scattering of Neutrons by Metals (T. J. Rowland and Paul A. Beck, eds). Gordon and Breach (1968).
5. A more detailed description of the general magnetic and crystallographic properties of the rare earth metals is given in Chapters 1, 2 and 4.
6. Details will be found in Chapters 4 and 7.
7. Cable, J. W., Wollan, E. O., Koehler, W. C. and Wilkinson, M. K., *Phys. Rev.*, **140**, 1896 (1965).
8. For a general description of the oscillatory structures, see Chapter 2.
9. Miwa, H. and Yosida, K. *Progr. Theoret. Phys.* (Kyoto), **26**, 693 (1961).
10. This compilation includes results from many sources. References are given in the appropriate paragraphs in the text.
11. Belov, K. P. and Ped'ko, A. V., *Zh. Eksp. Teor. Fiz.*, **47**, 87 (1962). [English Translation: *Soc. Phys. JETP*, **15**, 62 (1962)].
12. Graham, C. D. *J. Appl. Phys.*, **34**, 1341 (1963).
13. Will, G., Nathans, R. and Alperin, H. A., *J. Appl. Phys.*, **35**, 1045 (1964).
14. Cable, J. W. and Wollan, E. O., *Phys. Rev.*, **165**, 733 (1968).
15. Corner, W. D., Roe, W. C. and Taylor, K. N. R., *Proc. Phys. Soc.* (London), **80**, 927 (1962).
16. Graham, C. D., *J. Phys. Soc. Japan*, **16**, 1310 (1962).
17. Kuchin, V. M., Semenkov, V. A., Shil'shtein, S. Sh. and Patrikeev, Yu. B., *Zh. Eksp. Teor. Fiz.*, **55**, 1241 (1968) [English Translation: *Sov. Phys. JETP*, **28**, 649 (1969)].
18. Koehler, W. C., Child, H. R., Wollan, E. O. and Cable, J. W., *J. Appl. Phys. Suppl.*, **34**m 1335 (1963).
19. Koehler, W. C., in Transactions of the American Crystallographic Association (H. G. Smith, ed). Polycrystal Book Service, Pittsburgh (1967), Vol. 3, p. 53.
20. See, for instance, Elliott, R. J., *Phys. Rev.*, **124**, 346 (1961).
21. Dietrich, O. W. and Als-Nielsen, J., *Phys. Rev.*, **162**, 315 (1967).
22. Umebayashi, H., Shirane, G., Frazier, B. C. and Daniels, W. B., *Phys. Rev.*, **165**, 688 (1968).
23. Dietrich and Als-Nielsen observed a small difference in the turn angle measured from the two satellites of the 002 reflection. We give the average of the two.
24. Wilkinson, M. K., Koehler, W. C., Wollan, E. O. and Cable, J. W., *J. Appl. Phys.*, **32**, 48S (1961).
25. Jordan, R. G. and Lee, E. W., *Proc. Phys. Soc.*, **92**, 1074 (1967) have reported a very small (of the order of 0.1 β) spontaneous moment parallel to the c-axis for several specimens of Dy. It may be that the structures of Dy are not as simple as heretofore supposed. If this moment be produced by canting of the moments out of the basal plane, the semi-cone angle will be of the order of 89.5°.
26. Koehler, W. C., Cable, J. W., Wilkinson, M. K. and Wollan, E. O., *Phys. Rev.*, **151**, 414 (1966).
27. Koehler, W. C., Cable, J. W., Child, H. R., Wilkinson, M. K. and Wollan, E. O., *Phys. Rev.*, **158**, 450 (1967).

28. Stringfellow, M., private communication.
29. Koehler, W. C., unpublished.
30. Gray, W. J. and Spedding, F. H., IS-2044, February 1968, unpublished, have detected in weak fields an anomaly in the c-axis magnetization data in Er near 28 K.
31. Koehler, W. C., Cable, J. W., Wollan, E. O. and Wilkinson, M. K., *Phys. Rev.*, **126**, 16/2 (1962).
32. Davis, D. D. and Bozorth, R. M., *Phys. Rev.*, **118**, 1543 (1960).
33. Brun, T. O., Sinha, S. K., Wakabayashi, N., Lander, G. H., Edwards, L. R. and Spedding, F. H., *Phys. Rev.*, **B1**, 1251 (1970).
34. Cohen, R. L., *Phys. Rev.*, **169**, 432 (1968).
35. On reexamination of the older data in the light of the Mössbauer and new neutron results it was concluded that a wave vector as small as $0.275 \, b_3$ at T_N would have been detected. Unfortunately the crystal on which the measurements were made is no longer available for a repetition of the work.
36. Brun, T. O. and Lander, G. H., *Phys. Rev. Letters*, **23**, 1295 (1969).
37. Child, H. R., Koehler, W. C., Wollan, E. O. and Cable, J. W., *Phys. Rev.*, **138**, A1655 (1965).
38. Child, H. R. and Cable, J. W., *J. Appl. Phys.*, **40**, 1003 (1969).
39. Gray, W. J. and Spedding, F. H., IS-2044 (1968).
40. Bozorth, R. M. and Gambino, R. J., *Phys. Rev.*, **147**, 487 (1966).
41. Child, H. R., ORNL-TM-1063 (1965).
42. A brief description of these results appears in Millhouse, A. H. and Koehler, W. C., *Les Elements des Terres Rares*. Colloques Intern. C.N.R.S. Nº 180, Paris (1970), Vol. II, p. 214. A more complete account is in preparation for publication in the *Physical Review*.
43. Bozorth, R. M., *J. Appl. Phys.*, **38**, 1366 (1967).
44. Again, near T_N we cannot be certain that the ordering is helical and not a uniaxial oscillation along equivalent directions in the basal plane. What is certain is that Er atom moments, not just Dy atom moments, are ordered in the basal plane.
45. Bozorth, R. M., Gambino, R. J. and Clark, A. E., *J. Appl. Phys.*, **39**, 883 (1968).
46. Pickart, S., private communication.
47. Shirane, G. and Pickart, S., *J. Appl. Phys.*, **37**, 1032 (1966).
48. Lebech, Bente, *Solid State Comm.*, **6**, 791 (1968).
49. Bjerrum-Møller, H., Mackintosh, A. R. and Gylden Houmann, J. C., *J. Appl. Phys.*, **39**, 1078 (1965).
50. Spedding, F. H., Jordan, R. G., and Williams, R. W., *J. Chem. Phys.*, **51**, 509 (1969).
51. Spedding, F. H., Ito, Y. and Jordan, R. G., *J. Chem. Phys.*, **53**, 1455 (1970).
52. Spedding, F. H., Ito, Y., Jordan, R. G. and Croat, J., *J. Chem. Phys.*, **54**, 1995 (1971).
53. Evidence of a small c-axis component, $0.9 \, \beta$ was found for the alloy in the c-axis magnetization data. It would be difficult to detect in the neutron scattering experiments.
54. In a random domain sample the moment directions in the plane cannot be determined. Experiments in progress at the Oak Ridge National Laboratory (Q. Khan and W. C. Koehler) on $Dy_{0.50}Ho_{0.50}$ indicate that the easy direction at 4.2 K is the b-direction characteristic of Ho rather than the a-direction characteristic of Dy.
55. See, for example, Cooper, B. R., Solid State Physics. Academic Press, New York (1968), Vol. 21, p. 393.
56. Keeton, S. C. and Loucks, T. L., *Phys. Rev.*, **168**, 672 (1968).
57. Evenson, W. E. and Liu, S. H., *Phys. Rev.*, **178**, 783 (1969).
58. Nagamiya, T., Solid State Physics. Seitz, F. and Turnbull, D. (eds), Academic Press, New York (1967), Vol. 20, p. 305.
59. Feron, F. L. and Pauthenet, R., Proc. Seventh Rare Earth Research Conference. Coronado, California (1968).
60. Rhyne, J. J. and Clark, A. E., *J. Appl. Phys.*, **38**, 1379 (1967).
61. du Plessis, P. de V., *Physica*, **41**, 379 (1969).
62. Watson, R. E., Freeman, A. J. and Dimmock, J. P., *Phys. Rev.*, **167**, 497 (1968).
63. de Gennes, P. G., *J. Phys. Radium*, **23**, 510 (1962); *J. Phys. Radium*, **23**, 630 (1962).
64. de Gennes, P. G. and Saint James, D., *Solid State Comm.*, **1**, 62 (1963).
65. Elliott, R. J. and Wedgwood, F. A., *Proc. Phys. Soc.*, **84**, 63 (1964).
66. Miwa, H., *Proc. Phys. Soc.* (London), **85**, 1197 (1965).
67. Koehler, W. C., Child, H. R., Cable, J. W. and Moon, R. M., *J. Appl. Phys.*, **38**, 1384 (1967).
68. Child, H. R. and Koehler, W. C., *J. de Physique*, **32**, C1–1128 (1971).

69. Child, H. R. and Koehler, W. C., *J. Appl. Phys.*, **37**, 1353 (1966); *Phys. Rev.*, **174**, 526 (1968).
70. Wollan, E. O., *Phys. Rev.*, **160**, 369 (1967).
71. See Umebayashi *et al.*, Ref. 22, for references to the high pressure work.
72. Strandburg, D. L., Legvold, S. and Spedding, F. H., *Phys. Rev.*, **27**, 2046 (1962). These data are described in Chapter 4.
73. Herpin, A. and Meriel, P., *Compt. Rend.*, **250**, 1450 (1960); *J. Phys. Radium*, **22**, 337 (1961).
74. Enz, U., *Physica*, **26**, 69 (1960); *J. Appl. Phys.*, **32**, 22S (1961).
75. Nagamiya, T., Nagata, K. and Kitano, Y., *Progr. Theoret. Phys.*, (Kyoto), **27**, 1253 (1962); Kitano, Y. and Nagamiya, T., *ibid.*, **31**, 1 (1964).
76. Behrendt, D. R., Legvold, S. and Spedding, F. H., *Phys. Rev.*, **109**, 1544 (1958).
77. Schieber, M., Foner, S., Doclo, D. and McNiff, Jr., E. J., *J. Appl. Phys.*, **39**, 885 (1968).
78. Ofer, S., Segal, E., Nowick, I., Bauminger, E. R., Gradzins, L., Freeman, A. J. and Schieber, M., *Phys. Rev.*, **137**, A627 (1965).
79. Bozorth, R. M. and Van Vleck, J. H., *Phys. Rev.*, **118**, 1493 (1960).
80. Cohen, R. L., Hüfner, S. and West, K. W., *Phys. Letters*, **28A**, 582 (1969).
81. Nereson, N. G., Olsen, C. E. and Arnold, G. P., *Phys. Rev.*, **135**, A176 (1964).
82. Bjerrum-Møller, H. and Millhouse, A. H., private communication.
83. Wilkinson, M. K., Child, H. R., McHargue, C. J., Koehler, W. C. and Wollan, E. O., *Phys. Rev.*, **122**, 1409 (1961).
84. Moon, R. M., Cable, J. W. and Koehler, W. C., *J. Appl. Phys. Suppl.*, **35**, 1041 (1964).
85. Johansson, J., Lebech, B., Nielsen, M., Bjerrum-Møller, H. and Mackintosh, A. R., *Phys. Rev. Letters*, **25**, 524 (1970).
86. Parkinson, D. H., Simon, F. E. and Spedding, F. H., *Proc. Roy. Soc.* (London), **207A**, 137 (1951).
87. Lock, J. M., *Proc. Phys. Soc.* (London), **B70**, 566 (157).
88. Lounasmaa, O. V., *Phys. Rev.*, **133**, A211 (1964).
89. Cable, J. W., Moon, R. M., Koehler, W. C. and Wollan, E. O., *Phys. Rev. Letters*, **12**, 553 (1964).
90. Wedgwood, A., private communication.
91. Bleaney, B., *Proc. Roy. Soc.* (London), **A276**, 39 (1963).
92. Trammell, G. T., *Phys. Rev.*, **92**, 1387 (1953).
93. Odiot, Simone and Saint-James, D., *J. Phys. Chem. Solids*, **17**, 117 (1960).
94. Koehler, W. C. and Wollan, E. O., *Phys. Rev.*, **92**, 1380 (1953).
95. Koehler, W. C., Wollan, E. O. and Wilkinson, M. K., *Phys. Rev.*, **110**, 37 (1958).
96. Freeman, A. J. and Watson, R. E., *Phys. Rev.*, **127**, 2058 (1962).
97. Blume, M., Freeman, A. J. and Watson, R. E., *J. Chem. Phys*, **37**, 1245 (1962); **41**, 1874 (1964).
98. Child, H. R., Moon, R. M., Raubenheimer, L. J. and Koehler, W. C., *J. Appl. Phys.*, **38**, 1381 (1967).
99. Koehler, W. C., Moon, R. M., Cable, J. W. and Child, H. R., *J. de Physique*, **32**, C1-296 (1971).
100. Johnston, D. F., *Proc. Phys. Soc.* (London), **88**, 37 (1966).
101. Johnston, D. F. and Rimmer, D. E., *J. Phys. C* (*Solid St. Phys.*), **2**, 1151 (1969).
102. Lovesey, S. W. and Rimmer, D. E., *Reports on Prog. in Physics*, **32**, 333 (1969).
103. Lander, G. H. and Brun, T. O., *J. de Physique*, **32**, C1-571 (1971).
104. Halpern, O. and Johnson, M. H., *Phys. Rev.*, **55**, 898 (1939).
105. This conclusion differs from that of Kuchin, V. M. *et al.*, Ref. 17, who assert that the distribution of spin density in metallic gadolinium agrees well with that calculated for the Gd^{+3} ion.
106. Steinsvoll, O., Shirane, G., Nathans, R., Blume, M., Alperin, H. A. and Pickart, S. J., *Phys. Rev.*, **161**, 499 (1967).
107. For Gd the 4-f form factor is simply $\langle j_0 \rangle$.
108. Strictly speaking additional terms in $\langle j_6 \rangle$ and $\langle g_6 \rangle$ should be included. These have been omitted because they are negligibly small.
109. Brun, T. O., Thesis, Iowa State University, 1970.

Chapter 4

Bulk Magnetic Properties

James J. Rhyne

U.S. Naval Ordnance Laboratory, White Oak, Silver Spring,
Maryland 20910

4.1. INTRODUCTION

This chapter discusses the static magnetic properties of the rare earth metals. These, in the general sense, are the macroscopic average response of the material to the microscopic interactions discussed in other chapters of this volume. The study of bulk features complements atomic scale experiments such as neutron diffraction, Mössbauer and other resonance studies and is often required as a basic aid in their interpretation. In the areas of direct overlap, the data obtained from bulk results are usually more precise due to the relative simplicity of the measuring techniques. Some of these attributes are apparent in the material to follow.

The hexagonal unit cell found in a majority of the rare earths leads to considerable anisotropy in most of the crystal field, magnetic, elastic, and Fermi surface properties of these elements. In this chapter, due to the crucial effects of this anisotropy on many of the measurements, single-crystal results will be reviewed when available to the exclusion of polycrystalline data.

Unlike the elements heavier than Eu, there is still a paucity of single-crystal information available on the light rare earths. The complex crystal structures and symmetry transitions which occur well below the melting temperatures generally do not make the light rare earths amenable to traditional crystal growing techniques. The most successful method to date has been a recrystallization from the vapor phase.[1]

The chapter is divided into four basic sections: (4.2) magnetization and susceptibility, (4.3) magnetic anisotropy, (4.4) magneto-elastic properties, and (4.5) specific heat. These general sections are further subdivided by specific physical phenomena and experiments. Light and heavy rare earths are treated separately and in order of increasing atomic number.

4.2. MAGNETIZATION MEASUREMENTS

The discussion of magnetization measurements on the rare earths is in four parts: (4.2.1) the magnetic heavy rare earths; (4.2.2) the light rare earths and the paramagnetic elements Yb, Sc, Y, and Lu; (4.2.3) intra-rare earth alloy systems; and (4.2.4) magnetization properties at elevated pressure. Many of the results presented, especially in the heavy rare earths, depend heavily on neutron diffraction studies discussed in Chapter 3. The neutron diffraction experiments have elucidated the nature of the complex spin states found in these metals and of the transitions occurring between these states. Reference is made to Figure 3–2 in Chapter 3 which illustrates the intrinsic spin configurations for the magnetically ordered elements.

The existence of either the ferromagnetic or one of the several complex periodic moment states is determined by the minimum free energy for a particular temperature and applied field (if present). This free energy is derived from a Hamiltonian containing terms:

$$H = H_{\text{exchange}} + H_{\text{crystal field anisotropy}} + H_{\text{magneto-elastic}} + H_{\text{applied field}}.$$

The explicit forms of the first three terms are given respectively in Chapter 2 and sections 4.3.1 and 4.4.1 of this chapter.

The temperature dependences of the anisotropy and magneto-elastic energies are proportional to powers (greater than 1) of the magnetization and are thus relatively unimportant near the initial ordering temperature except to select a preferred moment direction. The initial moment configuration is then determined by the form of the exchange interaction. In particular, the material orders ferromagnetically (e.g., Gd) if the Fourier transform of the exchange integral $\mathscr{J}(\mathbf{q})$ exhibits a maximum at $\mathbf{q}=0$. A maximum occurring at some other point $\mathbf{q}=\mathbf{Q}_0$ of the zone leads to a periodic moment state. The stable $\mathbf{Q}_0$ is dictated by the $\mathbf{q}$ of the "nesting vector" of the Fermi surface as described in Chapter 6 and references 54–56.

At temperatures significantly lower than the ordering T, the energies associated with anisotropy and magnetostriction become sufficiently large to drive transitions between periodic states or from periodic to a ferromagnetic state of lower energy. Equivalent transitions can be effected with applied fields of appropriate magnitude as discussed in the following sections.

Table 4.1 summarizes information for the magnetic heavy rare earths on the experimental low-temperature saturation magnetic moment, ordering

Table 4.1. Moments and Ordering Temperatures for Heavy Rare Earth Single Crystals

	λJ (β)	$\mu_{sat}^{expt.}$ (β)	T_N (K)	T_C (K)	$T_{trans.}$[1] (K)	$\mu_{eff}^{theor.}$ (β)	$\mu_{eff}^{expt.}$ (β)	$\theta_\parallel^{para}$ (K)	$\theta_\perp^{para}$ (K)
Gd	7.0	7.55	—	293.2	—	7.94	7.98	317	317
Tb	9.0	9.34	229	221	—	9.72	9.77	195	239
Dy	10.0	10.33	178.5	85	—	10.64	10.64	121	169
Ho	10.0	10.34	132	20	—	10.6	11.2	73	88
Er	9.0	—	85	19.6	53	9.6	9.9	61.7	32.5
Tm	7.0	7.14	58	25	42	7.56	7.61	41	−17

(1) Intermediate spin transition temperatures from magnetic and neutron diffraction.

temperature, paramagnetic Curie temperatures, and effective moments from susceptibility. Theoretical moment values for the free tri-positive ions are given for comparison. The zero Kelvin saturation moments found for the elements heavier than Gd are generally greater than the theoretical λJ. The excess results from partial polarization of the p–d and s character conduction bands by the exchange field. The ordering and spin transition temperatures shown in Table 4.1 were generally derived from magnetic measurements but have been found to agree within experimental error with those obtained from neutron diffraction, transport and specific heat measurements. The light rare earths were omitted from the table since they have many more complexities in ordering temperatures and moment structures which are better left to the discussion in the text.

4.2.1. Magnetization—Heavy Rare Earths

The original single crystal magnetization measurements on the series Gd-Er made by Legvold and associates in the period 1958–63 at Iowa State University remain the essential basic data on these metals. Later results extended to higher fields have elucidated the field dependence of the complex spin structures particularly in Er and Tm.

Gadolinium[2] is the only one of the series with no orbital momentum contribution ($L=0$) to the total moment. Magnetization curves are typically characteristic of an almost isotropic ferromagnet. Small anisotropy effects (see next section) are seen at low field with a temperature-dependent variation in easy magnetic direction. The temperature dependence of the b axis magnetization curve is shown in Figure 4–1 and is reasonably well defined by the $S=7/2$ Brillouin function. The saturation magnetization is represented by a T^2 dependence from 0 K up to about 150 K as expected on a spin wave model.[3] The $T=0$ value of the saturation magnetic moment per gram is 268.2 emu/g corresponding to 7.55 Bohr magnetons and is the lowest of the series Gd-Er due to the S-state character of the ion. Theoretical saturation λJ equals 7.0β, and thus the excess moment of over one half Bohr magneton represents a polarization of the conduction electrons.

As shown in the figure, departures from Curie–Weiss behavior of the paramagnetic susceptibility were observed below about 370 K due to short

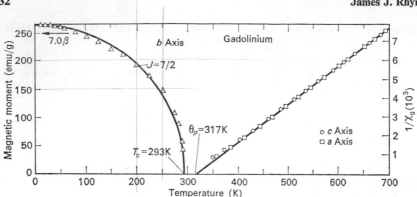

Figure 4–1. Temperature dependence of the spontaneous moment and reciprocal susceptibility in gadolinium. The solid curve is the $S=7/2$ Brillouin function. Departures from
Curie–Weiss susceptibility near θ_p result from short range ordering (after Ref. 2).

range ordering. Effective moment from the paramagnetic susceptibility was
7.98β in good agreement with the theoretical value of 7.94β. Ferromagnetic
and paramagnetic Curie temperatures were isotropic and are given in Table
4.1.

Terbium has the $b\langle 10\bar{1}0\rangle$ direction as the magnetically easy axis at all
temperatures. The spins are rigidly confined to the basal plane by a large two-
fold anisotropy (order of 10^8 ergs/cm³). Magnetization data at high fields[4, 5]
indicate that the susceptibility along $\langle 0001\rangle$ is constant with H and that the
spin direction is lifted from the basal plane only some 8.6° in fields of 70 kOe
at 4 K. Tb exhibits a helical spin structure between 221 K and 229 K which is
quite weak energetically. Magnetic fields of order 1 kOe applied in the basal
plane are sufficient to destroy the helical order and induce ferromagnetism.

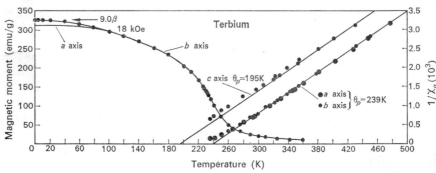

Figure 4–2. Temperature dependence of the magnetic moment in terbium at 18 kOe and of
the reciprocal susceptibility along basal plane and c axis directions. Basal plane
anisotropy prevents alignment of the moment along the a axis at 18 kOe below about
90 K (after Ref. 6).

Results of Hegland *et al.*[6] for magnetizations along $\langle 10\bar{1}0\rangle$ are shown in
Figure 4–2. The isofield data were extracted from a set of isothermal measurements made in fields up to 18 kOe. The raw data plotted as a function of $1/H$

and extrapolated to $1/H=0$ gave an absolute saturation moment of 328 emu/g or 9.34β with a temperature dependence proportional to $T^{5/2}$. It should be noted that a strict extrapolation of finite temperature moment data to infinite field $(1/H=0)$ would suppress the thermal fluctuations and yield the 0 K saturation moment for all temperatures. In practice, however, such a technique is experimentally useful at low temperature $(T\ll T_C)$ for eliminating the effects of small anisotropies and residual domain effects.

The unusual $T^{5/2}$ dependence found for Tb is modified to the more familiar $T^{3/2}$ when one includes the effect of the large ($\simeq 20$ K) anisotropy gap on the spin wave spectrum. This gives a temperature dependence of the form

$$\Delta M = M(T) - M(0) = AT^{3/2}e^{-\Delta/kT} \qquad (4.1)$$

as shown by Niira.[7] The power of T preceding the exponential is dependent on the spin wave dispersion relation (e.g. a two-dimensional spectrum yields a T^2 dependence—see reference 11 and Chapter 5). Experimentally the exponential dominates the behavior at low temperatures. Figure 4–3 shows the result

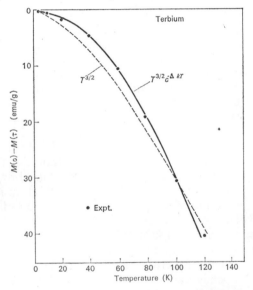

Figure 4–3. Experimental saturation moment of terbium compared to temperature dependence found from conventional spin wave theory with and without the effect of anisotropy gap Δ. The theoretical curves were matched to experiment at 100 K (after Refs. 6, 7).

(4.1) fitted to the experimental saturation moment of Tb and also the conventional $T^{3/2}$ spin wave dependence. The value found for the spin wave gap $\Delta/k=20$ K is in accord with direct observation by inelastic neutron scattering[8] and close to the value of 21 K calculated from anisotropy and magnetoelastic data[9] using the "frozen lattice" model as discussed in Chapter 5. The analysis shown in Figure 4–3 assumed Δ independent of temperature. In fact the gap is given to lowest order (neglecting magnetoelastic corrections) by

$$\Delta = \frac{C}{J}\sqrt{72 K_2 K_6{}^6} \qquad (4.2)$$

where K_2 and $K_6{}^6$ are axial and basal plane anisotropy constants and C is a conversion factor from energy/cm^3 to energy/ion (For Tb 10^8 ergs/cm^3 = 3.2×10^{-15} ergs/ion = 2.0 meV/ion). As discussed in Section 4.3, this dictates a twelfth-power dependence of Δ on the reduced magnetization which would modify the result shown in the figure somewhat.

Susceptibility data of Figure 4–2 show the effect of the extreme uniaxial anisotropy on the paramagnetic Curie temperature. A difference of some 44 K is found between θ_p measured with field parallel and perpendicular to the c axis. These results are analyzed for the effective 0 K anisotropy energy in a later section. Short range order effects on the susceptibility are evident for about 50 K above the Néel temperature.

Belov and Ergin[10] have shown by a critical phenomenon analysis of moment data near T_C that there is an effective lowering of the ferromagnetic Curie temperature for spins along the c axis direction of about 30 K due to the giant two-fold anisotropy. A similar analysis for Gd showed a difference of 1.5 K for T_C along a and c directions reflecting the much lower anisotropy.

Dysprosium exhibits the same general spin structure as Tb except that the helical state is stable over a much wider temperature range. The $\langle 11\bar{2}0 \rangle$ axis is the easy inplane magnetic direction and, as in Tb below T_C, the spins are constrained to the basal plane by a strong uniaxial anisotropy. The saturation magnetization at 0 K was found by Behrendt *et al.*[12] to be 350.5 emu/g corresponding to 10.2β. The temperature dependence as interpreted by Niira[7] is given from 0 to above 100 K by $T^{3/2}e^{-\Delta/kT}$ with $\Delta/k = 20$ K. This compares to a $\Delta/k = 26$ K calculated from anisotropy and magnetostriction data.

In zero field dysprosium undergoes a first-order transition from helical antiferromagnetic to ferromagnet states at 85 K accompanied by a discontinuous orthorhombic distortion of the crystal. Above 85 K this transition may also be induced by the application of a field in the basal plane as illustrated by the isofield curves of Figure 4–4. The critical field[12, 13] varies almost linearly from 0 at the Curie temperature to 11 kOe at 160 K as shown in Figure 4–5. Above about 135 K the moment collapse from helical to pure ferromagnetism is not fully complete at the indicated critical field, but rather proceeds by a series of field-dependent fan configurations as given originally by Enz,[14] Nagamiya *et al.*,[15] and others and discussed in Chapter 2. The moment vectors oscillate about the field direction from layer to layer in the form $\theta_n = B \sin kz$ (z is parallel to $\langle 0001 \rangle$) where B is field and exchange dependent. Such a fan-type structure in the presence of a finite hexagonal anisotropy is presumably responsible for the shift in easy direction (defined here as the direction along which maximum moment is measured at a field H) from $\langle 11\bar{2}0 \rangle$ to $\langle 10\bar{1}0 \rangle$ directions observed by Jew *et al.*[16] and Bly *et al.* [17] above approximately 130 K.

The inset of Figure 4–4 also shows the characteristic anomaly in the moment curves which, in the zero field limit, determines the Néel temperature. The paramagnetic susceptibility shown in Figure 4–4B again shows evidence of

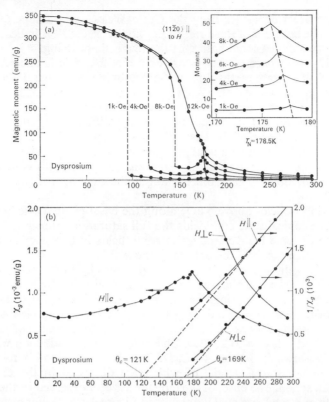

Figure 4–4. A. Isofield moment data in dysprosium showing the abrupt transitions from helical to ferromagnetic states at the critical temperature corresponding to the fields listed. The inset gives the behavior near the Néel temperature. The temperature intercept of the line connecting the peaks defines T_N. B. Susceptibility and reciprocal susceptibility for H parallel and perpendicular to the c axis. The small χ below T_N (178 K) for $H \| c$ reflects the essentially two-dimensional magnetic order resulting from the axial anisotropy (after Ref. 12).

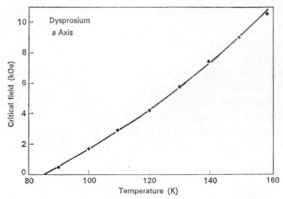

Figure 4–5. Critical fields and corresponding temperatures for the helical to ferromagnetic spin transitions in dysprosium (after Ref. 12).

short range order departure from the Curie–Weiss law at temperatures well above θ_p. The behavior of the c axis susceptibility below T_C is indicative of the essentially two-dimensional nature of the magnetic ordering.

The field dependence of the spin structure of *holmium* is one of the most complex of all the rare earths. In zero field holmium has a basal plane helical antiferromagnetic state existing from the Néel point (135 K) to 20 K. At 20 K the moments lift out of the plane onto a nearly flat cone with a ferromagnetic c axis component of 1.7β. As found from neutron diffraction results of Koehler *et al.*[18] the turn angle of the basal plane component of the moment below 20 K is 30° per layer. Such a configuration is energetically quite unfavorable in the presence of the large basal plane anisotropy, and to compensate, a preferential distribution of moments about the easy b $\langle 10\bar{1}0 \rangle$ directions is observed.

Application of fields above 5 kOe along the basal plane b axis below 20 K collapses the conical state and yields the full saturation moment of 10.3β.[19] Due to the extreme basal plane anisotropy, fields in excess of 300 kOe are required to align the full moment along the a axis.

The magnetization process for $T > 20$ K proceeds by a rather complex series of fan or incompletely collapsed helical phases dictated by anisotropy and exchange energy considerations as shown by Nagamiya *et al.*[15] The number and type of these states depend on both temperature and applied field directions as beautifully illustrated in the neutron diffraction results of Koehler *et al.* [20] taken in fields up to 20 kOe. The magnetic phase diagram which they obtained is shown in Figure 4–6C, D. The Roman numerals refer to the intermediate fan states, details of which are discussed in the chapter on neutron diffraction. The onset of the fan states is also observed in the magnetization data of Strandberg *et al.*[19] shown in Figure 4–6A, B. For example with $H\|b$, above 45 K the $H-T$ phase diagram indicates the transition to ferromagnetism proceeds by two intermediate steps. This is reflected in two distinct humps in the magnetization data, one occurring at about 100 emu/g and the other near 200 emu/g. Between 39 K and the zero field Curie point of 20 K only one transition field is apparent corresponding to the onset of the "helix + ferro →" state.

In the case of $H\|a$, the anisotropy energy stabilizes an additional intermediate state in the range 33 K to 40 K and this results in the kink near 100 emu/g not seen in the b axis data for this temperature range.

Magnetizations with $H\|c$[4, 21] have the effect of closing up the conical or helical configurations finally producing ferromagnetic alignment along $\langle 0001 \rangle$ for fields near 150 kOe. This process is illustrated in the 4 K data of Figure 4–7[21] which also show the large anisotropy of the basal plane directions. Some hysteresis was observed in the c axis data.[4] The sharp rise in moment about 120 kOe was also detected in pulse field studies by Flippen.[13]

Paramagnetic susceptibility data show Curie–Weiss behavior and give anisotropic θ_p values as given in Table 4.1.

A reversal in the sign of the $4f$ quadrupole moment for *Erbium* relative to Tb, Dy, and Ho causes the zero applied field spin direction to lie close to the c axis in this metal. Neutron diffraction results[22] indicate a sinusoidal order of c axis moment components from the Néel point (85 K) to about 53 K with no

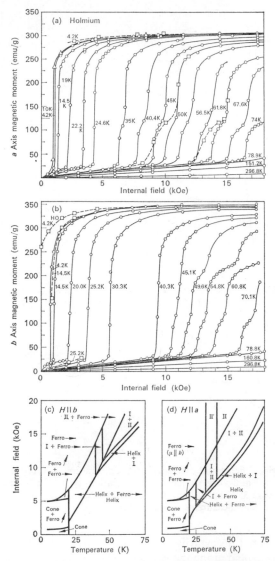

Figure 4–6. A and B. Magnetic moment of holmium measured along a and b axes as a function of internal field. Dotted curves are data taken in decreasing field sequence (after Ref. 19). C and D. The magnetic phase diagram determined by neutron diffraction. Above 20 K the moment curves reflect the appearance of intermediate "fan" states (I, II, etc.) before technical saturation (after Ref. 20).

detectable order of the basal plane component. At 53 K the basal plane orders into a helix and simultaneously a squaring up of the c axis component appears as shown in Figure 4–8B. Below 20 K the moments order onto a cone of half apex angle 30°. This equilibrium angle reflects competing effects of anisotropy and exchange on the lowest free energy state. As in the case of Ho, the stable

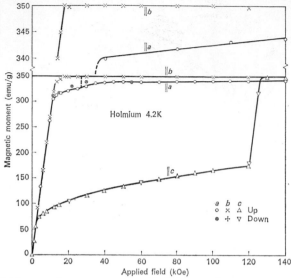

Fig. 4–7. High field magnetic moment curves of holmium at 4.2 K. Basal plane data expanded at the top of the figure illustrate the large basal plane anisotropy (after Ref. 21).

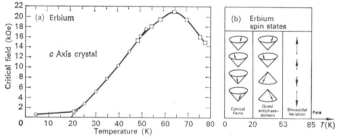

Figure 4–8. A. Critical fields and corresponding temperature for erbium required to transform the "quasi antiphase domain" configuration B to the conical ferromagnetic state. Data above 55 K correspond to transition fields in the sinusoidal moment region found by pulse field measurements (after Refs. 13, 23).

cone angle is thus not necessarily at the anisotropy potential energy minimum.

Application of a field along the c axis in Er serves to close up the cone angle. For $20\,K < T < 53\,K$ this is preceded by a first-order transition from the initial "anti-aligned cone" state of Figure 4–8B to a pure conical configuration (in effect T_C is raised). The critical field for this transformation is linear in temperature as shown in Figure 4–8A from the magnetization data of Green *et al.*[23] and Flippen.[13] The critical field shown for $T < 20\,K$ really corresponds only to the energy required to remove 180° domains and not to a spin state transition.

The magnetization process for H applied in the basal plane is quite complex as shown by the 4 K magnetization data of Rhyne and Foner[4] in Figure 4–9. This situation has been studied theoretically by Kitano and Nayamiya[24] for

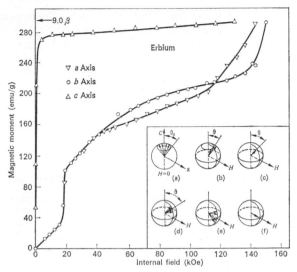

Figure 4–9. High field magnetization curves for erbium at 4.2 K (after Ref. 4). The inset
shows a set of theoretically determined stable spin configurations expected to occur in
the basal plane magnetization process (after Ref. 24).

several cases of varying relative strengths of exchange and anisotropy energies.
These calculations are described in detail in Chapter 2. The qualitative results
appropriate to Er are shown schematically in the inset of the figure. A
correspondence can be made between these calculated states and the moment
data. The collapse of the cone state to a fan lying at an angle near that of the
original cone (b) is manifested in the abrupt rise in basal plane moment at
17 kOe. The a and b axis moment curves are isotropic to 45 kOe. At this field
the moment is 150 emu/g or equal to $M_{sat} \sin 30°$ suggesting that the fan phase
has completely collapsed at this field (frame (c) of inset). The anisotropic
behavior observed for the a and b moment at higher fields is attributed to
directional dependent variations of the intermediate phases (d), (e). Final
ferromagnetic alignment (f) in the basal plane is not reached below 200 kOe
for the b axis direction.

   Thulium at low temperature has a c-axis magnetic order as determined from
neutron diffraction[25] in which the moments are directed up or down in a
three-four sequence. This yields a net moment in zero field of 1.0β at 4.2 K. At
higher temperatures the net unbalanced moment decreases somewhat (relaxation of the squared-up configuration) and above about 32 K the neutron
results of Brun[26] show a breakdown in the fundamental seven-layer periodicity. Above 56 K thulium shows no magnetic order.

   The magnetization results of Richards and Legvold[27] show that the
essentially ferrimagnetic structure below 32 K can be uncoupled by the
application of a field along $\langle 0001 \rangle$ yielding a ferromagnetic moment of 7.1β
with only a small high field susceptibility as shown in Figure 4–10. The critical
field is essentially temperature independent up to about 25 K, then exhibits a
small rise, and finally falls to zero near the Néel point. The high field suscepti-

bility above 25 K is large and shows considerable field dependence, indicating a field-induced variation of the moment structure.

The basal plane of Tm is magnetically very hard. Richards and Legvold's magnetization data show a moment of only 25 emu/g is developed along the b axis in a 100 kOe applied field. Early results of Foner *et al.* [28] indicated a transition field of less than 20 kOe to a state with a moment of 220 emu/g. This shift was found to be the result of an internal strain re-orientation of the crystal resulting from the high anisotropy torques on the lattice in fields over 100 kOe. Similar effects have been observed in c axis magnetizations on Tb, Dy,[4, 29] accompanied by permanent macroscopic distortions in shape amounting to several percent. Preliminary X-ray work on Dy[30] indicates a strain induced twinning mechanism may be responsible for the observed defects.

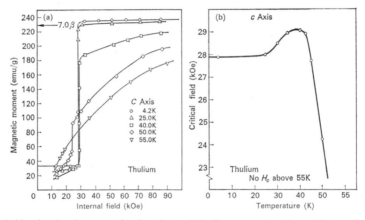

Figure 4–10. A. C axis magnetization data of thulium as a function of internal field for temperatures from 4.2 K to 55 K. B. The temperature dependence of the critical fields required to uncouple the 4↑3 spin structure (after Ref. 27).

Above the Néel temperature Tm has a Curie–Weiss susceptibility with an effective moment of 7.6β. The paramagnetic Curie temperature intercept is very anisotropic giving values of -17 K for H along $\langle 10\bar{1}0 \rangle$ and $+41$ K for H along $\langle 0001 \rangle$.

Thulium is the last magnetically ordered element of the heavy rare earth series. The light rare earths and non-magnetic elements are discussed in the next section.

4.2.2. Magnetization—Light Rare Earths, Yb, Sc, Y, and Lu

In some of the light rare earths the effect of the crystal electric field on the moment is appreciable. In this series the radius of the $4f$ electrons is larger and the resulting enhanced crystal field energy is of comparable magnitude to the exchange energy. This is in contrast to the heavy rare earths where crystal field effects are a small perturbation on the exchange and serve only as a source of magnetic anisotropy, small compared to the exchange.

Lanthanum does not order magnetically and its susceptibility results are complicated by the presence of both fcc and double hcp phases co-existing below room temperature. The purest lanthanum to date prepared by Finnemore et al.[31] was found by analyzing specific heat data[32] to be 96% d-hcp. The superconducting transition temperatures of the d-hcp and fcc phase differ by more than 1 K (4.87 K and 6.0 K respectively) and this makes it possible to use the relative height of the specific heat anomalies at T_C to determine the amount of each phase present. Susceptibility measurements by Wohlleben[33] shown in Figure 4–11 and by Finnemore et al.[31] showed a nearly temperature independent nature with a value $\chi = 0.7 \times 10^{-6}$ emu/g at 300 K. The latter study showed less than a 0.5% variation in χ from 5 K to above 80 K. Previous results[34] indicating a significant temperature dependence probably were the result of impurities. Approximately a 1% decrease in χ was observed in changing the measuring field from 2.3 kOe to 3.8 kOe.

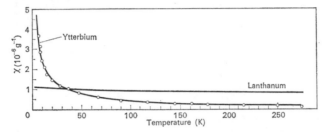

Figure 4–11. Temperature dependence of the magnetic susceptibility of single crystal ytterbium (after Ref. 46) and of polycrystalline lanthanum (after Ref. 33).

Neutron diffraction results on *Cerium*[35] show that below about 100 K the room temperature structure which is a combination γ-fcc and d-hcp cells partially transforms to a distorted α-fcc' crystal cell with a simultaneous promotion of the $4f$ electron to the $5d$ shell. The α-fcc' phase thus has no atomic moment. The $4f$–$5d$ electronic transition decreases the screening of the conduction electrons from the nucleus resulting in tighter outer electron orbits and an observed decrease in unit cell volume. The degree of completeness of the α-fcc' transformation is inhibited by repeated thermal cycling. The transition also exhibits considerable thermal hysteresis. These phenomena are reflected in the susceptibility measurements of Lock[34] shown in Figure 4–12. The Curie–Weiss susceptibility above 100 K from Lock's cooling curves gives an effective moment $\mu_{eff} = 2.51\beta$ almost exactly equal to the predicted free ion value. On the initial cool-down below 100 K the susceptibility is significantly reduced indicating partial transformation to the zero moment α-fcc' phase. The extrapolated paramagnetic Curie temperature varied from -46 K for the initial cooling to -38 K for the 100th cooling.

A time dependence of χ reflecting the metastable γ-fcc to α-fcc' transition was discovered by Burr and Ehara.[36] Over a period of 8 hours after initial cooling to 4.2 K, the susceptibility increased almost a factor of three. These researchers also measured the susceptibility at high temperatures through the melting point and found no evidence of a $4f$ to $5d$ electron transfer which had been proposed to explain the unique negative slope of the fusion curve with pressure in Ce.

The magnetic state of Ce metal can be described as a virtual $4f$ level lying close to the Fermi energy.[36a] Susceptibility results as a function of pressure by MacPherson *et al.*[36b] indicate that with increasing pressure this level approaches the Fermi level producing the observed gradual decrease in χ. At

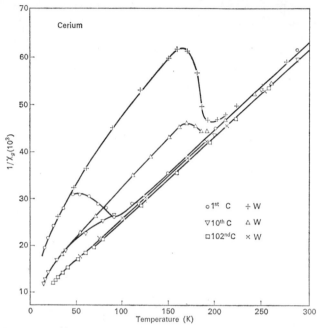

Figure 4–12. Reciprocal susceptibility of polycrystalline cerium as a function of temperature. The extreme hysterisis observed both between warming (W) and cooling (C) curves and between the 1st and 102nd runs is the result of an incomplete transformation to the α-fcc' crystal structure at low T (after Ref. 34).

the $\gamma - \alpha'$ transition ($\simeq 8$ kbars at room temperature) the $4f$ level abruptly becomes spin degenerate and the susceptibility drops to a value characteristic of an exchange enhanced quadravalent metal. On further increase of pressure the susceptibility is lowered approaching that of Th or Hf (4 valent) near 40 kbars.

Praseodymium is a singlet ground state ion with a cubic and hexagonal ion site symmetry[37] similar to Nd. Pr[34, 38] exhibits Curie–Weiss behavior of the susceptibility on a polycrystalline sample from above room temperature to below 100 K. The effective moment is 3.56β to be compared with a free ion theoretical value of 3.58β. Below 100 K the susceptibility is slightly higher than the Curie law value, and below 4 K is nearly constant at a value of 1.55×10^{-3} emu/g. Bleaney[39] calculated that a spontaneous moment would not occur in Pr because the ratio of exchange to crystal field parameters was below critical value required. However, Cable[37] *et al.* found on a polycrystalline sample that an antiferromagnetic state similar to Nd was developed below 25 K in which the maximum modulated moment was considerably less than the theoretical 3.2β.

In contrast to these results, very recent magnetization and neutron diffraction work on Pr[40] show that in a single crystal sample there is no net intrinsic moment (i.e., $\langle J_z \rangle = 0$) due to crystal field effects. The discrepancy in single and polycrystal data may arise from a slight change in the crystal field of the polycrystalline material due to strain or impurities. In addition, the application

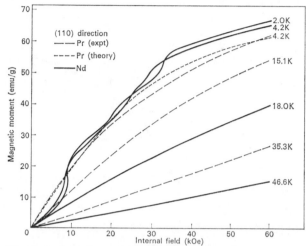

Figure 4–13. Magnetization data to 60 kOe at several temperatures for single crystal praseodymium and neodymium. The dotted line represents the Pr moment calculated on a crystal field model (after Ref. 40).

of only a small field to the single crystal produced an ordering of moments reaching 1.8β on the hexagonal site and 0.9β on the cubic site at 4.2 K and 46 kOe. Magnetization results in the $\langle 0001 \rangle$ direction (not shown in Figure 4–13) revealed that χ_c is about 10 times smaller than the susceptibility in the basal plane. This effect is opposite to that predicted by the crystal field work of Bleaney.

Susceptibility data of Nagasawa and Sugawara[41] on Pr, Pr-Ce, and Pr-La alloys indicate crystal field parameters corresponding to an overall splitting of the crystal field levels of order 50 K. Their polycrystalline susceptibility and transport results were consistent with the model in which the Pr^{3+} ions on the cubic sites carry no magnetic moment while those on the hcp sites order in a sinusoidal moment state below 25 K.

Neodymium has a hexagonal structure with a stacking sequence ABAC. In effect then atoms on alternate layers have either a hexagonal or face centred cubic environment. The Nd^{3+} is a Kramers ion and therefore the crystal field levels are doubly degenerate. Neutron diffraction studies of Moon *et al.*[42] show that the hexagonal sites order at 19 K in an antiferromagnetic arrangement. Cubic sites order at 7.5 K. The respective moments on the two sites are 2.3β and 1.8β. Such a complex ordering leads to somewhat confusing results for magnetization measurements. Inverse susceptibility data of Behrendt *et al.*[43] plotted vs. T showed a slight kink near 145 K and were fitted by two straight lines of slightly different slope. It is probable that the data are also representable by a single slope plus a temperature independent term (Pauli paramagnetism) as used in the polycrystalline work of Lock:[34]

$$\chi = \frac{9.47}{T-4.3} + 5.0 \times 10^{-6} \text{ emu/g.}$$

The temperature-independent term may be enhanced due to mixing of the closely spaced $J=9/2$ and $J=11/2$ $4f$ electron levels. In the single-crystal

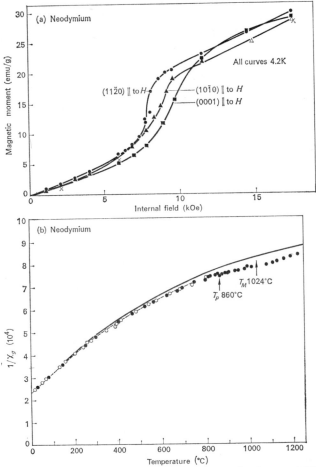

Figure 4–14. A. Single crystal magnetization data on neodymium at 4.2 K illustrating anisotrophy effects in the neighborhood of 8 kOe corresponding to a spin rearrangement described in the text (after Ref. 43). B. High-temperature susceptibility of neodymium showing no change in χ at the cubic phase change temperature (T_p) or at the melting temperature T_M. The solid curve is the result of a crystal field moment calculation (after Ref. 47).

results an anisotropy was found between basal plane and c axis susceptibility amounting to 5 ± 1 K in θ_p. The paramagnetic Curie temperatures found from the low temperature slope were $\theta_c = 0$ K, $\theta_a = +5$ K with identical effective moment $\mu_{eff} = 3.45\beta$. Theoretical effective moment for the ${}^4I_{9/2}$ ion is 3.68β. The reduced observed moment results from crystal field mixing of J levels. Magnetization data at 20 K showed some nonlinear variation of moment versus field up to 18 kOe. At 4.2 K clear evidence of a spin transformation near 8 kOe was present (see Figures 4-13 and 4-14). Moment anisotropy between c axis and basal plane was also present at all temperatures but none was observed in the basal plane above 20 K. The single crystal data of Behrendt did not show the dramatic rise in moment of some 20 times observed below 8 K in polycrystalline results of Lock.[34] This latter effect may have been associated with impurities.

Recent magnetization and neutron diffraction work on single crystals of Nd by Johansson et al.[40] show that at 4.2 K in low applied magnetic fields the moments on hexagonal sites turn parallel to those on the cubic sites saturating about 12 Oke. A ferromagnetic moment develops, which on cubic sites is roughly 2.5 times that on the hexagonal sites. At 23 kOe a sharp rise in ferro and periodic moments, resulting from a level crossing, brings the total moment on the hexagonal site close to the theoretical maximum. Above this field the metal becomes increasingly ferromagnetic with the periodic moment disappearing completely at 31 kOe. The field variation of bulk moment is shown in Figure 4-13.

High-temperature susceptibility results of Arajas and Miller[44] on Nd shown in Figure 4-14 show no anomalies at either the crystal phase change at 1133 K or at the melting point 1297 K indicating minimal interaction between $4f$ ions at this temperature. The χ data over most of the range from room temperature to 1500 K agree quite well with values calculated from a crystal field model as shown.

Samarium has a very complex hexagonal atomic stacking arrangement with a nine layer repeat sequence. Polycrystalline susceptibility results of Lock[34] and Jelinek et al.[45] revealed an anomalous kink near 106 K and a sharp decrease in susceptibility beginning at 14 K as shown in Figure 4-15. Recent single crystal data of Foner et al.[46, 46a] show that the anomalies occur only in the c axis susceptibility, while the basal plane susceptibility remains essentially constant from 4 K up to 14 K and then slowly decreases. Neutron diffraction results of Koehler et al.[46b] confirm that the spins on hexagonal sites only order antiferromagnetically at 106 K and at 14 K the cubic sites order into ferromagnetic layers. Sm has a $J = 7/2$ multiplet level lying only about 0.1 eV above the $J = 5/2$ ground state. Thus the magnetization departs from a Curie law at high temperatures. Susceptibility results from room temperature to 1400 K by Arajs[47] were compared to the results of a crystal field calculation. Significant departures were observed in contrast to the good agreement obtained for Nd.[44] Only a small change in χ was observed at either the 1190 K crystal phase change or at the melting point of 1325 K.

Europium is a body centered cubic structure helical antiferromagnet. The moment from neutron diffraction measurements[48] is 5.9β and lies in the $\{100\}$ planes with an inter-layer turn angle of $50°$. The Néel point is 89 K with the moment falling discontinuously to 0 from about 0.4 of the saturation value. Magnetization measurements on a single crystal by Schieber et al.[46] at 4.2 K show a linear variation of moment with applied field. The moment is identical along [100], [111], and [110] directions and has a value of 3β, at 140 kOe, less than half theoretical saturation of 7β for a Eu^{2+} ion in a $^8S_{7/2}$ state.

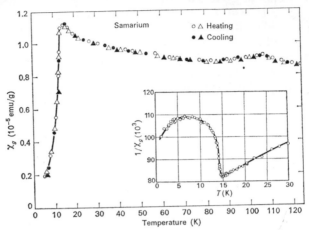

Figure 4–15. Susceptibility of polycrystalline samarium as a function of temperature. The inset shows the anomalous behavior of $1/\chi$ in the region below 14 K. As indicated by the data, no thermal hysteresis effects were observed (after Refs. 34 and 45).

Early paramagnetic susceptibility work by Bozorth and Van Vleck[49] on polycrystalline Eu gave an effective moment $\mu_{eff} = 8.3\beta$. This was about three times higher than the theoretical value for a free Eu^{3+} ion, which was the then indicated state on the basis of low temperature measurements, and led to a postulated electronic transition from Eu^{2+} to Eu^{3+} at the Néel point. Recent isomer shift measurements of Cohen et al.[50] do not substantiate the occurrence of any charge transfer and are consistent with the Eu^{2+} model at all temperatures. Their results attribute the discontinuous drop in moment at T_N to a first-order structure distortion.

Schieber et al.[46] determined in single crystal studies that *Ytterbium* was very weakly paramagnetic. Their results indicated no anisotropy in the moment and gave susceptibility results which were significantly smaller than those of Lock[51] and more nearly T independent. Even as low as 30 K χ was less than the value for pure La[31,33] (see figure 4–11). Thus Schieber et al. concluded that the paramagnetism of Yb could result from impurities and that completely pure Yb might be diamagnetic. Such is in fact the case in recent results by Bucher et al.[51a] on very high purity Yb metal which exhibited a first order paramagnetic to diamagnetic transition accompanying a fcc→hcp martensitic transformation. The transition occurred in the range 100 K to 360 K and exhibited a large hysteresis. The low-temperature phase was pure hcp, a result found also from recent X-ray investigations.[51b] The conventional fcc form could be recovered from the hcp phase by plastic deformation. The observed susceptibility and small volume change at the transition were consistent with a conversion of about 0.8% Yb^{2+} ($^1S_0$) to Yb^{3+}($^2F_{7/2}$). Lock[51] had originally proposed such a transfer model to explain the temperature dependence of his results in fcc Yb.

The magnetic susceptibilities of *Scandium* and *Yttrium* are small and quite sensitive to impurity levels. A very exhaustive study by Wohlleben[33] on Sc using more than twelve independent specimens showed that in general the purer samples exhibited the highest susceptibility, being some 20% higher than the best of the confusing assortment of earlier measurements. Figure 4–16 shows the small almost linear variation of χ from 5 K to room temperature. Below 5 K an anomaly develops in which χ rises almost 10% at 0.4 K, an effect shown to be intrinsic to Sc and not due to impurities. The value of χ at 0.4 K is 8.34×10^{-6} emu/g. The susceptibility of Sc is slightly anisotropic as found by Chechernikov *et al.*[52]

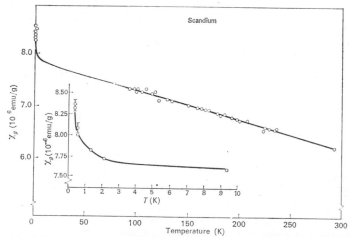

Figure 4–16. Susceptibility of scandium as a function of temperature. The inset illustrates the dramatic rise in χ below 3 K which is apparently not the result of impurities (after Ref. 33).

Yttrium susceptibility is also sensitive to impurities but not to the extent of Sc. The extensive work of Wohlleben[33] gave a value of 1.94×10^{-6} emu/g at 0 K with a weak linear temperature variation $d\chi/dT = +3 \times 10^{-10}$ emu/g K. Earlier results indicating a larger temperature variation were probably due to impurities. Wohlleben's results indicate that χ is lowest in the purest samples opposite to the case in Sc. The results of Gardner and Penfold[53] from 1 to 800 K agree well with the above results and show a broad maximum near 300 K reminiscent of the behavior in Pd.

Lutecium completes the rare earth series and thus has a filled $4f$ shell and is non-magnetic. The susceptibility at room temperature is of order 10^{-7}/g.

4.2.3. Magnetization—Intra-Rare Earth Systems

Due to the wide variety of spin structures and other effects found between elements of the rare earth series, the study of intra-rare earth alloys has natural appeal as a means of examining basic physical mechanisms. These researches can take several directions depending on the motivation. The concept that the long range magnetic interactions are mediated by the conduction electrons[54-56] leads to the natural investigation of ferromagnetically ordered rare earths diluted with non-magnetic metals which have similar electronic (Fermi surface) properties. Such is the work on RE-Y, RE-Sc, RE-La, and RE-Lu systems. These have led to the discovery of significant rare earth moment enhancement especially in the Sc alloys[33] and anomalies in the concentration dependence of the ordering temperatures as discussed later.

A second fertile area of investigation in rare earth alloys concerns the crystal field anisotropy and spin structures particularly in alloys of the heavy rare earths. For bulk studies this requires single crystals and thus progress has not been rapid. The major systems investigated by magnetization methods are Dy-Er,[21] Tb-Ho,[57] and Ho-Er.[21] In particular the latter system is of interest since it combines elements having opposite sign for the dominant Stevens' factor α (see magnetic anisotropy, Section 4.3), which is a measure of the degree of departure from spherical symmetry of the $4f$ charge cloud. As a consequence of this shape factor the moments in pure Ho favor alignment in the basal plane while Er moments align on a cone near the c axis. Thus, if single ion anisotropy is the dominant interaction in determining the equilibrium spin direction, one would expect in a HoEr alloy that spins on Er sites would align independently and at an angle approximately 60° from those on the Ho sites. Neutron diffraction[58] and magnetization[21] studies show that this is not the case and that all spins pick an intermediate angle of 52° independent of site. This indicates that it is a combination of exchange and anisotropy energies which determine the stable spin directions.

Magnetization measurements to 140 kOe on a single crystal specimen of HoEr at 4.2 K are shown in Figure 4–17, and illustrate the evolution of the magnetization process. One notes the considerable similarity between the general features of these curves and those for pure Er in Figure 4–9 indicating

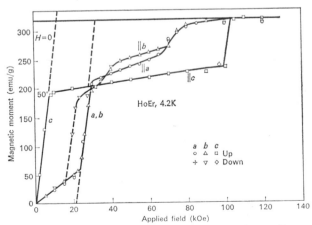

Figure 4–17. High field magnetization data for a single crystal $Ho_{0.5}Er_{0.5}$ alloy at 4.2 K. Evidence is present for several stable intermediate "fan" moment states as in pure Ho and Er (after Ref. 21).

that the magnetization sequence may be similar. The principal difference is that the initial configuration corresponds to the broader 52° cone angle instead of the 30° angle of Er. The respective spin transition fields are different in the alloy due to the averaged anisotropy, and a sharp moment rise to saturation at 100 kOe is observed along the c axis which is characteristic of Ho but not observed in Er to 200 kOe. Final theoretical saturation is achieved along all three axes in HoEr at 105 kOe indicating an overall anisotropy field only half that of pure Er.

The general investigation of intra-rare earth alloy systems is aided greatly by the "generous cooperation" of most of the elements of the series in forming complete solid solutions with one another. This enables the study of the magnetic transition temperatures of a wide range of alloys. Some of these results, largely on polycrystalline material, are shown in Figure 4–18. This monumental study is primarily the result of the work of Bozorth[59] and collaborators except for Tb-Sc[60] and Gd-Sc[61]. The highest magnetic ordering temperature (T_N, or T_C for Gd rich alloys) is plotted against the two-thirds power of the average effective spin as given by the DeGennes[62] factor (see (1.24) and Table 1.1). The ordering theory of DeGennes predicts that the Néel temperatures should vary linearly with G. The origin of the two-thirds

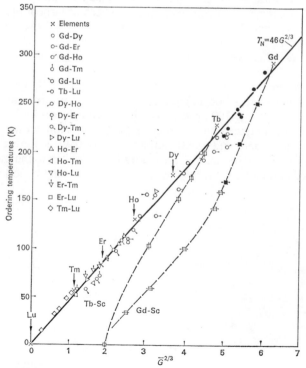

Figure 4–18. Ordering temperatures for a wide spectrum of intra-rare earth alloys plotted as a function of the two-thirds power of the average DeGennes factor (see text). All orderings are antiferromagnetic except for the ferromagnetic ordering of highly concentrated gadolinium alloys indicated by the solid symbols (after Ref. 59).

power relationship found in the moment experiments is obscure. High pressure experiments do show a linear dependence on G. For the alloys an average factor $\bar{G}$ was calculated from the expression:

$$\bar{G} = \sum_i C_i G_i, \qquad (4.3)$$

where C_i is the concentration of the element having DeGennes factor G_i. Note that significant departures from the $G^{2/3}$ relationship were observed in the Sc alloys. These are also the systems which exhibit large moment enhancement effects[33] as discussed below.

In contrast to the uniform behavior shown for T_N the Curie temperatures for the antiferromagnetic elements and alloys show no uniform relationship to $\bar{G}$, but are distinctly characteristic of the particular alloy system. These data are summarized in the paper by Bozorth.[59]

The magnetic interactions observed in alloys of Sc and Y with the heavy rare earths are extremely anomalous. In these alloys the basic RKKY mechanism would indicate a linear dependence of the ordering temperature on the concentration of the magnetic impurity. The interaction temperature would then vanish only in the limit of zero impurity concentrations. In fact, however, the experiments on GdSc by Nigh et al.[61] and TbSc by Child and Koehler[60] show that T_N for the alloys drops essentially to zero at large finite concentrations (15% for Gd and 25% for Tb) as shown in Figure 4–19A where the data are plotted against the average DeGennes factor. The vanishing of T_N at the same value of $\bar{G}$ indicates this anomaly is a property of the Sc host matrix and is independent of the particular rare earth impurity.

Gd-Sc alloys in the concentration range 15% to 69% Gd are antiferromagnetic and above this concentration range they are ferromagnetic. Very dilute alloys in the range 100 ppm to 1% have been shown by Wohlleben[33] to also exhibit antiferromagnetic order with a saturation moment which is concentration dependent and greatly enhanced over the free ion value of 7.0β. In a 79 ppm Gd alloy measured at 0.4 K and $H/T = 30$ kG/K the moment was fully saturated and had a value 9.9β; some 40% above the free ion value. This

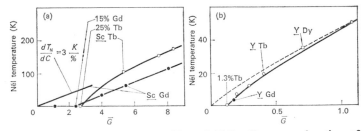

Figure 4–19. A. Néel temperature of GdSc and TbSc alloys as a function of average DeGennes factor. The line marked dT_N/dC was derived from saturation moment data at very dilute concentrations below the anomalous critical concentration for vanishing T_N (see text). (After Refs. 33, 60, 61.) B. Néel temperature of TbY, GdY, and DyY alloys also plotted as a function of average DeGennes factor. The dashed curve is the $\bar{G}^{2/3}$ dependence observed for a number of magnetic alloys in Figure 4–18 (after Refs. 64–66).

is a record enhancement for a rare earth impurity in any matrix. The study indicated a concentration dependence of the interaction temperature of 3 K/% Gd, exactly the same as given for T_N in the high concentration results of Nigh. Wohlleben has developed an oscillatory molecular field argument to explain the origin of the mysterious saturated moment behavior at a Gd concentration well below the critical value for vanishing T_N. Studies of Dy-Sc alloys by Isaacs et al.[61a] show similar moment enhancements to those in the GdSc system. The moment behavior is not affected by the non S-state character of the impurity.

Results on yttrium alloys are only slightly less anomalous. In this case, results in Figure 4.19B on Tb-Y by Nagasawa and Sugawara[63] and Child et al.,[64] on Dy-Y[65] and on Gd-Y[66] also show a vanishing of T_N at a finite but much lower concentration than in the Sc alloys. The difference in critical concentration for the vanishing of T_N in Sc and Y is related by Wohlleben to the relative range of the RKKY interaction in the two metals. As in Sc, the very dilute concentrations studied showed magnetic ordering with an effective moment enhanced by some 0.6β. The results of Nagasawa and Sugawara on a 3.7% Tb alloy gave $\theta_p = 0$ although a minimum in the χ^{-1} vs. T plot appeared corresponding to a $T_N = 19$ K. This lack of correlation in ordering temperatures is consistent with the Wohlleben model.

The effect on the susceptibility of the introduction of heavy rare earths to La has been studied extensively by Finnemore et al.[31] In addition to sharply suppressing the superconducting transition temperature, the inclusion of dilute amounts of Gd (order of 1 to 6 at. %) to La produced a Curie–Weiss behavior of the susceptibility for T well above θ_p and a linear dependence of θ_p on concentration. The effective moment per Gd ion at concentrations less than 0.5% was quite near the free ion value of 7.94β rising gradually to 8.5β at 1% Gd and above. This indicates a small band polarization effect of about 0.5β.

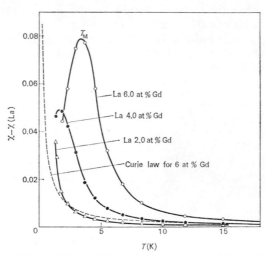

Figure 4–20. Susceptibility of dilute GdLa alloys at low temperature. The susceptibility of pure La has been subtracted. Strong enhancement effects in χ around $T \sim 0.5 \; \theta_p$ indicate magnetic order (after Ref. 31).

6

At a temperature about half θ_p a strong peak in χ was observed as shown in Figure 4–20 indicating some form of magnetic ordering. The peak height was quite field dependent. Application of a field of only 200 Oe reduced the maximum χ by one half and shifted it to lower temperatures. The introduction of Tb impurities to La[31] produced susceptibility effects qualitatively different from Gd. In this instance no evidence of long range order was observed by neutron diffraction.[31]

The preceding review of rare earth alloy systems is by no means intended to be comprehensive. Due to space limitations only a few representative examples were cited from the great wealth of alloy and compound data available in the literature.

4.2.4. Magnetization at Elevated Pressure

Magnetization measurements in the vicinity of the ordering temperatures have been made as a function of pressure on single crystals of Gd, Tb, and Dy by Bartholin and Bloch.[67] The pressure range studied was up to 5 kbars and the variation in moment was detected by a mutual inductance technique. The single crystal results were not in complete agreement with polycrystalline data which have shown a considerable variation depending upon material quality and experimenter. The general features of the results show that the transition temperatures (T_N and T_C) decrease in a completely linear fashion with temperature. Their analysis [67] suggests that the variation in ordering temperature arises from the variation of exchange and anisotropy interactions with lattice parameter, in particular the variation of exchange $\mathscr{I}(Q)$ with the c axis lattice constant. Figure 4–21 shows the observed moment change (proportion to mutual inductance) for Gd as a function of temperature for several pressures. Data are given for the a.c. magnetic driving field applied along b and c directions.

The slope $\partial T/\partial p$ of the linear change in T_N and T_C with pressure observed for the single crystals of Gd, Tb, and Dy is tabulated in Table 4.2. There was some thermal hysteresis in the values of the ordering temperatures in Tb and Dy but none was observed in Gd. Also included in the table are earlier values on polycrystalline Ho,[68] Er.[69]

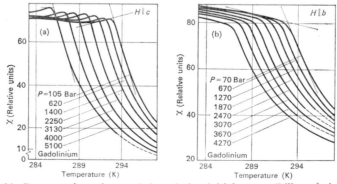

Figure 4–21. Pressure dependence of the relative initial susceptibility of single crystal gadolinium along c and b axes in the vicinity of the Curie temperature (after Ref. 67).

Except for Ce, Eu, and Tb, the crystal structure of the rare earth series La-Lu evolves in a regular progression from fcc (La for $T \geq 300$ K) through d-hcp (Pr, Nd) and Sm structure to hcp (Gd-Lu). The pressure studies to 85 kbar by McWhan and Stevens[68] and others indicate that the above fcc to hcp sequence occurs also with increasing pressure but in reverse order. It is postulated that the structure sequence is a result of a decrease in degree of localization of the 4f electrons with increasing pressures or decreasing volume.

Table 4.2. Linear Variation of Ordering Temperature with Pressure for Single and Polycrystal Rare Earth Elements

(After References 67–69)

Element	Sample	$\partial T_N/\partial p$ (K/kbar)	$\partial T_C/\partial p$ (K/kbar)
Gd	Single	—	−1.40
Tb	Single	−0.84	−1.24
Dy	Single	−0.41	−1.27
Ho	Poly	−0.48	—
Er	Poly	−0.26	−0.8

X-ray and initial susceptibility measurements on hcp Gd[68] indicate a metastable change to the Sm structure occurring in the range 25 to 52 kbars. The transition to the Sm structure is very sluggish and remnants of the low-pressure phase were observed some 30 kbars above the initial transition pressure. Likewise, on lowering the pressure, vestiges of the Sm structure remained down to 10 kbars in Dy ($P_{\text{trans}} = 75$ kbars). Other hcp rare earths and alloys show similar transitions but at widely varying pressures. Studies on Sm[70] and alloys of light and heavy rare earths indicate that the intrinsic Sm-type structure reverts to the dhcp structure at high pressures.

The effect of the structure changes on the magnetic susceptibility is quite pronounced. Figure 4–22 shows relative susceptibility data on Dy as a function of temperature for several pressures. The peak in the 29 kbar curve corresponds to the ordering temperature T_N. At higher pressures two additional transition temperatures become evident corresponding to the appearance of the Sm-type phase. The pressure variation of both the hcp and first (highest T) Sm-phase ordering temperatures are plotted in Figure 4–22B for the elements Gd-Ho. In all cases the T variation is linear with pressure. The heavy dotted lines indicate the hcp ordering temperature in the co-existence region. The transition to the Sm-type structure is accompanied by only a 1% volume change. The approximate 10% drop in ordering temperature of the Sm structure is thus apparently due to the symmetry change and is not just a volume effect.

It is noted that the appearance of the Sm type structure is accompanied by a disappearance of ferromagnetic order (e.g., in Gd or in $\text{Tb}_x\text{Y}_{1-x}$ alloys with $x > 0.8$) in favor of an antiferromagnetic order. It is further suggested by McWhan and Stevens' results that in the induced Sm (also d-hcp) structure the antiferromagnetic ordering occurs at different temperatures on the cubic and hexagonal sites in analogy with Nd.

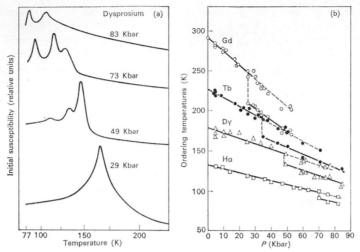

Figure 4–22. A. Relative initial susceptibility of dysprosium as a function of temperature at
several pressures. The additional peaks occurring at the higher pressures correspond to
ordering temperatures in the pressure induced Sm-type phase. B. Pressure dependence
of highest ordering temperature in the heavy rare earths. The vertical dotted line
indicates the onset of the Sm-type phase. At higher pressures ordering temperature are
shown both for the remnant hcp structure (dotted curves) and for the higher of the two
Sm-phase temperatures (after Ref. 68).

The ordering temperatures as a function of pressure (dT/dp) for the rare
earth elements and alloys[68] have been shown to be linear in the De Gennes[62]
factor $G=(\lambda-1)^2 J(J+1)$ which represents the square of the total effective
spin and which appears in the following molecular field expression for the Néel
temperature (see Chapter 1):

$$kT_N = \tfrac{2}{3}(\lambda-1)^2 J(J+1)\mathscr{J}(Q) + \tfrac{2}{5}K_2(J-\tfrac{1}{2})(J+\tfrac{3}{2})$$
$$= \tfrac{2}{3}G\mathscr{J}(Q) + \tfrac{2}{5}K_2(J-\tfrac{1}{2})(J+\tfrac{3}{2}), \tag{4.4}$$

where $\mathscr{J}(Q)$ is the Fourier transform of the real space exchange constant and
K_2 is the first-order anisotropy term. The linear dependence of the ordering
temperature on G found in the pressure studies is in marked contrast to the $G^{\frac{2}{3}}$
variation of T_N (or T_C) found in neutron diffraction and magnetization results
at 1 atmosphere on a variety of elements and alloys.

Dividing expression (4.4) by G and using data of dT_N/dp allows one to
examine the $\mathscr{J}(Q)$ exchange interaction as a function of volume (pressure).
Such has been done in Figure 4–23 where McWhan and Stevens[68] plot the
derived $\mathscr{J}(Q)$ versus the dimensionless ratio $\sqrt{\langle r_{4f}{}^2\rangle}/V^{\frac{1}{3}}$ for a series of rare
earths and alloys. The quantity $\langle r_{4f}{}^2\rangle$ is the average squared $4f$ wavefunc-
tion.[71] K_2 was assumed to vary as V^{-1}. The slopes shown in the figure are
quite consistent and correspond to an average variation of exchange integral
with lattice parameter

$$\frac{1}{k}\frac{d\mathscr{J}(Q)}{da} = 42 \pm 6 \text{ K/Å}.$$

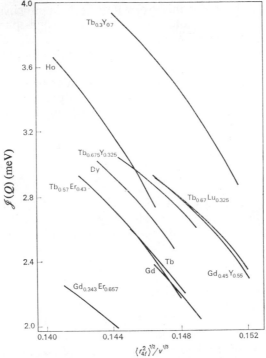

Figure 4–23. Volume dependence of the effective exchange interaction as determined from pressure studies on a variety of rare earth elements and alloys (after Ref. 68).

The variation in $\mathscr{J}(Q)$ is a reflection of the change in generalized susceptibility $\chi(Q)$ with pressure.

4.3. MAGNETIC ANISOTROPY

The magnetic anisotropy energy in many of the rare earth metals is extremely large and, in contrast to the $3d$ transition metals, is a significant perturbation on the exchange. The magnetic anisotropy energy appropriate for hexagonal symmetry can be written in the phenomenological form:

$$E_K = K_2 Y_2^0(\theta, \phi) + K_4 Y_4^0(\theta, \phi) + K_6 Y_6^0(\theta, \phi) + K_6^6 \sin^6 \theta \cos 6\phi. \quad (4.5)$$

Theta and phi are respectively angles of the magnetization with respect to the c (0001) and a (11$\bar{2}$0) axes. The spherical harmonics $Y_l^m(\theta, \phi)$ are normalized to unity for $\theta = \phi = 0$. The first three anisotropy coefficients K_l and accompanying harmonics describe the dominant uniaxial anisotropy, while the term K_6^6 refers to the six-fold basal plane component. In this convention positive values of K_l favor moment alignment in the basal plane and negative values favor the c axis. The minimum in this anisotropy potential energy expression determines the stable magnetization direction (ferromagnetic easy axis) except for the case of a conical ferromagnet where this angle is modified by the exchange energy as discussed later.

The various $l=2, 4, 6$ order anisotropy terms may in principal contain both single ion (i.e., interaction of Coulomb crystal field and non-spherical $4f$ charge cloud distribution) and two-ion (i.e., anisotropic exchange, or dipole-dipole) contributions. For all heavy rare earths but Gd the two-ion contributions are negligible.

The K_l's contain averages of a corresponding order spin polynomial (basis function), and these have a unique temperature dependence related to the magnetization. For the single-ion case this has been calculated by Callen and Callen[72] in the form:

$$K_l(T) = K_l(T=0) \frac{I_{l+\frac{1}{2}}[\mathscr{L}^{-1}(\sigma)]}{I_{\frac{1}{2}}[\mathscr{L}^{-1}(\sigma)]}$$

$$\equiv K_l(T=0)\hat{I}_{l+\frac{1}{2}}[\mathscr{L}^{-1}(\sigma)], \qquad (4.6)$$

valid for both molecular field and spin wave models.[73] $\hat{I}$ is a reduced hyperbolic Bessel function (modified first kind Bessel function of odd half integer order), $\mathscr{L}^{-1}$ is the inverse of the Langevin function and $\sigma = M(T)/M(0)$ is the reduced magnetization. At low T this expression predicts

$$K(T) \propto \sigma^{l/2[l+1]} \qquad (4.7)$$

in accord with conventional theories.[74] At high temperatures (also above T_C) $K(T)$ varies as σ^l.

Temperature dependences of two-ion anisotropy terms are difficult to calculate explicitly. Callen and Callen[75] have shown in a cluster model, that $K \propto \sigma^l$ over a fairly wide range of temperature below T_C. At the very low T limit K is again proportional to $\sigma^{l/2[l+1]}$.

Brooks[74a] has calculated that departures from the $\sigma^{l/2(l+1)}$ power law will occur at low temperature in Tb and Dy. The relationship (4.7) is based on the condition of cylindrical symmetry about the spin quantization axis. In the above metals this symmetry is destroyed by the large axial anisotropy which suppresses spin fluctuations out of the basal plane and leads, in Brooks' treatment, to a modification of the power laws. In particular for K_2 of Tb, the calculated power decreases monotonically below 100 K from 3 to 2 at 0 K. Conversely the σ^{21} conventional relationship for the basal plane K_6^6 rises to a σ^{36} law at 0 K. The experimental data available are not sufficiently accurate to adequately confirm these departures.

4.3.1. Single-Ion Anisotropy Theory

The giant single ion anisotropy of the heavy rare earth ions (except Gd) has its origin in the interaction between the large multipole moments of the $4f$ charge cloud and the low symmetry hexagonal crystalline field. Spin-orbit coupling strongly locks together spin- and orbital momentum making J a good quantum number. For the heavy rare earths only the lowest lying level of the J manifold is excited (i.e., $\langle J_z \rangle = J$).

The potential energy of a single $4f$ electron in spherical coordinates (r, θ, ϕ) can be expanded in spherical harmonics as

$$V = A_2^0 r^2 Y_2^0(\theta, \phi) + A_4^0 r^4 Y_4^0(\theta, \phi) + A_6^0 r^6 Y_6^0(\theta, \phi)$$
$$+ A_6^6 r^6 [Y_6^6(\theta, \phi) + Y_6^{-6}(\theta, \phi)]. \qquad (4.8)$$

The A's are the crystal field potentials derived from the tripositive ions on the crystal lattice sites. A_2^0 would be zero except for the departure of the c/a ratio in the rare earths from the ideal value. Calculations of the A_l^m potentials must include the significant effects of charge screening by the outer shell electrons as has been done by Kasuya.[3]

It is often convenient to remove the spatial dependence from the anisotropy energy (Eqn 4.8) and work in terms of the angular momentum, J. For a particular J manifold Elliott and Stevens[76] have shown that one can write

$$V = A_2^0 \langle r^2 \rangle \alpha Y_2^0(\mathbf{J}) + A_4^0 \langle r^4 \rangle \beta Y_4^0(\mathbf{J}) + A_6^0 \langle r^6 \rangle \gamma Y_6^0(\mathbf{J})$$
$$+ A_6^6 \langle r^6 \rangle \gamma [Y_6^6(\mathbf{J}) + Y_6^{-6}(\mathbf{J})]. \quad (4.9)$$

The terms

$$\langle r^2 \rangle \alpha Y_2^0(\mathbf{J}) = \sum_i (3Z_i^2 - r_i^2), \quad (4.10)$$

etc. represent the l-pole moments of the $4f$ charge distribution. The constants α, β, γ which effectively describe the degree of departure from spherical symmetry have been calculated by Stevens[77] and are tabulated in Table 1.3. The numbers $\langle r^2 \rangle$ etc. are averages over the $4f$ wave functions.[71] These averages and the potentials A_l^m show only relatively small variations (for $l=2$) across the heavy rare earth series. Comparison of Eqn (4.5) and Eqn (4.9) thus shows that the magnitudes of the anisotropy constant K_2 for different rare earths will scale approximately as the corresponding Stevens' factor α. Table 4.3 lists the values of $A_l^m \langle r^l \rangle$ and the total anisotropy energy per ion for the heavy rare earths using appropriate charge screening factors from Kasuya.[3]

Table 4.3. Theoretical Values of $\langle r^l \rangle A_l^m$ and of the 0 K Single-Ion Anisotropy Energies for the Heavy Rare Earth Series* (all values are $\times 10^{-15}$ ergs/ion)

	$-\langle r^2 \rangle A_2^0$	$-\langle r^4 \rangle A_4^0$	$\langle r^6 \rangle A_6^0$	$-\langle r^6 \rangle A_6^6$	K_2^0	K_4^0	K_6^0	K_6^6
Tb	26.9	5.16	1.93	1.07	9.77	−0.818	−0.101	0.056
Dy	29.8	5.09	2.52	1.38	10.63	0.954	0.464	−0.254
Ho	29.9	4.85	2.55	1.39	4.25	0.662	−0.863	0.471
Er	29.4	4.40	2.10	1.15	−4.20	−0.618	0.775	−0.423
Tm	28.2	3.71	1.52	0.83	−10.24	−0.785	−0.397	0.216

* Calculated from Kasuya.[3]

4.3.2 Anisotropy—Experimental Results

Experimental work on the anisotropy of the rare earths has largely been confined to the heavy series Gd-Tm due to the general lack of availability of single crystals of the light rare earths.

With the exception of gadolinium the two-fold anisotropy energy in these metals is predominately of single ion origin and is typically of order 10^8–10^9 ergs/cm^3 equivalent to 2–10 meV/ion using current values of density and atomic weight. The basal plane anisotropy is some two orders of magnitude smaller. The anisotropy of the S-state ion Gd is expected to be dominantly of two ion origin and is smaller by about a factor of 100. Gd is discussed last.

Dysprosium and *terbium* as expected from the theoretical values in Table 4.3 have almost identical two-fold anisotropy energies. Due to the large magnitude of this energy the magnetic moment can be pried out of the basal plane by only a few degrees (at 4 K) even in fields of over 100 kOe applied along the hard c axis direction. As a consequence conventional angular torque measurements are not of much utility. The better procedure is the use of magnetization measurements made as a function of H applied along the hard direction or use of a differential torque method.[9] From these data the angle of M relative to the crystal axes may be determined and the anisotropy constants found from the minimum energy expression:

$$\frac{\partial E}{\partial \theta} = 0 = |\mathbf{M} \times \mathbf{H}|_\theta + \frac{\partial E_K}{\partial \theta}, \tag{4.11}$$

where E_K is the anisotropy energy defined in Eqn (4.5).

It should be noted also that in measurements of extreme anisotropy the experimental uncertainty or scatter can be reasonably large. As a result the procedure of least squares fitting which is often used to determine the anisotropy constants from the moment or torque curves may produce non-unique values of K especially for the higher order terms. This may partially account for the discrepancies between anisotropy values reported by different experiments. The measurement of the field and temperature dependence of the $q=0$ spin wave energy gap by inelastic neutron scattering offers a promising method of determining the anisotropy. Recent results on terbium by Nielsen *et al.*[77a] are discussed in Chapter 5.

For *terbium* and *dysprosium* measurements to 80 kOe by Féron and Pauthenet[5] and to 150 kOe by Rhyne *et al.*[4, 9] all gave substantially the same results for values of the anisotropy constant $K_2 = 5.5 \pm 1 \times 10^8$ ergs/cm^3 at 4 K. Ferromagnetic resonance data at 100 GHz of Stanford and Young[78] yielded a 0 K value of $K_2 = 3.1 \times 10^8$ ergs/cm^3 also in fair agreement. Féron[5] determined values of K_4 which were approximately one order of magnitude smaller and are shown in Figure 4–24. The solid curve is the theoretical temperature dependence for a single ion anisotropy of order $l=2$ and $l=4$ (see Eqn 4.6). The agreement with theory is quite good especially considering the experimental difficulties involved.

Torque magnetometer measurements on dilute solutions of Dy (1.3 Wt%) and Tb (1.8 Wt%) in Gd have been made by Tajima and Chikazumi.[79] This method of determining the Dy and Tb ion anisotropy obviates the need for very high field measurements. Some question arises, however, concerning the linear extrapolation of the dilute results to 100% Dy or Tb (assumption of complete single ion behavior) and also the effect of the c/a ratio which is different between the alloy and the elements. The values of K_2 at 4 K for Dy and Tb which resulted were approximately half as large as those of Féron *et al.* and those of K_4 approximately equivalent. Tajima and Chikazumi also reported values of K_6 which were only about 30% smaller than those of K_4. It should be noted that the sign of K_4 and K_6 in both these results and those of Féron for Tb disagree with the theoretical predictions given in Table 4.3.

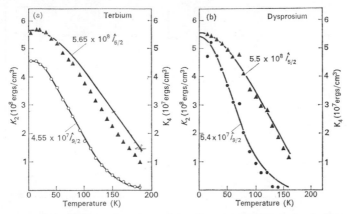

Figure 4–24. Second and fourth order axial anisotropy constants for terbium (A) and dysprosium (B). The solid curve represents the temperature dependence expected on the single ion model (after Ref. 5).

Values of K_2 for Dy and Tb were also obtained from torque measurements in pulsed fields up to 150 kOe by Levitin and Ponomarev.[80] These gave $K_2 = 2.7 \times 10^8$ ergs/cm³ for Tb at 105 K and $K_2 = 1.9 \times 10^8$ ergs/cm³ for Dy at 100 K somewhat smaller than those discussed above. Analysis of the classic magnetization data on Dy by Behrendt et al.[12] yielded

$$K_2 = 2.5 \times 10^8 \text{ ergs/cm}^3.[81]$$

Considering the data extended to only 18 kOe and the difficulties of using the Faraday balance technique for magnetization measurements in hard directions the agreement with later results is remarkable.

The six-fold (basal plane) anisotropy of Dy and Tb is approximately 100 times smaller than the axial anisotropy as predicted from the Stevens' factor analysis (Table 4.3). Further the value of $K_6{}^6$ for Tb is expected to be only about 25% of that in Dy as confirmed by the measurement shown in Figure 4–25. The values of Rhyne and Clark[9] were obtained from an analysis of the angular dependence of the basal plane magnetostriction. Also shown in Figure 4.25 for $K_6{}^6$ are results by Bly et al.[82] using a torque magnetometer. These results required considerable care in crystal alignment to effectively eliminate admixtures of the two-fold axial component into the six-fold torque data.

Torque results by Du Plessis[83] gave a 0 K extrapolated value for Tb of $K_6{}^6 = 2.9 \times 10^6$ erg/cm³. Féron and Pauthenet[5] obtained basal plane anisotropies from their magnetization data which were somewhat at variance with those shown in the figure. For Dy they found $K_6{}^6 = -11 \times 10^{+6}$ ergs/cm³ and for Tb, $K_6{}^6 = 1.85 \times 10^6$ ergs/cm³. Liu et al.[81] also using magnetization data of Dy obtained a value $K_6{}^6 = -10.2 \times 10^6$ ergs/cm³ at 4 K.

On preliminary inspection one would expect the temperature dependence of $K_6{}^6$ shown in Figure 4–25 to be given by the Callen $\hat{I}_{\frac{13}{2}}(\mathscr{L}^{-1}(m))$ expression for single-ion anisotropy of order $l = 6$. This is not strictly true, however, particu-

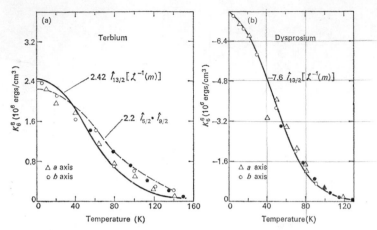

Figure 4–25. Basal plane anisotropy constant for terbium (A) and dysprosium (B). The solid
curve is the theoretical single ion dependence. The dotted curve for Tb is a calculated
result (after Ref. 84) for an anisotropy arising from the hexagonally symmetric magnetostriction term (data—open symbols, after Ref. 9; solid symbols, after Ref. 82).

larly in the case of Tb. The total angular dependent free energy contains in
addition to the terms given in Eqn (4.11), a contribution from the magneto-
elastic energy. Normally this term is negligible as in Co, for example. In the
rare earths however, this magneto-elastic energy is large, and in fact for Tb it
is of comparable magnitude to the six-fold anisotropy as shown by Cooper.[84]
One must properly consider the anisotropy constant to be made up of two
terms:

$$K_6^6(T) = L_6^6(T) - \tfrac{1}{4} C^{\gamma} \lambda^{\gamma,2} \lambda^{\gamma,4} \qquad (4.12)$$

where the term $L_6^6(T)$ is that originating from purely crystal field effects
(Eqn 4.9) while the second term is the magneto-elastic correction. The quan-
tity C^{γ} is an elastic constant in hexagonal basis and $\lambda^{\gamma,2}$ and $\lambda^{\gamma,4}$ are second
and fourth order magnetostriction constants (see Section 4.4). For Tb using
magnetostriction values given later the magneto-elastic term

$$\tfrac{1}{4} C^{\gamma} \lambda^{\gamma,2} \lambda^{\gamma,4} = -4.1 \times 10^6 \text{ ergs/cm}^3$$

which is actually larger than the reported total anisotropy at 0 K. Thus, taking
account of experimental uncertainty, one may at least state that the magneto-
elastic contribution to the measured K_6^6 in Tb is significant if not dominant.
This is confirmed in recent high field inelastic neutron scattering results.[77a]
The temperature dependence of the measured six-fold anisotropy for Tb
shown in Figure 4–25 should thus be a combination of the $\hat{I}_{13/2}$ ($\mathscr{L}^{-1}(\sigma)$)
dependence appropriate for the L_6^6 term of Eqn (4.12) plus a

$$\hat{I}_{5/2}(\mathscr{L}^{-1}(\sigma)) \times \hat{I}_{9/2}(\mathscr{L}^{-1}(\sigma))$$

dependence arising from the second (magneto-elastic) term. The temperature
dependence for these two expressions is not markedly different as shown in the

figure and it is hard to make a meaningful comparison. In Dy the crystal field energy is about 4 times larger than in Tb, while the magnetostriction is comparable, with the result that the magneto-elastic correction to $K_6{}^6$ should be less significant.

It is suggested that the magneto-elastic contribution to the anisotropy derived from static magnetization or torque measurements, in which the net moment is turned away from the easy direction, may be very different from the value appropriate for inelastic neutron or resonance results. Inelastic scattering results on Tb[85] (see Chapter 5) indicate for the dynamic case that the magnetostrictive crystal strain assumes an equilibrium value in response to spin wave disturbances; the so-called "frozen lattice approximation" formulated by Turov and Shavrov[86] and developed by Cooper.[84]

Anisotropy measurements on *holmium* have been made by at least three workers with rather wide variation of results. Analysis of the magnetization data to 200 kOe of Rhyne et al.[4] yielded 4 K values of $K_2 = 2 \times 10^8$ ergs/cm^3, $K_4 = 1.4 \times 10^7$ ergs/cm^3 and $K_6 = -1.6 \times 10^7$ with considerable uncertainty in the last parameter. In contrast, Féron and Pauthenet[5] reported values of $K_2 = 4.2 \times 10^8$ ergs/cm^3 and $K_4 = 1.8 \times 10^8$ ergs/cm^3 from magnetization measurements to only 60 kOe. These latter values are not consistent with the sharp increase in c axis magnetization observed above 100 kOe (see Figure 4–7). They also predict a saturation field of 985 kOe while the observed value[4] is only 155 kOe. The single-ion theory including charge screening of Kasuya[3] yields values $K_2 = 1.36 \times 10^8$ ergs/cm^3, $K_4 = 2.12 \times 10^7$ ergs/cm^3, and $K_6 = -2.77 \times 10^7$ ergs/cm^3. Experimental results by Tajima and Chikazumi[87] were again on dilute alloys of Ho and Gd with Ho concentrations of 1.3 and 2.5 weight per cent. Extrapolation to 100 per cent Ho (see previous remark on this point) gave values of $K_2 = 0.90 \times 10^8$ ergs/cm^3, $K_4 = 0.27 \times 10^8$ ergs/cm^3 and $K_6 = -0.13 \times 10^8$ ergs/cm^3. As in the case of dysprosium and terbium these extrapolated alloy values for the dominant K_2 term appear to be about 50% smaller than those measured on the non-dilute elements.

The basal plane anisotropy of Ho is anomalously large. In most of the heavy rare earths the basal plane anisotropy constant $K_6{}^6$ is approximately 1% of the K_2 term. However, due to the $4f$ charge configuration in Ho, the basal plane constant $K_6{}^6$ is about 10% of K_2 and is the largest of the series Tb, Dy, Ho. Magnetization results of both Féron and Pauthenet and Rhyne et al. gave similar 0 K values of 2.7×10^7 ergs/cm^3. The temperature dependence was examined in the former study and found to be only moderately well represented by a single ion expression.

The case of *erbium* is anomalous due to its non-planar spin structure. The anisotropy which is determined by magnetization or torque results is not totally representative of the crystal field energy but is expected to contain significant contributions from exchange as well. In the magnetization process with H applied along hard directions the angle between neighboring spins in the conical configuration is field dependent. Consequently, the exchange energy represented phenomenologically as

$$\sum_{ij} \mathscr{J}_{ij} S_i \cdot S_j$$

varies with H and is reflected as an additional anisotropy energy in the conventional analysis. No reliable determination may thus be made of the true crystal field anisotropy parameters in Er (or Tm) from bulk magnetization or torque measurements on the pure elements. Dilute alloy (1.3% Er) results of Tajima and Chikazumi[87] extrapolated to 100% Er indicate

$$K_2 = -1.2 \times 10^8 \text{ ergs/cm}^3,$$

$$K_4 = +0.39 \times 10^8 \text{ ergs/cm}^3,$$

and

$$K_6 = +0.06 \times 10^8 \text{ ergs/cm}^3.$$

The sign of only the K_2 term is in agreement with theory.

Gadolinium is an S state ion and as a consequence to first order its Steven's factors are 0 and the single-ion contribution to the anisotropy vanishes.

The magnetic anisotropy of Gd has received considerable attention. Torque measurements by Corner *et al.*,[88] Darby,[89] and Graham[90] indicate a strong variation in anisotropy energy with applied field particularly near the Curie temperature resulting from increasing magnetic saturation (paraprocess effect). In this case, it is not possible to obtain unique values of the K's for each temperature. It is thus appropriate to present the anisotropy constants† as a function of magnetization as shown in Figure 4–26A from Graham.[90] This removes the direct dependence on field. The parameters in the figure indicate the temperature at which the magnetization and torque measurements were

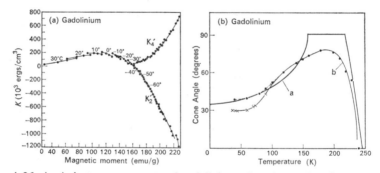

Figure 4–26. A. Anisotropy constants of gadolinium plotted as a function of magnetic moment. Corresponding temperatures are given as parameters. The anisotropy constants were strongly field dependent (after Ref. 90). B. Temperature dependence of easy direction of magnetization (cone angle) relative to c axis from two independent studies. The discrepancy in results probably arises from distinct intrinsic strains and impurities. The dotted curve gives the result extrapolated to high fields. (Curve a after Ref. 90, curve b after Ref. 89.)

† In the results shown the K's are anisotropy constants found from an energy expression of the form

$$E_K = K_2' \sin^2 \theta + K_4' \sin^4 \theta + K_6' \sin^6 \theta + K_{66}' \sin^6 \theta \cos 6\phi$$

These primed anisotropy constants are simple related to those of Eqn. (4.5) as follows:

$$-K_2' = 3/2 K_2 + 5 K_4 + 21/2 K_6$$
$$K_4' = 35/8 K_4 + 189/8 K_6$$
$$K_6' = -231/16 K_6$$
$$K_{66}' = K_6^6$$

made. Note that K_2' changes sign in the neighborhood of 238 K and K_4' essentially vanishes close to the same temperature. The sign reversal of K_2' is not a consequence of using the above expansion in $\sin^n \theta$ instead of the more physical Y_l^m of Eqn (4.5). Using the latter expression the anisotropy constants corresponding to $M = 220$ emu/g ($T \simeq 130$ K) are $K_2 = 245 \times 10^3$ ergs/cm³, and $K_4 = 137 \times 10^3$ ergs/cm³. At room temperature $K_2 \simeq -133 \times 10^3$ ergs/cm³. The coefficient K_6' was not determined by Graham; however, the data of Corner[88] show it to be approximately 0.25×10^6 ergs/cm³ at 80 K and to vanish also near 240 K. The behavior of the dominant K_2' term indicates a strong variation of the direction of easy magnetization with temperature as indicated in Figure 4–26B.[88, 90] Curves are shown for data at $H = 12.5$ kOe and also for results plotted vs. $1/H$ and extrapolated to $1/H = 0$. The c axis is easy down to 240 K with a shift toward the basal plane for lower temperatures. Neutron diffraction results by Cable and Wollan[90]a for the cone angle are in close agreement with those of reference 89.

Torque measurements with H applied in basal plane directions[90] indicate that the six-fold anisotropy coefficient K_{66}' is some 100 times smaller than the two-fold anisotropy. The value obtained by Graham at 4 K is

$$6.3 \times 10^{+3} \text{ ergs/cm}^3$$

with the energy minimum along the a $\langle 11\bar{2}0 \rangle$ direction. The temperature dependence is fairly well given by the σ^{21} power law. Departures are not surprising since the anisotropy predominantly is two-ion and this law is then strictly valid only at low temperatures.

Toyama et al.[91] have measured the pressure dependence of the anisotropy in Gd at 77 K to 4.4 kbars and find a drastic decrease in magnitude of K_2' with pressure. For a pressure of about 4 kbars K_2' is reduced to zero (their 0 pressure value is $-3.8 \times 10^{+5}$ ergs/cm³) and becomes positive for higher pressures. The effect on K_4' and K_6' is smaller and does not produce a sign change.

4.3.3. Anisotropy from $\Delta\theta_p$

As mentioned in a previous section the paramagnetic Curie temperature is a function of spin direction for the heavy rare earths (except Gd) as a consequence of the anisotropy energy. The molecular field expression for the anisotropy of θ_p is to first order:

$$k(\theta_{\parallel} - \theta_{\perp}) = -\frac{3[4J(J+1)-3]K_2^{\text{para}}}{10J(2J-1)} \tag{4.13}$$

assuming $J_z = J$. The quantities $\theta_{\parallel}$ and $\theta_{\perp}$ are the paramagnetic Curie temperatures for spin directions parallel and perpendicular to the $\langle 0001 \rangle$ direction. From this expression values of the effective 0 K two-fold anisotropy constant K_2^{para} can be calculated and are shown in Table 4.4. The absolute values of $|\theta_{\parallel} - \theta_{\perp}|$ for Ho and Er are anomalously different in striking contrast to the almost identical anisotropy magnitude expected from the single ion model. This may reflect an anisotropic exchange contribution.

Table 4.4. Comparison of Theoretical and Experimental Anisotropies for the Heavy Rare Earths

		K_2 (10^8 ergs/cm^3)	K_2^{para} (10^8 ergs/cm^3)	K_4 (10^7 ergs/cm^3)	K_6^6 (10^6 ergs/cm^3)
Tb	Theor.	3.1	—	−2.6	1.8
	Expt.	5.6	2.6	4.6	2.4
Dy	Theor.	3.4	—	3.0	−8.0
	Expt.	5.5	2.9	5.4	−7.6
Ho	Theor.	1.4	—	2.1	15.1
	Expt.	2.0	0.93	1.4	27.0
Er	Theor.	−1.4	—	−2.0	−13.8
	Expt.	—	−1.8	—	—
Tm	Theor.	−3.4	—	−2.6	7.2
	Expt.	—	−3.6	—	—

Table 4.4 gives a comparison between anisotropy constants for the series Tb-Tm calculated by the single ion method using the Elliott-Stevens factors and selected experimental results obtained (a) from the difference $\theta_\parallel - \theta_\perp$ (Eqn 4.13) and (b) from direct determination of the anisotropy by moment or torque measurements. Values obtained from dilute ion studies have been omitted due to difficulties in interpretation.

4.4. MAGNETOSTRICTION AND ELASTIC ENERGY

4.4.1. Brief Theory Review

The magnetostriction of a material is a dimensional distortion arising from the interdependence of magnetic and elastic energies. Such magnetostrictions occur to minimize the total free energy of the system. Phenomenologically the total energy of a magnetic system may be expanded in powers of the strain. The first order term is usually referred to as the magneto-elastic energy:

$$H_{\text{ME}} = \left(\frac{\partial E_{\text{anis}}}{\partial \epsilon_{ij}} + \frac{\partial E_{\text{exc}}}{\partial \epsilon_{ij}} + \ldots \right) \epsilon_{ij} \tag{4.14}$$

where the two dominant terms refer to strain derivatives of anisotropy and exchange energy respectively. Other terms such as dipole–dipole have been neglected. In general these magneto-elastic coupling terms arise from either single-ion (e.g., crystal field type) or multi-ion interactions (e.g., anisotropic exchange type).

Explicit expressions for the magneto-elastic Hamiltonian for various crystal symmetries have been written down by a number of authors. Callen and Callen[92] in a series of papers have obtained the most complete results based on the group theoretical representations for the allowed strains. For hexagonal symmetry their expression containing one and two-ion magneto-elastic coupling constants $B_j^{\mu l}$ and $D_j^{\mu l}$ is:

$$H_{\text{ME}} = - \sum_{\mu, j, l} B_j^{\mu l} \sum_i \epsilon_i^{\mu j} \sum_f K_i^{\mu l}(S_f)$$
$$- \sum_{\mu, j, l, i} \epsilon_i^{\mu j} \sum_{f, g} D_j^{\mu l}(f, g) K_i^{\mu l}(S_f S_g) \tag{4.15}$$

where $\epsilon_i{}^{\mu j}$ are irreducible hexagonal strains derivable from their Cartesian counterparts (see Ref. 92). The basis functions $K_i{}^{\mu l}(S_f)$ and $K_i{}^{\mu l}(S_f, S_g)$ are respectively one- and two-ion spin polynomials of order l (only terms with l even exist for crystals with a center of symmetry). Indices μ, j label the irreducible representations of the crystal group, i labels the basis functions, and f and g the lattice sites. Minimization of the above expression with respect to strain leads to the following expression[98] for the magnetostriction $\Delta l/l$ in hexagonal symmetry correct to order $l=2$.

$$\frac{\Delta l}{l} = [\lambda_1{}^{\alpha,0} + \lambda_1{}^{\alpha,2}(\alpha_z{}^2 - \tfrac{1}{3})](\beta_x{}^2 + \beta_y{}^2) + [\lambda_2{}^{\alpha,0} + \lambda_2{}^{\alpha,2}(\alpha_z{}^2 - \tfrac{1}{3})]\beta_z{}^2$$
$$+ \tfrac{1}{2}\lambda^{\gamma,2}[(\alpha_x\beta_x + \alpha_y\beta_y)^2 - (\alpha_x\beta_y - \alpha_y\beta_x)^2]$$
$$+ 2\lambda^{\epsilon,2}(\alpha_x\beta_x + \alpha_y\beta_y)\alpha_z\beta_z \quad (4.16)$$

The six magnetostriction constants λ are the experimentally determined quantities. Their explicit expression contains elastic constants, averages over the appropriate spin polynomial, and one and two ion magneto-elastic coefficients.[92] The two fully symmetric terms $\lambda_1{}^{\alpha,0}$ and $\lambda_2{}^{\alpha,0}$ contain only two ion coefficients and are related to the anomalous thermal expansion (or exchange magnetostriction of Section 4.4.3). The quantities α_i, β_i are direction cosines of the magnetization and strain (measuring) directions. The superscripts α, ϵ, γ describe the strain mode of the magnetostriction. As depicted in Figure 4–27 the α modes are symmetry preserving dilatations along the c axis ($\lambda_2{}^{\alpha,2}$) and in the basal plane ($\lambda_1{}^{\alpha,2}$), the γ mode is a distortion of the hexagonal symmetry into orthorhombic (circle to an ellipse in order $l=2$), and the ϵ mode is a c-axis shear.

The magnetostriction constants λ have a temperature dependence related (in the single ion case) to the magnetization. Callen and Callen[92] have obtained

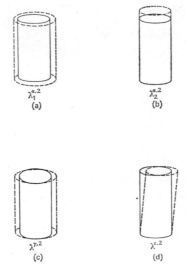

Figure 4–27. Magnetostriction strain modes in $l=2$ order for hexagonal symmetry (after Ref. 98).

expressions for this temperature dependence which is of the same form as for the anisotropy[72]

$$\lambda(T) = \lambda(T=0)\hat{I}_{l+\frac{1}{2}}[\mathscr{L}^{-1}(\sigma)] \qquad (4.17)$$

for the single ion case.

4.4.2. Experimental Results—Anisotropic Magnetostriction Constants

In this section we shall again confine our attention to single crystal results since polycrystalline data are often difficult to interpret due to possible preferential grain orientation in the sample. All of the reported single-crystal studies have been done on the heavy rare earths Gd, Tb, Dy, Ho, and Er due to the limitations of obtaining sufficiently large crystals of the other rare earths. Gd is discussed last due to its two ion character. The typical experimental procedure for determining the magnetostriction constants is to measure the linear strain with a resistance strain gage in a particular direction as a function of the angle of a saturating magnetic field. From these data the coefficients in Eqn (4.16) are then determined.

Of the highly anisotropic rare earths Tb-Tm, only the familiar twins Tb and Dy are very amenable to study. The complex spin states present in the other elements and the transitions between them give rise to large exchange-striction effects which cloud the determination of the single ion coefficients.

A complete determination of all six $l=0$ and $l=2$ magnetostriction constants has been performed only for Tb[93-95] and then only partially for temperatures below T_N. The giant anisotropy precludes measurements below T_N which require moving the magnetization out of the basal plane. This means essentially only $\lambda^{\gamma,2}$ of Eqn (4.16) can be directly determined over the entire temperature range. The other $l=2$ coefficients are obtainable in the paramagnetic temperature range, and, if they are of single ion origin only, can be extrapolated to low temperatures by the Callen theory.[92]

The success of this theory in predicting the temperature dependence of $\lambda^{\gamma,2}$ and an $l=4$ constant ($l=4$ order is not included in Eqn (4.16)) is shown in Figure 4–28. For Tb[93, 94] and Dy[96-98] the $l=2$ function ($\hat{I}_{\frac{5}{2}}$) represents the observed temperature variation of the strain through more than three orders of magnitude. In addition, the $l=4$ ($\hat{I}_{\frac{9}{2}}$) dependence represents $\lambda^{\gamma,4}$ well for Tb over most of the observed range. This indicates clearly that these coefficients are of only single ion source, and on this basis some confidence is gained is using the single ion theory to extrapolate the other $l=2$ constants $\lambda_1^{\alpha,2}$, $\lambda_2^{\alpha,2}$ and $\lambda^{\epsilon,2}$ from the paramagnetic region to low temperatures. Above T_N the constants are proportional to $M^2 = \chi^2 H^2$ as a consequence of the Maxwell relation

$$\left.\frac{\partial\lambda}{\partial H}\right|_\epsilon = \left.\frac{\partial M}{\partial\epsilon}\right|_H, \qquad (4.18)$$

where M is the magnetization and σ the stress. For $T > T_N$ the measured λ at the field H and the corresponding value of the $I_{\frac{5}{2}}$ function evaluated from the moment χH are plotted in Figure 4–29. The unity slope of the lines in the figure is a consequence of relation (4.17). In determining $\lambda^{\alpha,2}$ the requirement that the measurement be at constant moment dictated that the applied field H be changed in magnitude as a function of angle[94] due to the anisotropy in the susceptibility χ. From the figures the effective 0 K values of $\lambda_1^{\alpha,2}$, $\lambda_2^{\alpha,2}$, $\lambda^{\gamma,2}$,

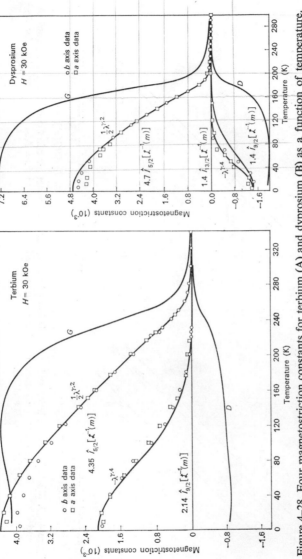

Figure 4–28. Four magnetostriction constants for terbium (A) and dyprosium (B) as a function of temperature. The theoretical single-ion dependence is shown for the γ mode constants. G and D are respectively combinations of c and a axis α mode constants as given in the text (after Ref. 93 and 96).

James J. Rhyne

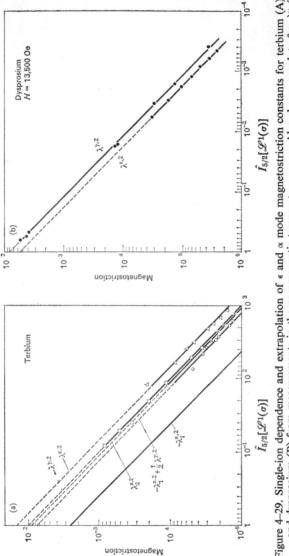

Figure 4-29. Single-ion dependence and extrapolation of ϵ and α mode magnetostriction constants for terbium (A) and dysprosium (B) from measurements in the paramagnetic temperature range. Also shown are data for $\lambda^{\gamma, 2}$ down to 77 K (after Refs. 94 and 98).

$\lambda^{\gamma,4}$, and $\lambda^{\epsilon,2}$ for Tb and $\lambda^{\gamma,2}$, $\lambda^{\epsilon,2}$ for Dy are obtained and are listed in Table 4.5. It is of interest to note that essentially all of the $l=2$ single ion coefficients for Tb and Dy have a 0 K value of order 1%. This represents a magnetostriction more than 1000 times larger than in the $3d$ transition elements and in Gd.

The coefficients G and D shown in Figure 4–28 represent the magnetic part of the thermal expansion (see Section 4.4.3) and are derived from the measured temperature dependence of c axis (for G) or a axis (for D) strain after subtraction of the normal lattice thermal expansion extrapolated from above T_N. The coefficients correspond to the following combination of magnetostriction constants which are not directly separable below T_N:

$$G = \lambda_2^{\alpha,0} - \tfrac{1}{3}\lambda_2^{\alpha,2}$$
$$D = \lambda_1^{\alpha,0} - \tfrac{1}{3}\lambda_1^{\alpha,2}. \tag{4.19}$$

The fact that D and G contain $l=0$ terms which can only be of two ion origin means that these constants will not have the $\hat{I}_{l+\frac{1}{2}}(\mathscr{L}^{-1}(\sigma))$ temperature dependence as is obvious from the figures.

The only anisotropic magnetostriction constant determined for the elements Ho[99] and Er[100] is $\lambda^{\gamma,2}$. Zero degree Kelvin values obtained by extrapolation of paramagnetic data are $\lambda^{\gamma,2} = 2.5 \times 10^{-3}$ for Ho and $\lambda^{\gamma,2} = -5.1 \times 10^{-3}$ for Er. The sign change for Er is in accord with the reversal in sign of the Stevens' factor α as discussed in the next subsection. Anisotropy and the presence of exchange magnetostriction effects prevent reliable measurements of even $\lambda^{\gamma,2}$ in erbium below T_N; in fact, a direct measurement of $\lambda^{\gamma,2}$ at 4.2 K produced a value $\lambda^{\gamma,2} = +7.2 \times 10^{-4}$ reversed in sign by the dominant effect of exchange magnetostriction in the fan conical moment state.

Gadolinium has received the most attention from at least four independent studies[101–104] of magnetostriction. The S-state character of the ion rules out single ion anisotropy and magneto-elastic contributions. The magnetostriction is of the same magnitude as in Co ($\Delta l/l$ order of 10^{-5}) but small compared to other rare earths with large single-ion contributions ($\Delta l/l$ order of 10^{-2}). Due to the two-ion origin of the magnetostriction the temperature dependence of the coefficients is therefore difficult to predict theoretically.

Figure 4–30 shows the temperature dependence of four $l=2$ anisotropic magnetostrticion coefficients. The coefficients used by Alstad[101] were derived by Mason[105] using a different set of basis functions than those employed in obtaining Eqn (4.16). The relationship of these coefficients to the group theoretical strain coefficients is as follows:

$$\lambda^{\gamma,2} = \lambda_A - \lambda_B$$
$$\lambda_2^{\alpha,2} = -\lambda_c$$
$$\lambda_1^{\alpha,2} = -\tfrac{1}{2}(\lambda_A + \lambda_B)$$
$$\lambda^{\epsilon,2} = 2\lambda_d - \tfrac{1}{2}(\lambda_A + \lambda_C). \tag{4.20}$$

As expected, the coefficients shown in the figure clearly have a temperature dependence different from that of the single-ion theory. In particular

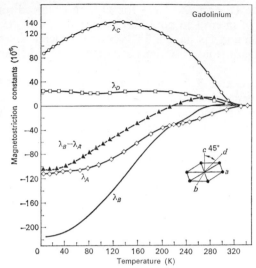

Figure 4–30. Temperature dependence of the four $l=2$ order anisotropic magnetostriction constants for gadolinium (after Ref. 101).

$\lambda^{\gamma,2}=\lambda_A-\lambda_B$ reverses sign near 200 K, an effect which can only be of two-ion origin. This behavior plus the kink in λ_A at the same temperature is associated with the change in sign of the dominant anisotropy constant.[90]

The microscopic origin of the large magneto-elastic strains in the heavy rare earths Tb-Tm has been examined by Tsuya et al.[106] The expression for the magnetostriction constants λ of order $l\geq 2$ is made up of a combination of one-ion and two-ion terms. The one-ion terms include a magneto-elastic coefficient B, elastic constants C_{ij} and a temperature and field-dependent spin correlation function. For example

$$\lambda^{\gamma,2}=\frac{B^{\gamma,2}}{2(C_{11}-C_{12})}\,\mathscr{L}(T,H),\qquad(4.21)$$

where $\mathscr{L}(T,H)$ is as defined in Ref. 92.

Tsuya et al. calculated the contribution to the one-ion magneto-elastic coefficients arising from two sources: (a) distortion of the crystal field and (b) strain rearrangement of the conduction electrons. This latter energy contribution considers the redistribution of conduction electrons near the $4f$ ion during lattice distortion and was calculated in a deformable ion model. The former crystal field contribution may be visualized by considering a very oblate $4f$ charge cloud (Stevens factor $\alpha>0$ as in Tb-Ho) residing in the crystal field produced by 3^+ ions on the hexagonal sites. Rotation of the $4f$ moment coupled to the charge distribution results in a large change in electrostatic energy. The crystal's response is an orthorhombic lattice distortion which lowers the total free energy. Ions having negative Steven's factors will experience the opposite sign dilatation since their charge clouds are prolate in form. This distortion of crystal field origin was shown[106] to result in a smaller magnetostriction than the conduction electron redistribution source and to

Table 4.5. 0 K Values of Magnetostriction Constants[a]

Element	$\lambda^{\gamma,2}$	$\lambda_{theor.}^{\gamma,2}$	$\lambda^{\gamma,4}$	$\lambda^{\varepsilon,2}$	$\lambda_1^{\alpha,0} - \frac{1}{3}\lambda_1^{\alpha,2}$	$\lambda_2^{\alpha,0} - \frac{1}{3}\lambda_2^{\alpha,2}$	$\lambda_1^{\alpha,2}$	$\lambda_2^{\alpha,2}$
Gd	0.11	0	—	0.062	—	—	0.162	-0.082
Tb	8.7[b]	8.7[d]	-2.1[b]	15.0[c]	-0.8	4.3[b]	-2.6[c]	9.0[c]
Dy	9.4[b]	8.4	1.5[b]	5.5[c]	-2.0[b]	7.3[b]	—	—
Ho	2.5[c]	3.0	—	—	-3.9[b]	7.1[b]	—	—
Er	-5.1[c]	-2.7	—	—	+0.3[b]	6.2[b]	—	—

[a] All values are in units of $\Delta l / l (10^{-3})$.
[b] Measured to 4.2 K.
[c] Extrapolated from paramagnetic range using single-ion theory.
[d] Theoretical values (Ref. 106) normalized to Tb experimental result.

yield the wrong sign for the ϵ mode constants. The inclusion of both mechanisms correctly predicts the sign for α, γ, and ϵ coefficients. The relative magnitudes of $B^{\gamma, 2}$ across the series Tb-Er scale reasonably well with experimental observations, although the absolute magnitude is too large by a factor of about 15. Screening factors which were not included in the calculation should improve the comparison.

Table 4.5 lists the 0 K extrapolated values of the magnetostriction constants which have been determined experimentally for the elements Tb-Er. Also given are the relative theoretical values of $\lambda^{\gamma, 2}$ normalized to the experimental value of Tb.

4.4.3. Anomalous Thermal Expansion and Magneto-elastic Energy

The observed thermal expansion of the rare earth metals is extremely large and has a very anomalous temperature dependence. This arises from the dominant contribution of the exchange magnetostriction to the measured lattice parameter. For example, the normal (phonon) part of the a axis thermal expansion in Ho from 4 K to room temperature accounts for only one third of the total dilatation.

In the rare earths having helical spin structures a giant c axis expansion anomaly of exchange origin is observed. In holmium this amounts to over 0.7% from the Néel temperature to the 20 K Curie temperature. The definitive microscopic calculation of the exchange magnetostriction of a helical antiferromagnet has been made by Evenson and Liu.[56] Phenomenologically this exchange magnetostriction may be represented in the molecular field approximation[107] as

$$\left.\frac{\Delta l}{l}\right|_c^{\text{exc}} = \frac{cM_s^2}{Y}\left[\frac{d\mathscr{J}_1}{dc}\cos\theta + \frac{d\mathscr{J}_2}{dc}\cos 2\theta + \ldots\right] \qquad (4.22)$$

where $\mathscr{J}_1$, $\mathscr{J}_2$ are effectively Fourier coefficients of the exchange energy between first and second neighbor layers. Theta is the interplanar helical turn angle, c the lattice parameter, Y the elastic constants, and M_s^2 the square of the magnetization. In practice the temperature dependence of the observed dilatation is dominantly M^2 with the angle factors providing small corrections.

The lattice distortions in the heavy rare earths are sufficiently large to be determined accurately by X-ray diffraction. Darnell has studied the temperature dependence of the a, b, c axis lattice parameters of Gd,[108], Tb,[109] Dy[108] Ho,[108] and Er.[109] Excellent agreement is found in the para and antiferromagnetic state between these results and those for the bulk thermal expansion obtained by strain gage methods. In the ferromagnetic regime, the X-ray results are the only reliable expansion data for the following reason: Below T_C in zero applied field the crystal relaxes into a domain configuration which is not unique from one measurement to the next. The bulk elastic energy is a function of this domain configuration and consequently the average strain as seen by a strain gage is not uniquely defined below T_C. Strain gage results below T_C have been shown[96] to vary upwards of 20% in different runs of the same sample. This difficulty can be obviated by measuring bulk strains in applied fields greater than the demagnetizing field and then extrapolating to zero field.

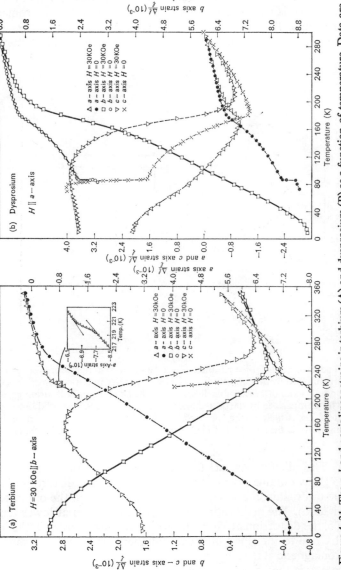

Figure 4-31. The a, b and c axis linear strains in terbium (A) and dysprosium (B) as a function of temperature. Data are shown for both 0 field and in a 30 kOe applied field. The inset of (A) shows the discontinuous lattice distortion in Tb at T_C which is small compared to that in Dy (after Refs. 93 and 96).

Figure 4–31 shows results for the a, b, and c axis bulk strain for Tb[93] and Dy[96, 98] obtained from strain gage experiments both at $H=0$ and in a saturating applied field. These results are compared to the microscopic X-ray lattice parameter results in Figure 4–32. The bulk results show that the basal plane strain ($H=0$) is isotropic in both paramagnetic and antiferromagnetic temperature ranges. At the Curie temperature, particularly in Dy, a large discontinuity is observed which corresponds to an orthorhombic distortion of

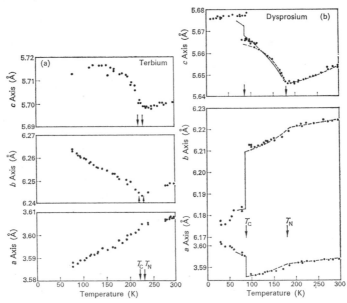

Figure 4–32. The a, b, c axis lattice parameter as a function of temperature for terbium (A) and dysprosium (B) from X-ray studies. The discontinuity in lattice constants for Dy at T_C corresponds to an orthorhombic distortion of the hcp lattice. Bulk strain gage results are given for comparison by the dotted line in the range $T>T_C$. The solid line shown for the c axis of Dy is a fit to the molecular field expression (4.22). The magnitude of the discontinuity is that calculated in Ref. 56. (X-ray data after Refs. 108 and 109.)

the lattice. The distortions at the transition are $\Delta a/a = +0.2\%$, and $\Delta b/b = -0.5\%$, as obtained from the X-ray results (bulk distortion data are not valid due to domain effects and do not reproduce even the opposite signs of $\Delta a/a$ and $\Delta b/b$). The orthorhombic distortion increases at lower temperatures. For Tb, the distortion at T_C is barely discernible as shown in the inset of Figure 4–31A due to the small moment and energy difference between ferro and antiferromagnetic states.

Along the c axis the crystals expand below T_N and in Dy a discontinuity $\Delta c/c = +0.3\%$ occurs at T_C corresponding to the abrupt drop in interlayer turn angle from 26.5° to 0°. The magnitudes of the discontinuities in lattice parameters were calculated by Evenson and Liu[56] from their expression for the differences in energy of the helical and ferromagnetic states (see below) and are in good agreement with experiment as shown in the Figure. Use of the recently published single-crystal Dy elastic constant data of Rosen and Klimer[110] would improve the comparison for the c axis.

The lowering of the magnetic elastic energy in passing from the helical to ferromagnetic state is the basic driving force behind the ferromagnetic transition.[111] Of the other possible mechanisms, the exchange energy continues to favor helical order below T_C, and the basal plane anisotropy energy is too small to be of major importance at this temperature. The total magneto-elastic energy in both helically ordered and ferromagnetic phases can be written as

$$E_{me} = -\tfrac{1}{2}C_{11}^{\alpha}(\epsilon^{\alpha,1})^2 - \tfrac{1}{2}C_{22}^{\alpha}(\epsilon^{\alpha,2})^2 - \tfrac{1}{8}C^{\gamma}(\epsilon^{\gamma})^2 \qquad (4.23)$$

from the work of Evenson and Liu.[56] This was a generalization of the previous work[92] to include non-uniform strains as found in the helical state. The C_{ij}^{μ} are elastic moduli in hexagonal basis and can be conveniently related to the conventional C_{ij}.[92] The $\epsilon^{\mu j}$ are the equilibrium strains appropriate to either ferromagnetic or helically ordered states.[56] In the helical state the ϵ^{γ} strain is 0 from the "lattice clamping" effect. Both above and below T_N the generalized strains can be deduced directly from the anomalous thermal expansion along a, b, and c axis. The total magneto-elastic energy may then be calculated over the entire temperature range of magnetic order. Using the X-ray thermal expansion data of Darnell,[108] Rosen and Klimer[110] have calculated E_{me} in Dy as shown in Figure 4-33. Note the abrupt drop in magneto-elastic energy at T_C of 1.36 J/cm^3 = 0.27 MeV = 3.2 K/atom. This lowering of energy favoring the ferromagnetic state is the source of the spin transition. Below T_C the magneto-elastic energy decreases at a more rapid rate with temperatures due to the effect the orthorhombic distortion ϵ^{γ} term.

In the antiferromagnetic range of both Dy and Tb as shown in Figure 4-31, the large difference between strains with and without an applied field reflects the exchange energy difference accompanying the forced destruction of the helical spin state. Below T_C no such transition occurs and only the curves with $H \neq 0$ have meaning as discussed above.

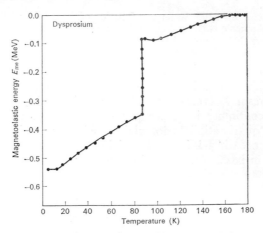

Figure 4-33. Magneto-elastic energy in dysprosium for both ferromagnetic and helical temperature ranges (after Ref. 110).

The energy associated with the helical state and the helical to ferro transition is large also in Ho^{99} as evidenced by the strain results shown in Figure 4–34A. The basal plane strain for $H=0$ is isotropic at all temperatures and no discontinuity or orthorhombic distortion is observed in the basal plane directions. A "lambda" type anomaly is observed in the calculated thermal expansion coefficients at 20 K. In applied fields the Ho strain reflects to some extent the onset of the complex fan states discussed in the magnetization section (see insert). Below the anomaly at 70 K the transverse magnetostriction (i.e., $G\|b$, $H\|a$; and $G\|a$, $H\|b$, where G is the gage direction) become anisotropic.

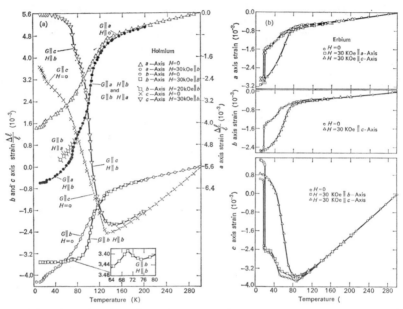

Figure 4–34. The a, b and c axis strain as a function of temperature for holmium (A) and erbium (B). Anomalous thermal expansion data $(H\doteq0)$, are shown as well as strain data taken in a 30 kOe applied field. G indicates the "gage" or strain measurement direction (after Refs. 99 and 100).

The effect of the change in exchange energy at the intermediate spin transition temperatures in Er is clearly seen in the strain results[93] of Figure 4–34B. The sharp discontinuity in lattice parameter at 19 K corresponds to the flipping of the "quasi-antiphase domain" state into the conical ferro configuration, an occurrence which is induced at higher temperatures in an applied field. Fields applied in basal plane directions have little effect except below 21 K where they induce a conical fan spin transformation as described in Section 4.2.1. As in Ho, no orthorhombic distortion[109] is observed below T_C since the basal plane moment component does not pick a preferred direction.

Gadolinium[104, 108] also shows the characteristic exchange-striction expansion of the c axis parameter below T_C accompanied by only a slight deviation of the basal plane lattice constants from normal thermal expansion behavior.

4.4.4. Elastic Constants

Due to the relatively low order symmetry of the rare earths there are five independent elastic constants. The complete specification of these as a function of temperature is an exacting task requiring data from three distinctly oriented disk shaped crystals. The discontinuous length changes encountered at some of the magnetic transitions present severe experimental problems in maintaining the bond to the transducer. Results have nevertheless been produced on Dy,[110, 112] Er,[112] and Gd[112] over a wide temperature range as shown in Figure 4–35. Recently elastic constant determinations made in a magnetic field to suppress domain effects have been reported for Ho and Dy by Palmer.[112a] Large fluctuations in the moduli are observed at the magnetic transition temperatures in all the elements studied but are most pronounced for Dy. The elastic coefficients C_{33} and C_{11} are proportional to the velocity of a longitudinal (compressional) sound wave propagated along the c axis and normal to it, respectively. Constants C_{44} and $C_{66} = \frac{1}{2}(C_{11} - C_{12})$ are the equivalent responses to shear wave propagations, while the cross coupling coefficient C_{13} is found from sound propagation perpendicular to a disk whose normal is $45°$ between c and a axes.

In Dy the degree of response at the transition temperature varies considerably among the C_{ij}'s. In particular for the longitudinal modes, the dip at T_N is much more pronounced in C_{33}, while at T_C C_{11} is most affected. These and the cross coupling constants C_{12} and C_{13} show much stronger effects than the shear constants C_{44} and C_{66}.

For erbium the elastic moduli show sharp anomalies at the 85 K sinusoidal moment ordering temperature and a fairly strong temperature dependence above T_N. The ratio of the shear coefficients C_{44}/C_{66} in both Dy and Er shows a strong increase with temperature in the range 300 K to 900 K. This behavior has been linked to the occurrence of the hcp-bcc phase transition just below the melting temperature as a result of studies on a wide range of hcp metals by Fisher and Dever.[112]

The elastic coefficients of Gadolinium are shown in Figure 4–35A. The curve marked QL is the "quasi-longitudinal" mode from which C_{13} is calculated. As in Er and Dy there is a very pronounced change in the c axis compressional mode constant C_{33} at the magnetic ordering temperature. C_{33} also shows effects of short range ordering in the paramagnetic range up to about 330 K. Near 220 K, C_{13} undergoes a "lambda" type anomaly while for C_{33} the acoustic attenuation becomes immeasurably large. These effects as well as the slope change in C_{66} are related to the rapid shift in moment direction which occurs in the vicinity of 220 K.

The attenuation of longitudinal and shear waves along the c axis has also been measured in Ho,[113, 114] Dy,[113, 114] and Tb[114] in the neighbourhood of the magnetic transition temperature. Typical results for the critical attenuation coefficient of Dy and Tb are shown in Figure 4–36 from the work of Pollina and Lüthi.[114] The sharp peak in longitudinal attenuation observed at T_N in Dy is broadened in Tb by the close proximity of the two transition points T_N and T_C. The spin-phonon coupling mechanism responsible for the longitudinal attenuation is largely of volume magnetostriction origin while that responsible for the shear wave attenuation is of single ion origin. The attenuation due to

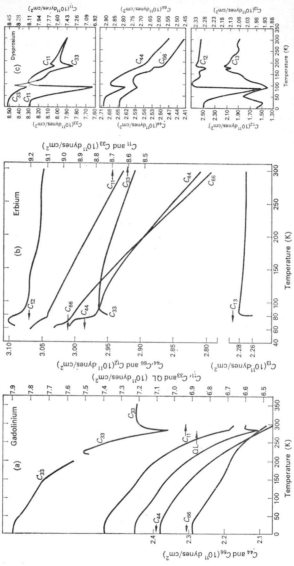

Figure 4-35. Temperature dependence of the five elastic constants for gadolinium (A), erbium (B), and dysprosium (C) single crystals. The curve labeled "QL" for Gd is a quasi-longitudinal mode from which C_{13} can be calculated (Dy data after Ref. 110; Gd, Er after Ref. 112).

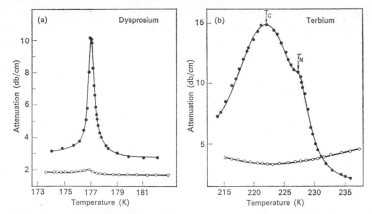

Figure 4–36. Critical ultrasonic attenuation near the Néel temperature in dysprosium (A) and in the vicinity of T_N and T_C for terbium (B). Solid symbols are data from longitudinal sound propagation and open symbols are for shear wave propagation (after Ref. 114).

this latter source is quite small at T_N. Values of the critical attenuation exponent η describing the temperature dependence of the attenuation (i.e., $\alpha \propto \omega^2 (T-T_N)^\eta$) have been compared with the predictions of scaling law theories.[114]

4.5. SPECIFIC HEAT

The specific heat measured at constant pressure in a magnetic material such as the rare earths consists of contributions from electronic C_E, lattice or phonon C_q, nuclear C_N, and magnetic C_M souces as follows:

$$C_P = C_N + C_E + C_q + C_M + \delta C, \qquad (4.24)$$

where the δC is a dilatation correction for the volume thermal expansion and is important only for measurements over an extended temperature range. The nuclear contribution is discussed in Chapter 8 and is obtainable from measurements at very low temperatures where all other excitations are frozen out. The other three terms C_E, C_q, and C_M are very difficult to separate in the magnetic rare earths. A complication is the appearance of an anomaly in the specific heat in the range 2–6 K arising from the excitation spectrum of the ever-present RE_2O_3 impurity. This makes difficult the determination of the electronic heat from the low-temperature data. At higher temperatures the lattice and magnetic spin wave terms dominate. Extreme difficulties are encountered in untangling these contributions since both may have a T^3 dependence over a considerable temperature range.

An exhaustive study of the specific heat of the rare earths has been done by Lounasmaa and co-workers[115] over the range 0.3 K to 25 K. Much of this research is involved in determining the explicit form of the spin wave and nuclear contributions and as such details will be left to the discussion of Chapters 5 and 8 respectively.

Heat capacity measurements over the approximate range 15 K to 300 K have been made by a group at Iowa State University.[116] Some of these results are shown in Figure 4–37, along with the results on Nd by Lounasmaa.[115] In general the specific heat shows a broadened "lamba" type anomaly at the paramagnetic to antiferromagnetic (or ferro in Gd) transition. In the helical to ferro transition in Dy a very sharp, presumably first order, anomaly is observed. In Nd the specific heat data clearly shows the separate ordering of cubic sites at 7.5 K and the ordering on hexagonal sites at 19 K.

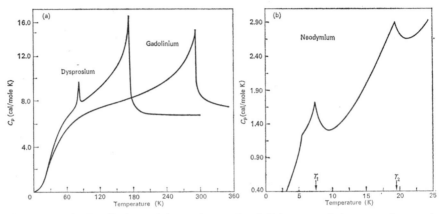

Figure 4–37. A. Specific heat of dysprosium and gadolinium over their range of magnetic order (after Ref. 116). B. Effect of distinct ordering temperatures for cubic (T_1) and d–hcp (T_2) sites on the low-temperature specific heat of neodymium (after Ref. 115).

The determination of the Debye temperature in the rare earths from specific heat data is subject to considerable uncertainty. The principal complication is the admixture of the large magnon specific heat having a similar temperature dependence as mentioned above. A procedure used by Lounasmaa[115] which appears most realistic is to assume a linear increase in θ_D with atomic number across the rare earth series. The end points La and Lu are non-magnetic and their θ_D can be determined relatively unambiguously. These values are $\theta_D = 142$ K for La[117] and $\theta_D = 210$ for Lu.[118] Debye temperatures determined by this interpolation method agree relatively well with those obtained from averaged sound velocity measurements[119] on both single and polycrystalline material which give probably the most reliable values. The θ_D values obtained by the Iowa State group from analyzing specific heat data are at variance with both of the above results and may be considered uniformly too low. Debye θ values are listed in Table 4.6 as derived from sound velocity averages for Gd–Er and by linear interpolation between La and Lu for the other elements.

The electronic component of the specific heat in the rare earths is a relatively small contribution except at low temperatures. In analyzing specific heat data the difficulty of isolating C_E has led to the use of a constant value of

$$C_E = 2.51 \times 10^{-3}\, T \text{ cal/mole-deg}$$

Table 4.6. Debye θ and Magnetic Entropy for the Heavy Rare Earths

	θ_D[a] (K)	S_{300}[c]	[d]S_{mag}expt. (cal/mole deg)	[d]S_{mag}theor.
Ce	147	16.68	3.74	3.56
Pr	152	17.49	4.67	4.37
Nd	157	17.54	4.94	4.58
Sm	166	16.68	4.41	3.56
Eu	—	18.64	—	4.13
Gd	$\begin{cases}173 \\ 181.5^{(b)}\end{cases}$	17.24	4.25	4.13
Tb	173	17.55	5.87	5.10
Dy	$\begin{cases}180 \\ 183^{(b)}\end{cases}$	17.91	6.41	5.51
Ho	188[b]	18.0	6.51	5.63
Er	191[b]	17.52	6.04	5.51
Tm	196	17.41	6.09	5.10

[a] Sound velocity results (polycrystal) for Gd–Er, rest interpolated between La and Lu.
[b] Single crystal result.
[c] Total Experimental entropy at 300 K except 360 K for Gd.
[d] Calculated magnetic contributions—see text.

found for La[117] for the first half of the series and the Lu[115] value

$$C_E = 2.69 \times 10^{-3}\, T \text{ cal/mole-deg}$$

for the second half. This approximation is reasonable in view of the assumed relative constancy of the electronic structure across the series.

The complication of the exact separation of the magnetic, elastic and electronic term encountered in analyzing the low temperature heat capacity is much less a problem in examining integrated quantities such as the entropy. Experimental values of the total entropy at 300 K have been provided by the work of Iowa State Group[116] and are given in Table 4.6. For Gd short range order persists above 300 K, thus the value given is for 360 K. This total entropy is made up of terms corresponding to those in Eqn (4.24). Using appropriate values of θ_D, the lattice contribution to the integrated entropy may be computed. Likewise the electronic contribution may be estimated under the assumptions discussed above. Subtraction of these terms from the total experimental entropy (the dilatation correction is negligible) yields the magnetic contribution to the entropy. These values are listed in Table 4.6, with corresponding values of θ_D.

Under the assumption that the metal goes from a completely disordered paramagnetic state to a fully ordered magnetic state in the temperature range covered, the magnetic entropy contribution should be

$$S_{mag} = R \ln (2J + 1) \tag{4.25}$$

This theoretical maximum is listed also in the Table. It is of interest to note that the derived experimental values of magnetic entropy are uniformly too large. This may indicate an additional contribution to either lattice or conduction electron entropy possibly of magnetic origin.

ACKNOWLEDGMENTS

The author wishes to thank B. R. Cooper, E. R. Callen, R. J. Elliott, W. C. Koehler and A. R. Mackintosh among others for many helpful discussions, and Mrs. E. Reidinger and Mrs. B. Gambel for typing the manuscript.

REFERENCES

1. Schieber, M., Crystal Growth, Suppl. *J. Phys. Chem. Solids* (H. S. Peiser, ed). Pergamon Press (1967), p. 271.
2. Nigh, H. E., Legvold, S. and Spedding, F. H., *Phys. Rev.*, **132**, 1092 (1963); also studied by Graham, C. D., Jr., Proc. Int. Conf. on Mag., Nottingham, England (1964), pp. 740. Féron, J. L. and Pauthenet, R., Proc. Seventh Rare Earth Research Conference, Coronado, Calif. 1968; and by Belov, K. P., Ergin, Yu. V., Levitin, R. Z. and Ped'Ko, A. V., *J. Expt. Theoret. Phys.* (USSR), **47**, 2080 (1964) [Engl. Trans.—*Sov. Phys-JETP*, **20**, 1397 (1965)].
3. Kasuya, T., in Magnetism (G. T. Rado and H. Suhl, eds). Academic Press, New York (1966), Vol. IIB.
4. Rhyne, J. J., Foner, S., McNiff, E. J. and Doclo, R., *J. Appl. Phys.*, **39**, 892 (1968).
5. Féron, J. L., Huy, G. and Pauthenet, R. Les Elements des Terres Rares II, 17 (Colloques Inter. du C.R.N.S. No 180, 1970).
6. Hegland, D. E., Legvold, S. and Spedding, F. H., *Phys. Rev.*, **131**, 158 (1963).
7. Niira, K., *Phys. Rev.*, **117**, 129 (1960); applied to terbium by A. R. Mackintosh, *Phys. Letters*, **4**, 140 (1963).
8. Møller, H. Bjerrum, Houmann, J. C. G. and Mackintosh, A. R., *J. Appl. Phys.*, **39**, 807 (1968).
9. Rhyne, J. J. and Clark, A. E., *J. Appl. Phys.*, **38**, 1379 (1967).
10. Belov, K. P. and Ergin, Yu. V., *J. Exptl. Theoret. Phys.* (*USSR*), **50**, 560 (1966). [Engl. Trans.—*Soviet Phys.-JETP*, **23**, 372 (1966).]
11. Kaplan, T. A., *Phys. Rev.*, **124**, 329 (1961).
12. Behrendt, D. R., Legvold, S. and Spedding, F. H., *Phys. Rev.*, **109**, 1544 (1958).
13. Flippen, R. B., *J. Appl. Phys.*, **35**, 1047 (1964).
14. Enz, U., *Physica*, **26**, 698 (1960).
15. Nagamiya, T., Nagata, K. and Kitano, Y., *Prog. Theoret. Phys.* (Kyoto), **27**, 1253 (1962); also Yoshimori, A., *J. Phys. Soc. Japan*, **14**, 807 (1959); Herpin, A., Meriel, P. and Villain, J., *C.R. Acad. Science* (Paris), **249**, 1334 (1959); and others.
16. Jew, T. T. and Legvold, S., U.S. Atomic Energy Commission Rept. IS-867, 1963.
17. Bly, P. H., Corner, W. D. and Taylor, K. N. R., *J. Appl. Phys.*, **40**, 4787 (1969).
18. Koehler, W. C., Cable, J. W., Wilkinson, M. K. and Wollan, E. O., *Phys. Rev.*, **151**, 414 (1966).
19. Strandburg, D. L., Legvold, S. and Spedding, F. H., *Phys. Rev.*, **27**, 2046 (1962).
20. Koehler, W. C., Cable, J. W., Child, H. R., Wilkinson, M. K. and Wollan, E. O., *Phys. Rev.*, **158**, 450 (1967).
21. Bozorth, R. M., Gambino, R. J. and Clark, A. E., *J. Appl. Phys.*, **39**, 883 (1968); Proc. Tenth Low Temperature Conference, St. Andrews, Scotland (1968).
22. Cable, J. W., Wollan, E. O., Koehler, W. C. and Wilkinson, M. K., *J. Appl. Phys.*, **32**, 49S (1961).
23. Green, R. W., Legvold, S. and Spedding, F. H., *Phys. Rev.*, **122**, 827 (1961).
24. Kitano, Y. and Nagamiya, T., *Prog. Theoret. Phys.* (Kyoto), **31**, 1 (1964).
25. Koehler, W. C., *J. Appl. Phys.*, **36**, 1078 (1965).
26. Brun, T. O., Sinha, S. K. and Wakabayashi, N., *Bull. Amer. Phys. Soc.*, **14**, 349 (1969).
27. Richards, D. B. and Legvold, S., *Phys. Rev.*, **186**, 508 (1969).
28. Foner, S., Schieber, M. and McNiff, E. J., Jr., *Phys. Letters*, **25A**, 321 (1967).
29. Chikazuma, S., Tanuma, S., Oguro, I., Ono, F. and Tajima, K., *IEEE Trans.*, **Mag-5**, 265 (1969).
30. Ernst, D. W. (private communication).
31. Finnemore, D. K., Williams, L. J., Spedding, F. H. and Hopkins, D. C., *Phys. Rev.*, **176**, 712 (1968).

32. Johnson, D. L. and Finnemore, D. K., *Phys. Rev.*, **158**, 376 (1967).
33. Wohlleben, D. K., Ph.D. Thesis, Univ. of Calif. (San Diego) 1968 (unpublished); also *Phys. Rev. Letters*, **21**, 1343 (1968).
34. Lock, J. M., *Proc. Phys. Soc.* (London), **B70**, 566 (1957).
35. Wilkinson, M. K., Child, H. R., McHargue, C. J., Koehler, W. C. and Wollan, E. O. *Phys. Rev.*, **122**, 1409 (1961).
36. Burr, C. R. and Ehara, S., *Phys. Rev.*, **149**, 551 (1966).
36a. Coqblin, B. and Blandin, A., *Adv. in Phys.*, **17**, 281 (1968).
36b. MacPherson, M. R., Wohlleben, D., Maple, M. B. and Everett, G. E. *Phys. Rev. Lett.*, **26**, 20 (1971).
37. Cable, J. W., Moon, R. M., Koehler, W. C. and Wollan, E. O., *Phys. Rev. Lett.*, **12**, 553 (1964).
38. Good agreement with reference 34 but with a slightly lower susceptibility was found below 20 K in the work of Nagasawa, H. and Sugawara, T., *J. Phys. Soc. Japan*, **23**, 701 (1967).
39. Bleaney, B., *Proc. Roy. Soc.*, **A276**, 39 (1963).
40. Johannsson, T., Lebech, B., Nielsen, M., Møller, H. Bjerrum and Mackintosh, A. R. *Phys. Rev. Lett.*, **25**, 524 (1970).
41. Nagasawa, H. and Sugawara, T., *J. Phys. Soc. Japan*, **23**, 701 (1967).
42. Moon, R. M., Cable, J. W. and Koehler, W. C., *J. Appl. Phys.*, **35**, 1041 (1964).
43. Behrendt, D. R., Legvold, S. and Spedding, F. H., *Phys. Rev.*, **106**, 723 (1957).
44. Arajs, S. and Miller, D. S., *J. Appl. Phys.*, **31S**, 325 (1960).
45. Jelinek, F. J., Hill, E. D. and Gerstein, B. C., *J. Phys. Chem. Solids*, **26**, 1475 (1965).
46. Schieber, M., Foner, S., Doclo, R. and McNiff, E. J., Jr., *J. Appl. Phys.*, **39**, 885 (1968).
46a. Foner, S., Voigt, C. and Alexander, E. J. *A.I.P. Conf Proc. Series* (1972 to be published).
46b. Koehler, W. C., Moon, R. M., Cable, J. W. and Child, H. R. *A.I.P. Conf. Proc. Series* (1972 to be published).
47. Arajs, S., *Phys. Rev.*, **120**, 756 (1960).
48. Nereson, N. G., Olsen, C. E. and Arnold, G. P., *Phys. Rev.*, **135**, A176 (1964).
49. Bozorth, R. M. and VanVleck, J. H., *Phys. Rev.*, **118**, 1493 (1960).
50. Cohen, R. L., Hüfner, S. and West, K. W., Les Elements des Terres Rares II, 101 (Colloques Inter. du C.N.R.S. Nº 180, 1970).
51. Lock, J. M., *Proc. Phys. Soc.* (London), **B70**, 476 (1956).
51a. Bucher, E., Schmidt, P. H., Jayaraman, A., Andres, K., Maita, J. P. and Dernier, P. D., *Phys. Rev.*, **132**, 3911 (1970).
51b. Tanuma, S., Datars, W. R. and Doi, H. *Solid State Comm.*, **8**, 1107 (1970).
52. Chechernikov, V. I., Pop, I. and Naumkin, O. P., *Sov. Phys.-JETP*, **17**, 1228 (1963).
53. Gardner, W. E. and Penfold, J., *Phys. Letters*, **26A**, 204 (1968).
54. Keeton, S. C. and Loucks, T. L., *Phys. Rev.*, **168**, 672 (1968).
55. Watson, R. E., Freeman, A. J. and Dimmock, J. P., *Phys. Rev.*, **167**, 497 (1968).
56. Evenson, W. E. and Liu, S. H., *Phys. Rev.*, **178**, 783 (1969).
57. Spedding, F. H., Jordan, R. G. and Williams, R. W., *J. Chem. Phys.*, **51**, 509 (1969).
58. Shirane, G. and Pickart, S. J., *J. Appl. Phys.*, **37**, 1032 (1966).
59. A summary of this work is found in Bozorth, R. M., *J. Appl. Phys.*, **38**, 1366 (1967).
60. Child, H. R. and Koehler, W. C., *J. Appl. Phys.*, **37**, 1353 (1966).
61. Nigh, H. E., Legvold, S., Spedding, F. H. and Beaudry, B. J., *J. Chem. Phys.*, **41**, 3799 (1964).
61a. Isaacs, L. L., Lam, D. J. and Fradin, F. Y., *J. Appl. Phys.*, **42**, 1458 (1971).
62. DeGennes, P. G., *C.R. Acad. Sci.* (Paris), **247**, 1836 (1958).
63. Nagasawa, H. and Sugawara, T., *J. Phys. Soc. Japan*, **23**, 711 (1967).
64. Child, H. R., Koehler, W. C., Wollan, E. O. and Cable, J. W., *Phys. Rev.*, **138**, A1655 (1965).
65. Weinstein, S., Craig, R. S. and Wallace, W. E., *J. Appl. Phys.*, **34**, 1354 (1963).
66. Nelson, D. and Legvold, S., *Phys. Rev.*, **123**, 80 (1961).
67. Bartholin, H. and Bloch, D., *J. Phys. Chem. Solids*, **29**, 1063 (1968).
68. McWhan, D. B. and Stevens, A. L., *Phys. Rev.*, **139**, A682 (1965); *Phys. Rev.*, **154**, 438 (1967); and Jayaraman, A., Sherwood, R. C., Williams, A. J. and Corenzwitt, E., *Phys. Rev.*, **148**, 502 (1966).

69. Milton, J. E. and Scott, T. A., *Phys. Rev.*, **160**, 387 (1967).
70. Jayaraman, A. and Sherwood, R. C., *Phys. Rev.*, **134**, A691 (1964).
71. Watson, R. E. and Freeman, A. J., *Phys. Rev.*, **131**, 250 (1963); *Phys. Rev.*, **132**, 706 (1963).
72. Callen, H. B. and Callen, E., *J. Phys. Chem. Solids*, **27**, 1271 (1966).
73. Callen, H. B. and Shtrikman, S., *Solid State Comm.*, **3**, 5 (1965).
74. Zener, C., *Phys. Rev.*, **96**, 1335 (1954).
74a. Brooks, M. S. S., *J. Phys. Chem. (Solid State Physics)*, **2**, 1016 (1969).
75. Callen, E. and Callen, H. B., *Phys. Rev.*, **139**, A455 (1965).
76. Elliott, R. J. and Stevens, K. W. H., *Proc. Roy. Soc.*, **A219**, 387 (1953).
77. Stevens, K. W. H., *Proc. Phys. Soc.* (London), **A65**, 209 (1952).
77a. Nielsen, M., Møller, H. B., Lindgard, P. A. and Mackintosh, A. R., *Phys. Rev. Letters* **25**, 1451 (1970).
78. Stanford, J. L. and Young, R. C., *Phys. Rev.*, **157**, 245 (1967); Wagner, T. K. and Stanford, J. L., *Phys. Rev.*, **184**, 505 (1969).
79. Tajima, K. and Chikazumi, S., *J. Phys. Soc. Japan*, **23**, 1175 (1967).
80. Levitin, R. Z. and Poromarev, B. K., *Soviet Phys.-JETP*, **26**, 1121 (1968).
81. Liu, S. H., Behrendt, D. R., Legvold, S. and Good, R. H., Jr., *Phys. Rev.*, **116**, 1461 (1959).
82. Bly, P. H., Corner, W. D. and Taylor, K. N. R., *J. Appl. Phys.*, **39**, 1336 (1968); *J. Appl. Phys.*, **40**, 4787 (1969).
83. Du Plessis, P. De V., *Physica*, **41**, 379 (1969).
84. Cooper, B. R., *Phys. Rev.*, **169**, 281 (1968).
85. Nielsen, M., Møller, H. Bjerrum and Mackintosh, A. R., *J. Appl. Phys.*, **41**, 1174 (1970).
86. Turov, E. A. and Shavrov, V. G., *Fiz. Tverd. Tela*, **7**, 217 (1965). [Engl. Trans. *Soviet Phys.-SS*, **7**, 166 (1965).]
87. Tajima, K. and Chikazumi, S., *J. Phys. Soc. Japan*, **24**, 1401 (1968).
88. Corner, W. D., Roe, W. C. and Taylor, K. N. R., *Proc. Phys. Soc.* (London), **80**, 927 (1962).
89. Darby, M. I. and Taylor, K. N. R., Proc. Intl. Conf. on Magnetism. Institute of Physics and The Physical Society, London, (1965), p. 742.
90. Graham, C. D., Jr., *J. Phys. Soc. Japan*, **17**, 1310 (1962); *J. Appl. Phys.*, **38**, 1375 (1967); and *J. Appl. Phys.*, **34**, 1341 (1963).
90a. Cable, J. W. and Wollan, E. O., *Phys. Rev.*, **165**, 733 (1968).
91. Toyama, K., Tajima, K., Chikazumi, S. and Sawaoka, A., *J. Phys. Soc. Japan*, **27**, 1070 (1969).
92. Callen, E. R. and Callen, H. B., *Phys. Rev.*, **129**, 578 (1963); Callen, Earl and Callen, Herbert B., *Phys. Rev.*, **139**, A455 (1965). An excellent review of the phenomenological theory of magnetostriction and experimental results is found in the paper by Earl Callen, *J. Appl. Phys.*, **39**, 519 (1968).
93. Rhyne, J. J. and Legvold, S., *Phys. Rev.*, **138**, A507 (1965).
94. DeSavage, B. F. and Clark, A. E., Proc. Fifth Rare Earth Research Conference, Ames, Iowa (1965) [lithographed].
95. Du Plessis, P. de V. and Alberts, L., *Phil. Mag.*, **18**, 151 (1968).
96. Rhyne, J. J., Ph.D. Thesis, Iowa State Univ., Ames, Iowa (1965).
97. Legvold, S., Alstad, J. and Rhyne, J., *Phys. Rev. Letters*, **10**, 509 (1963).
98. Clark, A. E., DeSavage, B. F. and Bozorth, R., *Phys. Rev.*, **138**, A216 (1965); *Phys. Letters*, **5**, 100 (1963).
99. Rhyne, J. J., Legvold, S. and Rodine, E. T., *Phys. Rev.*, **154**, 266 (1967).
100. Rhyne, J. J. and Legvold, S., *Phys. Rev.*, **140**, A2143 (1965).
101. Alstad, J. and Legvold, S., *J. Appl. Phys.*, **35**, 1752 (1964).
102. Belov, K. P., Levitin, R. Z. and Poromarjov, B. K., *J. Appl. Phys.*, **39**, 3285 (1968).
103. Coleman, W. E. and Pavlovic, A. S., *Phys. Rev.*, **135**, A426 (1964); *J. Phys. Chem. Solids*, **26**, 691 (1965).
104. Bozorth, R. M. and Wakiyama, T., *J. Phys. Soc. Japan*, **18**, 97 (1963).
105. Mason, W. P., *Phys. Rev.*, **96**, 302 (1954).
106. Tsuya, N., Clark, A. E. and Bozorth, R. M., Proc. Intl. Conf. on Magnetism. Institute of Physics and The Physical Society, London (1965), p. 250.

107. Lee, E. W., *Proc. Phys. Soc.* (London), **84**, 693 (1964).
108. Darnell, F. J., *Phys. Rev.*, **130**, 1825 (1963).
109. Darnell, F. J., *Phys. Rev.*, **132**, 1098 (1963).
110. Rosen, M. and Klimer, H., *Phys. Rev.*, B, **1**, 3748 (1970).
111. Calculated by Cooper, B. R., *Phys. Rev. Letters*, **19**, 900 (1967), also suggested by Jordan, R. G. and Lee, E. W., *Proc. Phys. Soc.* (London), **92**, 1074 (1967); and others. Helical spin regime treated in detail by Evenson and Liu (Ref. 56).
112. Fisher, E. S. and Dever, D., *AIME Trans.*, **239**, 48 (1967); *Proc. 6th Rare Earth Res. Conf.*, Gatlinburg, Tenn. 1967, p. 522 [lithographed].
112a. Palmer, S. B., *J. Chem. Phys. Solids*, **31**, 143 (1970) also Palmer, S. B., Ph.D. Thesis, University of Southampton (1969).
113. Tachiki, M., Levy, M., Kagiwada, R. and Lee, M. C., *Phys. Rev. Letters*, **21**, 1193 (1968).
114. Pollina, R. J. and Lüthi, B., *Phys. Rev.*, **177**, 841 (1969).
115. A summary of much of this work is contained in Lounasmaa, O. V. and Sundstrom, L. J., *Phys. Rev.*, **150**, 399 (1966) for the heavy rare earths and in *Phys. Rev.*, **158**, 591 (1967) for the light rare earths.
116. Results on the elements La, Ce, Pr, and Nd—Parkinson, Simon and Spedding, *Proc. Roy. Soc.* (London), **A207**, 137 (1951); Sm—Jennings, L. D., Hill, E. D. and Spedding, F. H., *J. Chem. Phys.*, **31**, 1230 (1959); Eu—Gerstein, B. C., Jelinek, F. J., Mullaly, J. R., Shickell, W. D. and Spedding, F. H., *J. Chem. Phys.*, **47**, 5194 (1967); Gd— Griffel, M., Skochdopole, R. E. and Spedding, F. H., *Phys. Rev.*, **93**, 657 (1954); Tb— Jennings, L. D., Stanton, R. M. and Spedding, F. H., *J. Chem. Phys.*, **27**, 909 (1957); Dy—Griffel, M., Skochdopole, R. E. and Spedding, F. H., *J. Chem. Phys.*, **25**, 75 (1956); Ho—Gerstein, B. C., Griffel, M., Jennings, L. D., Miller, R. E., Skochdopole, R. E. and Spedding, F. H., *J. Chem. Phys.*, **27**, 394 (1957); Er—Skochdopole, R. E., Griffel, M. and Spedding, F. H., *J. Chem. Phys.*, **23**, 2258 (1955); Tm—Jennings, L. D., Hill, E. and Spedding, F. H., *J. Chem. Phys.*, **34**, 2082 (1961).
117. Berman, A., Zemansky, M. W. and Boorse, H. A., *Phys. Rev.*, **109**, 70 (1958).
118. Lounasmaa, O. V., *Phys. Rev.*, **133**, A219 (1964).
119. Single crystal results are given in Section 4.4.4; polycrystal references are given in Gschneidner, K. A., Rare Earth Alloys. Van Nostrand, New York (1961), p. 38.

Chapter 5

Spin Waves

A. R. Mackintosh

H.C. Ørsted Institute, University of Copenhagen

and

H. Bjerrum Møller

Atomic Energy Commission Research Establishment,
Risø, Roskilde, Denmark

5.1 INTRODUCTION

In this chapter, we will review the knowledge which has been accumulated over the last few years on spin waves in the rare earth metals, and describe the way in which this information may be interpreted to elucidate the magnetic interactions which give rise to their characteristic magnetically ordered structures. Most of the discussion will be concerned with the hcp heavy rare earths, since the magnetic excitations in the light rare earths, which are complicated by strong crystal field effects, have not yet been extensively studied. Recent measurements on praseodymium will, however, be briefly described. The great majority of the information which has been obtained about the dispersion relations for the spin wave quanta, or magnons, has resulted from inelastic neutron scattering experiments since these allow, in principle and often even in practice, the determination of the magnon energy at any point in the Brillouin zone, as well as providing much valuable additional information about magnon interactions. An important supplementary technique is provided by the absorption of electromagnetic waves, which, however, only determines the magnon energies at certain high symmetry points within the zone. In addition, some information may be obtained from the low-temperature thermodynamic and transport properties, although the number of explicit features of the magnon dispersion relations which can be deduced from them is extremely limited. However, they do provide a useful check on the consistency of the measured or calculated magnon spectrum.

We shall begin with a brief review of the magnetic interactions which determine the magnon dispersion relations and discuss the various terms in the spin wave Hamiltonian which will later be used to interpret the experimental results. The most important of these is the indirect exchange interaction, and we shall initially derive the magnon dispersion relations and neutron scattering cross section for a ferromagnet in which the exchange is much greater than the anisotropy terms. The experimental results for gadolinium, in which the anisotropy is indeed very small, will be discussed within this framework. The diagonalization of the Hamiltonian for an anisotropic ferromagnet will be described more briefly, and applied to the ferromagnetic phase of terbium, for which extensive experimental results are available. The extraction of the exchange and anisotropy parameters from these results will be emphasized. The more difficult problems encountered in calculating and measuring the magnon spectra in the periodic structures, which result from the oscillatory nature of the indirect exchange, will then be described and the more limited amount of experimental information reviewed. The magnon spectrum determines the temperature variation of the magnetization and the magnetic contribution to the heat capacity, and we will discuss the consistency of these thermodynamic properties with the other experimental results. We will then review the information obtained from neutron scattering experiments on the interactions of the magnons with each other, with impurities, and with other excitations in the system. Finally we will discuss the results of recent measurements of the magnetic excitations in praseodymium.

5.2. THE SPIN-WAVE HAMILTONIAN

In this section we shall discuss the Hamiltonian in terms of which the experimental results on spin waves in the rare earths will be interpreted. The magnetic interactions of primary importance in determining the magnon dispersion relations are the indirect exchange interaction between the magnetic ions, which is mediated by the conduction electrons, and the magnetic anisotropy forces, due to the crystalline electric fields or an external magnetic field. These interactions are described extensively elsewhere in this volume, the former in Chapter 6 and the latter in Chapters 1 and 2, but it is convenient here to summarize the principal results of these discussions and emphasize the physical processes involved.

5.2.1. The Indirect Exchange Interaction

The direct overlap between the $4f$ electrons, which carry the ionic moments in the rare earths, is negligible but they are coupled together quite strongly through the conduction electrons. We can calculate the form of this coupling in a straightforward manner, provided that we assume the approximate form[1]

$$\mathcal{H}_{sf} = -\frac{1}{N} U(\mathbf{r} - \mathbf{R}_l) \mathbf{S}_l \cdot \mathbf{s} \qquad (5.1)$$

for the exchange interaction between the localized spin $\mathbf{S}_l$ on the site $\mathbf{R}_l$ and a conduction electron of spin $\mathbf{s}$. N is the number of ions in the system. This exchange may be viewed as an effective magnetic field acting on the conduction electron gas and giving rise to a moment at $\mathbf{r}'$, whose Cartesian components are given by

$$s^\alpha(\mathbf{r}') = \frac{1}{N} \int U(\mathbf{r} - \mathbf{R}_l) \chi^{\alpha\beta}(\mathbf{r} - \mathbf{r}') S_l^\beta d\mathbf{r} \qquad (5.2)$$

where $\chi^{\alpha\beta}$ is the nonlocal susceptibility tensor for the conduction electrons. This induced moment interacts through (5.1) with the spin $\mathbf{S}_{l'}$, producing a coupling of the form

$$\mathcal{H}_{ll'} = -\frac{1}{N^2} \int U(\mathbf{r} - \mathbf{R}_l) U(\mathbf{r}' - \mathbf{R}_{l'}) \chi^{\alpha\beta}(\mathbf{r} - \mathbf{r}') d\mathbf{r} \, d\mathbf{r}' S_l^\beta S_{l'}^\alpha. \qquad (5.3)$$

Taking the Fourier transforms

$$\chi^{\alpha\beta}(\mathbf{q}) = \int \chi^{\alpha\beta}(\mathbf{r}) e^{i\mathbf{q}\cdot\mathbf{r}} \, d\mathbf{r}$$

$$U(\mathbf{q}) = \int U(\mathbf{r}) e^{i\mathbf{q}\cdot\mathbf{r}} \, d\mathbf{r} \qquad (5.4)$$

and summing over lattice sites, we find

$$\mathcal{H}_{ff} = - \sum_{l>l'} j^{\alpha\beta}(\mathbf{R}_l - \mathbf{R}_{l'}) S_l^\beta S_{l'}^\alpha$$

$$= - \frac{1}{N} \sum_{l>l',\mathbf{q}}^{\alpha\beta} j^{\alpha\beta}(\mathbf{q}) e^{-i\mathbf{q}\cdot(\mathbf{R}_l-\mathbf{R}_{l'})} S_l^\beta S_{l'}^\alpha \tag{5.5}$$

where

$$j^{\alpha\beta}(\mathbf{q}) = \frac{2}{N} |U(\mathbf{q})|^2 \chi^{\alpha\beta}(\mathbf{q}) \tag{5.6}$$

and the sum is taken in such a way that the interaction between S_l and $S_{l'}$ appears only once.

If we neglect the spin-orbit coupling of the conduction electrons and the metal is unmagnetized, $\chi(\mathbf{q})$ is isotropic and the indirect exchange interaction takes the familiar isotropic Heisenberg form[2]

$$\mathcal{H}_{ff} = - \sum_{l>l'} j(\mathbf{R}_l - \mathbf{R}_{l'}) \mathbf{S}_l \cdot \mathbf{S}_{l'}. \tag{5.7}$$

It should be noted that the form (5.1) which we have assumed for the exchange interaction between the local moments and the conduction electrons is not generally strictly valid and this, together with the anisotropy of $\chi^{\alpha\beta}(\mathbf{q})$ in the magnetized state, implies that the true interaction between the ions is appreciably more complicated than (5.7) and contains components anisotropic in the spins. Furthermore the strong spin-orbit coupling in the $4f$ shell of the rare earths has the effect that $\mathbf{S}$ is not a constant of motion. The leading term in the interaction between the moments is then obtained by projecting $\mathbf{S}$ on $\mathbf{J}$ and still has the Heisenberg form

$$\mathcal{H}_{ff} = - \sum_{l>l'} \mathcal{J}(\mathbf{R}_l - \mathbf{R}_{l'}) \mathbf{J}_l \cdot \mathbf{J}_{l'} \tag{5.8}$$

where now

$$\mathcal{J}(\mathbf{q}) = \frac{2(\lambda-1)^2}{N} |U(\mathbf{q})|^2 \chi(\mathbf{q}) \tag{5.9}$$

and λ is the Landé factor, but there are anisotropic higher order terms,[3] which involve both space and momentum variables. In addition, the spin-orbit coupling of the conduction electrons introduces further terms which couple the spin on one site to the orbital moment on another.[4] The analysis of the experimental results which we shall present later shows, however, that the isotropic Heisenberg form (5.8) provides a good first approximation to the exchange in the rare earths and, for simplicity, we shall generally use it in our discussion.

5.2.2. Magnetic Anisotropy

The crystal field acting on a particular ion, which is the result of the anisotropic distribution of the other ions and conduction electrons, produces a splitting of the $4f$ quantum levels, as described in Chapter 1. The minimization of this crystal field energy causes a preferential orientation of the magnetic moments, which may be viewed classically as resulting from the action of the crystalline electric field on the anisotropic $4f$ charge distribution. The large spin-orbit coupling then ensures that the spin, as well as the orbital moment, follows the charge distribution. Thus the anisotropy is large, for example, in terbium which has a highly anisotropic $4f$ charge distribution, while it is small in gadolinium which has an almost spherical ion.

For a hexagonal metal, we may write the Hamiltonian for crystal field anisotropy in the form[5]

$$\mathcal{H}_{cf} = \sum \{V_2 J_{l\zeta}{}^2 + V_4 J_{l\zeta}{}^4 + V_6 J_{l\zeta}{}^6 - \tfrac{1}{2} V_6{}^6$$
$$\times [(J_{l\xi} + iJ_{l\eta})^6 + (J_{l\xi} - iJ_{l\eta})^6]\}. \qquad (5.10)$$

We have chosen Cartesian axes such that ξ is the easy direction of magnetization in the hexagonal plane, while ζ is the hexagonal axis.

Magnetic ordering may be accompanied by a magnetostrictive strain, which reduces the energy of the system by modifying the crystal fields. A slow rotation of the magnetization then results in a change in this strain, and this second order magnetoelastic effect makes an additional contribution to the magnetic anisotropy. As we shall see later, however, the experimental results show that the precession of the moments in a spin wave is sufficiently fast that the magnetoelastic strain is unable to follow it. It therefore remains static and the frozen lattice model of Turov and Shavrov[6] applies. In the case of a ferromagnet in which the moments are constrained to lie in the hexagonal plane, such as Tb or Dy, the magnetoelastic contribution to the spin wave Hamiltonian arising from distortions in the plane takes the form[7]

$$\mathcal{H}_{me} = -\frac{2Cc}{(J-\tfrac{1}{2})} \left[\frac{\bar{\epsilon}_1{}^\gamma}{2} \sum_l (J_{l\xi}{}^2 - J_{l\eta}{}^2) + \bar{\epsilon}_2{}^\gamma \sum_l (J_{l\xi} J_{l\eta}) \right]$$
$$-\frac{2Ac}{(J-\tfrac{1}{2})(J-1)(J-\tfrac{3}{2})} \left[\frac{\bar{\epsilon}_1{}^\gamma}{4} \sum_l (J_{l\xi}{}^4 - 6J_{l\xi}{}^2 J_{l\eta}{}^2 + J_{l\eta}{}^4) \right.$$
$$\left. - \bar{\epsilon}_2{}^\gamma \sum_l J_{l\xi} J_{l\eta}(J_{l\xi}{}^2 - J_{l\eta}{}^2) \right] \qquad (5.11)$$

where C and A are the first and second order magnetostriction coefficients[8] and c is a reduced elastic constant, defined by

$$c = c^\gamma / J. \qquad (5.12)$$

$\bar{\epsilon}_1{}^\gamma$ and $\bar{\epsilon}_2{}^\gamma$ are the equilibrium strains, related to the magnetostriction coefficients by

$$\bar{\epsilon}_1{}^\gamma = -C \cos 2\phi + \tfrac{1}{2}A \cos 4\phi$$
$$\bar{\epsilon}_2{}^\gamma = C \sin 2\phi + \tfrac{1}{2}A \sin 4\phi \qquad (5.13)$$

ϕ is the angle in the plane between the direction of magnetization and the hard axis. This angle will generally not be $\pi/6$ in the presence of a magnetic field, which contributes a term

$$\mathcal{H}_h = -\lambda\beta \sum_l \mathbf{H} \cdot \mathbf{J}_l \qquad (5.14)$$

to the Hamiltonian, and hence has a tendency to align the moments along the field direction.

5.3. THE FERROMAGNET WITH SMALL ANISOTROPY

The dominant term in the spin wave Hamiltonian is the indirect exchange coupling (5.8) which is responsible for the magnetic ordering. We shall first consider the properties of a system in which all other terms can be neglected, so that the magnetic anisotropy enters only by specifying the direction of magnetization, conventionally chosen to be the z-axis. The magnon dispersion relation for such an isotropic ferromagnet will be calculated for a general Bravais lattice, and the calculation then extended to the hcp structure. The symmetry properties of the magnons in this structure will be described both for the case of simple and more general interactions. The neutron scattering cross section will be calculated and the recent neutron measurements on gadolinium[9] used to illustrate some of the results of this section.

5.3.1. Ferromagnetic Magnons in a Bravais Lattice

The exchange Hamiltonian (5.8) involves the angular momentum operators, whose components satisfy commutation relations

$$[J_{lx}, J_{ly}] = iJ_{lz} \tag{5.15}$$

and are related by

$$\mathbf{J}_l \cdot \mathbf{J}_l = J(J+1). \tag{5.16}$$

In order to write the Hamiltonian in terms of boson creation and annihilation operators, we make the Holstein–Primakoff transformation[10]

$$J_l^+ = J_{lx} + iJ_{ly} = (2J)^{1/2}\left(1 - \frac{a_l^+ a_l}{2J}\right)^{1/2} a_l$$

$$J_l^- = J_{lx} - iJ_{ly} = (2J)^{1/2} a_l^+ \left(1 - \frac{a_l^+ a_l}{2J}\right)^{1/2} \tag{5.17}$$

where

$$[a_l, a_{l'}^+] = \delta_{ll'} \tag{5.18}$$

so that the commutation relation

$$[J_l^+, J_l^-] = 2J_{lz} \tag{5.19}$$

is satisfied.

From (5.16) we now have

$$(J_{lz})^2 = J(J+1) - \tfrac{1}{2}(J_l^+ J_l^- + J_l^- J_l^+) \tag{5.20}$$

so that, using (5.17) and (5.18) we have the exact relation

$$J_{lz} = J - a_l^+ a_l \tag{5.21}$$

a_l^+ is therefore a boson operator which creates a deviation of one quantum $\hbar$ of angular momentum from the completely aligned state at the site l. At low temperatures, the thermal average of the number of such spin deviations $\langle a_l^+ a_l \rangle$ is small, so we may neglect the terms $a_l^+ a_l/2J$ and write

$$J_l^+ \simeq (2J)^{1/2} a_l, \qquad J_l^- \simeq (2J)^{1/2} a_l^+. \tag{5.22}$$

The exchange Hamiltonian now becomes

$$\mathcal{H} = - \sum_{l>l'} \mathcal{J}(\mathbf{R}_{ll'})[\tfrac{1}{2}(J_l{}^+J_{l'}{}^- + J_l{}^-J_{l'}{}^+) + J_{lz}J_{l'z}]$$

$$= -J \sum_{l>l'} \mathcal{J}(\mathbf{R}_{ll'})[J - a_l{}^+a_l - a_{l'}{}^+a_{l'} + a_l{}^+a_{l'} + a_l a_{l'}{}^+]$$

$$= -J \sum_{l>l'} \mathcal{J}(\mathbf{R}_{ll'})[J - 2a_l{}^+a_l + 2a_l{}^+a_{l'}] \qquad (5.23)$$

where

$$\mathbf{R}_{ll'} = (\mathbf{R}_l - \mathbf{R}_{l'}).$$

This may readily be diagonalized by introducing magnon creation and annihilation operators, defined by

$$a_l{}^+ = N^{-1/2} \sum_{\mathbf{q}} e^{i\mathbf{q}\cdot\mathbf{R}_l} a_{\mathbf{q}}{}^+$$

$$a_l = N^{-1/2} \sum_{\mathbf{q}} e^{-i\mathbf{q}\cdot\mathbf{R}_l} a_{\mathbf{q}} \qquad (5.24)$$

where the sum is taken over the Brillouin zone. It follows from (5.18) that these operators satisfy boson commutation relations

$$[a_{\mathbf{q}}, a_{\mathbf{q}'}{}^+] = \delta_{\mathbf{q}\mathbf{q}'}. \qquad (5.25)$$

Transforming to these operators and using the Fourier transformed exchange function, defined by

$$\mathcal{J}(\mathbf{R}_{ll'}) = \frac{1}{N} \sum_{\mathbf{q}} \mathcal{J}(\mathbf{q}) e^{-i\mathbf{q}\cdot\mathbf{R}_{ll'}} \qquad (5.26)$$

we find

$$\mathcal{H} = -\frac{J^2}{N} \sum_{\mathbf{q}} \sum_{l>l'} \mathcal{J}(\mathbf{q}) e^{-i\mathbf{q}\cdot\mathbf{R}_{ll'}} + \frac{2J}{N^2} \sum_{\mathbf{q}\mathbf{q}'\mathbf{q}''} \sum_{l>l'} \mathcal{J}(\mathbf{q}) e^{-i\mathbf{q}\cdot\mathbf{R}_{ll}}$$
$$\times [a_{\mathbf{q}'}{}^+a_{\mathbf{q}''} e^{-i(\mathbf{q}'-\mathbf{q}'')\cdot\mathbf{R}_l} - a_{\mathbf{q}'}{}^+a_{\mathbf{q}''} e^{i\mathbf{q}'\cdot\mathbf{R}_l} e^{-i\mathbf{q}''\cdot\mathbf{R}_{l'}}]. \qquad (5.27)$$

Using the relation

$$\sum_{l>l'} e^{i(\mathbf{q}-\mathbf{q}')\cdot\mathbf{R}_{ll'}} = \tfrac{1}{2}N(N-1)\delta_{\mathbf{q}\mathbf{q}'} \simeq \tfrac{1}{2}N^2\delta_{\mathbf{q}\mathbf{q}'}. \qquad (5.28)$$

this reduces to

$$\mathcal{H} = -\tfrac{1}{2}NJ^2\mathcal{J}(o) + J \sum_{\mathbf{q}} [\mathcal{J}(o) - \mathcal{J}(\mathbf{q})]a_{\mathbf{q}}{}^+a_{\mathbf{q}}. \qquad (5.29)$$

The first term is the ground state energy and, from the second, the magnon dispersion relation is

$$\hbar\omega(\mathbf{q}) = J[\mathcal{J}(o) - \mathcal{J}(\mathbf{q})]. \qquad (5.30)$$

The magnon operator

$$a_{\mathbf{q}}{}^+ = N^{-1/2} \sum_{l} e^{-i\mathbf{q}\cdot\mathbf{R}_l} a_l{}^+$$

creates a collective excitation of the system, in which a small deviation from the completely ordered state propagates through the lattice. The z-component of the angular momentum for the whole system is given by

$$\sum_l J_{lz} = \sum_l (J - a_l{}^+a_l) = NJ - \frac{1}{N} \sum_{l\mathbf{q}\mathbf{q}'} e^{i(\mathbf{q}-\mathbf{q}')\cdot\mathbf{R}_l} a_{\mathbf{q}}{}^+a_{\mathbf{q}'}$$
$$= NJ - \sum_{\mathbf{q}} n_{\mathbf{q}} \qquad (5.31)$$

Each excited magnon therefore contributes a deviation of one unit of angular momentum from the fully ordered state.

5.3.2. The Hexagonal Close-Packed Lattice

The hcp lattice may be considered as two equivalent interpenetrating sublattices, as shown in Figure 5–1. The two types of sites will be denoted A

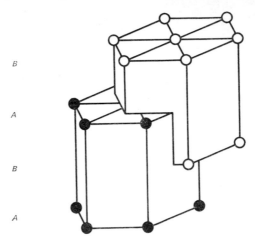

Figure 5–1. The hexagonal close-packed structure, composed of two interpenetrating simple hexagonal lattices.

and B, with corresponding indices l and m respectively. The exchange Hamiltonian for this lattice may then be written

$$\mathcal{H} = - \sum_{l>l'} \mathcal{J}(\mathbf{R}_{ll'})\mathbf{J}_l \cdot \mathbf{J}_{l'} - \sum_{m>m'} \mathcal{J}(\mathbf{R}_{mm'})\mathbf{J}_m \cdot \mathbf{J}_{m'} - \sum_{lm} \mathcal{J}(\mathbf{R}_{lm})\mathbf{J}_l \cdot \mathbf{J}_m \quad (5.32)$$

Performing the Holstein–Primakoff transformation as before, we find

$$\mathcal{H} = -J \left\{ \sum_{l>l'} \mathcal{J}(\mathbf{R}_{ll'})[J - 2a_l^+a_l + 2a_l^+a_{l'}] \right.$$
$$+ \sum_{m>m'} \mathcal{J}(\mathbf{R}_{mm'})[J - 2b_m^+b_m + 2b_m^+b_{m'}]$$
$$\left. + \sum_{lm} \mathcal{J}(\mathbf{R}_{lm})[J - a_l^+a_l - b_m^+b_m + a_l^+b_m + a_lb_m^+] \right\} \quad (5.33)$$

where a_l^+ and b_m^+ create a moment deviation on the A and B sites respectively. We now introduce the Fourier transformed exchange functions, defined by

$$\mathcal{J}(\mathbf{R}_{ll'}) = \frac{1}{N} \sum_{\mathbf{q}} \mathcal{J}(\mathbf{q})e^{-i\mathbf{q}\cdot\mathbf{R}_{ll'}}$$

$$\mathcal{J}(\mathbf{R}_{lm}) = \frac{1}{N} \sum_{\mathbf{q}} \mathcal{J}'(\mathbf{q})e^{-i\mathbf{q}\cdot\mathbf{R}_{lm}}$$

$$(5.34)$$

where N is now the number of unit cells in the system and, because of the lattice symmetry, $\mathcal{J}'(\mathbf{q})$ is generally complex. Introducing the Fourier transforms of the boson operators

$$a_{\mathbf{q}}^+ = N^{-\frac{1}{2}} \sum_l e^{-i\mathbf{q}\cdot\mathbf{R}_l} a_l^+$$

$$b_{\mathbf{q}}^+ = N^{-\frac{1}{2}} \sum_m e^{-i\mathbf{q}\cdot\mathbf{R}_m} b_m^+ \qquad (5.35)$$

(5.33) may be reduced to the form

$$\mathscr{H} = -NJ^2[\mathscr{J}(o)+\mathscr{J}'(o)]+J\sum_{\mathbf{q}}\{[\mathscr{J}(o)-\mathscr{J}(\mathbf{q})+\mathscr{J}'(o)][a_{\mathbf{q}}^+a_{\mathbf{q}}+b_{\mathbf{q}}^+b_{\mathbf{q}}]$$
$$-\mathscr{J}'(\mathbf{q})a_{\mathbf{q}}^+b_{\mathbf{q}}-\mathscr{J}'^*(\mathbf{q})a_{\mathbf{q}}b_{\mathbf{q}}^+\}. \qquad (5.36)$$

This Hamiltonian may be diagonalized by a canonical transformation[11] to new boson operators, defined by

$$\alpha_{\mathbf{q}}^+ = \left(\frac{\mathscr{J}'(\mathbf{q})}{4\mathscr{J}'^*(\mathbf{q})}\right)^{\frac{1}{4}}(a_{\mathbf{q}}^+ + b_{\mathbf{q}}^+)$$

$$\beta_{\mathbf{q}}^+ = \left(\frac{\mathscr{J}'(\mathbf{q})}{4\mathscr{J}'^*(\mathbf{q})}\right)^{\frac{1}{4}}(a_{\mathbf{q}}^+ - b_{\mathbf{q}}^+) \qquad (5.37)$$

in terms of which (5.36) becomes

$$\mathscr{H} = -NJ^2[\mathscr{J}(o)+\mathscr{J}'(o)]$$
$$+J\sum_{\mathbf{q}}[\mathscr{J}(o)-\mathscr{J}(\mathbf{q})+\mathscr{J}'(o)-|\mathscr{J}'(\mathbf{q})|]\alpha_{\mathbf{q}}^+\alpha_{\mathbf{q}}$$
$$+J\sum_{\mathbf{q}}[\mathscr{J}(o)-\mathscr{J}(\mathbf{q})+\mathscr{J}'(o)+|\mathscr{J}'(\mathbf{q})|]\beta_{\mathbf{q}}^+\beta_{\mathbf{q}}. \qquad (5.38)$$

For a given $\mathbf{q}$ we now have two excitations, created by $\alpha_{\mathbf{q}}^+$ and $\beta_{\mathbf{q}}^+$ and referred to as acoustic and optical magnons respectively. For $\mathbf{q}=o$ the moment precessions on the A and B sites are in phase for the acoustic mode and in antiphase for the optical mode. The two branches of the dispersion relation are given by

$$\hbar\omega_\alpha(\mathbf{q}) = J[\mathscr{J}(o)-\mathscr{J}(\mathbf{q})+\mathscr{J}'(o)-|\mathscr{J}'(\mathbf{q})|]$$
$$\hbar\omega_\beta(\mathbf{q}) = J[\mathscr{J}(o)-\mathscr{J}(\mathbf{q})+\mathscr{J}'(o)+|\mathscr{J}'(\mathbf{q})|]. \qquad (5.39)$$

The Fourier transformed exchanged functions may therefore be derived directly from the dispersions relations using the expressions

$$\mathscr{J}(o)-\mathscr{J}(\mathbf{q}) = \frac{1}{2J}[\hbar\omega_\alpha(\mathbf{q})+\hbar\omega_\beta(\mathbf{q})-\hbar\omega_\beta(o)]$$

$$\mathscr{J}'(o)-|\mathscr{J}'(\mathbf{q})| = \frac{1}{2J}[\hbar\omega_\alpha(\mathbf{q})-\hbar\omega_\beta(\mathbf{q})+\hbar\omega_\beta(o)]$$

$$\mathscr{J}'(o)+|\mathscr{J}'(\mathbf{q})| = \frac{1}{2J}[-\hbar\omega_\alpha(\mathbf{q})+\hbar\omega_\beta(\mathbf{q})+\hbar\omega_\beta(o)]. \qquad (5.40)$$

As may be seen from Figure 5–1, the c-direction in the hcp structure has a particularly high symmetry, since the two sublattices form alternate equally separated planes normal to the c-axis. $\mathscr{J}_c'(\mathbf{q})$ is therefore real and, from (5.34)

$$-\mathscr{J}_c'\left(\frac{2\pi}{c}-q\right)=\mathscr{J}_c'(q)$$

so that, from (5.39)

$$\hbar\omega_\alpha(q)=\hbar\omega_\beta\left(\frac{2\pi}{c}-q\right). \qquad (5.41)$$

We may therefore regard the dispersion relation in the c-direction as a single branch, provided that we translate each point in the optical branch from q to $(2\pi/c)-q$. In this double zone representation, we have from (5.34)

$$\mathscr{J}(o)-\mathscr{J}_c(q)=2\sum_{m=1}^{\infty}\mathscr{J}_m^c(1-\cos\tfrac{1}{2}mcq)\qquad\left(\frac{-2\pi}{c}<q<\frac{2\pi}{c}\right) \quad (5.42)$$

$\mathscr{J}_m^c$ is an interplanar exchange parameter and odd values of m correspond to exchange between planes in different sublattices.

$\mathscr{J}'(\mathbf{q})$ is also real in the a-direction, but since all planes perpendicular to the a-axis contain ions of both sublattices, a double zone representation cannot be used in this direction. From (5.34) we have

$$\mathscr{J}(o)-\mathscr{J}_a(q)=2\sum_{m=1}^{\infty}\mathscr{J}_m^a(1-\cos\tfrac{1}{2}maq)\qquad\left(\frac{-2\pi}{a}<q<\frac{2\pi}{a}\right)$$

$$\mathscr{J}'(o)-\mathscr{J}_a'(q)=2\sum_{m=1}^{\infty}\mathscr{J}'_m{}^a(1-\cos\tfrac{1}{2}maq)\qquad\left(\frac{-2\pi}{a}<q<\frac{2\pi}{a}\right)$$
$$(5.43)$$

while for the b-direction, in which $\mathscr{J}'(\mathbf{q})$ is complex

$$\mathscr{J}(o)-\mathscr{J}^b(q)=2\sum_{m=0}^{\infty}\mathscr{J}_m^b\left(1-\cos\frac{\sqrt{3}}{2}maq\right)$$
$$\left(\frac{-2\pi}{\sqrt{3}a}<q<\frac{2\pi}{\sqrt{3}a}\right)$$

$$|\mathscr{J}'(o)|^2-|\mathscr{J}'^b(q)|^2=\sum_{p=1}^{\infty}(\mathscr{J}'_p{}^b)^2\left(1-\cos\frac{\sqrt{3}}{2}paq\right)$$
$$\left(\frac{-2\pi}{\sqrt{3}a}<q<\frac{2\pi}{\sqrt{3}a}\right) \quad (5.44)$$

where

$$(\mathscr{J}'_p{}^b)^2=2\sum_{m=-\infty}^{\infty}\mathscr{J}_m'^b\mathscr{J}_{m-p}'^b.$$

It is convenient at this point to review briefly the symmetry properties of the magnon spectrum in a ferromagnetically ordered hcp lattice. This problem has been discussed from a group theoretical viewpoint by Brinkman[12] and Cracknell.[13] In general, the symmetry is described in terms of the magnetic space group, which contains the symmetry operations of the crystal in the ordered phase. However, those interactions which are most important in determining the spin wave spectrum may have a greater symmetry than that described by the magnetic space group, for instance if the moments can be rotated independently of the space coordinates. All of the interactions described in Section 5.2 fall into this category, as does the anisotropic exchange

$$- \sum_{l>l'} \mathscr{J}(\mathbf{R}_{ll'})(J_{xl}J_{xl'} + J_{yl}J_{yl'} + \alpha J_{zl}J_{zl'}) \qquad (5.45)$$

which has the effect of singling out the direction of ordering and would be expected to be significant, from the discussion of Section 5.2.1. The symmetry of the Hamiltonian is then described by the spin space group, and the increase in the number of symmetry operations produces extensive degeneracy between the acoustic and optical magnon branches. For a ferromagnetic structure with the moments in the plane, the spin space group predicts double degeneracy over the whole of the hexagonal AHL face of the Brillouin zone of Figure 5–2,

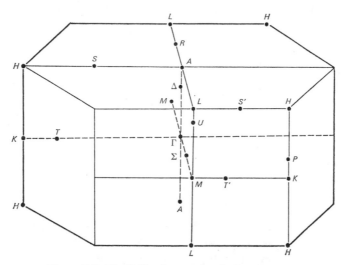

Figure 5–2. The Brillouin zone for the hcp structure.

and also along the line P. These degeneracies may be lifted by more complex interactions which do not have the symmetry of the spin space group, for instance the pseudo-dipolar exchange[3]

$$K_{ll'}[\mathbf{J}_l \cdot \mathbf{J}_{l'} - (3\mathbf{J}_l \cdot \mathbf{R}_{ll'} \mathbf{J}_{l'} \cdot \mathbf{R}_{ll'}/\mathbf{R}_{ll'}^2)] \qquad (5.46)$$

which mixes moment and space variables. If these interactions are important, the magnetic space group must be used and, for the structure mentioned above, the degeneracy is lifted, except along the lines S, S', and R. In addition, the hexagonal symmetry of the spectrum is reduced, since the direction of the moments is a unique axis in the plane. If a splitting can be observed between the acoustic and optical modes therefore, for instance along the line P, it provides direct evidence for more complex interactions than those normally considered. We will return to this point when discussing the magnon dispersion relations for terbium.

5.3.3. The Neutron Scattering Cross Section

Inelastic neutron scattering has proved to be an ideal technique for the study of the excitations of magnetic systems, since the neutron interacts through its magnetic moment with the moments in the system, undergoing changes in energy and momentum which can readily be measured and giving explicit information on the energies and lifetimes of the excitations. Accordingly we shall outline the derivation of the cross section for the scattering of a neutron through the creation of a magnon in a system, such as gadolinium, in which the ions are in S-states. More detailed derivations may be found in the literature,[14] and the cross sections for more complex situations will be introduced as they are required.

The probability per unit time that a neutron transfer momentum

$$\hbar\varkappa = \hbar(\mathbf{K}_0 - \mathbf{K}) \tag{5.47}$$

and energy

$$E = \frac{\hbar^2}{2M}(K_0^2 - K^2) \tag{5.48}$$

to the system is given in the Born approximation by

$$W(\varkappa, E) = \frac{2\pi}{\hbar} \sum_i \rho_i |\langle f|\mathcal{H}_i|i\rangle|^2 \delta[E - \hbar\omega(\mathbf{q})] \tag{5.49}$$

$\mathbf{K}_0$, σ_0 and $\mathbf{K}$, σ are the wavevectors and spins of the incident and scattered neutrons respectively, and $\mathcal{H}_i$ is the interaction Hamiltonian causing a transition from the initial state

$$|i\rangle = |\mathbf{K}_0\sigma_0\rangle |n_{\mathbf{q}}\rangle$$

with a state-density ρ_i, to the final state

$$|f\rangle = |\mathbf{K}\sigma\rangle |n_{\mathbf{q}}+1\rangle.$$

The differential neutron scattering cross section per unit solid angle per unit energy per atom is now given by

$$\frac{d^2\sigma}{d\Omega dE} = \frac{1}{N}\frac{M^2}{(2\pi\hbar)^3}\frac{K}{K_0} W(\varkappa, E). \tag{5.50}$$

The scattering interaction is that between the magnetic field produced by the neutron and the magnetic moment due to the incompletely filled shells of the scattering ions. The matrix element between neutron states of this interaction is[14]

$$\langle \mathbf{K}\sigma|\mathcal{H}_i|\mathbf{K}_0\sigma_0\rangle = \frac{4\pi\hbar^2}{M}\frac{a_m}{S}\sum_l e^{i\varkappa\cdot\mathbf{R}_l}\langle\sigma|\mathbf{S}_l\cdot(\sigma_n-(\mathbf{e}\cdot\sigma_n)\mathbf{e})|\sigma_0\rangle \tag{5.51}$$

where σ_n is the neutron spin operator, $\mathbf{e}$ is a unit vector in the direction of the scattering vector $\varkappa$, and the magnetic scattering amplitude is given by

$$a_m(\varkappa) = r_0\gamma Sf(\varkappa) \tag{5.52}$$

$f(\varkappa)$ is the magnetic form factor, defined as the Fourier transform of the distribution of unpaired electron spins, $r_0 = e^2/mc^2$ and $\gamma = -1.913$ is the value of the neutron magnetic moment in nuclear magnetons.

From (5.50), (5.49), and (5.51) we find, after averaging over spin orientations for an unpolarized neutron beam

$$\frac{d^2\sigma}{d\Omega dE} = \frac{K}{NK_0} \left(\frac{a_m}{S}\right)^2 \sum_{ll'} e^{-i\varkappa \cdot R_{ll'}} \sum_{\alpha\beta} (\delta_{\alpha\beta} - e_\alpha e_\beta)$$
$$\times \sum_i \rho_i \langle n_q | S_l^\alpha | n_q + 1 \rangle \langle n_q + 1 | S_{l'}^\beta | n_q \rangle \delta[E - \hbar\omega(\mathbf{q})] \quad (5.53)$$

where α, β are Cartesian coordinates, as before. If we consider a ferromagnetic Bravais lattice and transform to magnon operators, we find only the following non-vanishing matrix elements

$$\langle n_q | S_l^x | n_q + 1 \rangle = \frac{1}{2} \left(\frac{2S}{N}\right)^{\frac{1}{2}} (n_q + 1)^{\frac{1}{2}} e^{-i\mathbf{q}\cdot R_l}$$

$$\langle n_q + 1 | S_l^x | n_q \rangle = \frac{1}{2} \left(\frac{2S}{N}\right)^{\frac{1}{2}} (n_q + 1)^{\frac{1}{2}} e^{i\mathbf{q}\cdot R_l}$$

$$\langle n_q | S_l^y | n_q + 1 \rangle = \frac{1}{2i} \left(\frac{2S}{N}\right)^{\frac{1}{2}} (n_q + 1)^{\frac{1}{2}} e^{-i\mathbf{q}\cdot R_l} \qquad (5.54)$$

$$\langle n_q + 1 | S_l^y | n_q \rangle = -\frac{1}{2i} \left(\frac{2S}{N}\right)^{\frac{1}{2}} (n_q + 1)^{\frac{1}{2}} e^{i\mathbf{q}\cdot R_l}$$

For a static lattice, we have

$$\sum_{ll'} e^{-i(\varkappa+\mathbf{q})\cdot(R_l - R_{l'})} = \left| \sum_l e^{-i(\varkappa+\mathbf{q})\cdot R_l} \right|^2 = N \frac{(2\pi)^3}{\Omega} \sum_\tau \delta(\varkappa + \mathbf{q} - \tau) \qquad (5.55)$$

where Ω is the volume of a unit cell of the crystal and τ is a reciprocal lattice vector. The thermal vibrations of the lattice may be taken into account by multiplying (5.55) by the Debye–Waller factor $D(\varkappa)$[14]. Substituting (5.54) and (5.55) into (5.53), we obtain finally the one-magnon creation cross section

$$\frac{d^2\sigma}{d\Omega dE} = \frac{4\pi^3}{N\Omega} a_m^2(\varkappa) \frac{K}{K_0 S} (n_q + 1)(1 + e_z^2) D(\varkappa) \sum_\tau \delta(\varkappa + \mathbf{q} + \tau) \delta[E - \hbar\omega(\mathbf{q})]. \quad (5.56)$$

The one-magnon absorption cross section is identical with this, except that the population factor $(n_q + 1)$ is replaced by n_q.

For the hcp lattice, the use of the operators (5.37) leads to the same expression, but multiplied by a geometrical structure factor

$$F_\alpha(\varkappa) = 2 \cos^2 \tfrac{1}{2} [-\tau \cdot \rho + \phi(\mathbf{q})]$$
$$F_\beta(\varkappa) = 2 \sin^2 \tfrac{1}{2} [-\tau \cdot \rho + \phi(\mathbf{q})] \qquad (5.57)$$

for the acoustic and optical branches respectively. Here ρ is the vector between the two atoms in the unit cell and

$$\cos \phi(\mathbf{q}) = \frac{1}{2} \left[\left(\frac{\mathscr{J}'^*(\mathbf{q})}{\mathscr{J}'(\mathbf{q})}\right)^{\frac{1}{2}} + \left(\frac{\mathscr{J}'(\mathbf{q})}{\mathscr{J}'^*(\mathbf{q})}\right)^{\frac{1}{2}} \right] \qquad (5.58)$$

The natural variables in a neutron scattering experiment are therefore the energy transfer E and the vector $\mathbf{q} = \varkappa - \tau$. The experiental method normally consists of seeking peaks in the scattered intensity, either by fixing E and varying $\varkappa$ (constant-E scan) or by fixing $\varkappa$ and varying E (constant-$\mathbf{q}$ scan). Examples of such scans are shown in Figures 5–10 and 5–13 and as may be seen from (5.56) the positions of the peaks, or neutron groups, determine the dispersion relations.

5.3.4. Magnon Disperson Relations for Gadolinium

Natural gadolinium has a capture cross section for thermal neutrons which is so large that inelastic neutron scattering experiments are impracticable. However, Koehler *et al.*[9] have recently used a crystal highly enriched in the low capture isotope ^{160}Gd to study the magnon dispersion relations, and the

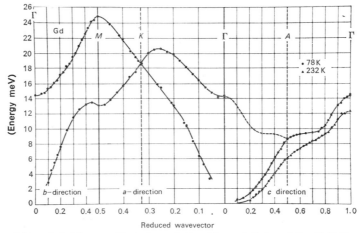

Figure 5–3. Magnon dispersion relations for gadolinium (after Ref. 9).

Table 5.1. Interplanar exchange parameters for terbium at 4.2 K and gadolinium at 78 K (after ref. 9). All values are in meV, with an accuracy of about ± 0.005 meV for terbium.

	a-direction		*b*-direction		*c*-direction	
	Tb	Gd	Tb	Gd	Tb	Gd
$\mathscr{J}_1$	0.183	0.713	0.223	0.789	0.269	0.893
$\mathscr{J}_2$	0.117	0.341	0.046	0.054	0.072	0.092
$\mathscr{J}_3$	0.043	0.106	0.009	0.040	−0.010	0.097
$\mathscr{J}_4$	0.018	0.027	0.006	0.008	−0.037	−0.131
$\mathscr{J}_5$	0.006	0.030	−0.001	0.007	0.008	0.032
$\mathscr{J}_6$	0.005		0.020			
$\mathscr{J}_1'$	0.197	0.773	0.146	0.626		
$\mathscr{J}_2'$	0.029	−0.146	0.214	0.704		
$\mathscr{J}_3'$	0.051	0.125	0.026i	0.109		
$\mathscr{J}_4'$	0.007	0.036	0.109	0.336		
$\mathscr{J}_5'$	0.010	0.029	0.037i	0.179		
$\mathscr{J}_6'$	0.003		0.062			

results of their measurements along the principal symmetry axes at 78 K are shown in Figure 5–3. Since gadolinium has a very small magnetic anisotropy the theory of Section 5.3.2 may be applied, and the interplanar exchange parameters deduced thereby are given in Table 5.1. These results will be discussed further and compared with those for ferromagnetic terbium in Section 5.4.2.

5.4. THE ANISOTROPIC FERROMAGNET

The occurrence of large magnetic anisotropy in the rare earths has a substantial effect on the spin wave spectrum, and we shall now consider the simple case of a ferromagnet in which the indirect exchange coupling (5.8) is augmented by the anisotropy terms (5.10) and (5.11). The magnon dispersion relations and neutron scattering cross section will first be derived on the assumption that the anisotropy is entirely due to crystal field effects, and this approximation will be used to discuss the dispersion relations in terbium. The indirect exchange functions will be deduced from the data and compared with those for other metals. The effect of the magnetic anisotropy will then be examined in more detail and it will be shown how measurements on long-wavelength magnons, by neutron scattering or electromagnetic absorption, can throw light on the mechanisms responsible for the anisotropy.

5.4.1. Magnon Disperson Relations and Neutron Scattering Cross Section

We begin by considering a planar ferromagnet, such as terbium or dysprosium, in which the magnetic anisotropy constrains the moments to lie along the ξ direction. We suppose initially that the anisotropy is dominated by the crystal field terms and neglect all axial terms higher than second-order, so that we may abbreviate (5.10) to

$$\mathscr{H}_{cf}= \sum_{l} \{V_2 J_{l\zeta}{}^2 - \tfrac{1}{2}V_6{}^6[(J_{l\xi}+iJ_{l\eta})^6+(J_{l\xi}-iJ_{l\eta})^6]\}. \qquad (5.59)$$

We wish to replace the angular momentum operators by the boson deviation operators (5.17) and, in the spirit of the one-magnon approximation, retain consistently only terms quadratic in these operators. We wish also to ensure that the matrix elements of (5.59) are invariant under this transformation for small deviations, since the one-magnon approximation is valid under such circumstances.

Recalling that the equilibrium magnetization is along the ξ direction, we have the relations

$$J_{l\xi}|m\rangle = m|m\rangle$$
$$J_{l\xi}|m\rangle = \tfrac{1}{2}[(J-m)(J+m+1)]^{\tfrac{1}{2}}|m+1\rangle + \tfrac{1}{2}[(J+m)(J-m+1)]^{\tfrac{1}{2}}|m-1\rangle \quad (5.60)$$
$$iJ_{l\eta}|m\rangle = \tfrac{1}{2}[(J-m)(J+m+1)]^{\tfrac{1}{2}}|m+1\rangle - \tfrac{1}{2}[(J+m)(J-m+1)]^{\tfrac{1}{2}}|m-1\rangle$$

from which we may deduce

$$J_{l\xi}{}^2|J\rangle = \tfrac{1}{2}J|J\rangle + \frac{1}{\sqrt{2}}[J(J-\tfrac{1}{2})]^{\tfrac{1}{2}}|J-2\rangle$$
$$\qquad (5.61)$$
$$J_{l\xi}{}^2|J-1\rangle = \tfrac{1}{2}(3J-1)|J-1\rangle + \sqrt{\tfrac{3}{2}}[(J-\tfrac{1}{2})(J-1)]^{\tfrac{1}{2}}|J-3\rangle.$$

Hence, as may easily be verified, the transformation

$$J_{l\zeta}^2 \equiv \tfrac{1}{4} + \tfrac{1}{2}(J-\tfrac{1}{2})(a_l{}^+a_l + a_l a_l{}^+ + a_l{}^+a_l{}^+ + a_l a_l) \qquad (5.62)$$

gives correctly all matrix elements involving states in which there is a deviation of zero or unit angular momentum on the site l, provided that we make the approximation

$$[J(J-\tfrac{1}{2})]^{\frac{1}{2}} \simeq J-\tfrac{1}{2} \simeq [(J-\tfrac{1}{2})(J-1)]^{\frac{1}{2}} \qquad (5.63)$$

which should be adequate for large angular momenta.

The hexagonal anisotropy may be treated in a similar manner. By a straightforward but tedious calculation, it may be shown using (5.60) that the transformation

$$(J_{l\xi}+iJ_{l\eta})^6 + (J_{l\xi}-iJ_{l\eta})^6 = -(J-\tfrac{1}{2})(J-1)(J-\tfrac{3}{2})(J-2)(J-\tfrac{5}{2})$$
$$\times [2J+21(1-a_l{}^+a_l-a_l a_l{}^+)$$
$$+15(a_l{}^+a_l{}^+ + a_l a_l)] \qquad (5.64)$$

also gives correctly all matrix elements for zero and one angular momentum deviation on the site l, again making use of the approximation (5.63). Using (5.62) and (5.64), we can now write the crystal field Hamiltonian.

$$\mathscr{H}_{cf} = \tfrac{1}{4}NV_2 J - \tfrac{1}{2}NV_6{}^6(2J+21)J_5$$
$$+ \tfrac{1}{2}\sum_l [A(a_l{}^+a_l + a_l a_l{}^+) + B(a_l{}^+a_l{}^+ + a_l a_l)] \qquad (5.65)$$

where

$$A = V_2 J_1 + 21 V_6{}^6 J_5$$
$$B = V_2 J_1 - 15 V_6{}^6 J_5 \qquad (5.66)$$

and we have used the abbreviation

$$J_n \equiv (J-\tfrac{1}{2})(J-1) \ldots \qquad (5.67)$$

for the n-term product of factors involving J. This expression was first obtained by Brooks et al.[15] by reordering the operators in (5.59). It is interesting to note that the truncated expansion of (5.59) first used by Niira[11] gives the same result, except that J_n in (5.66) is replaced by J^n.

Adding the indirect exchange (5.8) and transforming to creation and annihilation operators defined by (5.24), we obtain the Hamiltonian for a Bravais lattice

$$\mathscr{H} = -\tfrac{1}{2}NJ(J+1)\mathscr{I}(o) + \tfrac{1}{4}NV_2 - \tfrac{1}{2}NV_6{}^6(2J+21)J_5$$
$$+ \tfrac{1}{2}\sum_q \{(J[\mathscr{I}(o)-\mathscr{I}(q)]+A)(a_q{}^+a_q + a_q a_q{}^+) + B(a_q{}^+a_{-q}{}^+ + a_q a_{-q})\}. \qquad (5.68)$$

This expression may finally be diagonalized by the canonical transformation

$$\alpha_q{}^+ = u a_q{}^+ + v a_{-q}$$
$$\alpha_q = u a_q + v a_{-q}{}^+ \qquad (5.69)$$

with

$$u^2 = \frac{1}{2}\left(\frac{(J[\mathscr{I}(o)-\mathscr{I}(q)]+A)}{[(J[\mathscr{I}(o)-\mathscr{I}(q)]+A)^2 - B^2]^{\frac{1}{2}}} + 1\right) \qquad (5.70)$$

and
$$v^2 = u^2 - 1 \qquad (5.71)$$

giving
$$\mathscr{H} = -\tfrac{1}{2}NJ(J+1)\mathscr{J}(o) + \tfrac{1}{4}NV_2 - \tfrac{1}{2}NV_6{}^6(2J+21)J_5$$
$$+ \sum_{\mathbf{q}} [(J[\mathscr{J}(o)-\mathscr{J}(\mathbf{q})]+A)^2 - B^2]^{\frac{1}{2}}(\alpha_{\mathbf{q}}{}^+\alpha_{\mathbf{q}}+\tfrac{1}{2}). \quad (5.72)$$

The magnon energy is therefore
$$\hbar\omega(\mathbf{q}) = [(J[\mathscr{J}(o)-\mathscr{J}(\mathbf{q})]+A+B)(J[\mathscr{J}(o)-\mathscr{J}(\mathbf{q})]+A-B)]^{\frac{1}{2}}$$
$$= [(J[\mathscr{J}(o)-\mathscr{J}(\mathbf{q})]+2V_2J_1+6V_6{}^6J_5)(J[\mathscr{J}(o)-\mathscr{J}(\mathbf{q})]+36V_6{}^6J_5)]^{\frac{1}{2}} \quad (5.73)$$

The most striking feature of this result is that, because of the anisotropy forces, the magnon energy does not approach zero as the wavelength becomes very long. We shall discuss this magnon energy gap in more detail in Section 5.4.3. This dispersion relation is valid for the hcp lattice in the c-direction, when the double-zone representation may be used, but in other directions it must be modified in a manner analogous to that described in the last section. The dispersion relation then has acoustic and optical branches given by

$$\hbar\omega_\alpha(\mathbf{q}) = [(Jf_\alpha(\mathbf{q})+2V_2J_1+6V_6{}^6J_5)(Jf_\alpha(\mathbf{q})+36V_6{}^6J_5)]^{\frac{1}{2}}$$
$$\hbar\omega_\beta(\mathbf{q}) = [(Jf_\beta(\mathbf{q})+2V_2J_1+6V_6{}^6J_5)(Jf_\beta(\mathbf{q})+36V_6{}^6J_5)]^{\frac{1}{2}} \qquad (5.74)$$

respectively, where

$$f_\alpha(\mathbf{q}) = \mathscr{J}(o) - \mathscr{J}(\mathbf{q}) + \mathscr{J}'(o) - |\mathscr{J}'(\mathbf{q})|$$
$$f_\beta(\mathbf{q}) = \mathscr{J}(o) - \mathscr{J}(\mathbf{q}) + \mathscr{J}'(o) + |\mathscr{J}'(\mathbf{q})|. \qquad (5.75)$$

Provided that the anisotropy parameters V_2 and $V_6{}^6$ are known, we can therefore determine the $f(\mathbf{q})$ from the dispersion relations, and hence the Fourier transformed and interplanar exchange parameters by the method described after (5.39).

The neutron scattering cross section (5.56) must be modified to take account of the anisotropy and orbital angular momentum on the ion. Neutron scattering by an ion with orbital momentum has been considered by a number of authors[16] and a good approximation to the cross section is obtained by replacing S by J in (5.56) and re-defining the magnetic scattering amplitude as

$$a_m(\varkappa) = \tfrac{1}{2}r_0\gamma\lambda Jf(\varkappa) \qquad (5.76)$$

where now $f(\varkappa)$ is the Fourier transform of the total magnetization density.[17] The effect of the anisotropy may be determined by using J operators in (5.53) and transforming to the magnon operators $\alpha_{\mathbf{q}}$ for the anisotropic system. This has the consequence that $(1+e_z{}^2)$ in (5.56) is replaced by

$$(1-e_x{}^2)(u-v)^2 + (1-e_y{}^2)(u+v)^2$$
$$= (1+e_z{}^2)\left[1+\left(\frac{B}{\hbar\omega(\mathbf{q})}\right)^2\right]^{\frac{1}{2}} + (e_x{}^2-e_y{}^2)\frac{B}{\hbar\omega(\mathbf{q})} \qquad (5.77)$$

a result first derived by Lindgård et al.[18]

For the sake of completeness we will briefly discuss the axial ferromagnet, in which the moments lie along the (001) direction. This is the situation in thulium in an applied axial field, but no spin wave measurements have yet been made on this material. We consider a Hamiltonian consisting of the indirect exchange (5.8) and crystal field anisotropy (5.10), where now V_2 is negative, so that ζ is the magnetization direction. In the independent magnon approximation, the hexagonal anisotropy plays no role, while the axial terms may be treated by using the relation[18]

$$J_{l\xi}{}^n \equiv J^n + [(J-1)^n - J^n]a_l{}^+a_l \tag{5.78}$$

which, as usual, is correct for small moment deviations on the site l. Transforming to magnon operators, we readily find

$$\mathcal{H} = NJ^2[V_2 - \tfrac{1}{2}\mathcal{J}(o)] + NV_4J^4 + NV_6J^6$$
$$+ \sum_{\mathbf{q}} \{J[\mathcal{J}(o) - \mathcal{J}(\mathbf{q})] + (1-2J)V_2 + [(J-1)^4 - J^4]V_4$$
$$+ [(J-1)^6 - J^6]V_6\}a_{\mathbf{q}}{}^+a_{\mathbf{q}} \tag{5.79}$$

In this case, the axial anisotropy alone produces an energy gap and when, as in thulium, V_2 is dominant, it may be determined directly from the magnitude of the gap. The neutron scattering cross section for the axial ferromagnet is independent of the magnetic anisotropy, except in so far as it determines the magnon energies and hence the population factor $n_{\mathbf{q}}$ in (5.56).

5.4.2. Experimental Results for Anisotropic Ferromagnets

In this and the following section we will consider the experimental information which has been obtained on the magnon dispersion relations in anisotropic ferromagnetic rare earth metals. The most extensive measurements have been made on terbium,[19] which was the first rare earth to be systematically studied by inelastic neutron scattering, but we shall also refer to more recent measurements on holmium–10% terbium[20] and dysprosium.[21]

The magnon dispersion relations for terbium at 4.2 K along all the symmetry lines in the zone are shown in Figure 5–4.[22] They bear a close resemblance to those of Gd, except for the disturbance produced by the magnon–phonon interaction, to which we will return later, and the finite energy of the acoustic branch at Γ, which is a manifestation of the anisotropy forces. An interesting feature of these results is the small splitting between the acoustic and optical branches along the line KH. As discussed in Section 5.3.2, these branches should be degenerate along this line, according to the symmetry of the spin space group, and the observation of a maximum splitting of 0.35 meV, with an estimated uncertainty of 0.1 meV, might indicate that more complex interactions are present. The dipolar interaction between the moments would lead to such a splitting, but calculations[23] show that this falls monotonically from a value of almost 0.1 meV at K to zero at H, and is therefore too small to explain the observed effect. Further studies of this splitting, to confirm its existence and possibly furnish some information on higher-order exchange interactions, would clearly be desirable.

According to (5.73), the magnon energy gap is given to a good approximation by

$$\Delta^2 = 72V_2V_6{}^6J_1J_5 \tag{5.80}$$

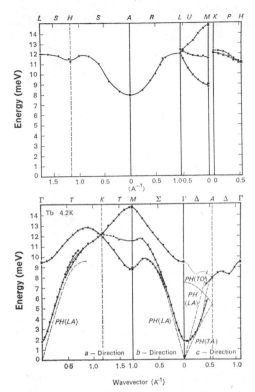

Figure 5–4. Magnon dispersion relations for terbium along the symmetry lines in the zone at 4.2 K. Certain phonon branches and the effects of magnon–phonon interactions are also shown.

since $V_2 \gg V_6^6$. V_2 may be deduced either from macroscopic measurements, or by the magnetic field experiments discussed in the next section, and the experimentally determined value of Δ then determines V_6^6. The Fourier transformed exchange function may then be deduced from (5.74) and (5.75). As we shall see, this procedure oversimplifies the description of the anisotropy forces by omitting magnetoelastic effects, but this simplification does not affect the values of $\mathscr{J}(o) - \mathscr{J}(\mathbf{q})$ deduced from the results and shown in Figure 5–5. As described in Section 5.3.2, the interplanar exchange parameters may readily be deduced from $\mathscr{J}(o) - \mathscr{J}(\mathbf{q})$, and these are compared with the corresponding values for gadolinium in Table 5.1. The similarity of the dispersion relations is reflected in the interplanar parameters, which have almost the same relative magnitudes in the two metals.

If $\mathscr{J}(\mathbf{q})$ is known throughout the Brillouin zone, it is clearly possible to deduce the interatomic exchange function $\mathscr{J}(\mathbf{R})$ from (5.26). In practice this means that dispersion relations must be obtained in a number of non-symmetry directions, by a combination of direct measurement and the type of interpolation method discussed in Section 5.6.2. The calculation is carried out most conveniently via an intermediate step, in which the exchange between a

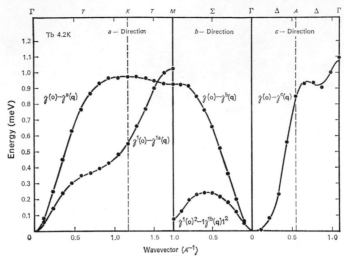

Figure 5–5. Fourier transformed exchange parameters for terbium at 4.2 K.

central atom and infinite lines of the other atoms is determined,[24] the interatomic exchange then being deduced from these interlinear parameters.[25] The interatomic exchange parameters so deduced for terbium are shown in Figure 5–6, and they have the long-range and oscillatory character anticipated for the indirect exchange interaction.

The magnon dispersion relations have also been measured in the c-direction in holmium–10% terbium[20] and in dysprosium.[21] The addition of 10% terbium to holmium stabilizes a simple ferromagnetic structure at low temperatures, whose magnon dispersion relations are presumably very similar to those of ferromagnetic pure holmium, stabilized by a small magnetic field.

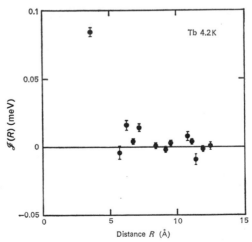

Figure 5–6. The spatial dependence of the exchange between ions on the same sublattice in terbium at 4.2 K.

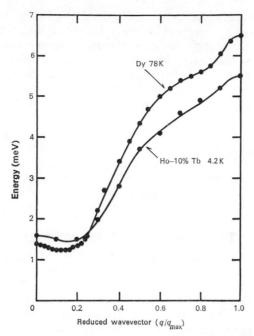

Figure 5–7. Magnon dispersion relations in the ferromagnetic phases of dysprosium (after Ref. 21) and holmium–10% terbium in the c-direction, in the double zone representation.

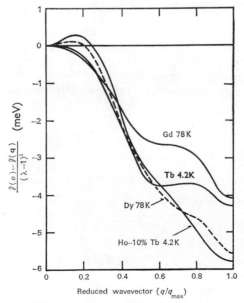

Figure 5–8. The exchange function $[\mathscr{J}(o)-\mathscr{J}(\mathbf{q})]/(\lambda-1)^2$ in the c-direction for a number of ferromagnetic heavy rare earths.

The dispersion relations for these metals are shown in Figure 5–7. They both have a small energy minimum at finite **q** in the *c*-direction, which is associated with the occurrence of stable periodic magnetic structures at higher temperature. The exchange function $[\mathcal{J}(\mathbf{q}) - \mathcal{J}(o)]/(\lambda - 1)^2$ is compared for gadolinium, terbium, dysprosium, and holmium–10% terbium in Figure 5–8. If the conduction electron band structure and 4*f* wavefunctions were identical, and if the effective coupling between the ions were due entirely to RKKY exchange, then according to the discussion of (5.2.1), this function should be identical for all of these metals. There are in fact notable differences in the **q**-dependences for the different metals, probably principally due to differences in the band structure and hence $\chi(\mathbf{q})$, and the average magnitude of the function also appears to increase with *L*. This may be an indication that there is some coupling between the orbital angular momenta on the ions, of the type suggested by Levy.[4]

5.4.3. Anisotropy and the Magnon Energy Gap

In the last section we discussed the way in which the **q**-dependence of the magnon energies gives information about the indirect exchange interaction between the magnetic ions. We shall now consider how the dependence on temperature and magnetic field of the magnon energy gap can give similarly detailed information on the mechanisms responsible for the single-ion anisotropy and the magnitude of its components.

The expression (5.80) for the magnon energy gap was derived on the assumption that the anisotropy is entirely due to crystal fields in the unstrained crystal. In order to take into account magnetoelastic effects in the frozen-lattice model, we must add to the exchange and crystal field contributions (5.8) and (5.59) the magnetoelastic Hamiltonian (5.11). In addition the external magnetic field contributes the Zeeman term (5.14).

When this field is applied in the easy (ξ) direction, the direction of the ordered moment does not change. The Hamiltonian may be diagonalized by expressing the angular momenta in terms of the deviation operators[26] and proceeding as before, giving the magnon energy in the long wavelength limit[27]

$$\Delta_e{}^2 = [2V_2J_1 + 6V_6{}^6J_5 + 2c(C^2 + A^2) + 3cAC + \lambda\beta H] \\ \times [36V_6{}^6J_5 + 4c(C^2 + A^2) + 10cAC + \lambda\beta H]. \quad (5.81)$$

If the field is applied in the hard (η) direction in the plane, the magnetization rotates so that it makes an angle ϕ with the η-axis, where

$$\lambda\beta H = 2\cos\phi(4\cos^2\phi - 1)(4\cos^2\phi - 3)(3cAC + 6V_6{}^6J_5). \quad (5.82)$$

It is then convenient to make the coordinate transformation

$$\begin{aligned} \zeta &= -x \\ \xi &= -y\cos\phi + z\sin\phi \\ \eta &= y\sin\phi + z\cos\phi \end{aligned} \quad (5.83)$$

so that the *z* axis is along the magnetization direction. This allows the use of the Holstein–Primakoff transformation in the new coordinate system, so that the Hamiltonian may be diagonalized by the usual techniques, giving

$$\Delta_h{}^2 = [2V_2J_1 + 6V_6{}^6J_5 + 2c(C^2 + A^2) + 3cAC + (6cAC + 12V_6{}^6J_5)$$
$$\times (\cos 4\phi + \cos 2\phi)]$$
$$\times [4c(C^2 + A^2) + 8cAC \cos 6\phi + (3cAC + 6V_6{}^6J_5)$$
$$\times (-5 \cos 6\phi + 2(\cos 4\phi + \cos 2\phi) + 1)]. \quad (5.84)$$

Above the critical field

$$\lambda\beta H_c = 36(\tfrac{1}{2}cAC + V_6{}^6J_5) \quad (5.85)$$

at which the magnetization turns into the hard direction, the gap is given by

$$\Delta_h{}^2 = [2V_2J_1 + 6V_6{}^6J_5 + 2c(C^2 + A^2) - 3cAC + \lambda\beta H]$$
$$\times [4c(C^2 + A^2) + 8cAC + \lambda\beta(H - H_c)]. \quad (5.86)$$

These expressions reduce to those of Cooper[7] when the second-order magnetoelastic effect is neglected, so that A is zero. The most striking difference between (5.84) and the corresponding expression in the free lattice model is that, in the latter case, Δ_h goes to zero at H_c,[7] while in the frozen lattice model the energy gap remains finite at all fields, although it has a minimum at H_c.

Although values of Δ had earlier been deduced from thermodynamic results[11, 28] the first explicit measurements of the magnon energy gap were made by inelastic neutron scattering in terbium.[19] In the analysis of these results, it was noted that an interpretation in terms of (5.80) required a hexagonal anisotropy substantially greater than the macroscopically measured value. Cooper[26] pointed out that this discrepancy could be explained by the frozen lattice model, but that the results of microwave absorption experiments apparently precluded the validity of this model. The microwave radiation couples to the magnetic system through its magnetic field, producing a perturbation of the form

$$\mathcal{H}(\omega) = -\lambda\beta \sum_l \mathbf{J}_l \cdot \mathbf{H}(\omega). \quad (5.87)$$

By expressing the angular momentum operators in terms of magnon variables, it may be shown[29] that an electromagnetic wave polarized in a plane perpendicular to the magnetization can excite the uniform mode of frequency $\omega(o)$ or energy Δ. Measurements on terbium and dysprosium at relatively low microwave frequencies[30] revealed peaks which were interpreted as magnon resonances. These observations were apparently inconsistent with the frozen-lattice model, which predicted that it would not be possible to reduce $\omega(o)$ to such low frequencies in the temperature range of the experiments. The temperature dependence of the energy gap in terbium–10% holmium, measured by neutron scattering,[19] showed that magnetoelastic effects are probably of importance in determining Δ, since the predicted temperature variation of Δ, if the planar anisotropy is entirely due to unstrained crystal fields, is much greater than that observed. Marsh and Sievers[31] concluded that the temperature dependences of Δ for terbium and dysprosium, measured by infra-red resonance and shown in Figure 5–9, are most plausibly interpreted in terms of

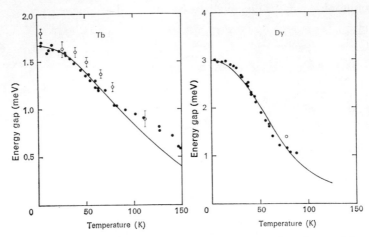

Figure 5–9. The magnon energy gap as a function of temperature in terbium and dysprosium, as determined from infra-red resonance (after Ref. 31). The results from neutron scattering experiments are shown as open circles. The full lines are calculated as described in the text.

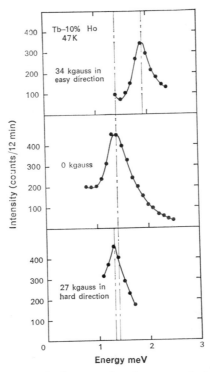

Figure 5–10. Neutron groups for long wavelength magnons in terbium–10% holmium in the ferromagnetic phase at 47 K, for different values of **H**.

the frozen-lattice model. Wagner and Stanford[32] observed a microwave resonance in terbium at around 100 GHz, which disappeared below about 190 K, and showed that their observations were also in qualitative agreement with the model. They later[33] explained the peaks observed at low frequencies as being due to domain alignment effects.

As may be deduced from the preceding discussion, a definitive test of the frozen-lattice model, and the determination of the anisotropy parameters as a function of temperature, require the measurement of Δ as a function of $\mathbf{H}$ and T in a single crystal. Such measurements were made on a small monocrystalline sphere of terbium–10% holmium by inelastic neutron scattering,[27] and the field dependence of the neutron groups is shown in Figure 5–10. The dependence of Δ on magnetic field is shown in Figure 5–11(a), from which it may immediately be concluded that the frozen-lattice model is valid, since a field just large enough to rotate the magnetization into a hard direction does not reduce the gap to zero. The full lines in Figure 5–11(a) are least-squares fits of (5.81), (5.84), and (5.86) to the experimental points, using $V_2 J_1$, $V_6^6 J_5$, cAC, and $c(C^2 + A^2)$ as adjustable parameters. The fit is quite good, provided that the magnetostriction coefficient A is included in the theoretical expressions, otherwise no satisfactory fit can be obtained. The least-squares analysis shows that the macroscopic hexagonal anisotropy is dominated by magnetoelastic effects,[27] especially at the higher temperatures, and the values of the magnetostriction coefficients obtained from this analysis are in good agreement with those obtained from macroscopic measurements.[8]

A comparison between the theoretical temperature dependences of Δ for terbium and dysprosium and those deduced from infra-red and neutron measurements is shown in Figure 5–9. As may be seen, the neutron results lie generally somewhat above the infra-red values. This may be associated with the difficulty of locating precisely the infra-red resonance.[31] The theoretical curve was calculated from (5.81), with $H = 0$, using macroscopic measurements to determine the values of the parameters cAC, $c(C^2 + A^2)$[8] and $V_2 J_1$.[34] $V_6^6 J_5$ was deduced from the value of Δ at 4.2 K, taken from the infra-red measurements. The values so determined are given in Table 5.2. The temperature dependence of the parameters was calculated according to the prescription of Callen and Callen.[35] If the magnetization is not too small, the dependence on the reduced magnetization is given by

$$V_2 J_1 \sim \sigma^3; \quad V_6^6 J_5 \sim \sigma^{21}; \quad cAC \sim \sigma^{13}; \quad c(C^2 + A^2) \sim \sigma^6 \qquad (5.88)$$

If these temperature dependences are used for the anisotropy parameters, the fit to the experimental results is quite good, as may be seen in Figure 5–9, except for terbium at high temperatures, where the theory[35] predicts a deviation from (5.88) in the observed direction. At low temperatures, the theory of Brooks,[36] which takes into account the elliptical spin precession in the spin waves, gives an equally good fit, but it deviates below the experimental points at higher temperatures, where it is not expected to be valid.[36] There is some evidence that the Brooks' theory gives a better fit at low temperatures to the measurements on terbium–10% holmium in the magnetic field,[27] but this point requires further investigation. The total macroscopic planar anisotropy deduced from Δ and given in Table 5.2 is substantially greater than that

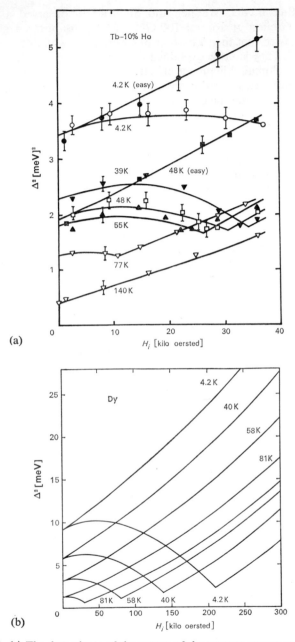

Figure 5–11. (a) The dependence of the square of the magnon energy gap on the internal
magnetic field in terbium–10% holmium. Except where indicated, the field is applied in
the hard direction in the hexagonal plane. The lines are the theoretical expressions
discussed in the text. (b) Theoretical field dependence of the magnon energy gap in
dysprosium, deduced from the expressions given in the text, with the parameters of
Table 5.2.

Table 5.2. Anisotropy parameters in terbium and dysprosium at 4.2 K in meV/ion, derived as described in the text.

	Tb	Dy
$2V_2J_1$	4.10	4.32
$36V_6{}^6J_5$	0.13	1.50
$18cAC$	0.43	0.31
$4c(C^2+A^2)$	0.28	0.25

derived from macroscopic measurements,[34] both for dysprosium and terbium. The ratio of the values of $V_6{}^6J_5$ for dysprosium and terbium in Table 5.2 is 5.8, which may be compared with the value 4.3 for the ratio of the J_5. This indicates that $V_6{}^6$ is not very different for terbium and dysprosium, as expected.[37]

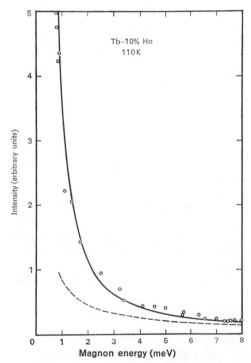

Figure 5–12. The integrated intensities of the neutron groups arising from magnon creation in terbium–10% holmium at 110 K. The dashed line is the predicted variation from (5.56), while the full line includes the term (5.77) which takes account of the magnetic anisotropy, with anisotropy parameters deduced from the field dependence of the magnon energy gap.

As may be seen from (5.77), the axial anisotropy also affects the neutron group intensity. In Figure 5–12 we show the corrected intensity as a function of $\mathbf{q}$ in the c-direction in terbium–10% holmium at 4.2 K.[38] The theoretical curve is calculated using the value of V_2J_1 deduced from the field dependence of Δ. The agreement is quite satisfactory, and it may be seen that the axial anisotropy increases the intensity substantially at low $\mathbf{q}$, which is very convenient from an experimental viewpoint.

5.5. PERIODIC STRUCTURES

Because of the long range and oscillatory nature of the exchange, the heavy rare earths frequently form periodic magnetic structures. In this section, we shall consider the spin wave excitations of such systems, with emphasis on the cone structure, in which the ordered moments have the form

$$J_{l\xi}=J\sin\theta\cos\mathbf{Q}\cdot\mathbf{R}_l; \quad J_{l\eta}=J\sin\theta\sin\mathbf{Q}\cdot\mathbf{R}_l; \quad J_{l\zeta}=J\cos\theta \quad (5.89)$$

where θ is the cone angle and $\mathbf{Q}$ the wavevector of the periodic structure. We shall first consider the simple case in which $\theta=\pi/2$ and derive the dispersion relations and neutron scattering cross section for the helical structure. These results will be compared with experimental measurements on terbium, dysprosium, and holmium. The general cone structure will then be discussed and illustrated by the experimental results on the conical phases of holmium and erbium. Finally a few comments will be made on other periodic structures.

5.5.1. Dispersion Relations and Neutron Scattering for the Helical Structure

In calculating the dispersion relations for the helix we shall neglect the hexagonal anisotropy, which is frequently relatively small in the temperature region in which this structure is stable. Furthermore, as we shall see later, V_4 and V_6 make no contribution to the excitation energy of a helix, so we may write the Hamiltonian in the simple form

$$\mathcal{H}=-\sum_{l>l'}\mathcal{J}(\mathbf{R}_{ll'})\mathbf{J}_l\cdot\mathbf{J}_{l'}+\sum_l V_2J_{l\xi}^2. \quad (5.90)$$

Since spin wave measurements in the helical phase have been performed almost exclusively in the c-direction, we will use the double zone representation throughout this section.

In order to apply the methods which we have previously used to obtain the magnon dispersion relations, we must first transform to a coordinate system in which the local z-axis is always along the equilibrium magnetization, so that

$$J_{l\xi}=J_{lz}\cos\mathbf{Q}\cdot\mathbf{R}_l-J_{lx}\sin\mathbf{Q}\cdot\mathbf{R}_l$$

$$J_{l\eta}=J_{lz}\sin\mathbf{Q}\cdot\mathbf{R}_l+J_{lx}\cos\mathbf{Q}\cdot\mathbf{R}_l$$

$$J_{l\zeta}=J_{ly}. \quad (5.91)$$

In this system, the Hamiltonian is

$$\mathcal{H}=-\sum_{l>l'}\mathcal{J}(\mathbf{R}_{ll'})$$
$$\times\{(J_{lz}J_{l'z}+J_{lx}J_{l'x})\cos\mathbf{Q}\cdot\mathbf{R}_{ll'}+(J_{lx}J_{l'z}-J_{lz}J_{l'x})\sin\mathbf{Q}\cdot\mathbf{R}_{ll'}$$
$$+J_{ly}J_{l'y}\}+\sum_l V_2J_{ly}^2. \quad (5.92)$$

Applying the Holstein–Primakoff and Fourier transformations as before, we obtain

$$\mathcal{H} = -\tfrac{1}{2}NJ(J+1)\mathcal{J}(\mathbf{Q})$$
$$+ \tfrac{1}{4}J\sum_{\mathbf{q}}[2\mathcal{J}(\mathbf{Q})-\mathcal{J}(\mathbf{q})-\tfrac{1}{2}[\mathcal{J}(\mathbf{q}+\mathbf{Q})+\mathcal{J}(\mathbf{q}-\mathbf{Q})]+2V_2]$$
$$\times(a_{\mathbf{q}}^{+}a_{\mathbf{q}}+a_{\mathbf{q}}a_{\mathbf{q}}^{+})$$
$$+ \tfrac{1}{4}J\sum_{\mathbf{q}}[\mathcal{J}(\mathbf{q})-\tfrac{1}{2}[\mathcal{J}(\mathbf{q}+\mathbf{Q})+\mathcal{J}(\mathbf{q}-\mathbf{Q})]-2V_2]$$
$$\times(a_{\mathbf{q}}a_{-\mathbf{q}}+a_{-\mathbf{q}}^{+}a_{\mathbf{q}}^{+}). \quad (5.93)$$

This Hamiltonian may finally be diagonalized by the canonical transformation

$$a_{\mathbf{q}} = \alpha_{\mathbf{q}}\cosh\theta(\mathbf{q}) - \alpha_{-\mathbf{q}}^{+}\sinh\theta(\mathbf{q})$$
$$a_{-\mathbf{q}}^{+} = \alpha_{-\mathbf{q}}^{+}\cosh\theta(\mathbf{q}) - \alpha_{\mathbf{q}}\sinh\theta(\mathbf{q}) \quad (5.94)$$

where

$$e^{4\theta(\mathbf{q})} = \frac{\mathcal{J}(\mathbf{Q})-\tfrac{1}{2}[\mathcal{J}(\mathbf{q}+\mathbf{Q})+\mathcal{J}(\mathbf{q}-\mathbf{Q})]}{\mathcal{J}(\mathbf{Q})-\mathcal{J}(\mathbf{q})+2V_2} \quad (5.95)$$

giving the result[39]

$$\mathcal{H} = -\tfrac{1}{2}NJ(J+1)\mathcal{J}(\mathbf{Q})$$
$$+ J\sum_{\mathbf{q}}\{[\mathcal{J}(\mathbf{Q})-\tfrac{1}{2}\mathcal{J}(\mathbf{q}+\mathbf{Q})-\tfrac{1}{2}\mathcal{J}(\mathbf{q}-\mathbf{Q})][\mathcal{J}(\mathbf{Q})-\mathcal{J}(\mathbf{q})+2V_2]\}^{\frac{1}{2}}$$
$$\times(\alpha_{\mathbf{q}}^{+}\alpha_{\mathbf{q}}+\tfrac{1}{2}). \quad (5.96)$$

It is of some interest that the ground state energy for this structure contains a contribution proportional to $J(J+1)$ and a zero point energy arising from the factor $\tfrac{1}{2}$ in the second term, whereas in the corresponding expression (5.29) for the isotropic ferromagnet, the ground state energy is contained entirely in the term proportional to J^2. This arises because, in the latter case, the zero point energy is independent of the form of the exchange. Using the relation

$$\sum_{\mathbf{q}}\mathcal{J}(\mathbf{q}) = \mathcal{J}(\mathbf{R}=o) = 0 \quad (5.97)$$

we can write (5.29) in the form

$$\mathcal{H} = -\tfrac{1}{2}NJ(J+1)\mathcal{J}(o) + J\sum_{\mathbf{q}}[\mathcal{J}(o)-\mathcal{J}(\mathbf{q})](a_{\mathbf{q}}^{+}a_{\mathbf{q}}+\tfrac{1}{2}) \quad (5.98)$$

Similarly, the zero point energy in the axial ferromagnet is independent of the exchange, so that the Hamiltonian may be written as in (5.79), whereas for the planar ferromagnet it must be included explicitly, as in (5.72). This has the corollary that there is a zero point deviation of the moments from their assumed equilibrium configuration in the helix and planar ferromagnet, whereas in the isotropic and axial ferromagnets, no such deviation occurs (see 5.6.1).

The neutron scattering cross section may readily be calculated by using J operators in (5.53) and transforming to the magnon operators $\alpha_{\mathbf{q}}$ for the helical structure. The result for the one-magnon creation cross section is[40]

$$\frac{d^2\sigma}{d\Omega\,dE} = \frac{4\pi^3}{N\Omega}a_m^2(\varkappa)\frac{K}{K_0 J}(n_q+1)D(\varkappa)$$
$$\times\sum_{\tau}\{\tfrac{1}{4}(1+e_Q^2)U(\mathbf{q})[\delta(\varkappa-\mathbf{q}-\mathbf{Q}-\tau)+\delta(\varkappa-\mathbf{q}+\mathbf{Q}-\tau)]$$
$$+ (1-e_Q^2)V(\mathbf{q})\delta(\varkappa-\mathbf{q}-\tau)\}\delta[E-\hbar\omega(\mathbf{q})] \quad (5.99)$$

where

$$U(\mathbf{q}) = V^{-1}(\mathbf{q}) = \left[\frac{\mathscr{J}(\mathbf{Q}) - \mathscr{J}(\mathbf{q}) + 2V_2}{\mathscr{J}(\mathbf{Q}) - \tfrac{1}{2}\mathscr{J}(\mathbf{q}+\mathbf{Q}) - \tfrac{1}{2}\mathscr{J}(\mathbf{q}-\mathbf{Q})} \right]^{\tfrac{1}{2}} \qquad (5.100)$$

and e_Q is the component of $\mathbf{e}$ along $\mathbf{Q}$.

5.5.2. Experimental Results for the Helical Structure

In a constant-E neutron scattering experiment, each branch of the dispersion relation in a helical structure in general gives rise to three peaks in the intensity as a function of $\varkappa$, as may be seen from (5.99). These peaks may be difficult to resolve, and the measurement of the magnon dispersion relation may therefore be considerably more difficult than for simple ferromagnets. The problem may be simplified somewhat if the scattering vector $\varkappa$ is parallel to $\mathbf{Q}$, in which case only two peaks are observed, as in Figure 5–13. Further difficulties may be

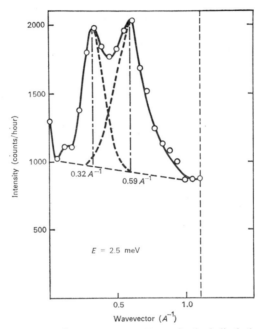

Figure 5–13. Neutron groups from a constant-E scan in the helical phase of terbium–10% holmium at 200 K. The scan is such that the scattering vector $\varkappa$ and $\mathbf{Q}$ are parallel.

encountered at large $\mathbf{q}$ values, where overlapping of branches originating from different τ vectors occurs and the natural broadening of the magnons is greater, but it is usually possible to determine the whole dispersion relation, at least in the c-direction, by a judicious combination of constant-E and constant-$\mathbf{q}$ scans.

The first measurements of spin waves in a helical structure[41] were performed on a terbium crystal to which 10% holmium had been added to stabilize the helical structure over a wider temperature range (195 K to 221 K).

The results of these measurements are shown in Figure 5–14. From (5.96) the magnon dispersion relation in the c-direction is given by

$$\hbar\omega(\mathbf{q}) = J\{[\mathscr{J}(\mathbf{Q}) - \tfrac{1}{2}\mathscr{J}(\mathbf{q}+\mathbf{Q}) - \tfrac{1}{2}\mathscr{J}(\mathbf{q}-\mathbf{Q})][\mathscr{J}(\mathbf{Q}) - \mathscr{J}(\mathbf{q}) + 2V_2]\}^{\frac{1}{2}}. \quad (5.101)$$

The experimental dispersion relations show the expected linear rise from zero at small $\mathbf{q}$, and the finite energy at $\mathbf{q}=\mathbf{Q}$ due to the axial anisotropy V_2. Because the energy at a particular $\mathbf{q}$ involves the four exchange parameters $\mathscr{J}(\mathbf{q})$, $\mathscr{J}(\mathbf{Q})$, $\mathscr{J}(\mathbf{q}+\mathbf{Q})$, and $\mathscr{J}(\mathbf{q}-\mathbf{Q})$, as well as V_2, the analysis of the results is much more complicated than in the ferromagnetic phase. In this case the value of V_2 was determined from the low-temperature results in the ferromagnetic phase,[19] renormalized according to the macroscopic measurements of Rhyne and Clark.[34] This procedure should be adequate for Tb, in which V_2 is much greater than V_4 and V_6. The $\mathscr{J}(\mathbf{q})$ function was then determined either

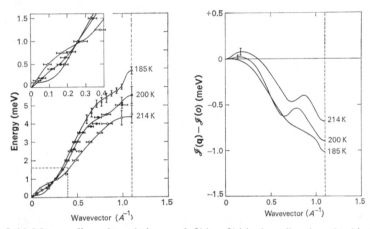

Figure 5–14. Magnon dispersion relations and $\mathscr{J}(\mathbf{q}) - \mathscr{J}(o)$ in the c-direction of terbium–10% holmium as a function of temperature in the helical and ferromagnetic (185 K) phases.

by a least-squares procedure using interplanar exchange parameters,[41] or by an iterative procedure in which it is assumed that $\mathscr{J}(\mathbf{q})$ has a maximum at $\mathbf{Q}$.[20] In either case there is an implicit assumption that $\mathscr{J}(\mathbf{q})$ is a smooth function, and the possible error in its determination is estimated by discovering within what limits it may be varied, while reproducing the dispersion relation within the experimental uncertainty. The values of $\mathscr{J}(\mathbf{q})$ resulting from such an analysis are also shown in Figure 5–14, which illustrates the interesting feature that the maximum at $\mathbf{q}=\mathbf{Q}$, which is necessary to stabilize the helical structure decreases in magnitude as the temperature decreases. This phenomenon may be observed directly in the dispersion relations, since from (5.101)

$$\hbar\omega(\mathbf{Q}) = J\{2V_2[\mathscr{J}(\mathbf{Q}) - \tfrac{1}{2}\mathscr{J}(o) - \tfrac{1}{2}\mathscr{J}(2\mathbf{Q})]\}^{\frac{1}{2}}$$

and $\quad\quad\quad\quad\quad\quad\quad\quad\quad\quad\quad\quad\quad\quad\quad\quad\quad\quad\quad\quad\quad\quad\quad\quad\quad$ (5.102)

$$\hbar\omega'(o) = J\{\tfrac{1}{2}\mathscr{J}''(\mathbf{Q})[\mathscr{J}(\mathbf{Q}) - \mathscr{J}(o) + 2V_2]\}^{\frac{1}{2}}.$$

A relative decrease in $\mathscr{J}(\mathbf{Q})$, and the concomitant decrease in the curvature $\mathscr{J}''(\mathbf{Q})$, therefore causes a decrease in the magnitude of $\hbar\omega(\mathbf{Q})$ and in the initial slope of $\hbar\omega(\mathbf{q})$, both of which are observed in the results. It should be mentioned that the increase in the effective V_2 with decreasing temperature due to renormalization effects has the opposite effect to the change in $\mathscr{I}(\mathbf{Q})$, and that the latter dominates in the temperature range of these measurements.

The magnon dispersion relations have also been measured in the helical phases of holmium[42, 43, 44, 21] and dysprosium[21] and the results, together with the values of $\mathscr{I}(\mathbf{q})$ deduced from them are shown in Figure 5–15. The

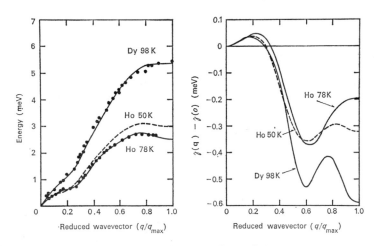

Figure 5–15. Magnon dispersion relations and $\mathscr{I}(\mathbf{q})-\mathscr{I}(o)$ in the c-direction for the helical phases of dysprosium and holmium (after Ref. 21).

analysis of the results for holmium is further complicated by the fact that the hexagonal anisotropy V_6^6 is still quite large at the temperatures of the measurements.[34] This does not affect the general form of the dispersion relation which still goes to zero in the long wavelength limit, due to the averaging of the hexagonal anisotropy in an undistorted incommensurable structure,[45] but the detailed $\mathbf{q}$ dependence may be affected by hexagonal anisotropy, and possibly magnetoelastic effects.[21] This may account for the relatively large difference between the exchange functions in the ferromagnetic and helical phases at large $\mathbf{q}$, and the strong temperature dependence in the helical phase. Yet a further complication for holmium arises because the higher order axial anisotropy is relatively large, so that the relation between V_2 and the macroscopic anisotropy parameters is complicated, and indeed the parameters deduced from spin wave and macroscopic measurements do not agree.[21] It is apparent from the results, however, that the exchange functions for both metals have maxima at $\mathbf{Q}$ which are substantially larger than those observed in the ferromagnetic phase.

The neutron scattering studies of the magnon dispersion relations have provided detailed information on the indirect exchange interaction and cast considerable light on the mechanisms which stabilize the different magnetic structures and lead to transitions between them. The exchange is, as expected, long range and oscillatory and, in the helical phases, $\mathscr{I}(\mathbf{q})$ has a maximum at the Q which characterizes the periodic structure. As discussed in Chapter 6, this peak is associated with details of the band structure, which result in a maximum in $\chi(\mathbf{q})$, the susceptibiilty of the conduction electron gas. The magnitude of the peak correlates qualitatively with the stability of the periodic structures. In holmium and dysprosium it is sufficiently pronounced that a residuum survives the transition to the ferromagnetic phase, and its relatively large size in the helical phase of holmium is consistent with the observation that this metal retains a periodic structure down to the lowest temperatures, in zero field. In contrast, gadolinium and terbium have no maximum at non-zero $\mathbf{q}$ in the ferromagnetic phase, although terbium–10% holmium, which presumably behaves similarly to pure terbium, has a pronounced peak in the helical phase which shifts and decreases in size as the temperature is lowered. The strong dependence of the band structure and hence $\chi(\mathbf{q})$ on magnetization is consistent with the large value of S in terbium. At the Curie temperature, the peak in $\mathscr{J}(\mathbf{q})$ has been reduced to the point where it is not sufficient to maintain the periodic structure against the magnetoelastic forces, which increase with decreasing temperature and ultimately drive the structure ferromagnetic.[46] The change in the band structure at the transition then eliminates the peak in $\mathscr{J}(\mathbf{q})$ in the ferromagnetic phase. On the other hand, in dysprosium and holmium–10% terbium, in which T_c is so low that the effect of the magnetization on the band structure is essentially fully developed, it is the rapid variation of the magnetoelastic and anisotropy forces with σ which determines the transition temperature. $\mathscr{J}(\mathbf{q})$ still changes abruptly at T_c, but not sufficiently to eliminate the maximum entirely. It is interesting that in the holmium–terbium system, the disordered alloys can be described in terms of an average exchange function which varies smoothly with the concentration of the constituents.

5.5.3. The Cone and Other Periodic Structures

The addition of a ferromagnetic component to the helical structure further complicates both the theoretical and experimental studies of the spin waves, so that the available information on this structure is still relatively limited. The magnon energies for the cone may be calculated by transforming to a coordinate system in which the ordered moment, given by (5.89), is always along the local z-axis. If the planar anisotropy is neglected, the usual Holstein–Primakoff and Fourier transformations, followed by an appropriate canonical transformation, then diagonalize the Hamiltonian, and the use of the condition for the stability of the cone structure allows the dispersion relation to be written in the form[29]

$$\begin{aligned}
\hbar\omega(\mathbf{q})/J = &\tfrac{1}{2}[\mathscr{J}(\mathbf{Q}+\mathbf{q})-\mathscr{J}(\mathbf{Q}-\mathbf{q})]\cos\theta \\
&+[\mathscr{J}(\mathbf{Q})-\tfrac{1}{2}\mathscr{J}(\mathbf{q}+\mathbf{Q})-\tfrac{1}{2}\mathscr{J}(\mathbf{q}-\mathbf{Q})] \\
&\times\{[\mathscr{J}(\mathbf{Q})-\tfrac{1}{2}\mathscr{J}(\mathbf{q}+\mathbf{Q})-\tfrac{1}{2}\mathscr{J}(\mathbf{q}-\mathbf{Q})]\cos^2\theta \\
&+[\mathscr{J}(\mathbf{Q})-\mathscr{J}(\mathbf{q})+2V_2+6V_4J^2\cos^2\theta \\
&+15V_6J^4\cos^4\theta]\sin^2\theta\}^{\frac{1}{2}}. \quad (5.103)
\end{aligned}$$

The neutron scattering cross section may be calculated by the usual method and is similar to that for the helix, except that the inelastic structure factors multiplying the momentum transfer δ-functions are modified.[40]

As may be seen from (5.103)

$$\hbar\omega(\mathbf{q}) \neq \hbar\omega(-\mathbf{q}) \qquad (5.104)$$

for the cone structure, so that the dispersion relation is non-centrosymmetric. In a multi-domain crystal therefore, six peaks generally arise from a single branch of the dispersion relation, in a constant-E neutron scattering experiment. This undesirably complicated situation may, however, be simplified if $\varkappa$ and $\mathbf{Q}$ are parallel, and the peaks may also be distinguishable by their inelastic structure factors.

Spin wave energies have been measured in the conical phase of Ho by Stringfellow et al.[44] Since the cone angle in Ho has the large value of $79°$, the splitting between $\hbar\omega(\mathbf{q})$ and $\hbar\omega(-\mathbf{q})$ was unobservable in these experiments, and the measured energies therefore correspond to an average value. The analysis of these results gives a $\mathscr{J}(\mathbf{q})$ which is quite similar to that observed in the helical phase, though there is again some disparity between the anisotropy parameters which give the best fit to the magnon energies and those determined from macroscopic measurements.

These experiments also give some information on the mechanism which causes the cone-helix transition in holmium. The requirement that the magnon energy of wavevector $\mathbf{Q}$ in the helical phase be real implies that

$$V_2(h) \geq 0 \qquad (5.105)$$

while a similar criterion for the cone structure, together with the condition that the total energy be stationary leads to[44]

$$V_2(c) < \tfrac{3}{4}[\mathscr{J}(o) - \mathscr{J}(\mathbf{Q})] + 3V_6 J^4 \cos^4 \theta +$$
$$+ \tfrac{1}{8} \cot^2 \theta [\mathscr{J}(\mathbf{Q}) - \tfrac{1}{2}\mathscr{J}(o) - \tfrac{1}{2}\mathscr{J}(2\mathbf{Q})]. \qquad (5.106)$$

Since the last two terms are small, V_2 should be negative in the conical phase and must therefore change sign at the transition. The rapid variation in magnitude of V_2 with temperature should lead to a similar rapid variation in $\hbar\omega(\mathbf{Q})$, and a considerable temperature variation of $\hbar\omega(0.2\pi/a, 0.2\pi/a, Q)$ was in fact observed in the experiments. The cone-helix transition is probably not the result of $\hbar\omega(\mathbf{Q})$ becoming zero, however, since such an instability would lead to a canted helical structure,[47, 48] which is not observed experimentally. The mechanism which leads to the transition is signalled by the rapid temperature dependence of a particular magnon mode, as it is in the ferromagnet-helix transition. It is not the instability of this "soft mode" which leads to the transition, however, but rather the abrupt change in sign of the difference between the total energies of the two phases, perhaps due to magnetoelastic effects, which produces the observed first-order phase change.

Magnon energies have also been studied in the conical phase of erbium,[49] in which the cone angle is small. This allows the splitting between $\hbar\omega(\mathbf{q})$ and $\hbar\omega(-\mathbf{q})$ to be clearly observed, as in the results of Nicklow et al.[49] shown in Figure 5–16. As may be seen from (5.103) the splitting between the two branches gives a rather direct determination of $\mathscr{J}(\mathbf{q})$, and its form in erbium is found to be rather different from that in the periodic structures of the other heavy rare earth metals. An especially interesting feature of these results is that they cannot be satisfactorily reproduced by a Hamiltonian

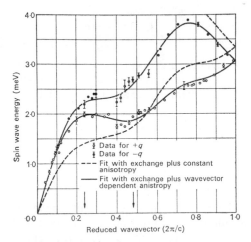

Figure 5–16. Magnon energies in the c-direction of the cone phase of erbium. The dashed line is an attempt to fit (5.103) to the experimental results, while the full lines are derived from an expression in which the anisotropy is allowed to depend on wavevector.

consisting of isotropic exchange and single-ion anisotropy only, as is illustrated in Figure 5–16. On the other hand, the inclusion of a very large anisotropic exchange of the form of (5.45) does allow the satisfactory fit shown in Figure 5–16 to be obtained.

The other, more complex, periodic structures in the heavy rare earths have not been systematically studied experimentally. The longitudinal wave structure, in which the component of the moment along the hexagonal axis orders according to

$$J_l^\zeta = J \cos \mathbf{Q} \cdot \mathbf{R} \qquad (5.107)$$

while those in the plane are disordered, is the stable high-temperature phase of thulium and erbium. The ordered moment in this structure varies from site to site, and spin waves are consequently not elementary excitations of the magnetic system.[29] On the other hand, the square wave ferromagnetic structure of thulium at low temperatures should exhibit magnon excitations and so should the canted helical structure of erbium, provided that it is not so distorted by the anisotropy that the total ordered moment varies significantly through the structure.

5.5.4. Applied Field Effects

The application of a magnetic field brings about complicated changes in the periodic magnetic structures of the heavy rare earths, and these complexities are reflected in the spin wave spectra. In a sufficiently high field, the ground state of the system is, of course ferromagnetic with the moments along the field, but it may reach this state by passing through several intermediate structures.[50] Since experimental evidence on these phenomena is very sparse, we will discuss only a single simple example; a helical structure with no planar anisotropy to which a field is applied in the plane. The ferromagnetic structure is reached at a field[50]

$$H_t = \frac{[\mathscr{J}(\mathbf{Q}) - \mathscr{J}(o)]J}{\lambda \beta} \qquad (5.108)$$

but there is an intermediate transition, occurring at approximately $H_t/2$, at which the helix transforms abruptly through a first-order transition to a fan structure, in which the moments make an angle θ with the field direction, given by

$$\sin \frac{\theta_l}{2} = 2\,\delta \sin \mathbf{Q} \cdot \mathbf{R}_l \qquad (5.109)$$

where

$$\delta^2 = \frac{\lambda \beta (H_t - H)}{2J[3\mathscr{J}(\mathbf{Q}) - 2\mathscr{J}(o) - \mathscr{J}(2\mathbf{Q})]}. \qquad (5.110)$$

The magnon dispersion relations in these structures have been studied theoretically by Cooper and Elliott,[51] and the field dependences of the modes of wave vector o and $\pm\mathbf{Q}$ are shown in Figure 5–17. In the helical phase, the normal modes in a magnetic field are symmetrical and antisymmetrical combinations of the magnons of wavevector $\mathbf{Q}$ and $-\mathbf{Q}$, with energies $\hbar\omega_S(\mathbf{Q})$ and $\hbar\omega_A(\mathbf{Q})$ respectively. The splitting between these modes increases with field until, at the transition to the fan phase, they both drop abruptly, with $\hbar\omega_S(\mathbf{Q})$ going to zero, while $\hbar\omega(o)$ rises to a finite value. A further increase in the field reduces $\hbar\omega_A(\mathbf{Q})$ further, and it is the instability of this mode at H_t which leads to the second order transition, above which both modes of wavevector $\pm\mathbf{Q}$ rise rapidly in energy. This behaviour may be studied by microwave resonance. The microwave field couples to the magnetic system through the Hamiltonian (5.87) and the modes which are excited with different polarizations may be determined by expressing the angular momentum operators in terms of magnon variables.[51] A microwave field polarized with its magnetic field

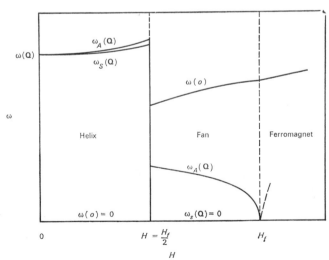

Figure 5–17. The variation of spin wave frequencies when a magnetic field is applied in the plane of a helical structure with negligible planar anisotropy (after Ref. 51).

along the c-axis excites $\omega(o)$, while polarizations parallel and perpendicular to the d.c. magnetic field $\mathbf{H}$ in the plane excite $\omega_A(\mathbf{Q})$ and $\omega_S(\mathbf{Q})$ respectively. In the fan phase, a microwave field perpendicular to $\mathbf{H}$ excites $\omega(o)$, while polarization parallel to $\mathbf{H}$ excites $\omega_A(\mathbf{Q})$ with an intensity proportional to the angular amplitude δ of the fan.

There have been a number of studies of microwave absorption in periodic structures as a function of the external magnetic field,[30, 52] but these have, been performed at such a low frequency that only effects associated with phase transitions have generally been observed. It would be interesting to study infra-red absorption in periodic structures, but the experimental difficulties associated with carrying out such measurements on single crystals are formidable and inelastic neutron scattering seems a more promising technique for studying applied field effects.

5.6. THERMODYNAMIC PROPERTIES

At low temperatures magnons are, to a very good approximation, independent boson excitations of the heavy rare earths, and the magnetic contribution to the thermodynamic properties can therefore be calculated from their energy spectrum. The partition function for the magnetic system is

$$Z = \mathrm{Tr}\left\{\exp\left(-\sum_{\substack{\mathbf{q} \\ i=\alpha, \beta}} \hbar\omega_i(\mathbf{q})n_{\mathbf{q}i}/\mathnormal{k}T\right)\right\} = \prod_{\mathbf{q}i}\{1 - e^{-\hbar\omega_i(\mathbf{q})/\mathnormal{k}T}\}^{-1} \quad (5.111)$$

where the subscript i represents the acoustic and optical branches. The thermal average of the occupation number for a state of wavevector $\mathbf{q}$ is then given by the Bose–Einstein distribution function

$$\langle n_{\mathbf{q}i}\rangle = (e^{\hbar\omega_i(\mathbf{q})/\mathnormal{k}T} - 1)^{-1} \quad (5.112)$$

and the thermodynamic properties can be calculated by the application of elementary statistical mechanics. As we shall see, the presence of the energy gap in the spectrum of the anisotropic ferromagnet has an important effect on the low-temperature properties.

As we shall discuss in (5.7), magnon interactions at finite temperatures cause a shift in the magnon energies, which may be described in terms of a renormalization of the exchange and crystal field interactions, and in addition limit the magnon lifetimes. It is still possible to write an approximate expression for the partition function[15] in terms of the renormalized magnon energies $E_i(\mathbf{q})$

$$Z = \mathrm{Tr}\left\{\exp\left(-\sum_{\mathbf{q}i} E_i(\mathbf{q})n_{\mathbf{q}i}/\mathnormal{k}T\right)\right\} \quad (5.113)$$

but this expression must be used with caution in computing the thermodynamic properties and few detailed calculations at higher temperatures have yet been attempted.

5.6.1. The Magnetization

The relative magnetization of an isotropic hcp ferromagnet at temperature T may readily be calculated from (5.31) as

$$\delta\sigma(T)=1-\sigma(T)=\sum_{\mathbf{q}i}\langle n_{\mathbf{q}i}\rangle/2NJ \qquad (5.114)$$

The quadratic dependence of energy on $\mathbf{q}$ at long wavelengths then gives rise to the familiar $T^{\frac{3}{2}}$ law for the low-temperature magnetization. The dispersion relation in the basal plane is approximately linear for gadolinium, as may be seen in Figure 5–3, and Kasuya[37] has shown how this can account for the temperature dependence of the magnetization at moderate temperatures.

To calculate the moment deviation due to the excitation of a magnon in an anisotropic ferromagnet, we require the expectation value of $a_{\mathbf{q}}^{+}a_{\mathbf{q}}$, which from (5.69) and (5.70) is given by

$$\langle a_{\mathbf{q}}^{+}a_{\mathbf{q}}\rangle=\langle(u\alpha_{\mathbf{q}}^{+}-v\alpha_{-\mathbf{q}})(u\alpha_{\mathbf{q}}-v\alpha_{-\mathbf{q}}^{+})\rangle$$
$$=(2u^{2}-1)\langle n_{\mathbf{q}}\rangle+v^{2}$$
$$=\frac{J[\mathscr{I}(o)-\mathscr{I}(\mathbf{q})]+A}{\hbar\omega(\mathbf{q})}\langle n_{\mathbf{q}}\rangle+\frac{1}{2}\left\{\frac{J[\mathscr{I}(o)-\mathscr{I}(\mathbf{q})]+A}{\hbar\omega(\mathbf{q})}-1\right\} \qquad (5.115)$$

remembering that the expectation value of non-diagonal terms is zero. The first term represents the departure from order with increasing temperature, while the second is the zero-point deviation, which goes to zero with the anisotropy. For the hcp structure, a straightforward generalization of (5.115), including a magnetic field in the easy direction, gives

$$\delta\sigma(T)=\frac{1}{2NJ}\left\{\sum_{\mathbf{q}i}\frac{Jf_{i}(\mathbf{q})+A+\lambda\beta H}{\hbar\omega_{i}(\mathbf{q},H)}[\langle n_{\mathbf{q}i}\rangle+\tfrac{1}{2}]-\tfrac{1}{2}\right\} \qquad (5.116)$$

where $f_{i}(\mathbf{q})$ is defined by (5.75).

The existence of the energy gap Δ in the magnon spectrum has a substantial effect on the low-temperature magnetization, as was first pointed out by Niira.[11] He found that the temperature dependence has the approximate form

$$\delta\sigma\sim T^{\frac{3}{2}}\sum_{n=1}^{\infty}e^{-n\Delta/kT}/n^{\frac{3}{2}} \qquad (5.117)$$

and that this expression, with Δ/k taking a value of about 30 K, fits the low-temperature magnetization of dysprosium much better than the conventional $T^{\frac{3}{2}}$ law. It was later shown[28] that (5.117) could also account for the low-temperature magnetization of terbium, with a Δ/k value of about 20 K.

At higher temperatures, the renormalization of the magnon energies must be taken into account. An expression similar to (5.117) may be used,[15] but $f(\mathbf{q})$, A, and $\hbar\omega(\mathbf{q})$ must be replaced by their renormalized values $f(\mathbf{q}, T)$, $A(T)$, and $E(\mathbf{q})$, which may be determined either theoretically or from experiment. Brooks et al.[53] used approximate theoretical renormalizations of both the exchange and crystal field terms to calculate the temperature dependence of the magnetization of terbium, which is compared with experiment[54] in Figure 5–18. The agreement is good at low temperatures, but less good at

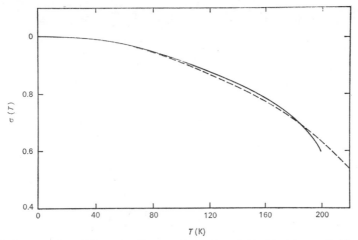

Figure 5–18. The reduced magnetization of terbium for a field of 18 kOe applied in the easy
    direction. The full curve is calculated as described in the text and the broken curve
    represents the experimental results (after Ref. 53).

higher temperatures, largely due to the approximations involved in the
renormalization procedure.

5.6.2. The Heat Capacity

At low temperatures, the magnetic contribution to the heat capacity is
readily calculated from

$$U = -\frac{\partial}{\partial(1/kT)} \ln Z = \sum_{qi} \hbar \omega_i(\mathbf{q})\langle n_{qi}\rangle \qquad (5.118)$$

and hence

$$C_m = \left(\frac{\partial U}{\partial T}\right)_v = Nk\hbar \int \frac{x^2 e^x g(\omega)d\omega}{(e^x - 1)^2} \qquad (5.119)$$

where $x = \hbar\omega/kT$ and $g(\omega)$ is the magnon density of states. The low-temperature
magnetic heat capacity in an anisotropic ferromagnet is dominated by the
energy gap, and it was shown[28, 55] to have the form

$$C_m = C(T, \Delta)e^{-\Delta/kT}. \qquad (5.120)$$

It is difficult experimentally to separate the magnetic contribution to C_v from
the lattice, electronic, and nuclear terms, but Lounasmaa and Sundström[56]
were able to accomplish this in a fairly satisfactory manner for a number of
heavy rare earths, by comparing with the results for non-magnetic lutetium.
They found that the assumption that C_m is proportional to $T^{\frac{3}{2}}e^{-\Delta/kT}$ is
consistent with their results for terbium and dysprosium, with Δ/k taking the
values 23 K and 31 K respectively.

The determination of the magnetic heat capacity at higher temperatures
presents severe theoretical and experimental problems. If the expression
(5.113) is used for Z, the contribution of the magnon interactions is counted

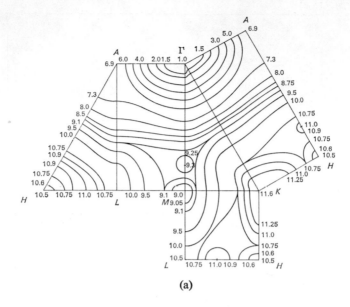

(a)

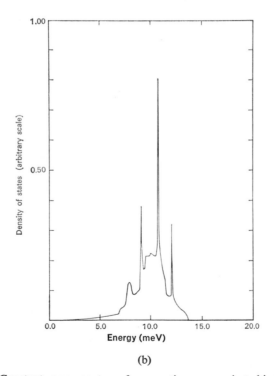

(b)

Figure 5–19. (a) Constant energy contours for acoustic magnons in terbium at 90 K on the faces of the irreducible zone. (b) The magnon density of states in terbium at 90 K (after Ref. 25).

twice.[53] This error may be approximately rectified by writing

$$U = \sum_{qi} \tfrac{1}{2}[\hbar\omega_i(\mathbf{q}) + E_i(\mathbf{q})]\langle n_{\mathbf{q}i}\rangle \tag{5.121}$$

and an even better approximation[53] is

$$U = \sum_{qi} \tfrac{1}{2}\{[\hbar\omega_i(\mathbf{q}) + E_i(\mathbf{q})]\langle n_{\mathbf{q}i}\rangle + (kT)^{-1}E_i(\mathbf{q})[\hbar\omega_i(\mathbf{q}) - E_i(\mathbf{q})]\langle n_{\mathbf{q}i}\rangle^2\}. \tag{5.122}$$

This expression has been used with an approximate magnon renormalization scheme to calculate C_m for terbium,[53] and it is found that the second term becomes quite important above 100 K.

In order to calculate the internal energy and heat capacity from experimentally determined magnon energies, it is necessary to determine the density of states from the dispersion relations. A suitable interpolation scheme for accomplishing this, taking into account the symmetry of the constant energy surfaces in the zone, was developed by Houmann.[25] Constant energy contours on the faces of the zone and the density of states for terbium calculated by this technique are shown in Figure 5–19.

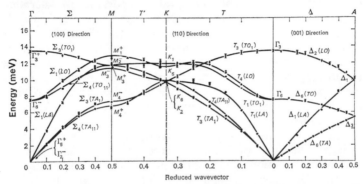

Figure 5–20. Phonon dispersion relations for terbium in the major symmetry directions at room temperature (after Ref. 57).

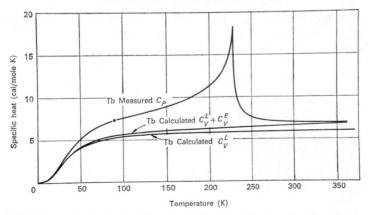

Figure 5–21. Measured and calculated heat capacities of terbium as a function of temperature. The point is the sum of the calculated lattice and electronic terms, together with the magnon contribution at 90 K, calculated as described in the text (after Ref. 57).

The experimental determination of C_m at elevated temperatures again requires the subtraction of the lattice and electronic terms. The latter is relatively small and, for terbium, the former was determined by Houmann and Nicklow[57] from the measured phonon dispersion relations, shown in Figure 5–20. The calculated and measured heat capacities of terbium are shown in Figure 5–21. The point at 90 K represents the sum of the lattice and electronic terms and a magnetic contribution calculated using the low-temperature expression (5.119), with the magnon energies and density of states determined from experiments at 90 K and the interpolation scheme. The agreement between experiment and theory is very good, although the use of the low-temperature approximation at 90 K is not strictly justified.

5.7. MAGNON INTERACTIONS

In the preceding discussion we have neglected, except incidentally, the interactions of magnons among themselves and with other excitations. Such interactions may, however, be of considerable importance in determining the behaviour of the magnetic system, and their effects may readily be observed experimentally, especially by neutron scattering. Interactions affect the magnon system in two ways, by modifying the dispersion relations and by limiting the magnon lifetimes. It appears that, in the rare earths, the latter process is dominated by magnon–magnon interactions, but other mechanisms have been considered theoretically.[58] A considerable variety of phenomena involving magnon interactions have been observed experimentally, but few systematic studies have yet been undertaken and a detailed comparison between theory and experiment is not therefore generally available. In the following, we shall therefore be satisfied with outlining the experimental evidence for the interactions of magnons among themselves and with phonons, impurities, and electrons, and with discussing qualitatively the mechanisms involved.

5.7.1. Magnon–Magnon Interactions

The modification of the magnon spectrum with increasing temperature due to magnon–magnon interactions may be illustrated by considering the isotropic ferromagnet. For small moment deviations, we may go beyond the approximation (5.22) by expanding the exact relations (5.17) in the form

$$J_l^+ = (2J)^{\frac{1}{2}}\left(a_l - \frac{a_l^+ a_l a_l}{4J}\right)$$

$$J_l^- = (2J)^{\frac{1}{2}}\left(a_l^+ - \frac{a_l^+ a_l^+ a_l}{4J}\right) \tag{5.123}$$

Using (5.21), we then find that the indirect exchange Hamiltonian (5.8) is the sum of (5.23) and the following magnon interaction Hamiltonian, involving products of 4 deviation operators

$$\mathscr{H}_{\mathrm{mm}} = J \sum_{l>l'} \mathscr{J}(\mathbf{R}_{ll'})[\tfrac{1}{2}(a_l^+ a_{l'}^+ a_l a_{l'} + a_l^+ a_{l'}^+ a_{l'} a_l) - a_l^+ a_{l'}^+ a_l a_{l'}]. \tag{5.124}$$

Fourier transforming in the usual way, we find

$$\mathscr{H}_{mm} = \frac{J}{4N} \sum_{q_1 q_2 q_3 q_4} [\mathscr{J}(q_1) + \mathscr{J}(q_3) - 2\mathscr{J}(q_1 - q_3)] \delta(q_1 + q_2 - q_3 - q_4)$$
$$\times a_{q_1}{}^+ a_{q_2}{}^+ a_{q_3} a_{q_4}. \quad (5.125)$$

This expression may be decomposed into two types of term, according as the wavevector pair (q_1, q_2) is the same as or different from (q_3, q_4). The first involves only magnon number operators and has the form

$$\mathscr{H}_{mm}{}^{(1)} = \frac{J}{4N} \sum_{qq'} [3\mathscr{J}(q) + \mathscr{J}(q') - 2\mathscr{J}(q - q') - 2\mathscr{J}(o)] n_q n_{q'}$$
$$= \sum_q \Delta E(q) n_q \quad (5.126)$$

where

$$\Delta E(q) = \frac{J}{4N} \sum_{q'} [3\mathscr{J}(q) + \mathscr{J}(q') - 2\mathscr{J}(q - q') - 2\mathscr{J}(o)] n_{q'}. \quad (5.127)$$

Hence the renormalized magnon energy is

$$E(q) = \hbar\omega(q) + \Delta E(q) \quad (5.128)$$

where, at low temperatures $\Delta E(q)$ varies[59] as $T^{\frac{5}{2}}$.

At higher temperatures, the expansion in powers of the moment deviation converges poorly, and the calculation of the exchange renormalization of the magnon energy becomes very difficult. A simple approximation is to assume that the phases of the deviations at different sites are uncorrelated. In this random-phase approximation[60] the effective exchange is proportional to the relative magnetization σ. It is in general also necessary to take into account the magnon interactions due to the crystal field terms in the Hamiltonian. Brooks et al.[15] showed that the principal effect of such interactions is to renormalize the anisotropy parameters in the way described by the Callen–Callen[35] theory, and hence to give rise to the temperature dependence of the energy gap discussed in (5.4.3).

There is no particular difficulty in studying the magnon renormalization experimentally and typical results for the c-direction of ferromagnetic terbium[19] are shown in Figure 5–22(a). The variation of the anisotropy parameters has been discussed earlier and the renormalization of the effective exchange is illustrated in Figure 5–22(b). According to the random-phase approximation, $[\mathscr{J}(o) - \mathscr{J}(q)]/\sigma$ should be independent of temperature. There is a rough proportionality between the exchange functions and the ordered moment, but there are noticeable differences between the shapes of the curves at different temperatures, probably reflecting the variation with σ of the electronic structure and $\chi(q)$.

The non-diagonal terms in (5.125) correspond to the mutual scattering of two magnons, with initial wavevectors q_3 and q_4, to final states q_1 and q_2. This process limits the magnon lifetime, or equivalently produces an energy broadening $\Gamma(q)$ of the magnon of wavevector q. The crystal fields may also lead to magnon–magnon scattering processes.[15] The magnon lifetimes may be measured by neutron scattering, since the energy transfer δ-function in

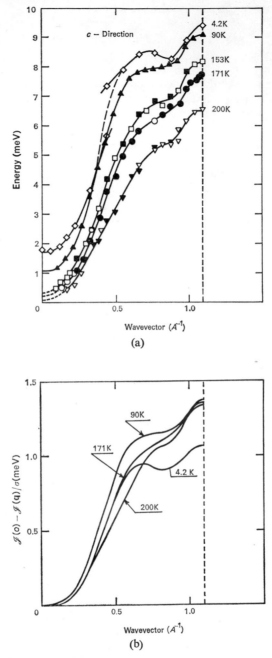

Figure 5–22. (a) Temperature dependence of the magnon energies in the c-direction in the ferromagnetic phase of terbium. The break in the curve for 4.2 K is due to the magnon–phonon interaction. (b) The exchange function $\mathcal{J}(o)-\mathcal{J}(q)$ in the c-direction, scaled by the relative magnetization σ, in the ferromagnetic phase of terbium.

(5.56) is modified by such processes to the Lorentzian form[61]

$$L(E)=\frac{\Gamma(\mathbf{q})}{[E-\hbar\omega(\mathbf{q})-\Delta E(\mathbf{q})]^2+\Gamma^2(\mathbf{q})}. \tag{5.129}$$

Provided that the instrumental resolution can be extracted,[62] the width of the neutron group in the constant-$\mathbf{q}$ scan therefore determines the magnon lifetime directly. Lifetimes so determined for magnons in the c-direction of terbium at a number of different temperatures are shown in Figure 5–23. At the higher

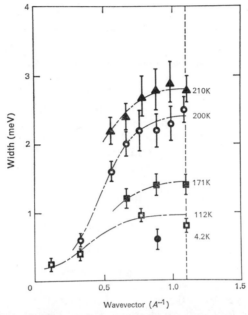

Figure 5–23. Magnons widths as a function of wavevector and temperature in the double zone representation in the c-direction, for the ferromagnetic phase of terbium.

temperatures the width is a substantial fraction of the magnon energy. The rapid temperature variation of these results and the correlation of the energy broadening with the magnetization indicates strongly that the magnon–magnon interaction is the dominant mechanism limiting the magnon lifetime, but no calculations are at present available with which the data can directly be compared. Preliminary measurements of magnon lifetimes in the ferromagnetic and helical phases of dysprosium at 78 K and 98 K have also been reported.[21] The energy broadening of the neutron groups is similar in the two phases, and comparable with that in terbium at the same temperatures.

5.7.2. Magnon–Phonon Interactions

The relative displacements of the ions associated with a lattice wave may modify both the exchange interactions among them and the crystal field, and thereby lead to a coupling between the magnon and phonon systems. Since no magnon–phonon interaction is observed experimentally in gadolinium, whereas pronounced effects appear in terbium (compare Figures 5–3 and 5–4) it is clear that the coupling is predominantly due to magnetoelastic effects, which are very small in gadolinium. The magnon–phonon interaction in terbium has been discussed on this basis by Jensen.[63] He finds that the magnon–phonon interaction may be expressed in the form

$$\mathscr{H}_{mp} = \sum_{q} \hbar\omega_m(\mathbf{q})(\alpha_q^+\alpha_q + \tfrac{1}{2}) + \sum_{k} \hbar\omega_p(k)(\beta_k^+\beta_k + \tfrac{1}{2})$$

$$+ \sum_{q} W(\mathbf{q})(\alpha_q^+ + \alpha_{-q})(\beta_q + \beta_{-q}^+)$$

$$+ \sum_{kq} [U(k, \mathbf{q})\alpha_{q+k}^+\alpha_q + \tfrac{1}{2}V(k, \mathbf{q})(\alpha_{q+k}^+ a_{-q}^+ + \alpha_q\alpha_{-q-k})]$$

$$\times(\beta_k + \beta_{-k}^+) \quad (5.129)$$

where α_q and β_k are respectively magnon and phonon annihilation operators, and the interaction functions $W(\mathbf{q})$, $U(k, \mathbf{q})$, and $V(k, \mathbf{q})$ may be expressed in terms of the exchange, crystal field, and magnetoelastic parameters of the crystal.[63] The last term in (5.129) represents the scattering of a magnon with absorption or emission of a phonon. It limits the magnon lifetime and renormalizes the phonon energies, which is of particular significance for acoustic phenomena, but we shall not consider it further here. The other interaction term mixes magnon and phonon modes and causes a substantial modification in the dispersion relations. In particular, it may cause a strong mode-mixing near points at which the magnon and phonon dispersion relations would cross in the absence of the interaction, resulting in a splitting apart of the branches at the virtual crossing point of magnitude

$$\Delta(\mathbf{q}) = |\{\hbar\omega(\mathbf{q})[\hbar\omega(\mathbf{q}) + 2W(\mathbf{q})]\}^{\frac{1}{2}} - \{\hbar\omega(\mathbf{q})[\hbar\omega(\mathbf{q}) - 2W(\mathbf{q})]\}^{\frac{1}{2}}| \simeq 2|W(\mathbf{q})|. \quad (5.130)$$

Near this $\mathbf{q}$-value, the normal modes of the system consist of a combination of lattice displacements and moment deviations.

It may be deduced from the form of $W(\mathbf{q})$ that, for propagation in the c-direction, neither longitudinal phonons nor transverse phonons polarized normal to the magnetization direction interact with the magnons. On the other hand, transverse phonons polarized along the magnetization do interact, giving rise to a splitting

$$\Delta_t^c(\mathbf{q}) = \frac{\hbar q c_{44} H_0}{2NJ} \left[\frac{J}{M(Jf(\mathbf{q}) + A + B)}\right]^{\frac{1}{2}} \quad (5.131)$$

where A and B are defined by (5.66) and $f(\mathbf{q})$ by (5.75), c_{44} is an elastic constant, M is the mass of the ions, and H_0 a magnetostriction coefficient.[64]

The interaction of magnons with transverse phonons propagating in the basal plane depends on the polarization direction. Polarization along the c-direction leads to a splitting given by (5.131), while the interaction for polarization in the plane is determined by the angle between the magnetization and polarization directions. Transverse phonons propagating perpendicular to the magnetization direction give rise to a splitting

$$\Delta_t^m(\mathbf{q}) = 2\hbar q c_{66}[C - A] \left[\frac{J}{M(Jf(\mathbf{q}) + A - B)}\right]^{\frac{1}{2}}/NJ \quad (5.132)$$

where c_{66} is an elastic constant and C and A are the magnetostriction co-efficients introduced previously. The splitting when **q** is along the other a-directions in the plane is one-half as great. Similar considerations apply to transverse phonon propagating along the b-direction.[63]

The interaction of magnons with longitudinal phonons propagating in the plane also depends upon the relation between the magnetization and **q**-vector. Longitudinal phonons perpendicular to the magnetization direction do not interact with the magnons, while those with **q** parallel to the other a-directions suffer a splitting

$$\Delta_l{}^a(\mathbf{q}) = \hbar q c_{66}[C-A]\left[\frac{3J}{M(Jf(\mathbf{q})+A-B)}\right]^{\frac{1}{2}}/NJ \qquad (5.133)$$

For a multi-domain crystal therefore, three magnon peaks will be observed in a constant **q** scan at the crossing of the magnon and longitudinal phonon dispersion relations in the a-direction; one propagating in the magnetization direction and unperturbed, and two split by the magnon–phonon interaction in the other a-directions. Again, similar considerations apply to longitudinal phonons in the b-direction.

These selection rules, which may also be deduced from group theoretical arguments,[48] are in accord with the experimental observations, as illustrated

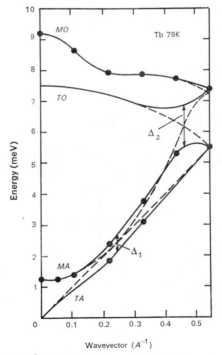

Figure 5–24. The magnon and transverse phonon (TA, TO) dispersion relations for terbium in the c-direction at 79 K. The magnon–phonon interaction causes a mixing of the modes and energy gaps Δ_1 and Δ_2 at the crossing points of the unperturbed dispersion relations, indicated by the dashed lines (after Ref. 63).

in Figure 5–4. Furthermore, the magnitude of the interaction between magnons and longitudinal phonons propagating in the basal plane is approximately correctly given by (5.133), while (5.131) also gives about the correct magnitude for the splitting Δ_1 observed between transverse acoustic phonons and magnons propagating in the c-direction at 79 K, which is illustrated in Figure 5–24. However, the theory does not predict the splitting Δ_2 caused by the interaction between transverse optic phonons and magnons, shown in Figures 5–4 and 5–24. This deficiency is ascribed[63] to the description of the interaction in terms of the strain, which represents the displacement of a single ion, whereas the important parameter is the relative displacement of neighbouring ions.

5.7.3. Magnon–Impurity Interactions

A magnetic impurity in a rare earth host is subjected to forces different from those experienced by the host atoms. If the impurity has a precession frequency in the exchange and crystal fields of its neighbours which lies outside the range of host magnon frequencies, it will form a localized magnon mode, in which the moment deviation decays rapidly away from the impurity.[65] Such localized modes have not yet been observed in rare earth metals. On the other hand a quasi-local or resonant magnon mode has been observed in terbium–10% holmium[19] and is illustrated in Figure 5–25 (a). In such a mode, the impurity frequency lies in the magnon band of the host, and the magnons close to this frequency therefore suffer a resonant scattering which strongly perturbs the dispersion relations.

The theoretical treatment of such modes in rare earths is complicated by the long range of the exchange, but Blackman[66] has performed an approximate

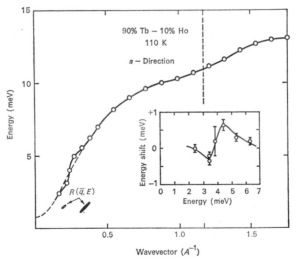

Figure 5–25. (a) The dispersion relation for magnons propagating in the a-direction of terbium–10% holmium at 110 K. The resolution functions used in obtaining these results are also shown. The insert shows the departure of the experimental points from the dashed straight line and has the characteristic shape corresponding to a resonant mode.

calculation in which he takes into account both the change in anisotropy forces acting on a holmium impurity, and the modification of the exchange interaction due to the difference in $(\lambda - 1)J$ between the host and the impurity. Using values of the anisotropy parameters based on the calculations of Kasuya,[37] he finds that the resonance at 110 K occurs at 4.25 meV with a width of 0.88 meV, in good agreement with the experimentally observed values of 4 meV and 1 meV respectively.

When the crystal is cooled to 4.2 K, the increase in the anisotropy raises the energy at which the resonance occurs, so that it falls in the vicinity of the crossing of the magnon and phonon dispersion relations in the c-direction. This has the effect of substantially amplifying the magnon–phonon interaction, as may be observed in Figure 5–25 (b), and the impurity mode appears as a small branch of excitations in the resulting energy gap, with some dispersion due to the finite concentration.

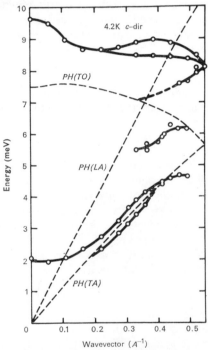

Figure 5–25. (b) The excitations in the c-direction in the same crystal at 4.2 K. The local modes now appear as a small branch in the gap produced by the magnon–phonon interaction.

5.7.4. Magnon–Electron Interactions

The localized magnetic moments and the conduction electrons interact through (5.1), which may be written more exactly in the form[67]

$$\mathcal{H}_{sf} = -\frac{(\lambda - 1)}{N} \sum_{kk'R_l} U(\mathbf{q}) e^{-i\mathbf{q}\cdot\mathbf{R}_l}$$
$$\times [(c_{k'\uparrow}{}^+ c_{k\uparrow} - c_{k'\downarrow}{}^+ c_{k\downarrow}) J_{lz} + c_{k'\uparrow}{}^+ c_{k\downarrow} J_l^- + c_{k'\downarrow}{}^+ c_{k\uparrow} J_l^+] \quad (5.134)$$

where $c_{k\uparrow}$ is the annihilation operator for the Bloch state $k\uparrow$. This interaction gives rise to the indirect exchange coupling between the ions and, in the ordered state, is also responsible for the modification of the conduction electron band structure, discussed in Chapter 6. In addition, it results in a magnon–conduction electron scattering which limits the lifetimes of both magnon and electron states.

The probability of an electron–magnon scattering process may be calculated from the interaction Hamiltonian (5.134) by expressing the angular momentum operators in terms of magnon operators and using an expression analogous to (5.49). The energy broadening of a particular magnon may then be calculated by summing over the electron states. The absorption of a magnon by a conduction electron requires a change in the spin of the latter, and conservation of energy and momentum may prohibit this process in ferromagnets, for magnons of small wavevector.[58] The interaction with the conduction electrons is the only mechanism which is expected to limit magnon lifetimes in pure crystals at low temperatures, and should therefore be susceptible to experimental study. In practice few such experiments have been performed, although the magnon broadening which has been observed in terbium at 4.2 K,[19] and is shown in Figure 5–23, is very likely due to interaction with the conduction electrons.

The most striking manifestation of the magnon–electron interaction is its effect on the transport properties, through the limitation of the mean free path of the conduction electrons. The relaxation time for a particular conduction electron may be calculated from the interaction Hamiltonian (5.134) by transforming to magnon operators and summing over all available magnon states, and the effect of the magnetic scattering on the transport properties thereby calculated.[68] In particular, the magnetic contribution to the low-temperature resistivity of an isotropic ferromagnet may be shown, for a simplified model[69] to vary as T^2. To a good approximation, this variation is modified to $T^2 e^{-\Delta/kT}$ in an anisotropic ferromagnet, which was shown[28] to account for the rapid rise in the low-temperature resistivity of terbium and dysprosium, compared with that of non-magnetic lutetium. The first attempt to study the effect of a magnetic field on the energy gap was made by measuring the temperature dependence of the resistivity of a terbium single crystal in magnetic fields as large as 50 kilogauss, applied along the hard direction.[70] Because of the small field dependence of Δ illustrated in Figure 5–11, however, only a very small effect was observed. More recent measurements on terbium single crystals[71] give a Δ/k value of about 21 K, in good agreement with other determinations.

The effect of magnon scattering on the electrical resistivity may also be studied by changing the magnetic structure through the application of a magnetic field.[72] If the field is applied in the b-direction of holmium at low temperatures, there is a transition from a conical to a simple ferromagnetic structure, which produces a very small increase in the resistivity at 4.2 K, as may be seen in Figure 5–26. At this temperature, magnon scattering is negligible and the change in conductivity is primarily due to the change in band structure at the transition. As the temperature is increased, however, the change in resistivity changes sign because the scattering by the low-lying

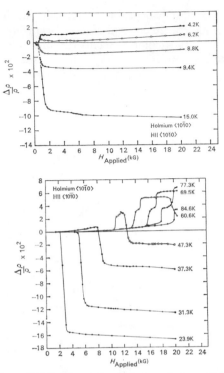

Figure 5–26. The relative resistivity change in a holmium single crystal due to the application of a magnetic field along the b-direction. The observed effects are principally due to changes in the magnon spectrum (after Ref. 72).

magnon states is eliminated in the ferromagnet, due to the energy gap, which has a value of about 20 K in holmium (see Figure 5–7). The effect is most pronounced around this temperature, and the relative change in resistance falls as the temperature increases and magnons are excited in the ferromagnetic phase. At higher temperatures, the application of a field brings about the helix–fan–ferromagnet sequence of structures discussed in (5.5.4). Because of the hexagonal anisotropy, there are actually two distinct fan structures in holmium.[73] On account of the greater density of low-lying magnon states in the fan structure (see Figure 5–17) the resistivity rises at the helix–fan transition, and then falls again at the ferromagnetic transition, when many of these low-lying states are eliminated. It would be of interest to carry out detailed calculations, to place these qualitative considerations on a more quantitative basis.

5.8. MAGNETIC EXCITATIONS IN PRASEODYMIUM

We have so far restricted our discussion to the magnetic excitations of the heavy rare earth metals, in which the exchange is generally much larger than the crystal field interactions. The latter then act as a source of magnetic anisotropy in these systems, in which the full moment $\lambda \beta J$ is ordered at low temperatures. In the light rare earths, on the other hand, the crystal fields may play a more fundamental role in determining the occurrence and nature of the magnetic ordering.[74] In particular, it seems that the crystal fields in praseodymium prevent the formation of a magnetically ordered state, at least at 4.2 K, and the magnetic excitations, which are discussed extensively in Chapter 2, are quite different from magnons.

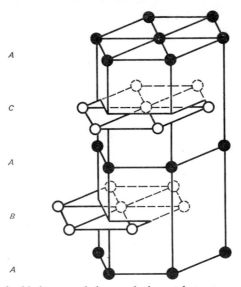

Figure 5–27. The double hexagonal close-packed crystal structure of praseodymium.

The double hexagonal close-packed structure of praseodynium is illustrated in Figure 5–27, and is composed of two types of site. The A sites lie on a simple hexagonal lattice and their immediate surroundings have approximately cubic symmetry, while the B and C sites form an hcp lattice with local hexagonal symmetry. The crystal field levels of the ions in praseodymium metal were first discussed in detail by Bleaney,[75] and his calculated level scheme is illustrated in Figure 2–8. Although this model gives good agreement with the experimentally measured heat capacities,[76] recent studies of the anisotropy of the magnetic susceptibility[77] indicate that the ordering and spacing of the levels may be somewhat different from the predictions of the calculation.

There seems to be no doubt, however, that the ground states of the ions at both the cubic and hexagonal sites are singlets, so that an exchange greater than a critical value is required for spontaneous ordering.[78] In the absence of exchange, the excitations of such a system just correspond to the excitation of a single ion from the ground state to an excited state, created by the operator

$$a_l^+ = c_{le}^+ c_{lg}. \qquad (5.135)$$

Exchange couples the ions together, so that the excitations have the form of magnetic excitons, created by an operator of the form

$$a_q^+ = N^{-\frac{1}{2}} \sum_l a_l^+ e^{iq \cdot R_l}. \qquad (5.136)$$

In the paramagnetic phase at low temperatures, the dispersion relation for these excitations then has the approximate form

$$\hbar\omega(\mathbf{q}) = \Delta \left[1 - \frac{2\alpha^2 \mathcal{J}(\mathbf{q})}{\Delta}\right]^{\frac{1}{2}} \qquad (5.137)$$

where Δ is the crystal field splitting between the two levels in the absence of exchange, and α is the matrix element of an angular momentum operator between them. The condition for spontaneous magnetic ordering then corresponds to the instability of one of these exciton modes. More sophisticated treatments, discussed in Chapter 2, change this picture quantitatively, but the qualitative features remain the same.

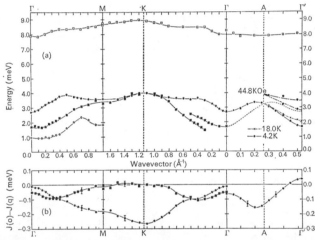

Figure 5–28. Magnetic excitons in praseodymium. The upper set of excitations corresponds to crystal field transitions on the cubic sites while the lower set corresponds to transitions on the hexagonal sites, plotted in the double zone representation in the c-direction. The temperature and magnetic field dependence of this branch is also shown. At the bottom of the Figure the values of $\mathcal{J}(\mathbf{q}) - \mathcal{J}(0)$ deduced from these results are given.

These excitations have recently been studied experimentally by Rainford and Houmann,[79] using inelastic neutron scattering, and some of their results are shown in Figure 5–28. The upper set of excitations in this figure correspond to transitions on the cubic sites. The $\mathbf{q}$-dependence of the excitation energy is small, but significant. The lower excitations correspond to transitions on the hexagonal sites between the ground state and an excited doublet. There are two branches, because of the hcp symmetry of the hexagonal sites. The dispersion of these branches is substantial, and provides a measure of the relative magnitudes of the exchange and crystal field energies. There is also evidence in the lower branch of an interaction between excitations. The dispersion relations in the c-direction are particularly interesting. The minimum excitation energy occurs in the optical branch at Γ, which corresponds to a mode in which equal and opposite moments are induced on neighbouring layers in the hcp lattice formed by the hexagonal sites. As may be seen in Figure 5–28, this mode decreases significantly in energy as the temperature is lowered, which is an indication of the incipient antiferromagnetism of praseodymium, since the instability of this mode would lead to an antiferromagnetic structure. A magnetic field results in a significant increase in the energy of the singlet–doublet excitations in the c-direction, as is also illustrated in Figure 5–28, and at the highest fields the doublet level is observed to split, as predicted theoretically.[79] These experiments promise to provide a wealth of detailed information on the exchange and crystal field interactions in the light rare earths.

5.9. CONCLUSION

The experimental information which has been accumulated during the last few years on the magnon dispersion relations in the rare earths has led to a substantial improvement in our understanding of the magnetic interactions in these metals. The rather detailed investigations which have been undertaken on the ferromagnets, especially terbium, have shown that indirect exchange of the RKKY form between the spins provides the dominant coupling between the magnetic ions, although a comparison between the different metals indicates that there may be an effective coupling between the orbital momenta, and some details of the dispersion relations suggest a small admixture of more complex forms of exchange.[80] Measurements in other phases, especially in the helix, have shown the relationship between the form of the exchange and the stability of the different magnetic structures found in the heavy rare earths. In particular, the role of the temperature dependence of both the exchange and crystal field interactions in inducing magnetic phase transitions has been clarified. The magnitudes of the maxima in $\mathcal{J}(\mathbf{q})$ in the c-direction correlate qualitatively with the stability of periodic structures in the different metals. The occurrence of such a peak can in turn be related to the band structures, as discussed in Chapter 6.

Unlike the exchange, the single-ion anisotropy parameters can readily be measured by macroscopic experimental techniques, but the field dependence of the magnon energy gap provides a convenient method for studying both the crystal field and magnetoelastic contributions. There is good agreement between the microscopically and macroscopically determined values in all cases except for the crystal field hexagonal anisotropy V_6^6. For reasons which are not at present clear, macroscopic measurements of the total hexagonal anisotropy appear to give results which are substantially too small. It is physically of great importance that the lattice strains cannot follow the magnetization precession in a spin wave, so that the frozen-lattice model is applicable and the energy gap is considerably enhanced. This affects all the magnetic properties, especially at low temperatures. The crystal fields affect the magnetic properties of the light rare earths in an even more dramatic way, as exemplified by praseodymium. The study of their magnetic excitations, which is still in an early phase, promises to reveal many new phenomena and to contribute in a crucial way to the understanding of their magnetic behaviour.

Although the main features of the magnon dispersion relations have been established, especially in the ferromagnetic and helical structures, there remain many details to be examined. Terbium has so far been the most extensively studied, principally because of its relatively favourable neutron properties, but the increasing availability of isotopically pure crystals of the other heavy rare earths opens the possibility of investigating their properties in similar detail. It would be of interest, for instance, to study the temperature dependence of the exchange in the helical structure over a wider range than has been possible in terbium, and to measure in more detail the dispersion relations in the cone structure and, perhaps eventually, in more complex periodic structures. Given sufficiently large magnetic fields, it would be possible to study the magnetic anisotropy in most of the heavy rare earths, by energy gap measurements in the ferromagnetic phase, but it would perhaps be of even greater interest to observe the effect of a magnetic field on the magnons in different phases, and thereby to investigate further the relation between the dispersion relations and magnetic phase transitions.

At a deeper level of complexity, the study of magnon interactions is still at a relatively rudimentary stage. A variety of interesting phenomena have been discovered, but they have so far been only cursorily investigated, and inter- preted, at best, semi-quantitatively. Considerable progress in both the experi- mental and theoretical understanding of these phenomena may be anticipated in the future. The nature and magnitude of the magnon interactions are largely determined by the exchange and crystal field interactions implicit in the dispersion relations, and they therefore provide a useful check on the con- sistency of the interpretation of these, but they also have a substantial effect on the macroscopic properties of the metals. In particular, the magnon–electron interaction profoundly affects the transport properties in the ordered phases, while the magnon–magnon renormalization is strongly reflected in the thermodynamic properties. The most straightforward of these to interpret is the magnetization, which is in principle entirely determined by the magnon dispersion relations and the field and temperature dependence of the magnetic interactions contained within them. A detailed calculation of the bulk magnetic properties in terms of the magnon spectrum now seems feasible, and it would be of interest to attempt such a careful comparison between the macroscopic and microscopic properties.

The rare earth metals exhibit a variety of magnetic phenomena as great as that of any class of magnetic materials, and our understanding of their behaviour has reached a considerable level of sophistication. The experimental study of the magnetic excitations has played a crucial part in this development and we may expect that, in conjunction with quantitative theoretical investiga- tions, it will have a no less significant role in the further extension of our knowledge of these important magnetic materials.

ACKNOWLEDGMENTS

Much of the experimental work described in this chapter has been carried out in collaboration with M. Nielsen and J. C. G. Houmann, and we are grateful to them for allowing us to present some of the results here for the first time. Dr. Houmann also carried out a number of the calculations for us. Valuable discussions with B. R. Cooper, R. J. Elliott, J. Jensen, P. A.

Lindgård, R. M. Nicklow, and B. D. Rainford are also gratefully acknowledged.

REFERENCES

1. Liu, S. H., *Phys. Rev.*, **121**, 451 (1961).
2. Ruderman, M. A. and Kittel, C., *Phys. Rev.*, **96**, 99 (1954); Kasuya, T., *Prog. Theor. Phys.* (Kyoto), **16**, 45 (1956); Yosida, K., *Phys. Rev.*, **106**, 893 (1957).
3. Kaplan, T. A. and Lyons, D. H., *Phys. Rev.*, **129**, 2072 (1963); Kasuya, T. and Lyons, D. H., *J. Phys. Soc. Japan*, **21**, 287 (1965).
4. Levy, P. M., *Solid State Commun.*, **7**, 1813 (1969).
5. See Chapters 1 and 2.
6. Turov, E. A. and Shavrov, V. G., *Fiz. Tverd. Tela*, **7**, 217 (1965). English transl.: *Soviet Phys. Solid State*, **7**, 166 (1965).
7. Lindgård, P. A. *J. de Physique*, **32**, C1-238 (1971). See also Chapter 2.
8. Rhyne, J. J. and Legvold, S., *Phys. Rev.*, **138**, A507 (1965).
9. Koehler, W. C., Child, H. R., Nicklow, R. M., Smith, H. G., Moon, R. M., and Cable, J. W., *Phys. Rev. Letters*, **24**, 16 (1970).
10. Holstein, T. D. and Primakoff, H., *Phys. Rev.*, **58**, 1098 (1940); It should be noted that the magnon energies can also conveniently be calculated by the equation of motion method. See Brooks, M. S. S., *Phys. Rev.*, **B1**, 2257 (1970); Goodings, D. A. and Southern, B. W., *Can. J. Phys.* **49**, 1157 (1971).
11. Niira, K., *Phys. Rev.*, **117**, 129 (1960).
12. Brinkman, W. F., *J. Appl. Phys.*, **38**, 939 (1967).
13. Cracknell, A. P., *J. Phys. C*, **3**, S175 (1970).
14. See e.g. Lomer, W. M. and Low, G. G., in *Thermal Neutron Scattering* (P. A. Egelstaff, Ed.). (Academic Press, London, 1965), p. 2.
15. Brooks, M. S. S., Goodings, D. A., and Ralph, H. I., *J. Phys. C*, **1**, 132 (1968).
16. Trammell, G. T., *Phys. Rev.*, **92**, 1387 (1953); see also Lovesey, S. W. and Rimmer, D. E., *Rep. Prog. Phys.*, **32**, 333 (1969).
17. Steinsvoll, O., Shirane, G., Nathans, R., Blume, M., Alperin, H. A., and Pickart, S. J., *Phys. Rev.*, **161**, 499 (1967).
18. Lindgård, P. A., Kowalska, A., and Laut, P., *J. Phys. Chem. Solids*, **28**, 1357 (1967).
19. Møller, H. Bjerrum and Houmann, J. C. G., *Phys. Rev. Letters*, **16**, 737 (1966); Møller, H. Bjerrum, Houmann, J. C. G., and Mackintosh, A. R., *J. Appl. Phys.*, **39**, 807 (1968); Møller, H. Bjerrum, in *Neutron Inelastic Scattering* (IAEA, Vienna, 1968), Vol. II, p. 3; Nielsen, M., Møller, H. Bjerrum, and Mackintosh, A. R., *J. Appl. Phys.*, **41**, 1174 (1970).
20. Møller, H. Bjerrum, Nielsen, M., and Mackintosh, A. R., in *Les Eléments des Terres Rares* (CNRS, Paris, 1970), Vol. II, p. 277.
21. Nicklow, R. M., *J. Appl. Phys.*, **42**, 1672 (1971); Nicklow, R. M., Wakabayashi, N., Wilkinson, M. K., and Reed, R. E., *Phys. Rev. Letters*, **26**, 140 (1971).
22. Møller, H. Bjerrum, Houmann, J. C. G., Nielsen, M., and Mackintosh, A. R., (to be published).
23. Lindgård, P. A., (private communication).
24. Stringfellow, M. W. and Windsor, C. G., *Proc. Phys. Soc.*, **92**, 408 (1967).
25. Houmann, J. C. G., *Solid State Commun.*, **6**, 479 (1968).
26. Cooper, B. R. *Phys. Rev.*, **169**, 281 (1968); Lindgård, P. A. *J. de Physique*, **32**, C1-238 (1971).
27. Nielsen, M., Møller, H. Bjerrum, Lindgård, P. A., and Mackintosh, A. R., *Phys. Rev. Letters*, **25**, 1451 (1970).
28. Mackintosh, A. R., *Phys. Letters*, **4**, 140 (1963).
29. Cooper, B. R., Elliott, R. J., Nettel, S. J., and Suhl, H., *Phys. Rev.*, **127**, 57 (1962).
30. Rossol, F. C. and Jones, R. V., *J. Appl. Phys.*, **37**, 1227 (1966); Bagguley, D. M. S. and Liesegang, J., *Proc. Roy. Soc.*, **A300**, 497 (1967).
31. Marsh, H. S. and Sievers, A. J., *J. Appl. Phys.*, **40**, 1563 (1969); Sievers, A. J., *J. Appl. Phys.*, **41**, 980 (1970).
32. Wagner, T. K. and Stanford, J. L., *Phys. Rev.*, **184**, 505 (1969).

33. Wagner, T. K. and Stanford, J. L., *Phys. Rev.*, **B1**, 4488 (1970).
34. Feron, J. L., Hug, G., and Pauthenet, R., in *Les Eléments des Terres Rares* (CNRS, Paris, 1970), Vol. II, p. 19; see also Chapter 4.
35. Callen, H. B. and Callen, E., *J. Phys. Chem. So.ids*, **27**, 1271 (1966).
36. Brooks, M. S. S., *Phys. Rev.*, **B1**, 2257 (1970) and private communication.
37. Kasuya, T., in *Magnetism* (G. T. Rado and H. Suhl, eds.). (Academic Press, New York 1966), Vol. IIB, p. 215.
38. Møller, H. Bjerrum, Thesis, University of Copenhagen (1968).
39. Yosida, K. and Miwa, H., *J. Appl. Phys.*, **32**, 8S (1961).
40. Baryakhtar, V. G. and Maleev, S. V., *Fiz. Tverd. Tela*, **5**, 1175 (1963). English transl.: *Soviet Phys. Solid State*, **5**, 858 (1963).
41. Møller, H. Bjerrum, Houmann, J. C. G., and Mackintosh, A. R., *Phys. Rev. Letters*, **19**, 312 (1967).
42. Nicklow, R. M., Mook, H. A., Smith, H. G., Reed, R. E., and Wilkinson, M. K., *J. Appl. Phys.*, **40**, 1452 (1969).
43. Nicklow, R. M., Houmann, J. C. G., and Mook, H. A., (to be published).
44. Stringfellow, M. W., Holden, T. M., Powell, B. M., and Woods, A. D. B., *J. Phys. C*, **2**, S189 (1970).
45. Elliott, R. J. and Lange, R. V., *Phys. Rev.*, **152**, 235 (1966).
46. Cooper, B. R., *Phys. Rev. Letters*, **19**, 900 (1967).
47. Sherrington, D., (to be published).
48. Elliott, R. J., (private communication).
49. Woods, A. D. B., Holden, T. M., and Powell, B. M. *Phys. Rev. Letters*, **19**, 908 (1967); Nicklow, R. M., Wakabayashi, N., Wilkinson, M. K., and Reed, R. E. *Phys. Rev. Letters* (1971). The magnetic excitations in the helix and cone phases of erbium-50% holmium have also been studied; see Holden, T. M., Powell, B. M., Stringfellow, M. W., and Woods, A. D. B. *J. Appl. Phys.*, **41**, 1176 (1970).
50. Nagamiya, T., Nagata, K., and Kitano, Y., *Prog. Theor. Phys.* (Kyoto), **31**, 1 (1964).
51. Cooper, B. R. and Elliott, R. J., *Phys. Rev.*, **131**, 1043 (1963). See also Cooper, B. R., in *Solid State Physics* (F. Seitz, D. Turnbull, and H. Ehrenreich, eds). Academic Press, New York (1968), Vol. 21, p. 393.
52. Rossol, F. C., Thesis, Harvard University (1966); Liesegang, J., Thesis, Oxford University (1966).
53. Brooks, M. S. S., Goodings, D. A., and Ralph, H. I., *J. Phys. C*, **2**, 1596 (1968).
54. Hegland, D. E., Legvold, S., and Spedding, F. H., *Phys. Rev.*, **131**, 158 (1963); see also Chapter 4 for a discussion of the magnetization.
55. Cooper, B. R., *Proc. Phys. Soc.*, **80**, 1225 (1962).
56. Lounasmaa, O. V. and Sundström, L. J., *Phys. Rev.*, **150**, 399 (1966).
57. Houmann, J. C. G. and Nicklow, R. M. *Phys. Rev.*, **B1**, 3943 (1970).
58. Elliott, R. J. and Stern, H., in *Inelastic Scattering of Neutrons* (IAEA, Vienna, 1961), p. 61.
59. Dyson, F. J., *Phys. Rev.*, **102**, 1217 (1956); Bloch, M., *Phys. Rev. Letters*, **9**, 286 (1962).
60. Tyablikov, S. V., *Ukr. Math. Zh.*, **11**, 287 (1959); see also Callen, H. B., *Phys. Rev.*, **130**, 890 (1963).
61. Kascheev, V. N. and Krivoglaz, M. A., *Fiz. Tverd. Tela*, **3**, 3167 (1961). English transl.: *Soviet Phys. Solid State*, **3**, 1117 (1961).
62. Nielsen, M. and Møller, H. Bjerrum, *Acta Cryst.*, **A25**, 547 (1969).
63. Jensen, J. *Intern. J. Magnetism*, **1**, 271 (1971).
64. Mason, W. P., *Phys. Rev.*, **96**, 302 (1954).
65. Wolfram, T. and Callaway, J., *Phys. Rev.*, **130**, 2207 (1963).
66. Blackman, J. A., *J. Phys. Chem. Solids*, **31**, 1573 (1970).
67. Kasuya, T., *Prog. Theor. Phys.* (Kyoto), **16**, 45 (1956).
68. Kasuya, T., *Prog. Theor. Phys.* (Kyoto), **22**, 227 (1959).
69. Mannari, I., *Prog. Theor. Phys.* (Kyoto), **22**, 335 (1959).
70. Brun, T. O., (1965—unpublished).
71. Sze, N. H., Rao, K. V., and Meaden, G. T., *J. Low Temp. Phys.*, **1**, 563 (1969).
72. Mackintosh, A. R. and Spanel, L. E., *Solid State Commun.*, **2**, 383 (1964).
73. Koehler, W. C., Cable, J. W., Wilkinson, M. K., and Wollan, E. O., *Phys. Rev.*, **151**, 414 (1966).
74. See Mackintosh, A. R. *J. de Physique*, **32**, C1-482 (1971) for a recent review.

75. Bleaney, B., *Proc. Roy. Soc.*, A276, 39 (1963).
76. Parkinson, D. A., Simon, F. E., and Spedding, F. H., *Proc. Roy. Soc.*, A207, 137 (1951).
77. Johansson, T., Lebech, B., Nielsen, M., Bjerrum Møller, H. and Mackintosh, A. R. *Phys. Rev. Letters*, 25, 524 (1970); Johansson, T., *J. de Physique*, 32, C1-372 (1971).
78. Trammell, G. T., *J. Appl. Phys.*, 31, 362S (1960).
79. Rainford, B. D. and Houmann, J. C. G. *Phys. Rev. Letters*, 26, 1254 (1971).
80. Bjerrum Møller, H., Houmann, J. C. G., Jensen, J. and Mackintosh, A. R. in *Neutron Inelastic Scattering* (IAEA, Vienna, to be published) have studied the anisotropy of the exchange in terbium by measuring the magnetic field dependence of the magnon energy, as a function of q in the c-direction. They find that, if the exchange is written as in (5.45), with a different parameter multiplying each Cartesian component, its isotropic and anisotropic components have the same order of magnitude.

Chapter 6

Energy Band Structure, Indirect Exchange Interactions and Magnetic Ordering

A. J. Freeman

Racah Institute of Physics, The Hebrew University of Jerusalem, Jerusalem, Israel, and Physics Department, Northwestern University, Evanston, Illinois, U.S.A.

6.1. INTRODUCTION

The great variety of unusual phenomena exhibited by the rare earth metals has made them a subject of much interest to experimentalists and theorists alike. In addition to their somewhat exotic magnetic structures, they also possess unusual (often "anomalous") electrical and optical behavior, as is clear from other chapters in this book, which is associated with and related to their magnetic ordering properties. The response of theory to the challenge posed by this wealth of experimental results has been slow and late in coming. In the absence of any detailed knowledge about the electronic band structure of these metals, theoretical workers focused their efforts within the framework of a free-electron model.[1, 2] From this early work there emerged a body of fundamental ideas regarding the origin of magnetic ordering in these systems and the basic interactions responsible for the anomalous properties observed below the magnetic ordering temperature. That these theories were so successful in providing a qualitative understanding of some of the fundamental properties of the rare earth metals is due mostly to the ingenuity of the physical methods employed, and to some degree, as we shall see, to fortuitous circumstances.

From the first determinations of the electronic band structure of the heavy rare earth metals by Dimmock, Freeman, and Watson[3] in 1964, there has emerged a picture of conduction bands which resemble closely those of the transition metals, but differ drastically from the free electron model. Subsequent calculations, by these workers[4-6] and by Loucks and associates,[7-10] have confirmed this view and have yielded detailed results for the energy bands, density of states and Fermi surfaces of both the heavy and light rare earths. The importance of these theoretical studies lies in the fact that they provide a means for undertaking quantitative correlations and interpretations of the electric, magnetic, and optical properties of these metals. Although experimental determinations of the electronic structures are still sparse (and here theory appears to have gotten ahead), what there is appears to agree with the more detailed predictions provided by band structure calculations. While still strictly within the framework of the one-electron model, the band results are also meaningful in revealing the nature and extent of many-body interactions in this interesting and important part of the periodic table. This development must, however, await the availability of high purity single crystals without which a wide variety of experiments cannot be made.

This chapter reviews the present state of our understanding of the electronic properties of the rare earth metals as derived from band theory. Following a rather detailed description of the electronic band structures and Fermi surfaces determined for the heavy and light rare earth metals, we discuss the Ruderman–Kittel–Kasuya–Yosida (RKKY) interaction and its relation to real systems and to magnetic ordering, the effects of magnetic superzones on the Fermi surface and the resultant contribution to the resistivity. Extensions of RKKY theory are considered in some detail. The relation of the observed magnetic ordering to the actual band structures and Fermi surfaces is described with emphasis on those "nesting" features thought to be responsible for the magnetic ordering and compared with results obtained from wave vector dependent susceptibilities calculated from the RAPW band structures. The effects of the magnetic periodicity on the band structure, Fermi surface and some observable properties such as magnetization and optical anomalies are treated using the realistic band structures and the s-f Hamiltonian. Finally, the origin of crystal fields in the rare earths is discussed briefly.

6.2. ELECTRONIC STRUCTURE OF THE RARE EARTH METALS

6.2.1. Atomic Properties (Localized vs. Itinerant Electrons) and Band Expectations

The atomic structure of the 4f rare earth elements has been discussed in Chapter 1. Outside the xenon core there is a partially filled 4f shell and several electrons in the atomic 5d and 6s levels. Most of the rare earth metals (with the exception of Eu, Yb, and one phase of Ce) are thought to have three valence electrons and this is the configuration assumed in all band studies done to date. The relative positions in energy of these atomic states[11] are shown in Figure 6–1. The atomic 5d and 6s states lie close to

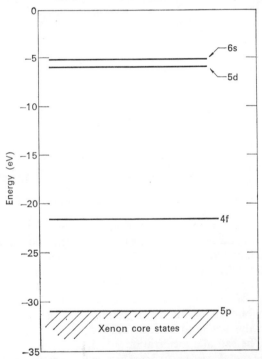

Figure 6–1. The relative positions in energy of the outer atomic electrons in atomic Gd.

one another in energy and are widely separated from the 4f level. The xenon core states all lie at considerably lower energies and so need not be considered further. From the figure one expects the 5d and 6s electrons to form the conduction bands in metals and hence contribute to the conduction processes independently of the 4f electrons.

It is further instructive to examine the relative outer radial extent of the

9

atomic electrons in a rare-earth atom. This is shown for Gd in Figure 6–2 where we have also indicated the Wigner–Seitz radius appropriate for Gd in its metallic state. [11] (Not shown are the various orthogonality oscillations of the wave functions at small radial distances.) As can be seen from Figure 6–2 the 5d and 6s atomic functions on different atom sites do overlap one another to a considerable extent. Consequently, they will form an s–d conduction band of considerable width. The contribution to this band width from the 5d electrons is expected to be somewhat narrower than that of the 6s electrons since the spatial overlap of the 5d functions is somewhat less than that of the 6s functions. Note that this is expected to bear a strong similarity to the situation found to exist in the case of transition metals.

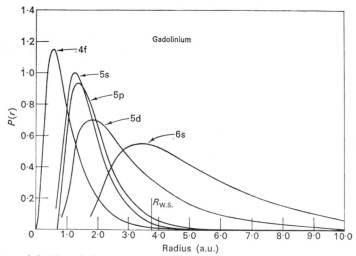

Figure 6–2. The relative outer radial extent of the atomic electrons in atomic Gd.

Whereas the 5d and 6s electrons are very extensive, the 4f electrons are tightly bound to the atom and so do not overlap neighboring atoms appreciably. Note that their maximum occurs well inside the outer maxima of the closed 5s and 5p shells. As a result, the 4f electrons will form a very narrow energy band in the metal. The band calculations [3–6] do in fact yield a 4f band with a width of only about 0.05 eV located well below (by approximately 10 eV from) the bottom of the 5d–6s bands. This separation, however, is very sensitive to the potential used in the calculations and is consequently unreliable. [3–6] The extreme localization which the atomic 4f orbitals retain upon entering the metal arises in large measure from the large (kinetic energy) centrifugal potential term [12] $l(l+1)/r^2$. The effective crystal potential acting on an electron (i.e. the sum of the crystal potential and the centrifugal potential term) is plotted in Figure 6–3 for several different values of the orbital angular momentum quantum number along with the atomic potential, V, and the crystal potential, labeled V_{MT}. The effect of the centrifugal repulsive barrier is striking in localizing the 4f electrons spatially to the same region near the nucleus in the metal as in the free atom. The d electrons are similarly, but less strongly, confined and they form energy bands of narrow to moderate widths because, while mostly localized, they can tunnel through the potential barrier. Their quasi-band nature accounts for the fact that the d band electrons exhibit both localized and itinerant properties and has resulted, particularly for the 3d bands, in the familiar controversy between the proponents of the localized vs. itinerant models. Little needs to be said for the s–p electrons which lie well above the maximum in V_{MT} and so form essentially free electron energy bands.

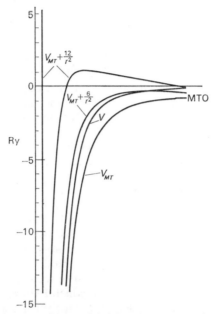

Figure 6–3. The effective crystal potential (after O. Jepsen, private communication) for several different values of the orbital angular momentum quantum number showing the effect of the d and f centrifugal potentials. The muffin tin potential V_{MT} lies below the atomic potential V because of the contribution from neighboring atoms in the metal.

We have seen the large separation in energy and in space between the atomic 4f and the 5d–6s electrons and have presented arguments for the sharp localization of the 4f orbitals even in the metal. What may be said about the location in energy of the 4f states in the metal? As a result of this spatial localization, the open 4f shell of electrons experiences large intra-atomic Coulomb and exchange correlation interactions which are entirely omitted from an energy band calculation.[6] This makes the energy band description inappropriate for the 4f states; they must be treated as localized electrons— not as band electrons—and so do not fit the energy band picture at all. Their calculated position in energy is therefore even more unrealistic than is indicated by the variation obtained from band calculations using different potentials. We shall see later, as in other chapters, that the completely localized description of the ionic 4f shell (Hund's rule coupling of S, L, and J) interacting with its environment through the weak crystalline electric field and the exchange interactions via the conduction electrons is substantiated by a number of experiments. In what follows in this section we shall focus our attention on the conduction bands in the metals as determined by energy band methods.

6.2.2. Energy Bands in Solids: The Augmented Plane Wave Method

The energy band method for calculating electronic eigenstates in crystals is based on a number of simplifying assumptions and approximations which reduce the many-body problem, involving the interaction between all the particles—electrons and nuclei—in the system, to a one-electron or independent electron model.[13] (The reason for these approximations is clear: the many-body problem for the crystal entails the solution of Schrödinger's equation for 10^{23} nuclei and electrons and is a completely hopeless task.) It is important to keep these approximations in mind as they are not completely justifiable and may affect seriously some of the physical results obtained.[6]

The first of these is the Born–Oppenheimer approximation which essentially amounts to neglecting the electron–phonon interaction and reduces the problem to that of an interacting electron system only. Actually, electron–phonon interactions in metals can cause an enhancement of the measured electron mass and the measured oscillator strength for optical transitions by a factor of up to 2.5 for some polyvalent metals. The neglect of the electron–phonon interactions does not appear to be so bad for alkali metals as it is for polyvalent metals and in general it does not appear seriously to affect Fermi surface dimensions even though it modifies the density of states obtained from specific-heat measurements. The second assumption is the use of the Hartree–Fock (HF) approximation which reduces the problem to that of an independent electron model and which neglects electron–electron correlations among electrons of opposite spin. Even with this assumption, the problem is still a formidable one and its exact solution is beyond our present capabilities. It is uncertain at this time how important a role electron–electron correlations play in the electronic band structure of metals. It is assumed that these effects are least important for the broad s–p band and much more important for the narrow d band metals. As already discussed, the Coulomb and exchange correlation effects are so large for the 4f electrons, that they may not be treated by the band model.

The third approximation is that of averaging the exchange term which arises in the Hartree–Fock equations. In order to obtain an effective single-electron local potential, it is necessary to average the nonlocal exchange term in one manner or another. It can be averaged over atomic orbitals leading to an l-dependent exchange term. However, the more common approach is to use Slater's $\rho^{\frac{1}{3}}$ approximation for the free electron gas.[14] Serious questions have been raised as to the applicability of this approximation to the core (including rare earth 4f) electrons in metals. The principal justification of this approach is that it is simple, relatively easy to use and yields results in good agreement with experiment, especially for the Fermi surface.

With these approximations, each electron is assumed to experience, as it moves through the crystal, an *average* but spatially periodic potential due to all the other electrons and nuclei in the crystal. All the band methods thus reduce the problem into two parts : the choice of the crystal potential and the accurate solution of the Schrödinger (or, in the relativistic case, the Dirac) equation with the assumed potential. High speed computers, good mathematical analysis and programming, have made the second part of the problem relatively easy to solve and have led to considerable confidence in the numerical accuracy of the solutions obtained. The crystal potential problem still remains to introduce uncertainties into the reliability of the resulting solutions.

The Augmented Plane Wave (APW) Method has been used in all energy band calculations for the rare earth metals. Central to this method, as originally proposed by Slater,[15] and to the parallel Kohn–Korringa–Rostoker Method,[16] is the subdivision of the crystal potential into the so-called "muffin-tin"-form, i.e. the potential is spherically symmetric within a sphere about each atomic site and constant in the region outside these spheres. Although seemingly a crude approximation, this form of potential is more appropriate for metals than for semiconductors or insulators and this has been borne out by extensive calculations on quite a large number of different metals.

The construction of the crystal potential, which is then approximated by the muffin-tin-form, follows a procedure introduced by Matthiess[17] and now generally used : superposition of atomic potentials and charge densities consisting of Coulombic and exchange parts which are calculated separately. The Coulombic potential arises from all the atoms in the crystal and is calculated in a straightforward way, limiting the summation to a certain number of nearest neighbors. The exchange contribution to the potential is commonly approximated by the Slater free electron exchange term

$$V_{ex}(r) = -6 \left[\frac{3}{8\pi} \rho(r) \right]^{\frac{1}{3}} = c[\rho(r)]^{\frac{1}{3}} \tag{6.1}$$

where $\rho(r)$ is the electronic charge density at the spatial position r arising from superposition of the charge density of the central atom and neighboring shells of atoms. [Recently Kohn and Sham[18] questioned the Slater derivation of $V_{ex}(r)$, as did Gáspár[19] previously, and using a variational procedure have derived the same exchange potential as Eqn (6.1) only reduced by two-thirds in magnitude. Since it is not clear at this time which form of $V_{ex}(r)$ is more appropriate, many workers are carrying out calculations using the magnitude of the exchange term as an empirical parameter.]

The atomic charge densities are obtained from the appropriate atomic wave functions calculated for the free atoms or ions. The potential between the muffin tin spheres is flat and is usually taken as an average of the potential over this region as obtained from the superposition of atomic charge densities. This potential contains a number of adjustable parameters. In the case of compounds, one can vary the assumed ionicity of the various components and their Madelung energies, since both of these quantities are usually known only within broad limits. In obtaining the atomic charge densities, one can vary the atomic configuration and the state of ionization assumed in the

free atom calculations. One can also vary the radii of the APW muffin tins
and the potential between the spheres.

Having constructed the potential, the next step in solving the Schrödinger
equation is to choose a set of basis functions—augmented plane waves in
the APW scheme. An APW is a solution of the spherically symmetric poten-
tial inside the muffin tin spheres and a plane wave $e^{i\mathbf{k}\cdot\mathbf{r}}$ in the flat (constant)
part of the potential in the region between the spheres. Inside the muffin
tin the solution is expanded in terms of products of radial wave functions and
spherical harmonics; this facilitates the matching of solutions on the APW
sphere. Eigenvalues (and most recently eigenfunctions) are determined at
selected $\mathbf{k}$ values in the Brillouin zone. Details and procedures involved in
carrying out these calculations in either the non-relativistic or the relativistic
cases are available in the review articles by Loucks,[20] and by Matthiess,
Wood, and Switendick.[21] A review of the APW method and the results
which have been obtained using it for different solids may be found in the
forthcoming review article by Dimmock.[22]

6.2.3. Electronic Band Structure of the Rare Earth Metals

The last half-decade has witnessed the determination from first principles
of accurate band structure, density of states and Fermi surfaces of non-
magnetic rare earth metals. Whereas nothing was known about their electronic
band structure prior to 1964, today we are confronted with a wealth of
results both for the heavy metals (Gd to Lu) and the light metals (Ce to
Eu). The calculations were done for their different crystal structures and in
some instances when these metals are subjected to external hydrostatic pressures. This section reviews the results of calculations to date obtained from
the non-relativistic and relativistic methods and discusses their accuracy
and reliability with a view towards using them in a later section where the
magnetic properties of these metals are considered.

6.2.3.1. Non-relativistic Band Structure and Fermi Surface of the hcp Metals
A. Electronic Bands

The energy band structure of the 4f rare-earth metals was first determined
by Dimmock, Freeman and Watson[3, 4] using the non-relativistic APW
method for hexagonal close packed gadolinium, thulium, lanthanum, and
lutecium metals. We shall discuss these non-relativistic results first because
they lack some of the complications of the relativistic calculations, and so
simplify the presentation and discussions. As we shall see, there are, surprisingly, only small differences between the two types of calculations.

The calculations for Gd were carried out with a number of starting potentials. These were all generated by superpositions of atomic charge densities
obtained for free Gd atoms in which the assumed atomic configuration for
the free atoms was varied. Atomic charge densities were obtained by using
the Hartree–Fock–Slater (HFS) calculational procedure of Herman and
Skillman[11] applied to the atomic Gd configurations $4f^7\,5d^1\,6s^2$, $4f^7\,5d^2\,6s^1$,
$4f^7\,5d^3\,6s^0$, and $4f^8\,5d^0\,6s^2$. Also used was the atomic Hartree–Fock charge

density obtained by Freeman and Watson[23] for the singly ionized Gd configuration $4f^7 6s^2$ to which was added a 5d function. As expected, the differences in the potentials occurred in the outer regions of the atom since only the configuration of the outer electrons was varied. In these calculations, the full value of the Slater free-electron exchange potential was used. All of these potentials gave qualitatively the same results with a maximum variation in the relative energies of the conduction band states of about 0.5 eV. The results which we present are those obtained using the HFS $4f^7 5d^1 6s^2$ potential. These are representative of results obtained using the other potentials.

Figure 6–4 presents the results of the electronic energy band calculations for Gd metal in the paramagnetic state using the HFS atomic results for the configuration $4f^7 5d^1 6s^2$. The striking feature of these results is the important, indeed dominant, role played by the 5d bands which strongly hybridize with

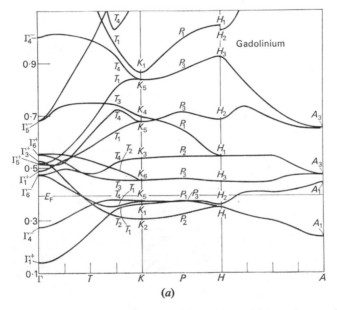

(a)

Figure 6–4. Non-relativistic energy band results for Gd metal (after references 3 and 6).

the 6s–p bands and give rise to an unusually high density of flat bands near the Fermi energy. This type of band structure for the hcp heavy rare earth metals obviously differs from the free electron bands previously used for these systems and instead is almost identical to the band structure of the hcp transition metals.

To illustrate the difference between these rare earth bands and those of a free-electron metal, we compared in Figure 6–5 the bands calculated for[3, 4] Gd with those for hcp[24] Mg along the $\Gamma - K$ symmetry direction. The 5d bands which at Γ lie above the Fermi energy are seen to drop sharply below E_F and to hybridize strongly with the s–p bands, resulting in flat bands as

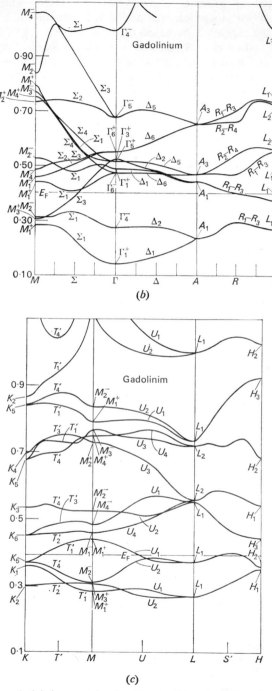

Figure 6–4. Non-relativistic energy band results for Gd metal (after references 3 and 6).

one approaches K. Furthermore, the $\Gamma_3{}^+$ state, which is pure p, is separated from the lowest $\Gamma_1{}^+$ state (mostly s-like) in Gd by approximately the same amount as in Mg. As is typical of s–p bands, this separation between the $\Gamma_3{}^+$ and $\Gamma_1{}^+$ is also found to be much less sensitive to changes in potentials in the different Gd calculations than are the d energy levels. The similarity of the Gd results with those of a free-electron-like hcp transition metal arises from the hybridization of the 5d and 6s–p bands in much the same way as the 3d and 4s–p bands hybridize in the 3d transition metals.

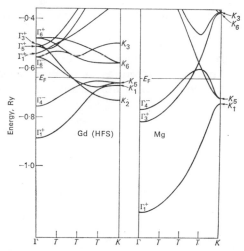

Figure 6–5. Comparison of the energy bands for a free electron hcp metal, Mg, with those calculated for Gd, along the Γ–K symmetry direction (after references 3 and 24).

The 4f levels appear to play no role in the formation of the conduction bands. As already mentioned, the 4f electrons form a very narrow band well below the bottom of the 5d and 6s–p conduction bands. The position of the 4f bands, really localized atomic-like states, was found to be very sensitive to the potentials used. (One should not forget that any band description of the 4f level is totally meaningless because of the neglected intra-atomic Coulomb and exchange correlation terms.)

The d-bands originating from the Gd 5d states contribute a high density of states in the vicinity of the Fermi energy with a width of about 0.5 Rydbergs (6 eV). This width is nearly the same as that obtained by Wood[25] for the 3d-bands in iron. The Fermi energy, for three electrons per atom, is $E_F = 0.25$ Ry measured from the bottom of the band as compared with a value of 0.54 Ry for the free-electron model. At the Fermi energy, the calculated density of states is large, $N(E_F) = 1.8$ electrons per atom per eV compared with the free electron value of 0.6 (in the same units). (This is due to the fact, discussed above, that the electron bands in the vicinity of the Fermi surface are of mixed s–d character and are consequently much flatter than would be expected from a free-electron model.) This large density of states accounts for a number of experimental observations such as the large specific heats

observed for the heavy rare earth metals and the large saturation magnetization of Gd metal. It also predicted the large peak in the optical density of
states subsequently observed by Blodgett and Spicer[26] in photoemission
from Gd.

We must emphasize that the bands in Figure 6–4 are unusually flat. In
some important directions, the pairs of bands (the third and fourth) which
intersect the Fermi energy have a total width of only about 0.5 eV, which is
of the order of the spin–orbit splitting. This implies large effective masses
and substantial contributions to the density of states $N(E)$. There will also
be, as we shall see, large effects due to the exchange field perturbations which
introduce magnetic band gaps of the order of 0.1 to 0.6 eV. These perturbations will be of primary interest to us later on.

Soon after their first work on gadolinium, Freeman, Dimmock and Watson[27, 4] determined the band structure (and Fermi surface) of thulium
metal, again using the conventional non-relativistic APW method. The

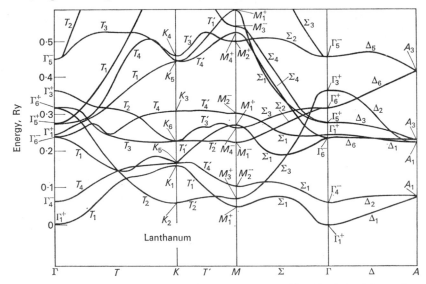

Figure 6–6. Non-relativistic electronic band structure of hcp La (after Dimmock,
Freeman, and Watson, unpublished results).

assumed free atom configuration for the HFS calculations was $4f^{12}\ 5d^1\ 6s^2$;
the full value of the Slater exchange potential was used. The energy band
structure and Fermi surface were found to be quite similar to that of Gd.
These authors[28] also obtained the band structure of lanthanum and lutecium
metals with assumed configurations $4f^0\ 5d^1\ 6s^2$ and $4f^{14}\ 5d^1\ 6s^2$ respectively.
Their results for hcp La are shown in Figure 6–6; those for Lu are shown
in Figure 6–7. By comparing the results of Figures 6–6, 6–4, and 6–7 one
obtains a view of the general trend across the 4f series for the hcp rare earth
metals. Not surprisingly, as one proceeds to the heavier metals the 5d bands
increase in energy with respect to the 6s–p bands.

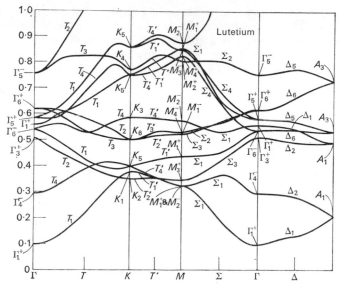

Figure 6–7. Non-relativistic electronic band structure of hcp Lu (after Dimmock, Freeman, and Watson, unpublished results).

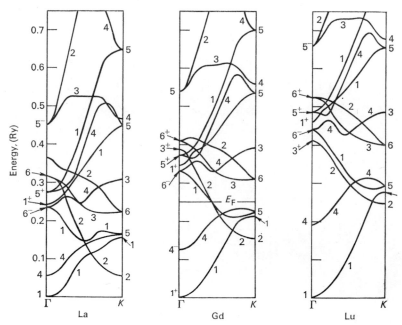

Figure 6–8. Comparison of bands in La, Gd, and Lu along the Γ–K direction in the hcp zone (after Dimmock, Freeman, and Watson, unpublished results). The band numbers indicate symmetry type (e.g. Γ_5^-, T_2, K_4).

For a detailed comparison, we show in Figure 6–8 the bands in La, Gd, and Lu along the $\Gamma - K$ direction. In addition to the general similarity one might have expected, there is a striking similarity in the detailed structure of the computed energy bands, with many aspects of the band structure unchanged as one goes from La to Lu. The general trend of increasing the 5d energy relative to the 6s–p bands is probably a reliable feature of the band calculations. Those details which vary from metal to metal are sensitive to the assumed potential and hence are probably not particularly reliable. These results should be remembered when making comparisons between theory and experiment.

Detailed results of all band calculations are uncertain due to the inaccuracies described above. In addition, the band results discussed so far were

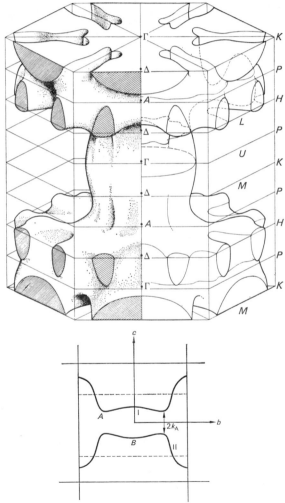

Figure 6–9. Complete non-relativistic Fermi surface of non-magnetic Gd (after reference 3). A section of the free electron Fermi surface (after reference 2) is shown below.

obtained from non-relativistic APW calculations. It is clear that relativistic effects should play an important role in determining band structures of heavy metals like the rare earths. Estimates of the magnitude of these effects may be made by using the HFS free atom calculations to calculate relativistic corrections as has been done by Herman and Skillman.[11] For Gd they find a relative shift of the 5d state with respect to the 6s state of 0.027 Ry, which is less than the difference observed[3, 5, 6] for the Gd band calculations using the HFS potential and the analytic HF potential. The relativistic shifts are thus seen to be within the uncertainties of the non-relativistic calculations. The spin–orbit splitting of the 5d states, found to be[11, 6] 0.028 Ry, is expected to change qualitatively the structure of the conduction bands. We shall examine these changes in the next sub-section where we discuss the relativistic energy band calculations.

B. Fermi Surface

The Fermi surface is of particular interest to readers of this volume because many of the features of the magnetic properties (both static and dynamic) as well as the more conventional transport properties of the rare earth metals are determined by its topology. The Fermi surface for non-magnetic Gd, determined by Dimmock, Freeman, and Watson[3] using the non-relativistic APW bands results of Figure 6–4 is shown in Figure 6–9 in the double zone scheme. This is the complete Fermi-hole surface as there appear to be no additional pockets of holes or electrons. The surface consists of a single volume with a trunk, contributed by bands of largely s–p character, and arms in the vicinity of the ALH plane, which arise mainly from bands of largely d character. The trunk provides the Fermi surface parallel to the c-axis while the arms are responsible for that perpendicular to it. The E vs. k plots of Figure 6–4 show the relatively steep s–p bands contributing to the trunk. The d bands contributing to the arms which are normal to the c-axis are important for understanding the resistance anomalies due to the magnetic ordering which occur along the c-axis direction. We shall return to these matters in a later section.

The Fermi surface permits open orbits both along the c-axis of the crystal and in the plane perpendicular to this axis. There are also closed electron and hole extremal orbits in planes normal to the c-axis. Several horizontal Fermi surface sections are shown in Figure 6–10 in order to present a clearer view of some of the symmetry of the Fermi surface.

It should be pointed out that this Fermi surface bears no resemblance at all to the Fermi surface of the free-electron model. The Fermi surface of Gd derived by Kasuya[2] on the basis of the nearly free electron model is plotted, using the double zone scheme, as an insert to Figure 6–9. This surface is shown on the plane which includes the c- and b-axes. The dotted lines denote the ALH planes which in the absence of spin–orbit coupling are not Bragg scattering planes.

Given the importance of the Fermi surface to the magnetic ordering and other properties, we must ask, how believable are the results? In the absence of detailed experimental results we must answer this question within the theoretical framework alone. We have seen that the unusual characteristics

of the band structure of the rare earth metals were due to the flatness of the
d bands. This flat character of the d bands makes it difficult to determine the
details of the Fermi surface normal to c. For example, raising the bands
shown in Figure 6–4 by ∼0.2 Ry at the point L causes the pairs of arms of
Figure 6–9 to merge into single larger arms. Such a shift is within the uncertainties of the band calculations. A number of factors, quite aside from
numerical accuracy, contribute to these uncertainties, and several deserve
mention here. As stated earlier, a traditional problem associated with APW

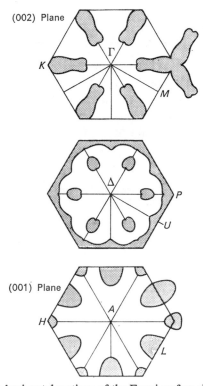

Figure 6–10. Some horizontal sections of the Fermi surface shown in Figure 6–9.

calculations for transition metals is the choice of atomic starting potential
(in particular its effect on the relative placement of the s and d bands). Gd
band calculations[6] employing various non-relativistic potentials suggest
that this uncertainty is ∼0.02 Ry in the placement of one band with respect
to another. The approximate exchange potential[15–17] employed in such
band calculations introduces another source of uncertainty. Comparing
results obtained[6] with the Slater[15] and the Gáspár[19] approximations
suggests the uncertainty from this source to be ∼0.04 Ry. Finally, the results
just discussed are non-relativistic. Comparison of these with the relativistic
results of Keeton and Loucks,[7, 8] to be presented later, indicates that the
differences are within the uncertainties of either calculation.

6.2.3.2. Relativistic Calculations of the Band Structure and Fermi Surface of the hcp Metals

The important effects of the spin–orbit interaction on the non-relativistic band structures are : (1) to eliminate many of their degeneracies throughout the Brillouin zone ; (2) to prevent many of their bands from crossing each other ; and (3) to eliminate the degeneracy across the ALH plane perpendicular to the c-axis (except along the line AL). This last effect is most important for the rare earth metals ; it will be seen to introduce discontinuities (energy gaps) in the Fermi surface in the ALH plane in the Brillouin zone.

As shown by Loucks,[29, 30] relativistic effects may be included in the APW method in a straightforward way. Following a procedure analogous to that of the APW method, energy eigenvalues are determined as solutions of the single particle Dirac equation with a muffin tin potential. Relativistic energy band calculations for the rare earth metals were first carried out by Keeton and Loucks[7] for Lu metal using the relativistic APW method[20] (RAPW). Calculations were later reported by them for Gd, Dy, Er, and Lu. In all these calculations, potentials were constructed from the relativistic HFS calculations of Liberman, Waber, and Cromer[30] and the full value of the Slater exchange. The assumed atomic configurations were Gd ($4f^7 5d^1 6s^2$), Dy$^{(I)}$ ($4f^9 5d^1 6s^2$), Dy$^{(II)}$ ($4f^{10} 6s^2$), Er$^{(I)}$ ($4f^{11} 5d^1 6s^2$), Er$^{(II)}$ ($4f^{12} 6s^2$), and

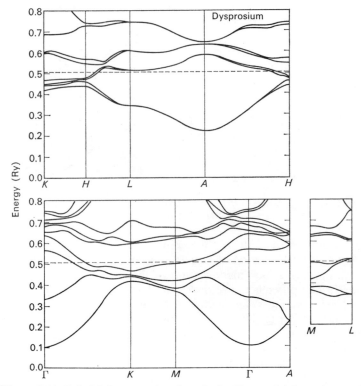

Figure 6–11. Relativistic energy band results for Dy metal (after reference 8).

Lu ($4f^{14}\ 5d^1\ 6s^2$). Energy bands, density of states histograms and Fermi surfaces were determined for each of the metals and for the two different configurations in Dy and Er.

As in the non-relativistic case, the energy bands resulting from all the different potentials were again quite similar to each other, differing only in details. Figure 6–11 presents the RAPW calculated bands for dysprosium metal as a representative example. It is seen that the major effect of relativistic corrections is to remove some of the degeneracies and level crossings in the non-relativistic bands (cf. Figure 6–4). This similarity is seen more clearly in comparing in Figure 6–12 the non-relativistic and relativistic bands obtained by Keeton and Loucks[7] for lutetium metal. (The non-relativistic

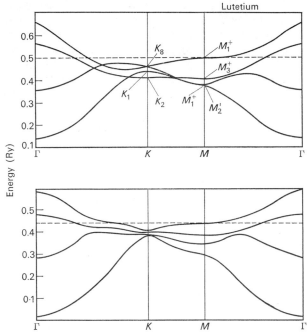

Figure 6–12. Comparison of the non-relativistic and relativistic energy bands for Lu metal (after reference 7).

bands are identified in the figure by comparison with the results of Freeman *et al.*;[28] the RAPW method of Loucks does not make use of the symmetry of the states and so makes identification of the calculated states exceedingly difficult.[31]) The differences are, however, large enough to show up in the calculated Fermi surfaces. The Fermi surface obtained by Keeton and Loucks for Gd is very similar to those obtained by Freeman, Dimmock, and Watson for both Gd and Tm; their results for the other metals—Dy, Er, and Lu— resemble the Fermi surface obtained earlier by Loucks[32] for yttrium metal.

The most important difference in details of the band structure occur in the bands between M and L in the zone and this difference causes a qualitative difference in the Fermi surface of the different heavy hcp rare earth metals.

These differences are shown in Figure 6–13 which compares the intersections of the surfaces with the symmetry planes of the Brillouin zone for Gd with those for Dy, Er, and Lu. Note that unlike the case of Gd, the Lu electron surface encloses the point M and the hole surface extends past $A-L$ almost to the point H. Note the flat sections between the electron and hole parallel sheets over a large region of the Brillouin zone. Although initially thought

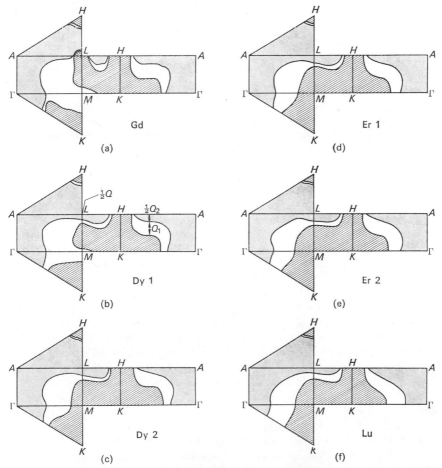

Figure 6–13. Intersections of the RAPW Fermi surfaces of Gd, Dy, Er, and Lu with the symmetry planes of the Brillouin zone (after reference 8).

by Williams, Loucks, and Mackintosh[33] to be important for characterizing the magnetically ordered phases of the rare earths, it was shown by Lomer[34] that this parallelism ("nesting") between electron and hole sheets labeled as Q_1 on the Dy I diagram in Figure 6–13 does not result in a large contribution to the susceptibility because it requires a coupling between two occupied regions. Keeton and Loucks[8] have proposed that the proper "nesting" feature is that labeled by $\frac{1}{2}Q_2$ in Figure 6–13. This "webbing" of the arms of the Fermi surface, and its absence in Gd, is shown in Figure 6–14 for Y, Lu, Er, and Dy. On the basis of their results for these metals, these authors postulated a "webbing" feature for Tb metal. (We shall return to these Fermi surfaces later on when we discuss the magnetic ordering properties of the rare earth metals.)

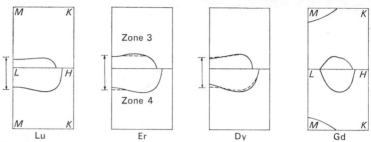

Figure 6–14. Cross sections of various Fermi surfaces showing webbing feature (after reference 8).

The band structure of Tb metal has been calculated by Mackintosh[35] using the RAPW method and a potential generated from relativistic Hartree–Fock–Slater solutions for the $4f^8\ 5d^1\ 6s^2$ configuration. These calculations gave energy bands which are qualitatively similar to those in the other hcp rare earth metals, except that the d bands are substantially lower than those in, for example, Dy (cf. Figure 6–11). The position of the second lowest doubly degenerate level at L relative to the Fermi level is found to be below E_F in Tb as in Gd (which is ferromagnetic) but unlike the other metals discussed above where the degeneracy occurs above the Fermi level and both bands cut E_F. Jackson[36] has given a more complete description of the energy bands, density of states and relationship between the Fermi surface and the periodicity of the magnetic ordering in Tb metal. Using the same RAPW method, the calculation differs from Mackintosh's only in the potential which was obtained from relativistic Hartree–Fock wave functions for the configuration $4f^9\ 6s^2$. The two results are found to be in very close agreement with each other in almost all details. Jackson's Fermi surface for terbium agrees very closely with that found for dysprosium (their Dy I calculation) by Keeton and Loucks[8] with the "webbing" modified in the way these authors predicted would occur for Tb. The position of the second doubly degenerate band is found to be just above the Fermi level. Figure 6–15 shows a comparison of the energy bands (3 and 4) between M and L evaluated by Mackintosh et al.,[35] Keeton and Loucks,[8] and Jackson.[36]

The energy bands responsible for these key features of the Fermi surface are flat and the surface they produce could be very sensitive to changes in the potential. The remarkable agreement between the two sets of results indicates that these features are not very sensitive to small changes in the potential. However, it should be recalled from the earlier discussion of the Dimmock, Freeman, and Watson[3, 6] calculations on Gd that increasing the d character in the potential tended to shift the 5d bands to higher energy relative to the 6s–p bands and that this same sort of shift occurred when using the Hartree–Fock potential for the same atomic configuration. It is thus possible that in going from the relativistic Hartree–Fock–Slater ($\rho^{\frac{1}{3}}$) potential of Mackintosh for the $4f^8\ 5d^1\ 6s^2$ configuration to the relativistic HF wave functions to $4f^9\ 6s^2$ used by Jackson there exists a cancellation of effects to produce the nearly identical results observed.

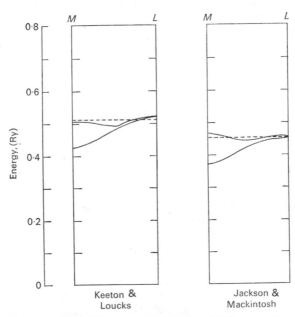

Figure 6–15. Comparison of the energy bands (3 and 4) between M and L in Tb from results of references 35, 8, and 36.

6.2.3.3. Band Structure and Fermi Surface of the Light Rare Earth Metals

As discussed in Chapter 1, the light rare earth metals have crystal structures which are considerably more complex than those of the heavy rare earth metals. The first two elements in the series, La and Ce, form in both face centered cubic and double hexagonal close-packed (dhcp) crystal structures and the next two, Pr and Nd, also form in the dhcp structure. Promethium has no stable isotope and the next element, samarium, forms the most complex structure of any of the rare earth metals (see section 1.2.1). Europium forms a body centered cubic structure and is unique among the metals which order magnetically for being divalent (two outer conduction electrons) rather than trivalent. (We discuss Eu and divalent fcc Yb in the next sub-section.) The magnetic properties of the dhcp rare earths are also found to be considerably different from the other rare earth metals. Details of the magnetic structure of these metals are given in Chapters 1, 3, and 4.

The dhcp structure shown in Figure 5–17 of Chapter 5 consists of stacking close packed hexagonal layers in the sequence ABACABAC . . . compared with the hcp sequence ABABAB . . .. There are two inequivalent sites in the lattice (the B and C layers have the same nearest neighbors as does an atom in an fcc lattice) and this makes the dhcp lattice unique among the rare earth crystal structures. The unit cell of the dhcp structure contains four atoms and is twice as long in the *c*-axis direction as the single hcp unit cell. The *c*-axis dimension in the Brillouin zone, shown in Figure 6–16, is reduced by this factor of two and the number of energy bands is doubled over that of the hcp structure because it is necessary that there be twice as many states per unit volume of reciprocal space.

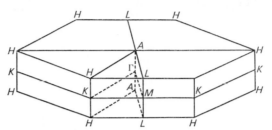

Figure 6–16. Brillouin zone for the dhcp structure.

Fleming, Liu, and Loucks[9] have calculated the energy band structure and Fermi surfaces in dhcp La, Nd, and Pr using the RAPW method. (From the electronic band structure Fleming *et al.* also calculated generalized electronic susceptibilities for Nd and Pr, as we shall describe later.) Relativistic Hartree–Fock–Slater solutions for the atomic configurations $4f^0$ $5d^1$ $6s^2$ in La, $4f^4$ $5d^0$ $6s^2$ in Nd and $4f^3$ $5d^0$ $6s^2$ in Pr were superposed to obtain crystal potentials in the way described above. (Note that the divalent free atom configurations $5d^0$ $6s^2$ were assumed for the starting potentials for Pr and Nd.) The similarity of the results obtained shows again the relative insensitivity of the computed band structures to the choice of potential. Their

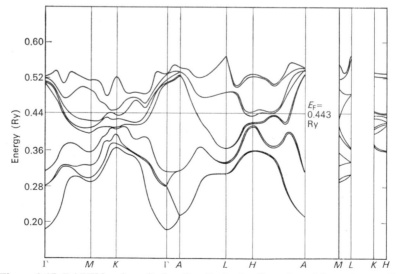

Figure 6–17. RAPW band results for dhcp La in the dhcp phase (after reference 9).

energy band results for La, shown in Figure 6–17, are similar to the results obtained for Pr and Nd and are similar to those of the hcp structure heavy rare earth metals. The major difference arises from the doubling of the number of energy bands (since for a trivalent metal six bands must be filled instead of three) and twice as many bands which intersect the Fermi energy as for the simple hcp metal. There are thus four bands which contribute to the Fermi surface rather than two as in the hcp metals. A four zone scheme similar to the double zone scheme used by Freeman et al.[4] cannot be used because the inclusion of spin–orbit coupling in the band calculations causes splitting of the bands at all AHL zone boundaries and at alternate ΓKM faces and hence a Fermi surface in the four zone scheme would have discontinuities at these locations. The Fermi surface becomes more complicated if the four pieces of the Fermi surface due to the four bands are presented in the first zone, since there are regions in the zone which are occupied by electrons in all four bands.

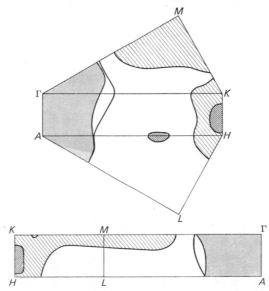

Figure 6–18. Intersection of the Fermi surface of dhcp La with the faces of 1/24 of the zone (after reference 9).

Fleming et al.[9] simplify the representation by showing the hole regions from the fifth and sixth bands and the electron regions from the seventh and eighth bands. The intersections of the Fermi surface of lanthanum with the faces of 1/24 of the zone are given in Figure 6–18. The surfaces for Pr and Nd differ only in detail.[9]

A distinguishing feature of the calculated Fermi surfaces is the absence of large, very flat pieces perpendicular to the c direction. In the simple hcp metals, described above, these flat Fermi surface sections occur about halfway between the center (the ΓKM face) and the top face (AHL) of the Brillouin zone. Fleming et al.[9] have proposed a possible connection between this missing feature of the Fermi surface and the occurrence of the dhcp structure in these metals. A periodic potential with a propagation vector lying along the c axis and having a period twice the hcp periodicity would introduce an additional Brillouin zone surface and resulting energy gaps just at this place in the zone and thereby eliminate these flat pieces of Fermi surface and lower the total electronic energy of the crystal. Since in La, Pr, and Nd with the dhcp structure the flat surfaces are eliminated, this indicates the relative stability of the dhcp structure over the simple hcp structure.

Recently, the energy bands of La and Pr in the *face-centered cubic* structure have been determined by Myron and Liu.[37] Interest in these structures (and in Nd which also orders in fcc form) derives from the fact that both phases of La are superconducting at low temperatures with slightly different superconducting and normal state parameters and both fcc Pr and fcc Nd order ferromagnetically instead of antiferromagnetically as for the dhcp structures. The RAPW method was used along with a muffin tin potential, constructed from atomic densities with configuration $5d^1 6s^2$ for La and $5d^0 6s^2$ for Pr, and with the full value of the Slater $\rho^{\frac{1}{3}}$ exchange.

Singularities in the relativistic analogue of the logarithmic derivative of the wave function, indicative of the presence of 4f states, were removed from the range of energy of interest in order to remove the 4f bands. In addition to the energy bands, a density of states and Fermi surface were determined for each of the metals. The densities of states at the Fermi energy were found to be high (24.8 states/Ry/atom for La and 17.8 for Pr) and in good agreement with experiment.

The Fermi surfaces of fcc La and Pr are multiply connected and difficult to depict graphically. Several cross-sections in the (100), (110), and (111) planes are shown in Figure 6–19 as a partial representation of the Fermi surface of Pr. The occupied regions are shown shaded. The Fermi surface of La has an additional electron pocket between Γ and W. From the band structure, a wave vector dependent susceptibility $\chi(\mathbf{q})$, was computed for Pr in order to determine possible mechanisms for the observed ferromagnetic ordering of this metal and differences in magnetic ordering of the two phases of Pr and Nd.

The electronic band structure of metallic Ce in its two allotropic fcc forms (γ and α) has been extensively studied by Waber and Switendick[38] using the non-relativistic APW method. As discussed in Chapters 3 and 7, interest in this metal derives particularly from the α–γ phase transition (which is thought to be related to a change in occupation of the 4f level), and from the associated anomalous behavior of the resistivity for the metal (by itself or as an impurity in other metals).[39] Since no proposed configuration for cerium metal was generally accepted, five different atomic configurations $4f^{2-x} 5d^x 6s^2$ (with $x = 0$, 0.5, 1.0, 1.5, and 2.0) determined from relativistic HFS wave functions and the full value of Slater $\rho^{\frac{1}{3}}$ exchange were used to generate muffin tin potentials for each phase. Waber and Switendick[38] found a great sensitivity of the position of the 4f band to the assumed atomic configuration. Both the results for α– and γ–Ce are qualitatively similar with a variation of the band structure and position of the 4f bands which follows from the amount of occupied 4f electrons added to the assumed potential. In the $4f^0 5d^2 6s^2$ configuration, the 4f levels are well below the 5d–6s levels, but the more 4f in the potential the higher in energy the calculated 4f levels fall, thereby overlapping and strongly hybridizing with the 5d–6s bands and becoming broader.

Recently, Mukhopadhyay and Majumdar[40] reported on both relativistic and non-relativistic APW calculations of the α and γ-phases of Ce. The full value of Slater exchange was used together with muffin tin potentials generated from relativistic atomic HFS calculations[30] for the RAPW computations and non-relativistic HFS atomic calculations[11] for the APW computations.

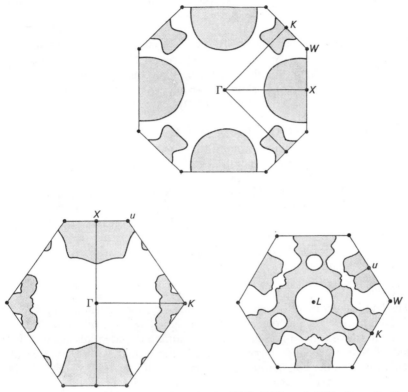

Figure 6–19. Fermi surface cross-sections in the (100), (110), and (111) planes for fcc Pr (after reference 37). The occupied regions are shown shaded.

For both calculations the 4f band width is very narrow (0.01 Ry for γ–Ce and 0.02 Ry for α–Ce) and the 4f bands lie well below the conduction band (0.36 Ry below for γ–Ce and 0.24 Ry below for α–Ce), in disagreement with the earlier suggested model[39] based on optical and magnetic data, which placed the 4f band only 0.076 eV below ϵ_F. They suggested that their inconclusive results might be resolved by self-consistent calculations. However, as discussed earlier in this chapter, the 4f levels cannot be described by a band picture as the strong Coulomb correlation energy has not been included. It is, therefore, not possible to determine the occupancy of the 4f bands by use of the usual procedure of filling up energy bands with electrons to some maximum energy; viz. the Fermi energy. Because of this occupation number dependence, Koopman's theorem does not hold and only total energy calculations which take account of the electron–electron (Coulomb) repulsion can be used to determine the 4f state occupancy. Waber et al.[41] carried out relativistic HFS self-consistent field calculations for atomic cerium confined to a sphere and subject to boundary conditions corresponding to those of the conventional Wigner–Seitz treatment but appropriate to the relativistic differential equation. The total energy was computed using the Slater determinant expression for four different configurations and ten different sphere radii (to simulate compression of the lattice) in order to determine which configuration was the most stable for each radius. They found that their calculations, which depend critically on the sphere radius and therefore on the lattice parameter, gave a minimum energy at the spacing corresponding to α–Ce in the configuration $4f^{0.75} 5d^{1.25}$. While these calculations are not a good model of the metallic state the results are suggestive of the sort of calculations required to elucidate this complicated system.

6.2.3.4 The Divalent Cubic Metals Eu and Yb

Both Eu and Yb are unique among the rare earth metals in that they form in a simple crystallographic structure, bcc for Eu and fcc, bcc and hcp for Yb. Of all the rare earths only these two metals have a 4f occupation ($4f^7$ for Eu and $4f^{14}$ for Yb) such that only two electrons per atom are available for the conduction bands and hence are called divalent. Europium is antiferromagnetically ordered whereas fcc Yb is non-magnetic but shows a transition to a semiconductor under moderate pressures and transforms from an fcc to a bcc structure at about 40 kbar. Highly pure Yb transforms to the hcp phase below room temperature.

Eu Metal

The band structure of bcc europium metal was first determined by Freeman and Dimmock[42] using the non-relativistic APW method along the lines described above for the hcp metals. Two different potentials using the full value of Slater exchange were employed, generated from HFS free atom calculations for the configurations $4f^7 6s^2$ and $4f^6 5d^1 6s^2$ corresponding to the "divalent" and "trivalent" configurations. The energy bands for the divalent configuration are shown in Figure 6–20 and are seen to resemble closely the band structure of the bcc metals of the first transition series—V, Cr, and Fe. The d-band width (H_{12}–H_{25}') is <0.4 Ry which makes it comparable with, or a bit smaller than V (0.5 Ry), Cr (>0.5 Ry), and Fe (0.4 Ry); it is at least one-half the band width for tungsten metal. The total width of the s–p bands (Γ–H_{15}), about 0.6 Ry, is about one-half that of V, Cr, Fe, and W.

Figure 6–21 presents a comparison of the bands obtained for the two configurations along the Γ–H direction. While it is expected that removing a 4f electron and replacing it with a 5d electron would tend to raise the 5d bands, just the opposite effect is observed. The 5d bands in the $4f^6$ calculation shift to lower energy with respect to the 6s–p bands because the 4f electrons, which are deep inside the atom (cf. Figure 6–2) shield the 5d electrons more

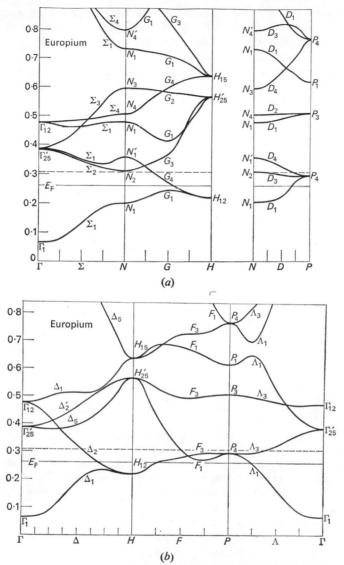

Figure 6–20. Non-relativistic energy bands for bcc Eu obtained (from a starting potential with configuration $4f^7 6s^2$—after reference 42).

effectively than do the 5d electrons themselves. In both configurations, the shielding experienced by the 6s–p electrons is about the same and these states shift a good deal less. Otherwise the two sets of results resemble each other and are what one expects for transition metals.

The density of states histogram calculated for the $4f^7 6s^2$ results is shown in Figure 6–22. The two E_F's shown correspond to the Fermi energy determined by using a rigid band model (not at all correct) and filling the conduction bands with 2 or 3 electrons/atom. The divalent E_F gives a density of states at the Fermi energy, $N(E_F)$, of about 2.5 states/atom-eV, in good agreement with specific heat measurements in the antiferromagnetic phase after allowing for electron–phonon enhancement. The $N(E_F)$ for 3 conduction band electrons is much too high, giving a very unreasonable result. Since one might expect the $4f^6\ 5d^1\ 6s^2$ calculation to yield a similar density of states histogram to that shown in Figure 6–22, we may conclude that the

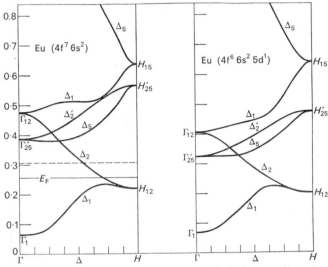

Figure 6–21. Comparison of bands for the divalent and trivalent configurations in bcc Eu along the Γ–H direction (after reference 42).

trivalent calculation is not at all likely to agree with experiment. (In this regard, the Bozorth–Van Vleck model for europium metal, with Eu divalent at high temperatures and trivalent at low T, must be viewed with scepticism.) The low value of $N(E_F)$ resembles the case of Cr which is also antiferromagnetic and is consistent with simple notions of magnetic ordering based on the Ruderman–Kittel–Kasuya–Yosida theory (to be discussed later).

Detailed reflectance measurements on Eu (and Ba, Sr, and Yb) have been carried out by Endriz and Spicer.[43] These allowed some detailed comparisons to be made with the results of band calculations especially since the Freeman and Dimmock[42] results were given at high enough energies. Figure 6–23 presents their comparison of theory and experiment for Eu using the Freeman and Dimmock band structure (cf. Figure 6–20) and density of states (cf. Figure 6–22) and assuming constant transition matrix elements. Also included is a theoretical plot resulting from the nearly free electron model. The Freeman and Dimmock curve shown in Figure 6–23 is to arbitrary scale so that quantitative comparisons with the experimental interband conductivity ($\omega\sigma$) should not be made. Nevertheless, the positions of the two large pieces of structure in this convolved density of states qualitatively confirms the experimentally observed presence of a strong onset of interband transitions near 1 eV followed by a second broad piece of structure in the ultraviolet. Some further analysis suggested to Endriz and Spicer that a systematic contraction of the energy scale associated with the Freeman and Dimmock band calculation would give better agreement with experiment. The nearly free electron picture cannot be adequately reconciled with the experimental data (i.e. the strong piece of structure in the ultraviolet region which is prominent in the band calculations at higher energy). These results give the best experimental evidence available to date for the role of 5d electrons in the four metals studied.[43]

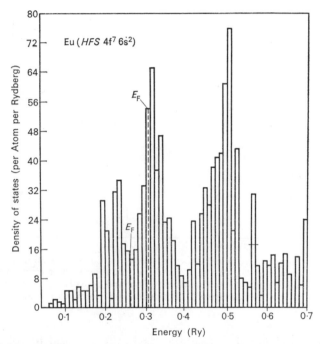

Figure 6–22. Density of states for bcc Eu ($4f^7\,6s^2$ configuration). The two E_F's correspond to filling the conduction bands with 2 or 3 electrons/atom (after reference 42).

A comparison of the interband optical conductivities for the rare earth (Eu and Yb) and the alkaline earth metals (Ba and Sr) failed to indicate strong structure or increased oscillator strength in the rare earth metals which could be unambiguously attributed to transitions from 4f electrons in Eu and Yb. The existence of such 4f levels in both Eu and Yb was found by Brodén et al.[44] by means of ultraviolet photoemission studies. Transitions were observed to occur from what they believed to be 4f states lying between 2.0 and 3.0 eV below the Fermi level of both metals. Endriz and Spicer discuss the paradox between theirs and Brodén et al.'s results in terms of atomic oscillator strengths but no final explanation has emerged.

Andersen and Loucks[10] used the RAPW method to obtain energy bands and a Fermi surface for Eu metal from a potential generated with full Slater exchange and a relativistic HFS atomic calculation from the configuration $4f^7\,6s^2$. These relativistic results agree very well with the non-relativistic results just described except that the RAPW 6s–p bands are shifted downward (by ~ 0.08 Ry) with respect to the 5d bands. The energy bands for Eu were found to resemble closely the band structure of tungsten metal.[45] The Fermi surface is found to differ from the free-electron picture and both the

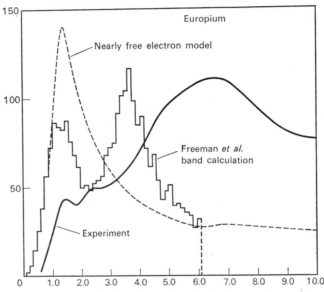

Figure 6–23. Comparison of interband conductivity in Eu metal determined by Endriz and Spicer (reference 43), with that calculated in the nearly free electron theory, and that calculated by ignoring matrix element effect and convolving the filled and empty densities of states shown in Figure 6–22 calculated by Freeman and Dimmock (reference 42).

energy bands and Fermi surface are practically identical to Johansen's[46] calculations for Ba (to which Eu is isoelectronic). The Fermi surface is found[10] to consist of two pieces: an electron surface at the symmetry point H, which has the shape of a rounded-off cube—called the "superegg"—and a hole surface at the point P, which is also a rounded-off cube (half the size of the one at H) but with ellipsoids tetrahedrally positioned on four of the corners—called the "tetracube". These two pieces are shown in Figure 6–24; each has the same volume, as required by compensation of electron and hole volumes. (However, Andersen and Loucks[10] point out the sensitivity of these results to slight changes of the potential or the Fermi energy. A slight change can result in the appearance of tiny pieces of Fermi surface inside the tetracube near where the ellipsoids join onto the rounded off cube.) These authors also discussed the observed helical magnetic order in europium metal as arising from the separation ("nesting") between opposing faces of the nearly cubial part of the hole surface at P; this feature is demonstrated in Figure 6–25 and discussed in Section 6.4.1.

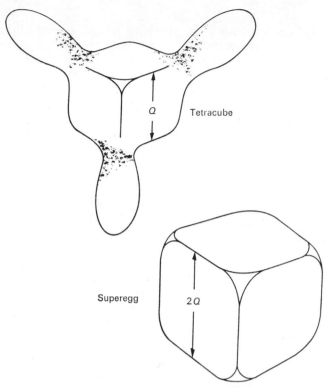

Figure 6–24. "Superegg" and "tetracube" Fermi surfaces for Eu metal determined from the RAPW calculations of Andersen and Loucks (reference 10).

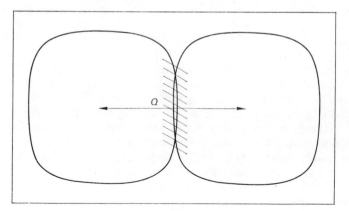

Figure 6–25. Hypothetical nearly square Fermi surface in Eu showing one edge nesting into the opposite one (after reference 10).

Yb Metal

The energy band structure of fcc ytterbium metal was studied theoretically as a function of volume by Johansen and Mackintosh[47] using the RAPW method. A muffin tin potential was constructed from a $4f^{14} 6s^2$ (divalent configuration) atomic change density and the Slater approximation for exchange. Their energy band results are shown in Figure 6–26. A noteworthy feature is the occurrence of two extremely narrow sets of 4f bands near the bottom of the conduction band (indicated by thick lines in the figure) corresponding to the $j=7/2$ and $j=5/2$ single 4f electron states split apart by the spin–orbit coupling. The 4f bands lie unusually high for the rare earths

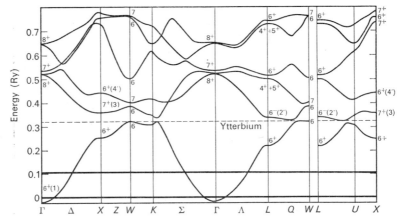

Figure 6–26. The calculated energy band structure of Yb at zero pressure. The thick lines indicate the 4f bands and the dashed line the Fermi level. Relativistic symmetry labels are shown at the symmetry points together with non-relativistic labels, in brackets, in some cases (after reference 47).

because of the effective screening of the nuclear charge in the divalent Yb^{2+} ion. The calculated spin–orbit splitting (1.45 eV) agrees remarkably well with the ultraviolet photoemission measurements of Brodén *et al.*[44] (1.30 ± 0.05 eV). The widths of the 4f bands are very narrow (2.5 mRy and 0.12 mRy respectively at Γ) and show that the crystal field splitting is much less than the splitting caused by the spin–orbit interaction.

It appears at first sight surprising that the calculated spin–orbit splitting of the bands is about the right magnitude for Yb^{3+} whereas the metal's divalent configuration gives it a closed shell which has no spin–orbit splitting.[48] Actually it is probably just a simple accident that the one-electron band Hamiltonian, which treats exchange only approximately, and which does not distinguish in its potential between open and closed shells, gives one-electron spin–orbit split energies $\epsilon_{\frac{7}{2}}$ and $\epsilon_{\frac{5}{2}}$ which agree with experiment. The ultraviolet photoemission experiments measure the difference in *total energies* between the ground state (like Yb^{2+}—4f shell filled) and excited state (Yb^{3+}—with one-hole in the 4f shell). Two peaks are then observed corresponding to the two possible *j*-states for the hole.[49] The one-electron

Hamiltonian cannot be used to calculate this many-electron property because the re-arrangement of the electron shells requires Coulomb correlation and exchange energy contributions not considered in the one-electron local description.

As noted by Johansen and Mackintosh, the position and widths of the 4f bands are uncertain in these calculations because the 4f wave functions are largely confined to the ion core where the charge density is large, so that the corresponding energy levels are very sensitive to the approximation used for exchange. The same sensitivity was found in determining the position and width of the d-bands leading to predictions that Yb should be a semiconductor with a small energy gap ($\sim 10^{-3}$ Ry) in contradiction with Fermi surface experiments.[50]

Recent calculations by Koelling and Harmon[50a] show that a semimetal is obtained if one includes the warped muffin tin (WMT) potential terms, i.e., the actual potential in the interstitial region. The electron pockets then occur along the Σ line near K and result from the hybridization of the plane wave and d-band states (which is why the WMT is important). These features are so potential-sensitive that they are affected by the small readjustment of the 4f-state charge within the crystalline environment. Therefore, Koelling Harmon have allowed the 4f-states to become self-consistent within the crystal environment.

At low temperatures, the stable structure of Yb is hcp and the band structure of this phase has been calculated by Jepsen and Anderson,[51] using a $4f^{14}$ 5d 6s configuration and Slater exchange. The band structure resembles that of the trivalent hcp rare earths (see Figure 6.11) except that the d-bands near the Fermi level are extremely flat, leading to a very high peak in the density of states, just above E_F. It is suggested[51] that this peak may help to account for the hcp–fcc phase transformation. The calculated de Haas–van Alphen frequencies are in quite good agreement with preliminary experimental results.[50] Using $\alpha = 2/3$ exchange, Harmon and Koelling[51a] have obtained a similar band structure also with a pronounced flat band. However, they find this feature to be very sensitive to changes in the potential such as use of the WMT, variation of α, etc., for their calculations.

6.2.3.5. The Effects of Hydrostatic Pressure on Band Structures

Many experiments have been performed to study the effects of hydrostatic pressure on the magnetic ordering of some of the heavy rare earths.[52–55] The magnetic ordering temperature of Gd, Tb, Dy, and Ho was found to decrease linearly with pressure by order of -1 K/kbar for pressures below a critical value where a crystallographic transition occurs. However, although the type of initial magnetic ordering is unchanged by pressure[53, 55] just below the ordering temperature, a reduction of helical turn angle in Tb and Ho when pressure is applied has been observed in neutron diffraction experiments.[55]

The pressure shift of the electronic energy bands of Gd, Tb, and Dy metals have been calculated by Fleming and Liu[56] in order to calculate the effects of pressure on the observed magnetic ordering properties through the effects of this shift on the indirect exchange interactions (to be discussed later in this chapter). The RAPW method was used with a crystal potential of the muffin tin form constructed from a superposition of atomic potentials using the full Slater exchange. The lattice parameters under pressure (0 and 20 kbar) were deduced from the elastic constants of Gd, Tb, and Dy but the same RAPW sphere radius was used for both the zero pressure and the 20 kbar calculations. From the energy bands calculated at 147 points in 1/24 of the zone a spline interpolation method was used to interpolate the bands over a mesh of 450,000 points in the full zone. From these interpolated bands a wave vector dependent susceptibility $\chi(\mathbf{q})$ was calculated for $\mathbf{q}$ along $\Gamma A \Gamma$ over a mesh of 60 points in the double zone scheme.

Fleming and Liu[56] do not present the energy band results but do discuss some of the limitations of the calculation. Foremost among the difficulties is that there is no way to assess the accuracy of the results because of the absence of Fermi surface data for these metals. Also it is not clear that the pressure shift can be entirely separated from the noise level. By choosing a high enough pressure, they hoped to overcome this latter difficulty; by taking great care to do both calculations in an identical manner they hoped to minimize random error. Some of their $\chi(\mathbf{q})$ calculations will be discussed further in a later section.

We have described earlier the energy band calculations[47] on ytterbium metal and noted the interest in the pressure effects on this metal. Johansen and Mackintosh[47] have done the RAPW calculations as a function of volume (compressions of 0.8 and 0.6 of the zero pressure volume) and they were able to make some qualitative arguments about the effect of pressure on the band structure. We quote their results in some detail here as a guide to viewing pressure effects in other metals, although it should be borne in mind that these results are still somewhat tentative speculative. They find that decreasing the volume increases the overlap between neighboring atoms and hence reduces the energy of the muffin-tin zero. The increase in electron density raises the p-like $L_6^-(L_2')$ level relative to $\Gamma_6^+(\Gamma_1)$, the bottom of the band. The mean d-band energy, determined as the mean of the two doubly-degenerate Γ_8^+ levels, drops somewhat on an absolute scale, but changes very little compared with the p-level. However, the width of the d-band, measured as the difference between the upper and lower X_7^+ levels, increases substantially, so that the lower d-levels move down relative to the p-level and the hybridization near the Fermi level increases. This has the effect of increasing the energy gap, as is observed experimentally. The spin–orbit splitting which lifts the accidental degeneracy between L and W, is essential for the semi-conducting behavior. When the volume is reduced further, the decrease of the $X_7^+(X_3)$ energy level first reduces the energy gap and finally produces a metallic structure with electron pockets at X.

It is believed that strong sp–d hybridization near the Fermi level favors the bcc structure in the alkaline earth metals since Ba, which has low-lying d-bands, is bcc at zero pressure, while Sr and Ca, in which the d-bands lie higher, are fcc. By comparing some of their results, particularly the 0.6 compression bands to results on fcc Ba, Sr, and Ca, Johansen and Mackintosh ascribe the fcc–bcc phase transition in Yb (and also in Sr) to the broadening of the d-resonance on compression. Compression of the lattice causes the 4f bands to broaden and to rise relative to the muffin tin zero, as expected.

We have also described in an earlier section of this chapter the pressure shift band studies on the properties of Ce metal. The normal room temperature fcc phase (γ–Ce) transforms at low temperatures (~ 120 K and 1 atm) or at high pressure (~ 7.5 kbar and 298 K) to another fcc modification (γ–Ce) with a large contraction in volume (17 to 12% depending on temperature). [57] This large volume change has been suggested by many workers to be due to the transfer of the 4f electron to the 5d band. There are a number of models used to explain the band structure of Ce in both these phases [58] and some calculations [41]—as we have earlier noted. Still Ce remains the least well understood rare earth metal today.

6.2.4. Experimental Evidence on Fermi Surface Properties

Whereas there exists today a wide variety of experiments on the rare earth metals, there is at present little direct experimental evidence on their conduction band structures and Fermi surfaces. The difficulty of preparing samples of sufficiently high purity has prevented, until now, the study of Fermi surfaces by the conventional techniques. Recently, the de Haas–van Alphen effect has been observed in fcc and hcp ytterbium metal [50] and reported for Lu metal [59]—both non-magnetically ordered structures. The magnetic ordering in the other metals at low temperatures severely restricts the number of methods which may be used to study the electronic structure in the paramagnetic phase. The resulting situation is unfortunate in that direct comparison with the band structure calculations described above is difficult.

The measurement of the angular correlation of the photons emitted when positrons annihilate is a technique which does not suffer from these limitations. Details of the procedure plus a discussion of the application of this technique to the study of the heavy rare earth metals Gd, Tb, Dy, Ho, and Er and the related metal Y (configuration $4d^1\ 5s^2$ and also hcp in structure) have been given by Williams and Mackintosh. [60] In the independent particle model, the number of coincidences in the conventional parallel slit geometry at an angle θ is proportional to

$$N(\theta) = N\left(\frac{\hbar p_z}{mc}\right) = \int\limits_{-\infty}^{\infty}\!\!\int dp_x\, dp_y \sum_k F(\mathbf{p}, \mathbf{k})$$

where the sum is over all occupied states, p_x and p_y are components of the momentum and

$$F(\mathbf{p}, \mathbf{k}) = \left| \int \psi_\mathbf{k}(\mathbf{r})\, \psi_+(\mathbf{r})\, e^{-i\mathbf{p}\cdot\mathbf{r}}\, d^3 r \right|^2$$

where $\psi_\mathbf{k}$ is the electron wave function with wave vector $\mathbf{k}$ and ψ_+ is the positron wave function (determined from the same potential as ψ_k but of opposite sign and without the exchange contribution).

Williams, Loucks, and Mackintosh [33] measured the positron annihilation in single crystal discs of Ho and Er and also Y. They found that the coincidence distributions in these three metals was similar and highly anisotropic—

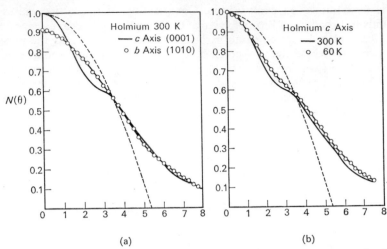

Figure 6–27. (a) Photon coincidences as a function of angle in Ho at 300 K. The dashed
curve is the parabola corresponding to three free electrons per atom. (b) The temperature dependence of the coincidence distribution in the c-axis crystal (after reference
33).

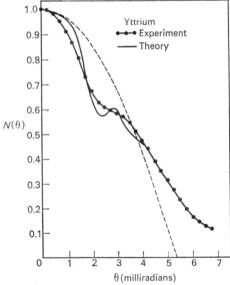

Figure 6–28.* The angular distribution of photon coincidences from position annihilation
in a c-axis single crystal of Y. The points are the experimental results and the full
line is the calculation of Loucks (reference 32); the dashed parabolic curve shows the
free electron prediction (after reference 33).

* Later calculations by Gupta and Loucks[32a] using a less approximate expression give
somewhat different results. The hump is much more prominent and the curve lies well
below the experimental results at small momenta. This may be due to a relatively larger
inaccuracy is the s part of their wave function as compared to the d part.

in disagreement with predictions of free-electron theory. The results for Ho, shown in Figure 6–27, show pronounced structure and a hump at 3 mrad, showing evidence of Fermi surface anisotropies such as found from the band calculations. The first calculation of Loucks[32] using his APW results for yttrium metal reproduces quite well the qualitative features of the experimental results [32a] (cf. comparison shown in Figure 6–28). The differences between theory and experiment are attributed to Coulomb correlation contributions which are neglected in the independent particle model description. The rapid drop in coincidences at low angles and the hump in the c direction seen in Figure 6–28 are both attributed to the rapid increase in the hole surface area near the hexagonal face of the zone. [See footnote on previous page.]

The later results of Williams and Mackintosh[60] on the other hcp rare earth metals show similar results with differences between them attributable to the differences in the Fermi surface features; Tb and Gd were found to have a much less distinct hump than the other metals. This effect is seen in the band calculations (cf. Section 6.3) and is very pronounced for Y and the heavy rare earths other than Tb and Gd (where it is less pronounced) because of the thick webbing near the point L in the zone (cf. Figure 6–14). In Tb and Gd, this webbing is either very thin or absent.

These experiments thus give qualitative confirmation of the theoretical band structures and Fermi surfaces described earlier. Hopefully, the formidable problem of obtaining high purity single crystals will soon be overcome and more detailed Fermi surface studies undertaken for the magnetically ordered rare earth metals. From the Fermi surface viewpoint, La is probably the best candidate to be undertaken first because it does not order magnetically (thereby simplifying the analysis) but has the characteristic band structures of the rare earths.

6.3. INDIRECT EXCHANGE: THE s–f INTERACTION AND RUDERMAN–KITTEL–KASUYA–YOSIDA THEORY

It may at first sight appear rather remarkable that thus far only very little has been said about the role played by the 4f electrons in the rare earth metals other than that they are highly localized and may not be described as band electrons. In considering the magnetic properties of the rare earth metals (cf. Chapters 3 and 4) it is clear that the open 4f shell contributes almost the entire observed bulk magnetization and paramagnetic moment well above the ordering temperature, giving in both cases a moment per ion which is close to the isolated ion values. The direct exchange interaction between 4f electrons on different sites is clearly too small to play a significant role in accounting for the strong magnetic ordering observed because the 4f wave functions on different atoms overlap only negligibly (cf. Figure 6–2).

As the carriers of the net magnetization, the open 4f shell electrons play an obviously essential role in promoting magnetic ordering but through the mechanism of indirect exchange between the localized 4f electrons and the conduction band electrons. This indirect exchange mechanism produces an effective 4f–4f coupling which is responsible for the various observed magnetic properties. First proposed by Ruderman and Kittel[61] to explain the effective

long range coupling between nuclear spins interacting via the Fermi hyperfine interaction, this mechanism was later extended by Kasuya[62] and
Yosida[63] to treat local moment–conduction electron, s–d (and s–f), interactions in magnetic materials and is now referred to as the RKKY interaction.

6.3.1. The "s–d", "s–f" Interaction

The Hamiltonian of the exchange interaction between the localized and
conduction electrons has been treated by a number of authors[1, 2, 64–68]
following the work of Kasuya.[62]

The s–d interaction is derived from the original Hamiltonian which unlike
the simple band or Bloch Hamiltonian includes the electron–electron interaction

$$\sum_{i>j} \frac{1}{r_{ij}}$$

explicitly rather than through some average field approximation. The electronic energy band description of electrons in metals is an approximate
treatment of the $N=10^{23}$ body problem using a simplified average of the
true (non-local) interactions between electrons to obtain a periodic potential
Hamiltonian (cf. Section 6.2). The complete non-relativistic Hamiltonian is
given by

$$\mathscr{H} = \sum_i -\nabla_i^2 + \sum_{i,l} V(\mathbf{r}_i - \mathbf{R}_l) + \sum_{i>j} \frac{2}{r_{ij}} \tag{6.2}$$

in Rydberg units. The energy band approximation averages the electrostatic
interaction between electrons (i and j) and results in a periodic (local) potential from which Bloch's theorem about the translational properties of the
solutions follows. For magnetic interactions in which the spin orientations on
a given site are disturbed by say the scattering of conduction electrons, the
exchange interaction must be considered in detail, as is done in the following.

In second quantized notation this interaction term may be written as

$$V(\mathbf{r}_1\mathbf{r}_2) = \tfrac{1}{2} \sum_{k_1 k_2 k_3 k_4} \langle \mathbf{k}_4, \mathbf{k}_3 | V | \mathbf{k}_2, \mathbf{k}_1 \rangle \, c_{\mathbf{k}_4}^+ c_{\mathbf{k}_3}^+ c_{\mathbf{k}_2} c_{\mathbf{k}_1} \tag{6.3}$$

with

$$\langle \mathbf{k}_4, \mathbf{k}_3 | V | \mathbf{k}_2, \mathbf{k}_1 \rangle = \iint d^3\mathbf{r}_1 d^3\mathbf{r}_2 \, \psi^*_{\mathbf{k}_4}(\mathbf{r}_1) \, \psi_{\mathbf{k}_3}^*(\mathbf{r}_2) \, V(\mathbf{r}_1 - \mathbf{r}_2)$$
$$\times \psi_{\mathbf{k}_2}(\mathbf{r}_1) \, \psi_{\mathbf{k}_1}(\mathbf{r}_2) \tag{6.4}$$

Here $V(\mathbf{r}_1 - \mathbf{r}_2)$ can denote the bare electron interaction $1/r_{ij}$ or more generally
an effective two body interaction which includes screening of this Coulomb
interaction. The c^+ and c are respectively creation and annihilation operators.

Among the many processes which occur when one considers this additional
term in the Hamiltonian as a perturbation, one is interested solely in those
in which a conduction electron scatters and a localized (d or f) state is
occupied both before and after. Thus one each of the states $\mathbf{k}_4$, $\mathbf{k}_3$ and $\mathbf{k}_2$, $\mathbf{k}_1$
must be a localized state and the others conduction band states. Since
$V(\mathbf{r}_1 - \mathbf{r}_2)$ is independent of spin, the spin of the conduction bands states
must be the same as that of the localized state.

If one assumes that the localized orbitals centered at R_l are approximated by non-overlapping atomic-like functions $\phi_L(r - R_l)$ and the conduction electron orbitals as Bloch type wave functions, $\psi_k(r) = e^{ik \cdot r} u_k(r)$ where $u_k(r)$ is periodic, we can substitute the spin operators S_n for the creation and annihilation operators of unfilled shell electrons and obtain the exchange term of the interaction Hamiltonian

$$\mathcal{H}_{sf} = -N^{-1} \sum_{k, k', l} j_{sf}(k, k') \, e^{i(k-k') \cdot R_l}$$

$$\times [(c_{k\uparrow}^+ c_{k\uparrow}' - c_{k\downarrow}^+ c_{k\downarrow}') \, S_l^z + c_{k\uparrow}^+ c_{k\downarrow}' \, S_l^- + c_{k\downarrow}^+ c_{k\uparrow}' \, S_l^+] \quad (6.5)$$

where N is the number of magnetic lattice sites per unit volume and

$$j_{sf}(k, k') = N \int d^3r_1 d^3r_2 \phi_L^*(r_1) \, \psi_k^*(r_2) \, V(r_1 - r_2) \, \phi_L(r_2) \, \psi_{k'}(r_1) \, e^{i(k-k') \cdot R_l}$$

$$(6.6)$$

is the generalized exchange integral and does not depend on the lattice position R_l because of the translational symmetry of the Bloch functions.

In writing Eqn (6.5) we have written in a spin dependence ($\uparrow$ or $\downarrow$) for the c_k operators and S_l denotes the spin operator of the localized (open) shell electrons located at R_l with S^+ the step-up and S^- the step-down operators. It is assumed that the localized orbital $\phi_L(r - R_l)$ is always occupied by just one electron so that $c_{l\uparrow}^+ c_{l\uparrow} + c_{l\downarrow}^+ c_{l\downarrow} = 1$. In real cases where the localized orbital $\phi_L(r - R_l)$ may be degenerate, Eqn (6.5) is used with S_l taken to be the total spin operator provided the intra-atomic exchange interaction is strong enough to retain the Hund coupling even in the metal. For more than one localized electron, the exchange integral must be determined as a suitable average over the wave functions of the open shell electrons. If, as in the rare earths, the orbital angular momentum is not zero and $J = L + S$ is a good quantum number then we obtain, as a good approximation, the s–f interaction Hamiltonian by replacing S_l in Eqn (6.5) by the projection of the spin along J, namely by $(\lambda - 1)J$ where λ is the Lande splitting factor.

In many applications $j_{sf}(k, k')$ is assumed to be independent of k and k' and replaced by a constant term, j_{sf}. As a better approximation, since $j_{sf}(k, k')$ depends sensitively on $|k - k'|$, it is assumed that $j(k, k')$ depends only on $k' - k = q$. (We shall describe later the actual dependence of $j_{sf}(k, k')$ on k and k' and on q.) The $j_{sf}(q)$ approximation allows Eqn (6.5) to be written in a different, more familiar and often used form. Introducing the well-known spin–wave operators

$$S_q = N^{-\frac{1}{2}} \sum_l S_l \, e^{iq \cdot R_l} \quad (6.7)$$

spin–wave operators

$$S_q = N^{-\frac{1}{2}} \sum_i S_i \, e^{iq \cdot r_i}$$

where r_i and s_i are respectively the coordinate and spin operator of the ith electron, results in

$$\mathcal{H}_{sf} = -\sum_q j_{sf}(q) \, S_q \cdot s_{-q} \quad (6.8)$$

or in spatial coordinates

$$\mathscr{H}_{sf} = -\sum_i \sum_l j_{sf}(|\mathbf{r} - \mathbf{R}_l|)\, \mathbf{S}_l \cdot \mathbf{s}_i \qquad (6.9)$$

where

$$j_{sf}(|\mathbf{r} - \mathbf{R}_l|) = N^{-1} \sum_{\mathbf{q}} j_{sf}(\mathbf{q})\, e^{i(\mathbf{r}_i - \mathbf{R}_l)\cdot \mathbf{q}} \qquad (6.10)$$

The exchange interaction between the (4f) localized electrons in the rare earths and the conduction electrons is basic for understanding the electronic, magnetic, and optical properties of the rare earth metals (and for a variety of physical problems such as local moments in dilute alloys, Kondo effect,[68, 64] etc.[69]). Here we shall concern ourselves with the effects produced by this s–f interaction Hamiltonian. Unlike the Coulomb interaction which is periodic and does not scatter the conduction electrons, the nonlocal exchange interactions are not periodic and do scatter the conduction electrons when the spin directions of the open shell electrons are disturbed. This spin disorder scattering gives rise to the resistivity in first order; the second order processes are responsible for the effective spin–spin interaction between the open shell electrons which is responsible for the magnetic ordering.

6.3.2. Ruderman–Kittel–Kasuya–Yosida Interaction: Indirect Exchange Coupling

Ruderman and Kittel[61] first considered the indirect coupling of nuclear spins by calculating the second-order perturbation of the energy using the hyperfine interaction analogy of the $\mathscr{H}_{sf}$ interaction. Kasuya[62] and de Gennes[70] proposed that it was the important interaction for the case of the rare earths and Yosida[63] used it for the electron spin coupling in transition metal alloy systems. The relation between the R–K and the K–Y types of calculations was given by Van Vleck.[71] A detailed discussion of the exchange coupling parameter was given first by Liu[72] and more recently by Watson and Freeman[73, 74] who also investigated the long-range conduction electron spin polarization induced by the s–f interaction. The 4f electrons overlap the conduction electrons strongly and their net spin polarizes the conduction electrons via the s–f interaction. This polarization has an oscillatory component due to the Fermi distribution which restricts the wave vector of the conduction sea electrons that carry the polarization. This resultant polarization carried over to the vicinity of other ions will then interact with the moment of their 4f shells and produce an alignment of the moments.

Consider the interaction between two localized spin moments S_l and $S_{l'}$ located at sites $\mathbf{R}_l$ and $\mathbf{R}_{l'}$ and the conduction electrons. If we use the $\mathscr{H}_{sf}$ interaction (Eqn (6.5)) the shift in energy of the system to the *lowest* order in $\mathscr{H}_{sf}$ which gives an interaction is of second order. [The coupling may be obtained by considering the virtual excitation of an electron $\mathbf{k}$ (with spin s) into an empty state $\mathbf{k}'$ (with spin s) and taking account of the Fermi distribution function which accounts for the Pauli Principle.] In second-order perturbation theory the shift in energy is given by using Eqn (6.5) with one matrix element from each site l and l'

$$\delta E^{(2)} = \sum_i \frac{\langle 0|\mathcal{H}_{sf}(\mathbf{R}_l)|i\rangle\langle i|\mathcal{H}_{sf}(\mathbf{R}_{l'})|0\rangle}{\epsilon_0 - \epsilon_i} + \sum_i \frac{\langle 0|\mathcal{H}_{sf}(\mathbf{R}_{l'})|i\rangle\langle i|\mathcal{H}_{sf}(\mathbf{R}_l)|0\rangle}{\epsilon_0 - \epsilon_i}$$

where $|0\rangle$ is the initial state and $|i\rangle$ are the excited (intermediate) states. (Since both sums give identical results we consider the first only and multiply the result by a factor of two.) Substituting $\mathcal{H}_{sf}$ from Eqn (6.5), using the properties of closure on the intermediate states and retaining only terms dependent on the spin orientation yields in operator form

$$\mathcal{H}_{ff'} = -\frac{1}{N^2}\sum_{\mathbf{k},\,\mathbf{k}'} |j_{sf}(\mathbf{k},\mathbf{k}')|^2 \frac{f_{\mathbf{k}}(1-f_{\mathbf{k}'})}{\epsilon_{\mathbf{k}'} - \epsilon_{\mathbf{k}}} e^{i(\mathbf{k}'-\mathbf{k})\cdot(\mathbf{R}_l - \mathbf{R}_{l'})}$$
$$\times (2S_l^z S_{l'}^z + S_l^- S_{l'}^+ + S_l^+ S_{l'}^-) \quad (6.11)$$

where $f_{\mathbf{k}}$ is the Fermi occupation function. This expression was derived for the interaction of two localized sites. For many lattice sites with localized spins we may write $\mathcal{H}_{ff'}$ as

$$\mathcal{H}_{ff'} = -\sum_{l,\,l'} j(\mathbf{R}_l - \mathbf{R}_{l'})\, \mathbf{S}_l \cdot \mathbf{S}_{l'} \quad (6.12)$$

where the sum is over all *pairs* of spins in the system. Here the indirect exchange coupling constant is given by

$$j(\mathbf{R}_l - \mathbf{R}_{l'}) = \frac{1}{N^2}\sum_{\mathbf{k},\,\mathbf{k}'} |j_{sf}(\mathbf{k},\mathbf{k}')|^2 \frac{f_{\mathbf{k}}(1-f_{\mathbf{k}'})}{\epsilon_{\mathbf{k}'} - \epsilon_{\mathbf{k}}} e^{i(\mathbf{k}-\mathbf{k}')\cdot(\mathbf{R}_l - \mathbf{R}_{l'})} \quad (6.13)$$

As stated earlier, this effective isotropic exchange interaction has been derived without taking account of the orbital contribution to the ionic moment. For the rare earths, when $\mathbf{J}$ is a good quantum number, one can rewrite Eqn (6.12) in terms of the $\mathbf{J}$ manifold by replacing $\mathbf{S}_l$ by its projection onto $\mathbf{J}$ namely by $(\lambda - 1)\mathbf{J}_l$ giving for Eqn (6.12)

$$\mathcal{H}_{ff'} = -(\lambda-1)^2 \sum_l \sum_{l'} j(\mathbf{R}_l - \mathbf{R}_{l'})\,\mathbf{J}_l \cdot \mathbf{J}_{l'} = -\sum_{l,\,l'} \mathcal{J}(\mathbf{R}_l - \mathbf{R}_{l'})\,\mathbf{J}_l \cdot \mathbf{J}_{l'} \quad (6.14)$$

As in Chapter 5, it is convenient to define a wave vector dependent susceptibility of the conduction electron system, $\chi(\mathbf{q})$, which yields the response of the electron gas to the exchange field of the localized spin. We may rewrite Eqns (6.12) and (6.13) to obtain a useful approximate expression which depends on the wave vector dependent susceptibility $\chi(\mathbf{q})$ of the non-interacting gas. Writing the usual expression for $\chi(\mathbf{q})$

$$\chi(\mathbf{q}) = N^{-1}\sum_{\mathbf{k}} \frac{f_{\mathbf{k}} - f_{\mathbf{k}+\mathbf{q}}}{\epsilon_{\mathbf{k}+\mathbf{q}} - \epsilon_{\mathbf{k}}} \quad (6.15)$$

and replacing $j_{sf}(\mathbf{k},\mathbf{k}')$ by $j_{sf}(\mathbf{q})$ for convenience and simplification gives

$$\mathcal{H}_{ff'} = -\sum_{l,\,l',\,\mathbf{q}} j(\mathbf{q})\, e^{i\mathbf{q}\cdot(\mathbf{R}_l - \mathbf{R}_{l'})}\, \mathbf{S}_l \cdot \mathbf{S}_{l'} = -\sum_{l,\,l} j(\mathbf{R}_l - \mathbf{R}_{l'})\, \mathbf{S}_l \cdot \mathbf{S}_{l'} \quad (6.16)$$

where

$$j(\mathbf{q}) = \frac{2}{N} |j_{sf}(\mathbf{q})|^2 \chi(\mathbf{q}) \quad (6.17)$$

For rare earth ions we use Eqn (6.14) but with

$$\mathscr{J}(\mathbf{q}) = 2N^{-1}(\lambda - 1)^2 |j_{sf}(\mathbf{q})|^2 \chi(\mathbf{q}) \tag{6.18}$$

and

$$\mathscr{J}(\mathbf{R}_l - \mathbf{R}_{l'}) = \sum_{\mathbf{q}} \mathscr{J}(\mathbf{q}) \, e^{i\mathbf{q} \cdot (\mathbf{R}_l - \mathbf{R}_{l'})} \tag{6.19}$$

The evaluation of the RKKY interaction is possible only if one makes some *extreme* approximations. In addition to considering a free-electron like band structure for the conduction electrons and non-overlapping of the localized electrons on different sites (a rather good approximation for the rare earths), it is common to set $j_{sf}(\mathbf{k}, \mathbf{k}') = \text{constant} = j$ (corresponding to a delta function interaction) and to use the free electron approximation for $\chi(\mathbf{q})$. For a free electron gas in three dimensions,[2, 75] $\chi(\mathbf{q})$ is proportional to the functional

$$f(q) = 1 + \frac{4k_F^2 - q^2}{4k_F q} \ln \left| \frac{2k_F + q}{2k_F - q} \right| \tag{6.20}$$

where k_F is the radius of the conduction–electron Fermi surface, which is taken to be spherical. The function $f(q)$ arises from the energy denominator and density of states in the second order perturbation mixing of $\mathbf{k}$ and $\mathbf{k}'$. Figure 6–29 shows a plot of $f(q)$ for one, two, and three dimensions. As can be seen for the three-dimensional case, $f(q)$ slowly decreases between 0 and $2k_F$ from the value 1 to $\tfrac{1}{2}$. At $q = 2k_F$ the slope is infinite and for $q > 2k_F$,

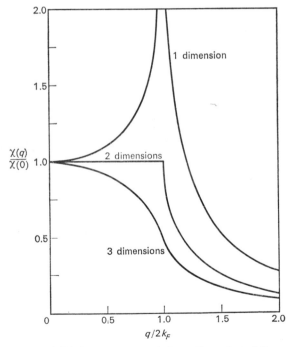

Figure 6–29. Plots of $f(q)$ in one, two, and three dimensions (after reference 2).

$f(q)$ falls rapidly to zero because the energy denominator in Eqn (6.15) becomes large.

These approximations permit the summations in Eqn (6.13) (or in Eqn (6.15)) to be carried out and result in the familiar Ruderman–Kittel[61] expression

$$j(\mathbf{R}_l - \mathbf{R}_{l'}) = 9\pi (j^2/\epsilon_F) \, F(2k_F |\mathbf{R}_l - \mathbf{R}_{l'}|) \qquad (6.21)^*$$

where

$$F(x) = \frac{x \cos x - \sin x}{x^4} \qquad (6.22)$$

The interaction is thus of long range, decreasing as R^{-3} for large R so that it is closely centered about the ion site and oscillates with the period $(2k_F)^{-1}$. It is the infinite slope of $\chi(\mathbf{q})$ when $q = 2k_F$ (or twice the diameter of the Fermi sphere) which causes its Fourier transform to have this oscillatory behavior. Thus, in this model, finite temperatures or finite relaxation times of the conduction electrons will smear out the singularity at $\chi(q = 2k_F)$, so the oscillations in $j(\mathbf{R}_l - \mathbf{R}_{l'})$ will be smeared out also for sufficiently large R.

An exchange interaction of this oscillating form (alternating in sign) with increasing interionic separation has the potential for accounting for many different kinds of order. It is known from the molecular field model that the presence of maxima in the Fourier transform of a long-range interaction determines the form and periodicity in a helical ordering pattern. Note that while $\chi(\mathbf{q})$ has its maximum at $q = 0$ showing that ferromagnetic alignment is preferred, the maximum in q for real systems need not be at $q = 0$. This is the reason for the great interest in applications of RKKY theory to the rare earths with their variety of observed magnetic orderings.

The **q**-dependent susceptibility $\chi(\mathbf{q})$ measures the linear response of the conduction electron system to the effective fields of the local moments. Relating the indirect exchange interaction to $\chi(\mathbf{q})$ as in Eqns (6.17) and (6.18) thus separates the interaction pictorially into two parts, the driving exchange polarization interaction given by $j_{sf}(\mathbf{q})$, arising from the s–f interaction, and the response function of the system given by $\chi(\mathbf{q})$. Although $\chi(\mathbf{q})$ was taken here as that for the non-interacting free-electron gas it has been considered more generally from the band picture of metals and from the many-body point of view which includes interactions between electrons. Detailed and extensive discussions of the generalized susceptibility problem are given in the volume by Herring.[75]

As has been emphasized by Overhauser,[76] Kasuya,[20] Herring,[75] Kittel,[77] and others, the form of $\chi(\mathbf{q})$ is of central importance for determining a variety of phenomena associated with the indirect exchange mechanism. Of primary concern to us here is how the indirect exchange and $\chi(\mathbf{q})$ relate to the most stable configuration in a crystal. Clearly the most stable spin configuration is that which minimizes the energy of interaction (6.12) or (6.14). Nagamiya's[78] treatment, outlined below, was to minimize the Fourier transform of the exchange energy expression. (Earlier, Villain[79] had obtained the relation between stable configurations and indirect exchange using a statistical approach.)

We now consider the simple helical spin ordering at $T = 0$ for a monoatomic lattice. As in the discussion centering around Eqns (6.7)–(6.10) it is readily seen that the total exchange energy Eqn (6.14) for $\mathscr{H}_{ff'}$ can be written as

$$\mathscr{H}_{ff'} = -\sum_{\mathbf{q}} j(\mathbf{q}) \, \mathbf{S_q} \cdot \mathbf{S_{-q}} \qquad (6.23)$$

where $\mathbf{S_q}$ is defined in Eqn (6.7). The most stable configuration results when the summation is minimized subject to the constraint that

$$S_n^2 = \text{constant} = S^2$$

for lattice sites n. A milder form of this constraint is

$$\sum_n S_n^2 = \text{constant}$$

which can be written in Fourier component form as

$$\sum_{\mathbf{q}} \mathbf{S_q} \cdot \mathbf{S_{-q}} = \text{const.} \qquad (6.24)$$

Under this weaker constraint $\mathscr{H}_{ff}$ is readily minimized when one takes only one Fourier component of spin configuration, i.e. that $\mathbf{q} = \mathbf{Q}$ for which $j(\mathbf{q})$ is the highest maximum and the minimum value of $H_{ff'}$ is (with $\mathbf{q} = -\mathbf{Q}$ also allowed)

$$\mathscr{H}_{ff'} = -j(\mathbf{Q}) \, [\mathbf{S_Q} \cdot \mathbf{S_{-Q}} + \mathbf{S_{-Q}} \cdot \mathbf{S_Q}] \qquad (6.25)$$

From the definition of $\mathbf{S_q}$ (Eqn 6.7) it follows that

$$\mathbf{S}_l = N^{-\frac{1}{2}} \, [\mathbf{S_Q} \, e^{i\mathbf{Q} \cdot \mathbf{R}_l} + \mathbf{S_{-Q}} \, e^{-i\mathbf{Q} \cdot \mathbf{R}_l}] \qquad (6.26)$$

which in component form gives for the spin arrangement of minimum energy

$$S_{lx} = A \cos (\mathbf{Q} \cdot \mathbf{R}_l + \alpha)$$

$$S_{ly} = B \cos (\mathbf{Q} \cdot \mathbf{R}_l + \beta)$$

$$S_{lz} = C \cos (\mathbf{Q} \cdot \mathbf{R}_l + \gamma) \qquad (6.27)$$

where A, B, α, β, and γ are arbitrary constants.

As discussed by Nagamiya in great detail,[78] there are many spin configurations which can result from the general conditions described by Eqn (6.27) (cf. Chapters 2, 3, and 5). It should be emphasized that a characteristic feature of the helical spin-arrangement is that the period of the arrangements is, generally, incommensurate with the lattice period since the magnitude of $\mathbf{Q}$ is determined solely by the exchange coefficients. As discussed in Chapter 2, the spin-configuration that a particular crystal will assume depends on anisotropy of the crystal field and magnetoelastic effects, but for any configuration there is the common property that the wave vector $\mathbf{Q}$ which denotes the periodicity is the same wave vector for which $j(\mathbf{q})$ is a maximum. From the relation (6.17) or (6.18) it is seen that, within the approximation $j_{sf}(\mathbf{q}) = \text{constant}$, the statement that $j(\mathbf{q})$ (or $\mathscr{J}(\mathbf{q})$) be a maximum simply means that $\chi(\mathbf{q})$ is a maximum. The more realistic case of considering the $\mathbf{q}$ dependence of $j_{sf}(\mathbf{q})$ or of $j_{sf}(\mathbf{k}, \mathbf{k}')$ itself has not been treated to date.

6.3.3. Extensions of the RKKY Theory

Although popular with both theorists and experimentalists because of its simplicity and ease of applicability, the simple RKKY model is not an overly realistic description for systems like the rare earth metals. As discussed extensively in Section 6.2, the conduction bands are mostly comprised of the 5d-like electrons which are hybridized strongly with the 6s-p like electrons near the Fermi energy. A proper treatment of s–f exchange must include this multiple sub-band structure, the actual density of states, the highly anisotropic Fermi surface in these metals, magnetic ion multiplicity and, of course, a realistic calculation of the wave vector dependent susceptibility, $\chi(\mathbf{q})$, determined from the actual band structure and taking account of interactions between the electrons. The drastic assumption that $j_{sf}(\mathbf{k}, \mathbf{k}') = j = \text{constant}$ must also be dropped, as model calculations have shown this to be invalid.[73, 74]

6.3.3.1. Non-spherical Fermi Surfaces

The usual assumption of spherical Fermi surfaces for the conduction electrons has been questioned by Blandin,[80] Gauthier,[81] and by Roth, Zeiger, and Kaplan.[82] By dropping this requirement and considering non-spherical Fermi surfaces they have attempted to generalize the RKKY interaction and have presented some applications by way of results. Roth et al.[82] find that in general $j(\mathbf{R}_l - \mathbf{R}_{l'})$ is found to fall off as $1/R^3$ and to oscillate with a period corresponding to a calipering of the Fermi surface in the $\mathbf{R}_l - \mathbf{R}_{l'}$ direction. For the special cases of parallel or cylindrical regions of a Fermi surface which they used as a model, they found a slower fall-off of $j(\mathbf{R}_l - \mathbf{R}_{l'})$ (R^{-1} and R^{-2} respectively.) This gives a considerable increase to the range of the interaction. They showed that $j(\mathbf{q})$ displays Kohn[83]-type anomalies and constructed a locus of these singularities forming what they call a Kohn anomaly surface. The form of these anomalies depends on the detailed shape of the Fermi surface in the vicinity of pairs of calipering points. In this way, for more general Fermi *surfaces*, one can look for maxima in $j(\mathbf{q})$ in the vicinity of the Kohn anomaly surface and relate these to stable configurations for the spin system. The infinite slope in $j(\mathbf{q})$ for spherical Fermi surfaces arising from $\chi(\mathbf{q})$ at $q = 2k_F$ (cf. Eqn (6.20)) changes for the more general case and becomes a discontinuous slope in the case of a "waist" in the Fermi surface. For parallel regions of the Fermi surface, $j(\mathbf{q})$ has a logarithmic singularity and this large maximum is important in real systems like the rare earth metals where a "nesting" between electron and hole Fermi surfaces is thought to be responsible for the observed magnetic ordering (cf. Sections 6.2.3 and 6.4). Recently, Bambakidis[84] has attempted to calculate the indirect exchange interaction between near-neighbor spins in Gd metal using the method of Roth et al.[82] and the APW bands determined for Gd by Dimmock et al.[3, 6] He concluded from his results that the present accuracy of the band calculations did not permit a realistic determination of indirect exchange energies.

While this work by Roth et al. has had little further direct application, its importance lay in pointing up the need for considering real, i.e. non-spherical

Fermi surfaces, within RKKY theory. As we shall see (cf. Section 6.4), recent emphasis of theory has been on the calculation of $\chi(q)$ using the actual band structure (described in section 6.2.3) and looking for maxima with which to determine the most stable magnetic ordering.

6.3.3.2. Correlation and Exchange Effects on $\chi(q)$

As emphasized by Wolff,[85] Overhauser,[76] and Herring[75] among others, the Coulomb interactions between the conduction electrons play an essential role in determining the wave vector dependence of the spin susceptibility, $\chi(\mathbf{q})$. The use of a non-interacting gas approximation for $\chi(\mathbf{q})$, such as discussed so far, leaves out important correlation and exchange effects which can significantly effect $\chi(\mathbf{q})$ and any predictions based on it.

The various calculations[45] indicate that the electron–electron interaction greatly *enhances* the spin susceptibility of the non-interacting gas (of the order of 10–15 for the case of Pd metal[86] at $q=0$). Here we sketch a simple but informative derivation[73] of the modification of $\chi(\mathbf{q})$ due to the inclusion of conduction electron–conduction electron exchange. The electron gas is perturbed with a magnetic field $\mathbf{H_q}$ giving the response

$$M(\mathbf{q})=\chi(\mathbf{q})\,H(\mathbf{q})$$

Now if the response itself induces a field proportional to $M(\mathbf{q})$, say $vM(\mathbf{q})$, where v is a constant which may depend on $\mathbf{q}$, this results in

$$M(\mathbf{q})=\chi(\mathbf{q})\,[H(\mathbf{q})+vM(\mathbf{q})] \tag{6.28}$$

Solving for $M(\mathbf{q})$ gives

$$M(\mathbf{q})=H(\mathbf{q})\,[\chi(\mathbf{q})/1-v\chi(\mathbf{q})]=\chi_v(\mathbf{q})\,H(\mathbf{q})$$

where the exchange enhanced susceptibility is given by

$$\chi_v(\mathbf{q})=\chi(\mathbf{q})/[1-v\chi(\mathbf{q})] \tag{6.29}$$

The susceptibility is thus seen to be enhanced when v is positive (as is certainly the case for a delta function type of interaction). From the non-interacting gas $\chi(\mathbf{q})$ depicted in Figure 6–29, $\chi(\mathbf{q})$ decreases monotonically as $\mathbf{q}$ increases so that $\chi(\mathbf{q})$ is enhanced more at low values of $\mathbf{q}$ than at high $\mathbf{q}$. The range of the RKKY interaction is increased by this selective enhancement of $\chi(\mathbf{q})$ at low $\mathbf{q}$ as is seen from Figure 6–30 taken from the work of Giovanni, Peter, and Schrieffer.[87]

The many difficulties besetting determinations of the susceptibility $\chi(\mathbf{q})$ have been discussed at length by Herring.[75] Quite aside from the difficulties inherent in any explicit calculation in terms of the energies and wave functions of the Bloch bands, there is the uncertainty even for the free-electron gas at metallic densities, whether $\chi(\mathbf{q})$ is a monotonically decreasing function of $\mathbf{q}$ (as for the non-interacting gas of Eqn (6.20)), or has a maximum near $q=2k_F$, as suggested by unscreened Hartree–Fock theory. The actual result is determined by the correlation energy between electrons (not treated in the H–F picture and much less well understood for d bands than for free electrons).

The calculation of the wave vector dependent susceptibility (even for the simpler Hartree–Fock approximation) in the case of real (non-free-electron) overlapping bands (such as the 5d bands in the rare earth metals) would be very messy. The reason lies in the fact that here one uses the Hamiltonian given in Eqn (6.2) which introduces matrix elements over the operator $V(\mathbf{r}_1 - \mathbf{r}_2)$ (cf. Eqn (6.4)) coupling the multiple bands, and this gives a strong dependence on the wave vectors $\mathbf{k}$ and $\mathbf{k}'$ (cf. Eqn (6.13)) and a dependence on the band indices as well.[75]

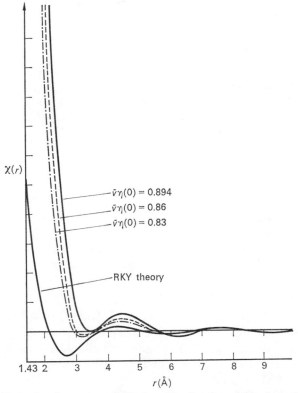

$\chi(r)$

$\bar{v}\eta(0) = 0.894$
$\bar{v}\eta(0) = 0.86$
$\bar{v}\eta(0) = 0.83$

RKY theory

1.43 2   3   4   5   6   7   8   9

$r(\text{Å})$

Figure 6–30. Effect of exchange interaction in enhancing the magnitude and increasing the range of the indirect exchange interaction (after reference 87). The quantity $\bar{v}\eta(0)$ on the figure corresponds to v as used in the text.

Let us look again briefly at the derivation of the $\mathscr{H}_{ff'}$ interaction given in Eqns (6.12) and (6.13). The more simplified expression derived for the free electron conduction band problem made the assumption (cf. discussion preceding Eqn (6.15)) that these matrix elements could be approximated instead by a dependence on only $\mathbf{q} = \mathbf{k}' - \mathbf{k}$. This permitted the separation of the expression (6.13) into two parts: one part is a purely $\mathbf{q}$ dependent term and the other a summation over $\mathbf{k}$ (cf. Eqn (6.15)) which was then identified as the non-interacting gas susceptibility.

The multi-band problem has been discussed by Yosida[88] who gave a greatly simplified formulation but called attention to the need for taking interband transitions into account in the perturbational calculations of the susceptibility. However, Yosida wrote a single equation for an averaged $\tilde\chi(\mathbf{q})$ whereas, as pointed out by Herring,[75] for a multi-band situation a set of simultaneous equations for quantities like $\chi(\mathbf{q}, \mathbf{q}+\mathbf{K})$, where $\mathbf{K}$ is a reciprocal lattice vector, would have to be solved. (In a gas (no periodic potential), $\chi(\mathbf{q}+\mathbf{K}, \mathbf{q})$ is of course diagonal and $\chi(\mathbf{q}, \mathbf{q})=\chi(\mathbf{q})$ as before.) In all calculations on the rare earths to date, some of which we discuss later, these matrix elements have been entirely neglected or severely approximated and the more simplified form (cf. Eqn (6.15)) has been used. As a consequence, all such predictions should be viewed with some caution.

We should point out the consequences of the possible non-linearity of the susceptibility. Overhauser[76] has called attention to some of the simpler effects which may occur due to the higher order effects of the interaction of conduction electrons with localized spins. As discussed by Herring,[89] "peculiar shapes of the Fermi surface may cause $\chi(\mathbf{q})$ to have a sharp peak at the $\mathbf{q}$ of the perturbation and the effective susceptibility may change considerably as soon as the antiferromagnetic gap (induced by the magnetic ordering) gets big enough to disturb the particular features of the Fermi surface responsible for the peak in $\chi(\mathbf{q})$." Peculiar things will happen in this case, including the significant excitation of certain modes, other than the modes $\pm\mathbf{q}$ which are excited to near saturation.

Finally, Kim[90] has shown that an enhanced effective exchange interaction between the conduction electrons arises when one eliminates the exchange interaction between localized and conduction electrons. This new interaction *enhances* the susceptibility over that caused by the Coulomb interaction and must be considered wherever exchange enhancement effects are expected to exist.

6.3.3.3. *The Exchange coupling $j_{sf}(k, k)$ and Induced Conduction Electron Spin Densities*

The assumption that $j_{sf}(\mathbf{k}, \mathbf{k}')=j=$constant allows analytic results to be obtained for the RKKY theory, and seemed reasonable for the nuclear coupling problem originally considered by Ruderman and Kittel (because a delta function interaction between a nucleus and the conduction electrons yields $j=$constant). This resulted in an oscillating spin density whose well-known asymptotic form $(\cos 2k_Fr/r^3)$ unfortunately goes to infinity as $r\to0$. This is due to keeping $j(\mathbf{q})=$constant as $q\to\infty$. To overcome this non-physical result, Yosida[63] assumed instead that

$$j_{sf}(\mathbf{q})=\text{const} \quad \text{for} \quad q\leqslant2k_F$$

$$j_{sf}(\mathbf{q})=0 \qquad \text{for} \quad q>2k_F \qquad\qquad (6.30)$$

and obtained a spin density which is rk_F times the Ruderman–Kittel result. This avoids the infinite value at $r=0$ but falls off as $\cos(2k_F)/r^2$ for large r.

A more reasonable approximation for $j_{sf}(\mathbf{q})$ was obtained by Overhauser[76] by assuming that the Coulomb interaction in Eqn (6.2) was strongly shielded and could be replaced by a delta function. This gives for the exchange integral

(Eqn 6.6), using plane waves for the Bloch functions,

$$J_{FF}(\mathbf{q}) = \text{const.} \times \int |\phi_L(r)|^2 \, e^{i\mathbf{q}\cdot\mathbf{r}} \, dv \qquad (6.31)$$

which is simply the form factor of the local moment density.[91]

In several investigations, Watson and Freeman calculated first[73] the behavior of $j_{sf}(\mathbf{k}, \mathbf{k}')$ when assumed to be of the approximate form $j_{sf}(\mathbf{q})$ and then[74] the $(\mathbf{k}, \mathbf{k}')$ dependence of $j_{sf}(\mathbf{k}, \mathbf{k}')$ for a simplified model. They found some important differences between the two sets of results for $j_{sf}(\mathbf{k}, \mathbf{k}')$ and for the induced conduction electron spin polarization. For their calculations, they used the illustrative case of a spherical[92] local moment which consisted of the half-filled shell of Gd^{3+} ($4f^7$) and conduction electrons which were treated as plane waves orthogonalized to both the closed shell and 4f electrons (OPW's). Analytic atomic Hartree–Fock wave functions[93] were used for the Gd^{3+} ion.

Figure 6–31 presents some of the $j_{sf}(\mathbf{q})$ results obtained for several values of k_F (0.5 and 1.0 a.u.) and a choice of magnitudes and relative orientation of $\mathbf{k}$ and $\mathbf{k}'$. The solid curves, labelled J_Q, conform to the choice that $\mathbf{k}$ and $\mathbf{k}'$ each trace the Fermi surface for $q < 2k_F$ and are antiparallel for $q > 2k_F$ in which case $\mathbf{k}'$ must leave the Fermi surface. (This choice is associated with the minimum energy denominator for given $|\mathbf{q}|$ in the perturbation theory expressions for the spin density.) It is found that $j_{sf}(\mathbf{k}, \mathbf{k}')$ can be reasonably well approximated by a suitably chosen $j_{sf}(\mathbf{q})$ for the case of Gd when $k_F = 0.5$ (but not for the case of Fe, also considered by Watson and Freeman but not discussed here).

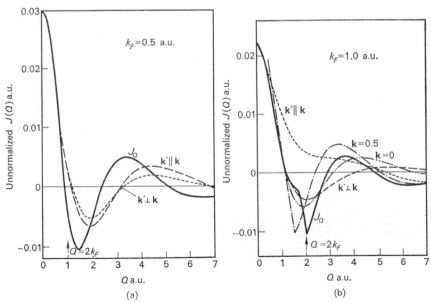

Figure 6–31. Un-normalized OPW-Gd-Gd^{3+} exchange integrals, $J(Q)$, for various choices of $\mathbf{k}$ and $\mathbf{k}'$: (a) with $k_F = 0.5$ a.u. $= |\mathbf{k}|$ and, of necessity, $|\mathbf{k}'| \geqslant k_F$; and (b) with $k_F = |\mathbf{k}| = 1.0$ a.u. (after reference 73).

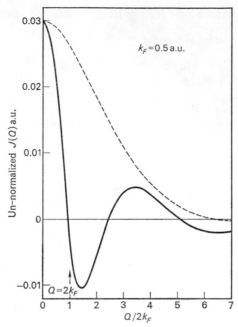

Figure 6–32. Un-normalized exchange couplings: the OPW J_Q (solid line) and J_{FF} (dashed line) for Gd^{3+} and $k_F = 0.5$ a.u. (after reference 73).

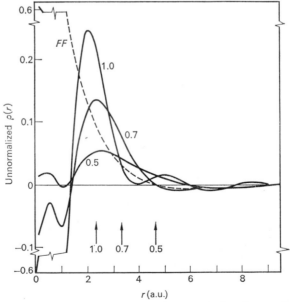

Figure 6–33. Un-normalized spin-density predictions employing the unenhanced $\chi(\mathbf{q})$ with J_Q for $k_F = 0.5$, 0.7, and 1.0 a.u. and with J_{FF} for $k_F = 0.5$ a.u. (after reference 73).

A comparison of the Overhauser form factor result and the J_Q approximation is given in Figure 6–32. Here the $J_{FF}(\mathbf{q})$ has been scaled so that the integrals match at $\mathbf{q}=0$. The large disagreement depends on the question of whether one expects substantial Coulomb shielding over distances short with respect to the dimensions of the local moment. While there is evidence for such strong shielding over distances as short as the inter-atomic lattice spacings in metals,[75] this is not likely for distances on the intra-atomic scale. Some resultant induced conduction electron spin density results obtained with both J_Q and for different k_F values and J_{FF} and perturbation theory are shown in Figure 6–33. (These results are also of interest to the interpretation of hyperfine interaction experiments such as are discussed in Chapter 8.) All these display the characteristic Friedel[67] type oscillations at large r with periods proportional to $(2k_F)^{-1}$. The outer region shown is

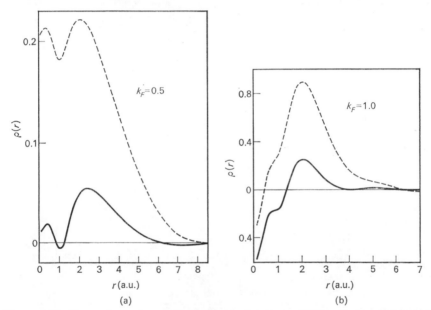

Figure 6–34. Comparison of un-normalized spin density predictions for the unenhanced (solid line) and seven-fold enhanced (dashed line) susceptibility for k_F values equal to (a) 0.5 a.u. and (b) 1.0 a.u. (after reference 73).

not the asymptotic region; all the plotted results have a positive net spin associated with positive values of the exchange integral for $q=0$. The same net spin, i.e.

$$\int \rho(r)\, r^2 \, dr,$$

is obtained for both the $k_F=0.5$ a.u. result and the form factor result. The J_Q results show an outward shift of the region where the spin density is concentrated (away from the nucleus to the region some 2 a.u. away) and indicate that the spin at the origin may be parallel or anti-parallel to the net spin, and of the order of, larger than, or smaller than $\rho(r)$ at the nearest neighbor distance. By contrast, the $\rho(r)$ result predicted with the use of J_{FF} peaks at the origin; this peaking becomes more (unrealistically) pronounced in the Yosida[63] and, of course, the Ruderman–Kittel[61] approximations.

The effects of exchange enhancement in the simple case of enhancing just $\chi(0)$ were also considered by Watson and Freeman.[73] The results are in keeping with the Giovanni et al.[87] results in that there is a raising of the curves shown in Figure 6–30 sharply at lower r values. Some of their results are presented in Figure 6–34.

The general $(\mathbf{k}, \mathbf{k}')$ dependence of $j_{sf}(\mathbf{k}, \mathbf{k}')$ was determined for the simplified OPW model by Watson and Freeman.[74] For Gd, many of the features of the results obtained for $j_{sf}(\mathbf{q})$ were found for the more exact estimates of $j_{sf}(\mathbf{k}, \mathbf{k}')$. As was found for the $j_{sf}(\mathbf{q})$ results, the range of the "main peak" of the induced spin distribution is significantly greater than that yielded by more traditional estimates of the RKKY model. Near neighbor sites are more likely than not to lie within the main peak. Exchange enhancement of the response, of course, further increases this "range". In addition, they found that

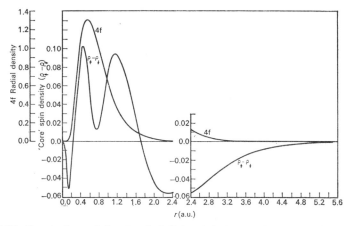

Figure 6–35. Comparison of the closed shell (core) induced spin polarization with the 4f density in the free Gd^{3+} ion. Note the change of scale to a common scale in the right side of this figure (after reference 94).

the amplitude and phase of the near oscillations are sensitive to details of the coupling. Lower amplitudes are obtained with $j_{sf}(\mathbf{k}, \mathbf{k}')$ than with the $j_{sf}(\mathbf{q})$ approximation. In their words, "this result and questions regarding the range of the main peak would suggest that the frequent usage of asymptotic density expressions in what are near neighbor regions is at best doubtful and at worst simply wrong."

These same authors[74] also emphasized the role of the core electrons and discussed the relation between hyperfine fields and neutron magnetic scattering experiments. By including the core electrons they showed that the bulk-spin density as measured by neutron magnetic scattering need not be faithfully reflected in the density distribution seen by nuclei in a hyperfine measurement.

This result suggests that spin distributions obtained by neutron diffraction and those inferred from hyperfine field measurements may differ significantly. As pointed out by these authors,[74] the presence of core effects in metals has been widely recognized in other phenomena but somehow has been almost overlooked and its implications ignored, for the s–f interaction. Future work particularly on the rare earths will have to account for these core effects as they must certainly play a role.

As an example of these effects (as shown earlier[94]), the exchange inter-action even in free ions produces a large spin polarization of the closed 5s–5p shells (which lie spatially outside the 4f shell) in the rare earths, which has a much larger range and hence a greater overlap with neighboring ion shells than does the 4f spin density itself. Note in their result for the Gd^{3+} ion, shown in Figure 6–35, that the induced large spin density at large r is negative. There are certain striking similarities between the atomic exchange polarization results seen for Gd^{3+} in Figure 6–35 and the induced conduction electron spin densities described above. The atomic results, of course, do not display Friedel oscillations; instead they show the piling up of majority spin density in the vicinity close to the local moment (the open 4f shell) which is exactly the same tendency as displayed by the conduction electrons.

Finally, as stressed by these authors, the rare earth metals are not describ-able as free electron (or simple OPW) bands. Further screening effects are likely to be important and to play a significant role in the determinations of $j_{sf}(\mathbf{k}, \mathbf{k}')$ and its interplay with the conduction electron bands themselves.

6.3.3.4. Interband Mixing (Hybridization)

A. Role of Interband Mixing on $j_{sf}(\mathbf{k}, \mathbf{k}')$

The assumption that $j_{sf}(\mathbf{k}, \mathbf{k}')$ is defined solely over the electrostatic exchange integral, as in Eqn (6.6), means that $j_{sf}(0)$ is positive and this immediately implies that the net induced moment is parallel to that of the local moment. And yet anti-parallel net spin polarizations are observed.[95] Further, to explain a variety of effects associated with the Kondo effect,[64] such as resistance minima, the phenomenological Kondo theory uses the s–d inter-action Hamiltonian but requires that $j =$ constant < 0, i.e. antiferromagnetic coupling. This means that the theory developed so far is incomplete and that there exist other exchange effects which give negative contributions to a total effective exchange coupling and that these effects can be so large as to domi-nate over the electrostatic contribution. Such an effective exchange inter-action can arise out of the mechanism of interband or covalent mixing between the conduction and local moment electrons.

While the implications of interband mixing for a variety of properties of metals were recognized in the 1930s, its significance for the problem of the mixing between the local moment and the conduction bands and the resulting exchange coupling, was emphasized relatively recently.[96] The interband mixing mechanism may be described within the s–f model in terms of absorp-tion and emission processes. These can readily occur because the theory, used earlier to develop the RKKY model, started with a non-polarized conduction band of either spin which does not represent exact eigenstates of the H–F Hamiltonian. Consider a model[97] depicted in Figure 6–36, in which the localized occupied 4f state of spin up lies below the Fermi level of the occupied free electron parabolic band representing the conduction electrons and the 4f state of spin down lies above the Fermi energy. In an absorption process a Bloch electron of wave vector **k** is absorbed into one of the local moment hole states and re-emitted as a Bloch electron with wave vector **k′**. This process tends to lower the energy of the occupied conduction electron state (consider when **k**=**k′**). Now, since all, or at least a majority, of the local moment holes available to such mixing are of spin antiparallel to that of the moment proper, this mixing stabilizes the energy of minority spin conduction electrons relative to those of spin parallel to that of the moment.

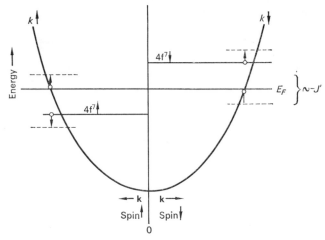

Figure 6–36. Model for interband mixing in Gd metal showing a simple unmagnetized conduction band of either spin occupied up to the Fermi energy E_F and a half-filled localized 4f shell of spin ↑ below E_F and an empty shell of spin ↓ above E_F (after reference 97).

As a result, this mixing acts like a negative exchange coupling. (The conventional electrostatic exchange interaction, now denoted as $j_0(\mathbf{k}, \mathbf{k}')$, of course stabilizes the majority spin electron energy.) Omitting constants, the absorption process has the form of an exchange coupling

$$J_{abs}(\mathbf{k}, \mathbf{k}') \sim -\frac{\langle \mathbf{k}|H|\text{loc}\rangle\langle\text{loc}|H|\mathbf{k}'\rangle}{\epsilon_{4f\downarrow} - \epsilon_{\mathbf{k}}} \qquad (6.32)$$

where $\langle|H|\rangle$ are hybridization matrix elements. Similarly, an emission term $J_{em}(\mathbf{k}, \mathbf{k}')$ arises from a process in which an electron in an occupied local moment level (with energy $\epsilon_{4f\uparrow}$) is emitted into a Bloch state **k′** and the created hole absorbs a Bloch electron **k**. Neglecting higher order effects, we may approximate the *total* exchange integral for the s–f interaction as

$$j_{tot}(\mathbf{k}, \mathbf{k}') = j_0(\mathbf{k}, \mathbf{k}') + J_{abs}(\mathbf{k}, \mathbf{k}') + J_{em}(\mathbf{k}, \mathbf{k}')$$

$$= j_0(\mathbf{k}, \mathbf{k}') + J_{inter}(\mathbf{k}, \mathbf{k}')$$

The interband contributions for the case of rare earth local moments and free-electron-like conduction bands was calculated by Watson et al.[97] Their predictions for the variation in $J_{\text{inter}}(\mathbf{k}, \mathbf{k}')$ across the row of rare earths was found to be in crude agreement with the then available experimental data. Rather good agreement was later found between these predictions for the k_F dependence of $J_{\text{inter}}(k_f, k_f)$ and experiments on Knight shifts and susceptibilities.[98] It should be recognized, however, that the exchange integrals inferred from experiment necessarily involve approximations which are as severe as those which arise in the model calculations.

Watson, Freeman, and Koide[99] have done more detailed calculations on the effect of interband mixing using the OPW model described above for the electrostatic exchange calculations. An orthogonalized plane wave was used to represent the conduction electrons and free atom Hartree–Fock functions were assumed to represent the local moment 4f and core electrons. They found that the induced conduction–electron spin density polarization has a very diffuse main peak and that the phase of the Friedel-type oscillations is shifted outwards. For the case of rare earth interband mixing, these oscillations serve to enhance those already due to electrostatic exchange. Thus, while it is often assumed that the amplitude of the s–f Friedel oscillations faithfully reflects the net induced spin, this is not the case if s–f interband mixing occurs.

Further, Watson et al.[93] emphasized that if one assumes that the perturbing potential, responsible for the matrix elements in Eqn (6.32), is spherically symmetric, the result is that the interband mixing contribution to the effective exchange interaction is given as

$$J_{\text{inter}}(\mathbf{k}, \mathbf{k}') = P_3 (\cos \Omega)\, G(\mathbf{k}, \mathbf{k}') \qquad (6.34)$$

Here Ω is the angle between $\mathbf{k}$ and $\mathbf{k}'$ and G is a function of the magnitudes of the wave vectors. This is to be compared with the form of the free-electron-band exchange integral arising from a spherical local moment.[73, 74]

$$j_0(\mathbf{k}, \mathbf{k}') = \sum_L P_L (\cos \Omega)\, F_L(\mathbf{k}, \mathbf{k}') \qquad (6.35)$$

The one L term in Eqn (6.34) and sum of L terms in Eqn (6.35) are each associated with the Lth term in the expansions of the Bloch functions $\psi_k(\mathbf{r})$ about the local moment site $\mathbf{R}$, i.e.

$$\psi_{\bar{k}}(r) = V^{-\frac{1}{2}} U_k (\mathbf{r} - \mathbf{R})\, e^{i\mathbf{k}\cdot\mathbf{R}} \qquad (6.36)$$

$$= \frac{4\pi}{V}\, e^{i\mathbf{k}\cdot\mathbf{R}} \sum_{l,\, m} (i)^l\, \Xi_l{}^k (|\mathbf{r} - \mathbf{R}|)\, Y_l{}^{-m} (\theta_k, \phi_k)\, Y_l{}^m (\theta, \phi)$$

and V is the volume of an atomic site.

The hybridization matrix elements of Eqn (6.32) deserve some discussion since this will perhaps help to explain the origin of the hybridization. Hybridization has been treated in different ways by various authors. The model dependence arises from the fact that evaluation of the hybridization matrix elements is model dependent in that a choice of basis functions ϕ_L and ψ_k is required. This matter is more subtle than the traditional question as to the choice between Bloch and Wannier functions in a solid, for it involves the

Hamiltonian used to obtain the orbitals in the ground state. Watson et al.[99] first assumed that the occupied and hole 4f states, ϕ_L, have common spatial character which presumes that they are common eigenfunctions of some Hamiltonian H_0. Secondly, they orthogonalized the conduction electron ψ_k to these local wave functions. This implies that a ψ_k is an eigenfunction which is different from ϕ_L or some linear combination of different eigenfunctions of H_0. Hybridization then arises from the difference H' between H_0 and the Hartree–Fock Hamiltonian H_{HF} for the system.

Other models which can be chosen include, for example, that of Kasuya[24] who took a set of ϕ_{loc} and simply required them to be properly orthogonalized to the occupied ϕ_{loc}. Hybridization and a negative $J'(k_F, k_F)$ then occurs to the extent that 4f admixture arises due to lack of orthogonalization to the hole ψ_{loc}. Variationally, this is incomplete since it does not allow the full admixture yielded by the Hamiltonian.

B. Hybridization Effects in Ce

The importance of interband mixing (or hybridization) has come to some prominence as a consequence of the anomalous behavior of cerium metal and its alloys.[100, 39, 57] This behavior is thought to be associated[101] with the proximity of the cerium 4f level to the Fermi level, E_F. Alloys with cerium impurities are magnetic when the 4f level is below E_F and non-magnetic when it is above[102] E_F. The closeness of ϵ_{4f} to E_F makes interband effects large in the case of Ce (cf. Eqn (6.32)). In a series of dilute alloys with rare earth impurities in lanthanum or yttrium, the alloys with cerium impurities are the only ones that show a resistivity minimum (Kondo effect) at low temperatures; all the other rare earth alloys, though magnetic, do not show a resistivity minimum and this is contrary to expectations. Further, since the s–f Hamiltonian for constant $j_{sf}(\mathbf{k}, \mathbf{k}')$ is usually written as

$$\mathscr{H}_{sf} = -2(\lambda - 1)j\,\mathbf{J}\cdot\mathbf{s} \tag{6.37}$$

(where $\mathbf{s}$ is the spin of the conduction electron) this leads to the puzzling result that since $(\lambda - 1)$ is negative there will be a Kondo effect only if j were positive, in contrast with the case of transition metal alloys.

From the above discussion, [since it follows that the interband mixing (hybridization) leads to a negative j, at least for S state ions] it is hard to reconcile this with usual presumption that the strong s–f hybridization in cerium is responsible for the Kondo effect in dilute cerium alloys.

Coqblin and Schrieffer[103] have studied the role of interband mixing on the exchange interaction in alloys with Ce impurities. Treating the interband mixing matrix elements as constants and neglecting the direct exchange they started with the Anderson[66] Hamiltonian and performed the Schrieffer–Wolff[104] transformation taking into account combined spin and orbit exchange scattering. They found a resultant interaction Hamiltonian which differs qualitatively from the conventional s–f exchange interaction. The RKKY interaction was worked out as a first approximation with other terms showing that strong anisotropic contributions to RKKY theory exist. They showed that values of the interband mixing J_{inter} and the ϵ_0, the energy of the localized state, could be determined from experiments. For different assumed values of the density of states at the Fermi energy and effective mass they listed values of J_{inter} and ϵ_0 for Y–Ce and La–Ce alloys (derived from their spin disorder resistivity) and for La–Ce alloys (derived from their superconducting temperatures).

6.4. RELATIONSHIP OF MAGNETIC ORDERING TO BAND STRUCTURE AND FERMI SURFACES

In the previous section we discussed the theory of indirect exchange in the rare earth metals, reviewed the approximations inherent in the RKKY formulation, and described extensions of this simpler theory. We showed how the magnetic ordering in the rare earth metals is thought to arise from this indirect exchange interaction between localized moments as mediated by the conduction electrons. The interaction $\mathscr{I}(\mathbf{q})$ was related to a $\mathbf{q}$-dependent susceptibility $\chi(\mathbf{q})$ which is the linear response of the conduction electron system to the effective exchange field of the localized moments. The most stable magnetic spin structure, determined by the minimum in the free energy, was seen to be determined by a maximum in $\mathscr{I}(\mathbf{q})$ when exchange is the dominant term in the free energy. This maximum in $\mathscr{I}(\mathbf{q})$ was shown, under certain approximations, to be proportional to $\chi(\mathbf{q})$ and hence the stable magnetic configuration would be determined by the maximum in the susceptibility.

It is seen from Eqn (6.15) that a large contribution to $\chi(\mathbf{q})$ results when the energy of an occupied state, $\epsilon(\mathbf{k})$ is approximately equal to that of an unoccupied state, $\epsilon(\mathbf{k}+\mathbf{q})$. Because of the Fermi distribution function, essentially all states below E_F are occupied at low temperature and all above E_F are empty. This requires $\epsilon(\mathbf{k})$ to be slightly less than E_F and $\epsilon(\mathbf{k}+\mathbf{q})$ slightly greater; hence $\mathbf{k}$ must lie just inside the Fermi surface. Similarly, $\mathbf{k}+\mathbf{q}$ must be slightly outside the Fermi surface.

Lomer,[105] Roth et al.,[82] and Kasuya,[2] among others, have discussed the relation between the geometry of the Fermi surface and the features of the susceptibility. While there are many pairs of points $\epsilon(\mathbf{k})$ and $\epsilon(\mathbf{k}+\mathbf{q})$ which satisfy the above requirements for any Fermi surface, there must be a large number of such pairs of points separated by the same $\mathbf{q}$ in order to obtain a large $\chi(\mathbf{q})$. Lomer[105] first pointed out that the existence of large nearly flat parallel areas separated by $\mathbf{q}$ was the Fermi surface feature required to produce a maximum in $\chi(\mathbf{q})$ and showed that this "nesting" feature of the Fermi surface section in chromium metal was responsible for its magnetic ordering. Kasuya[2] generated free-electron bands in one, two, and three dimensions to produce spherical, cylindrical, and cubic Fermi surfaces and showed that only the cubic surface resulted in a maximum for $\chi(\mathbf{q})$ (it became infinite for the $\mathbf{q}$ vector separating two parallel faces of the cube). Roth et al.[82] showed that the occurrence of nearly parallel pieces of the Fermi surface satisfying the "nesting" requirement with wave vector $\mathbf{Q}$ gives a logarithmic divergence in $\chi(\mathbf{q})$ at the nesting $\mathbf{Q}$. The nesting of points, or of lines, of Fermi surface was shown to give other characteristic anomalies without producing a maximum in $\chi(\mathbf{q})$ at any point related to the nesting $\mathbf{Q}$.

Observations of such nesting features of the Fermi surface of the rare earth metals were first made in the calculations reported by Dimmock and Freeman,[3] Anderson and Loucks,[10] Keeton and Loucks,[8] Watson *et al.*,[4] and more recently others, as mentioned in Section 6.2. The main thrust of these efforts has been to determine, as we have seen in Section 6.2.3, highly accurate energy bands and Fermi surfaces and to identify nesting features which can be considered to be responsible for maxima in $\chi(\mathbf{q})$ and hence responsible for the magnetic ordering observed in the different metals.

It is clear that one can give only qualitative discussions of magnetic ordering by means of Fermi surfaces alone. For quantitative descriptions one must calculate maxima in $\mathscr{J}(\mathbf{q})$, or (to proceed more approximately) to calculate instead $\chi(\mathbf{q})$ from the computed band structure and to look for maxima in this function. In recent years a considerable amount of effort has gone into both the determinations of nesting features in the Fermi surface and into the $\chi(\mathbf{q})$ calculations. We discuss both in what follows.

6.4.1. Nesting Features in the Fermi Surfaces

It is informative to discuss first the prediction of the free-electron picture, as the early body of work was done solely within this framework. The Fermi surface for Gd is shown in Figure 6.9 in the reduced zone scheme derived from the nearly-free-electron approximation.[2] The horizontal portion of the Fermi surface in the central part of the figure is assumed to be flatter for the heavier metals. The $\chi(\mathbf{q})$ determined from this Fermi surface has a maximum at $\mathbf{q} = 2k_A$ (as indicated in the figure) corresponding to the nesting of the two parallel sections.[106] This maximum in $\chi(\mathbf{q})$ has been interpreted to mean that a helical order having a wave vector $\mathbf{q} = 2k_A$ stabilizes itself by producing energy gaps at these planes (with one plane produced by one kind of spin, the other by the opposite spin). The subsequent distortion of the Fermi surface was then held responsible for the strong anomalies observed in electrical resistivity[1, 2, 107–109] and in optical absorption.[110–113]

The Fermi surfaces determined from band structure calculations were described at length in Section 6.2.3 and the important nesting features were pointed out when these results were presented. Here we describe briefly some of these features and their relation to the observed magnetic ordering.

The features of the Fermi surfaces of the heavy (hcp) rare earth metals were described in Section 6.2.3 with emphasis on the differences predicted by the non-relativistic and relativistic band structures. As discussed in Section 6.2.3, the major differences between the details of the two sets of band struc-tures lies in the bands between M and L, which gives rise to a qualitative difference in the Fermi surfaces of the different heavy hcp metals. The nesting feature proposed by Keeton and Loucks[8] as important for characterizing the magnetically ordered state is the "webbing" of the arms of the Fermi surface shown in Figure 6–13. This webbing feature, shown in detail in Figure 6–14, was found in the RAPW calculation for Lu, Er, Dy, and later for[36] Tb but not for Gd, which orders ferromagnetically. The correlation between the thickness of the webbing and the magnetic Q vector was found to be very good[8] (cf. Figure 6–14). This rather remarkable agreement with experiment thus appears to confirm the simple view that Fermi surface geometry is the dominant cause for the occurrence of periodic moment arrangements in these metals.

For dhcp Nd and Pr the principal flat pieces of Fermi surface are in the sixth and seventh zones. (The very flat pieces of Fermi surface perpendicular to the c direction which distinguish the hcp Fermi surface are absent in the case of the dhcp metals.) Fleming et al.[9] have emphasized that it is the flat region parallel to the $KMHL$ zone face (seventh zone electron surface) which will determine the magnetic ordering in dhcp Nd and Pr and which has the correct magnitude for the observed wave vector. However, there are several additional contributions which come from pieces of Fermi surface separated by other wave vectors (different in magnitude from the principal component wave vectors). More study of these metals is necessary in order to elucidate the dominant contributions of the Fermi surface to the observed magnetic ordering.

The interesting Fermi surfaces of europium metal were described earlier in Section 6.3.2. Here again the actual Fermi surface[10] is nothing like that predicted from the free-electron picture (which gives multiply connected holes centered at the points H and electrons at N). The "tetracube" and "superegg" surfaces were shown in Figure 6.24. In europium the Q vector is observed to be perpendicular to a cube face (along [001]) and has a magnitude at T_N equal to $4\pi/7a$. In the Brillouin zone, the wave vector is along the edge ΓH (or equivalently NP and $N\Delta$, where Δ is the midpoint of ΓH) and has a magnitude equal to $(\frac{2}{7})\ \Gamma H$ or $(\frac{4}{7})\ NP$. While in the antiferromagnetic hcp metals (and in Cr) the wave vector of the spin density wave can be associated with approximately parallel electron and hole portions of the respective Fermi surfaces, Andersen and Loucks found that this cannot be the case in Eu. The superegg (at H) and tetracube (at P) are separated along the [111] disection instead of along [001] as required by experiment. Instead they point to the nesting that occurs between pieces of the Fermi surface into themselves. This is illustrated in Figure 6–25 for a two-dimensional nearly square Fermi surface; for a nearly cubical Fermi surface it is possible for one face to nest into the opposite one. This could occur for the tetracube in Eu which is essentially a rounded-off cube with ellipsoids tetrahedrally positioned on four of the corners. Thus the base of this object is nearly cubical and opposing faces can nest into one another. The calculations give 0.238 a.u. for the distance between the faces of the tetracube, in good agreement with the observed wave vector (0.209 a.u. at T_N). This suggests that this nesting feature fixes the observed Q vector.

Andersen and Loucks[10] point out another interesting feature of the Eu Fermi surface. The electron surface at H, the "superegg", is itself a rounded-off cube with parallel faces separated by almost exactly twice the observed Q vector (0.388 vs. 0.418 a.u.). Thus second-order contributions will assist in maintaining the maximum at $q=Q$ whereas first-order effects from the superegg will produce a second maximum in $\chi(q)$ at $q=2Q$. Since the second order effects become increasingly important below T_N, they argue that this superegg contribution will help to stabilize the Q of the magnetic ordering as a function of temperature (but they caution that the peak at $2Q$ could become dominant at some low temperature). Experimentally the Q is found to be insensitive to temperature and no effects of the second maximum at $q = 2Q$ have yet been observed. Further experimental and theoretical work on this system is clearly necessary.

6.4.2. Calculations of $\chi(q)$ from the Electronic Band Structure

The first calculation of $\chi(q)$ for the rare earths using realistic energy bands was performed by Evenson and Liu[113, 114] along the line Γ to A. They used the energy bands for the heavy rare earth metals, Gd, Dy, Er, and Lu previously calculated by Keeton and Loucks.[8] Their first calculations were done using a generalization of (6.15)

$$\chi(q) = \frac{1}{N} \sum_{k, n, n'} \frac{f_{k, n} (1 - f_{k+q+K_0, n'})}{\epsilon_{n'} (k + q + K_0) - \epsilon_n(k)} \tag{6.38}$$

derived for all the approximations described earlier. Here n and n' are band indices and K_0 is the reciprocal lattice vector necessary to reduce $k + q$. The calculations were confined to q along the line Γ to A in the Brillouin zone because all the magnetic ordering structures observed in the heavy rare earths can be described by a wave vector in that direction. For hcp crystal structures *and* for q in the $\Gamma - A$ direction, $\chi(q)$ can be treated in the double zone representation.

Since the RAPW bands are for the non-magnetic form of the metals, the results are for the initial ordering of the metals before the bands are too greatly perturbed by the magnetic interactions.[113, 114] The temperature dependence of the Fermi functions was neglected and a mesh containing 27,216 points in the *full* Brillouin zone was used in a three-dimensional extension of the trapezoidal rule to perform the k integration. This finite number of mesh points may introduce spurious peaks in the susceptibility and as pointed out by these authors this is the principal source of noise in their calculation.[114a] They followed the procedure of comparing with the Fermi surface geometry and thereby eliminating peaks due only to the numerical methods since the Fermi surface is critical in determining the shape of $\chi(q)$.

Their results for $\chi(q)$ are shown in Figure 6–37 along with the appropriate Fermi surface cross-sections for the directions involved. Evenson and Liu[114] focus on the peaks found only for Dy, Er, and Lu around 0.6 π/c whereas the actual magnetic wave vectors are (in unit of π/c) 0.49 for Dy, 0.57 for Er, and 0.53 for Lu. The beginning of the peaks in $\chi(q)$ for Dy, Er, and Lu is suggested by Evenson and Liu to correspond quite well to the Fermi surface separation between the two arms called the "webbing" by Keeton and Loucks[8] and shown in Figures 6–13 and 6–14. The $\chi(q)$ results appear to confirm this feature as being important in determining the exact position of the peak in Lu, Er, and Dy; $\chi(q)$ in Gd where the webbing feature is absent appears to be rather flat with a large range of q's contributing. Evenson and Liu argue that inclusion of q dependent matrix elements[73, 74] in the Gd calculation would probably pull down the right side of the curve shown in Figure 6–37 and establish $q = 0$ quite firmly as the maximum of $\chi(q)$. This would, of course, bring about agreement with the ferromagnetism observed in Gd at its initial ordering point. The webbing in the heavier metals is viewed

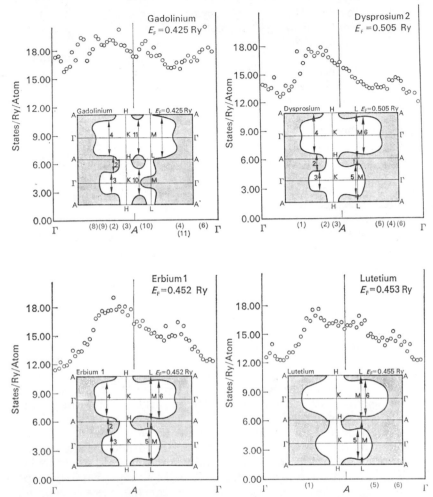

Figure 6–37. Generalized susceptibilities and Fermi-surface cross sections for some hcp rare earth metals (after references 113, 114a).

as tending to allow a cluster of $\mathbf{q}$'s in a small range around the webbing $\mathbf{q}$ to dominate the susceptibility.

Evenson and Liu[114] have also made a start at estimating the role played by the matrix elements neglected in writing Eqn (6.38). These are a generalization of $j_{sf}(\mathbf{k}, \mathbf{k}')$ defined in Section 6.3.3 and have the form $I_{nn'}^{2}(\mathbf{k}, \mathbf{k}+\mathbf{q}+\mathbf{K}_0)$, where n, n' are band indices. Assuming that they are functions only of $\mathbf{q}, n,$ and n' they may be factored out completely. The generalized form of (6.18) gives

$$\mathcal{J}(\mathbf{q}) \sim I^2(\mathbf{q}) \, \chi(\mathbf{q})$$

one looks for maxima in $\mathcal{J}(\mathbf{q})$. Writing simple Gaussian expressions for $I^2(\mathbf{q})$ allowed them to evaluate the $\mathcal{J}(\mathbf{q})$ expression again using the $\chi(\mathbf{q})$

results found earlier. These slowly varying matrix elements are found to shift the maxima found for $\chi(\mathbf{q})$ to smaller q values and to smoothe out the curves shown in Figure 6–37.

The susceptibility function for Gd, Tb, and Dy in the $\Gamma A\Gamma$ direction was calculated as a function of pressure by Fleming and Liu[56] using the band structures described in Section 6.2.3. The peaks of $\chi(\mathbf{q})$ for Tb and Dy shift to smaller q values when the pressure is applied and the sizes of the peaks are reduced. The small peak in the susceptibility function for Gd is argued away as probably being wiped out by a q-dependent matrix element. In view of all the uncertainties in the calculation, Fleming and Liu conclude that the agreement with the experimental values must be termed highly satisfactory. The good results are taken as further support that the initial magnetic ordering in these metals is mainly determined by the energy band structure through the indirect exchange mechanism.

A similar calculation of $\chi(\mathbf{q})$ for dhcp Nd and Pr was carried out by Fleming, Liu, and Loucks[9] using the expression given in Eqn (6.38) only with a structure factor $\sum_\nu e^{-i\kappa\cdot\rho_\nu}$ included to account for the necessary sum over the various atoms of the unit cell. (This structure factor serves to put a restriction on the bands between which transitions can occur.) The summation over bands was restricted to those four bands which determine the Fermi sufrace. As in the Evenson and Liu[113, 114] calculations the matrix elements were approximated by a Gaussian. The susceptibility was calculated for wave vectors lying in the ΓALM plane and found to have sharp peaks for both elements near $q_x=0.125b_1$ and $q_z=0$. In addition a wide flat peak was found extending out from near the point A of the first branch of the $\chi(\mathbf{q})$ curve. This peak has a maximum value near A for Pr and at $q_x=0.075b_1$ and $q_z=0.58b_3$ for Nd. Fleming et al.[9] ascribe the stable magnetic structure to the sharp peak. The periodicity of the moment in the basal plane indicated by this maximum agrees closely with the observed experimental ordering determined by neutron diffraction[115, 116] to be $0.13b_1$ for Nd and Pr. Their results, however, indicate that the moments in adjacent hexagonal layers should be aligned to each other in contrast to the antiparallel arrangement in the experimental model.[117] As they point out, it thus appears that there is some additional interaction which must be included in order to explain the observed[117] c-axis periodicity.[114a]

Myron and Liu[37] have calculated a $\chi(\mathbf{q})$ for fcc Pr and compared it with that for dhcp Pr. Whereas for dhcp Pr there is a peak at $q=0.15b_1$, there are two peaks located at Γ and X for the fcc structure. Since the peak at X corresponds to a sizeable wave vector, they argue that it may be strongly suppressed by a wave-vector dependent matrix element[73, 74, 114] (cf. discussion in Section 6.3.3) in which case the peak at Γ will dominate and the ordering of the moments in fcc Pr will be ferromagnetic. If this is the case then this would explain the difference in magnetic ordering properties of the two phases of Pr and Nd.

Using the non-relativistic energy bands for tungsten calculated by Matthiess,[45] Evenson, Fleming and Liu[118] have calculated $\chi(q)$ along ΓH for europium metal. Two peaks were found at the diameters of the two Fermi surfaces tetracube and superegg found by adjusting the Fermi energy to match the RAPW results of Andersen and Loucks[10] described earlier. As in the earlier cases cited above, it is argued that the peak at the smaller wave vector (tetracube) will dominate because of the effect of the wave-vector dependent matrix elements. This would lend support to the idea that the nesting between the opposite faces of the tetracube is responsible for the helical structure of Eu. (The superegg nesting anomaly, it is pointed out, should best be observed in the phonon spectrum of europium metal.)

The agreement with experiment obtained in these first calculations is encouraging although they are still somewhat crude. The sensitivity of the $\chi(\mathbf{q})$ results to different starting potentials has not yet been investigated, nor has the actual role of the neglected matrix elements. In this regard, it should be emphasized that the matrix elements reflect information contained in the electronic wave functions and differ significantly from the free-electron

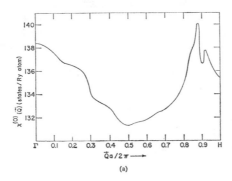

(a)

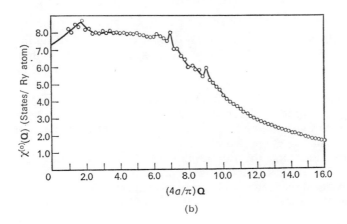

(b)

Figure 6–38. The unenhanced wave vector dependent susceptibility for chromium metal calculated (a) by neglecting the matrix elements, and (b) including the matrix elements for $T=0$ using a mesh of 128,000 points in the Brillouin zone (after reference 119).

approximation (as is expected from the emphasis given to the strong deviations of the energy from the free-electron limit). The inconsistency of using the energy bands from the APW calculations but not the wave functions is obvious. It will also greatly affect the final results, as has been found by Gupta and Sinha [119] for the classic case of paramagnetic chromium. They found that the inclusion of matrix elements (both interband and intraband) made a dramatic difference in the susceptibility: the peak in the calculated susceptibility due to the "nesting" feature of the Fermi surface is reduced by the inclusion of matrix elements so that it is no longer an actual maximum of $\chi(\mathbf{q})$. They conclude that the Fermi surface nesting is not enough to cause the spin density wave to choose the observed wave vector, if allowance is made for the matrix elements and stress the importance of the local field corrections in greatly increasing the exchange enhancement and in determining the spin density wave. The striking difference between these sets of calculations is shown in Figure 6–38 which is included here as a dramatic cautionary note on accepting results which ignore matrix elements.

Recently, much attention has been given to the problem of numerical accuracy in the calculation of Brillouin zone integrations such as $\chi(\mathbf{q})$. The possible errors introduced by a limited number of integration mesh points was recognized by Evenson and Liu [114] and described in great detail by Gilat and Raubenheimer [120] and by Mueller et al. [121] These authors speak of the need for using 10^6 independent points in the zone. More recently the problem of accurate $\chi(\mathbf{q})$ calculations was discussed by a number of authors. [122] These possible errors are not to be minimized, particularly when coupled with the problem of the inclusion of matrix elements into the generalized susceptibility calculations.

6.5. EFFECTS OF MAGNETIC ORDERING ON BAND STRUCTURE, FERMI SURFACE AND OBSERVABLE PROPERTIES

We have thus far considered the band structures, Fermi surfaces and other properties in the absence of magnetic ordering, and described the theory of indirect exchange interactions and how the magnetic ordering is thought to result from this indirect coupling between localized moments as mediated by the conduction electrons. Those geometrical features of the Fermi surface which give rise to nesting of sections of the Fermi surface were related to the onset of magnetic order and descriptions of their contributions to maxima in the wave vector susceptibility, $\chi(\mathbf{q})$, and the resultant lowering of the free energy of the system, were presented.

In order to understand the various interesting magnetic-order-dependent phenomena, such as optical and anisotropic resistance anomalies, we must examine the metal in the magnetically ordered state. In this section we examine the nature of the magnetic perturbation of the band structure and Fermi surface by the periodic magnetic ordering and consider the relationship between these perturbations, the 4f ordering itself, and the observed anomalies. The magnetic perturbation of the 5d conduction bands by the localized ordered 4f moments will be shown to be quite large; they introduce gaps in the energy bands largely at or near the Fermi surface sections normal to the hexagonal axis, thereby destroying or perturbing large segments of that surface. The energy gaps themselves are quite large, especially in view of the flat d bands involved, and approach the important bandwidths.

6.5.1. Magnetic Ordering and Superzones

The magnetic ordering observed for the rare earth metals described in Chapters 1, 3, and 4 shows some exotic structures. Consider the two basic types represented by Tm and Ho, as examples. The spin ordering is given by the average spin values of the types:

$$\text{Tm type:} \quad \langle S_n^z \rangle = SM \cos (\mathbf{Q} \cdot \mathbf{R}_n + \phi) \tag{6.39}$$

where $\mathbf{Q}$ is parallel to the hexagonal c axis (taken to be the z axis) so that (6.39) gives a longitudinal wave.

$$\text{Ho type:} \quad \langle S_n^x \rangle = SM' \cos \mathbf{Q} \cdot \mathbf{R}_n$$
$$\langle S_n^y \rangle = SM' \sin \mathbf{Q} \cdot \mathbf{R}_n \tag{6.40}$$

which gives a helical structure, with the moments in the hexagonal planes varying direction but not magnitude. Here M and M' where $0 < M, M' < 1$, are temperature dependent factors associated with the sublattice magnetization. Both these structures may, of course, occur simultaneously, higher harmonics may be present, and finally $\mathbf{Q}$ may be zero in which case the magnetization has a ferromagnetic component.

When a coupling of the s–f type exists (cf. Eqn (6.5)), the effect of these magnetic structures is to perturb the conduction bands. For an ordering like Eqns (6.39) and (6.40), there is now added to the crystallographic translational symmetry a new magnetic translational symmetry which comes from the exchange field acting on the conduction electrons. This magnetic periodicity modifies the conduction band structure and affects various properties.

In the simplest case of the occurrence of a ferromagnetic array of local moments an exchange splitting of the conduction bands is induced at $\tau = 0$ given by

$$\Delta_{\text{ferro}} = 2SM j_{sf}(0) = 2M(\lambda - 1) J j_{sf}(0) \tag{6.41}$$

where $S[=(\lambda - 1)J]$ is the expectation value of the local spin moment and $j_{sf}(0)$ is the diagonal ($\mathbf{k} = \mathbf{k}'$) matrix element. The assumption of a j_{sf} which is a function of only $|\mathbf{k} - \mathbf{k}'|$ implies a simple rigid, $\mathbf{k}$-independent, exchange splitting of spin parallel and spin antiparallel bands when the 4f moments are ordered ferromagnetically.

The periodic structures described by the orderings expressed in relations (6.39) and (6.40) result in much more interesting cases in that the periodic magnetic ordering causes new energy gaps to appear in the conduction band structure rather than the rigid splitting just seen for the ferromagnetic case. Inserting either of the spiral structures into Eqn (6.5) yields off-diagonal exchange matrix elements between Bloch states for $\mathbf{k} - \mathbf{k}' = \pm \mathbf{Q} + \tau$, τ being any reciprocal-lattice vector. These matrix elements arise between states of like spin for the Tm spiral and between states $\mathbf{k}$ of spin down and $\mathbf{k} + \mathbf{Q} + \tau$

of spin up for the Ho helix. The new perturbed energies for wave vectors
near degenerate bands separated by $\tau + Q$ may be determined by diagonalizing
a 4×4 matrix including the unperturbed states $\epsilon_k{}^0$ and $\epsilon_{k+\tau+Q}{}^0$ of *both*
spins.[108] Band gaps occur when $Q + \tau$ connects a degenerate pair of band
states. These gaps are

$$\Delta_{\text{Tm}} = M S j_{sf}(Q + \tau) \mid F(\tau) \mid \qquad (6.42)$$

$$\Delta_{\text{Ho}} = 2M'S j_{sf}(Q + \tau) \mid F(\tau) \mid \qquad (6.43)$$

where $F(\tau)$ is the structure factor

$$F(\tau) = (1/n) \sum_{i=1}^{n} \exp{(i R_i \cdot \tau)} \qquad (6.44)$$

summed over n atomic sites in a unit cell in real space. The presence of $F(\tau)$
causes the mixing and associated gaps to vanish for certain values of τ.
These gaps are known as the magnetic superzone energy gaps[107, 108, 110]
and give rise to the resistivity anomalies observed for the rare earth metals
below the ordering temperature. Miwa[110] had proposed that these energy
gaps would give rise to optical (absorption and reflectivity) anomalies as
well. We discuss these superzone effects later on in this section and shall
distinguish between the cases when the magnetic symmetry is commensurate
and incommensurate with the lattice periodicity.

6.5.2. Effect on Band Structure and Fermi Surfaces

The calculation of the band structures of the rare earths in their magnetically ordered phases is still a very complex problem which remains to be
solved. If the magnetic symmetry is commensurate with the lattice symmetry
the Hamiltonian can be diagonalized explicitly and the exchange perturbed
bands may be determined. Such an approximate calculation was carried out
for Tm by Watson, Freeman, and Dimmock[27] using their calculated non-
relativistic APW energy bands. We shall discuss their work in detail here as
it represents the only such calculation performed to date. (For an incommensurate structure, perturbation theory must be used, in which case the state k
mixes with $k + \tau + nQ$ in nth order.)

In order to estimate the effect on the bands of the spiral perturbations, one
requires a knowledge of the exchange coupling $j_{sf}(Q)$. The value $j_{sf}(0)$ is
readily estimated utilizing the conduction–electron moment $\Delta\mu$, observed
in ferromagnetic Gd. Assuming no conduction-electron–conduction-electron
enhancement of the response, $\Delta\mu$ is given by[27]

$$\Delta\mu = \tfrac{1}{4} g \mu_B N(E_F) \Delta_{\text{ferro}} \qquad (6.45)$$

For Gd the total moment is observed to be 7.55 μ_B per atom, of which
0.55 μ_B is considered to be due to conduction-electron polarization. Taking
the calculated value[3, 5] of $N(E_F) = 1.8$ electrons per atom per eV, Eqns
(6.41) and (6.45) yield $j_{sf}(0) \sim 0.087$ eV, assuming $g = 2$ for the conduction
electrons. This value of $j_{sf}(0)$ appears to agree well with estimates based on
both plane wave and orthogonalized plane wave (OPW) approximations.[3, 6]

There is one simple but important consequence of the relation Eqn (6.45) in that it shows clearly the origin of the extra magnetization observed in the rare earth metals. The key is the high density of states at the Fermi energy arising from the flat 5d bands. This high density of states also accounts for the high electronic specific heats observed for the rare earths.

Given $j_{sf}(0)$ for Gd and assuming that this has the same value for the heavier rare earths Δ_{ferro} becomes proportional to $S[=(\lambda-1)J]$. The values obtained for the heavy rare earths are listed in Table 6.1.[5] The moderate variation in band structure and 4f-shell character, with varying nuclear charge Z, suggest this to be a reasonable first approximation. (We shall discuss later the remarkable agreement of these values of Δ_{ferro} with the energies of the optical anomalies observed for Ho and Dy.)

Table 6.1. Observed Spiral Δ's Compared with the Δ_{ferro} for the Heavy Rare Earth Metals as Predicted by Eqns (6.41) and (6.45) Employing the Observed $\Delta\mu$ for Gd (after Reference 27)

	S	Δ_{ferro} (eV)	Δ
Gd	7/2	0.61	
Tb	3	0.52	
Dy	5/2	0.44	0.44
Ho	2	0.35	0.35
Er	3/2	0.26	
Tm	1	0.17	

6.5.2.1. Exchange Perturbed Bands in Tm

The energy bands and Fermi surface obtained from the non-relativistic APW calculations were described earlier in Section 6.2.3. Watson *et al.*[27] use these results rather than the RAPW results because they differ only in the few details discussed earlier. Some of their Fermi surface cross-sections

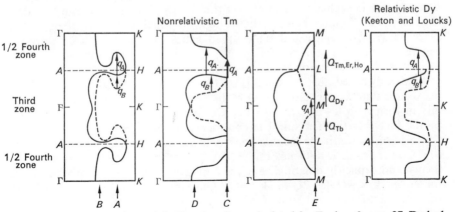

Figure 6–39. Cross-sections of the Fermi surface calculated for Tm in reference 27. Dashed curves show the fourth-zone Fermi surface folded into the third zone. The q_A and q_B vectors indicate typical Fermi-surface calipers.

are shown in Figure 6–39, and energy vs. k plots are plotted in Figure 6–40 for the c-axis lines A through E indicated in Figure 6–39.

The bands intersecting the Fermi surface lie, in the extended zone scheme, in the 3rd and 4th Brillouin zones and $\tau=0$ couplings may occur within a band in a single zone and between bands across the (001) boundary between zones. A number of such inter- and intraband Q vectors, designated as q_A, are shown in Figure 6–39. The case of $\tau=(001)$ is also of interest since it includes interband terms such as the q_B of Figure 6–39. The structure factor is zero for $\tau=(001)$, indicating that this mixing is zero in first approximation. Some mixing and associated gaps will occur because spin–orbit coupling,

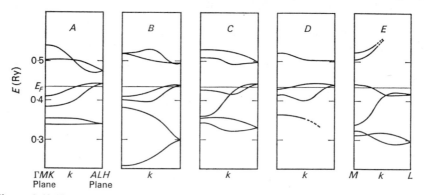

Figure 6–40. Energy versus k, for nonmagnetic Tm, plotted for the c-axis lines A through E indicated in Figure 6–39. For all cross-sections but D, the lower bands are displayed. Similarly, nearest lying higher bands are shown, except in E (after reference 27).

which was neglected in the APW calculations, will mix the bands and render Eqn (6.44) invalid. Watson $et\ al.$ nevertheless expect such q_B gaps to be smaller than the $\tau=0$, q_A gaps. Spin–orbit mixing will also cause initially zero-valued Ho spiral-matrix elements to take on small nonzero values. For bands to mix in such a way as to lower the energy of the system in first order, their slopes at E_F must be opposite in sign, in which case the mixing takes place between occupied and unoccupied states. For the bands shown in Figure 6–40 the q_A vectors of Figure 6–39 meet this criterion whereas the q_B vectors do not. Watson $et\ al.$ have neglected the q_B mixing.

As already noted, the Tm spiral (Eqn (6.39)) introduces exchange coupling between a state $\mathbf{k}$ and states $\mathbf{k}\pm\mathbf{Q}$ of like spin. Of course a state $\mathbf{k}+\mathbf{Q}$ will mix in turn with $\mathbf{k}+2\mathbf{Q}$, and so on. Computational matters are simplified by the fact that the Tm spiral is commensurate with the lattice, with a Q which is one-seventh the double-zone c-axis dimension (see $\mathbf{Q}_{Tm}$ in Figure 6–39). Then, since $\mathbf{k}+7\mathbf{Q}$ has returned to $\mathbf{k}$, we may abandon perturbation theory and explicitly diagonalize a 7×7 matrix for a given choice of $\mathbf{k}$, i.e. for states $\mathbf{k}, \mathbf{k}+\mathbf{Q}, \mathbf{k}+2\mathbf{Q}\ldots\mathbf{k}+6\mathbf{Q}$. Once one has the energies of the states involved, obtained from a nonmagnetized metal band calculation, and has assumed a value for the exchange constant $j_{sf}(\mathbf{Q})$, the process is trivial to carry out. Inspection of the resulting eigenvectors normally shows that one $\mathbf{k}$ component, of one band, dominates. Plotting the energies of the perturbed bands as a function of these dominant **k** components, Watson *et al.* obtain the results shown in Figure 6–41. These results should be considered to be only qualitative because of uncertainties primarily in the nonmagnetic bands.

Their results show the appearance of energy gaps of up to 0.085 eV at, or close to, all but one intersection of the bands with the Fermi energy and this band, in *D*, would have such a gap if $\tau = (001)$ mixing had been included. Watson *et al.* have not calculated the shift in E_F due to the band perturbations, but expect the shift to be small compared to the uncertainties in placement of the nonmagnetic bands with respect to one another and with the placement of the nonmagnetic E_F.

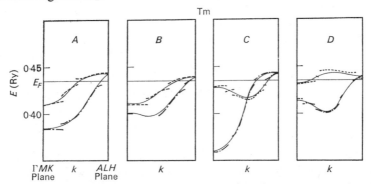

Figure 6–41. The exchange perturbation induced by the Tm 4f spiral in conduction-electron bands *A* through *D* of Figure 6–40 (after reference 27).

Even in the presence of the magnetic perturbation, one will still have a volume of hole states in **k** space much like that plotted in Figure 6–9 but the volume is now bounded by energy gaps, as well as by Fermi-surface segments. The gaps will largely replace Fermi-surface sections normal to the *c* axis since the Fermi surface associated with the flat d bands is most severely perturbed while the "s–p" band Fermi surface of the hollow trunk, involving steeper bands, is rather less perturbed. The effect of this is to produce resistance anomalies parallel, but not perpendicular, to the *c*-axis, as are seen experimentally. Watson *et al.* also emphasize that details, such as whether the gaps do or do not hit E_F in particular band section *A*, *B*, or *C*, are within the uncertainty in position of bands with respect to one another and to E_F. However, the correlation between spiral length and Fermi-surface calipers is not accidental and such gaps will occur at or near E_F.

6.5.2.2. Exchange Perturbed Bands for Ho Metal

We have seen that the Ho helix (Eqn (6.40)) exchange couples Bloch states of wave vector **k** and spin ↓ only with states **k**+**Q** and spin ↑. Inclusion of spin–orbit coupling, however, mixes the spin character, causing the states of wave vector **k** to be coupled with both states **k**+**Q** and this in turn with **k**+2**Q**, etc. This leads to chains of coupled states of the type seen for Tm. These are, however, more weakly coupled, since this process depends on the presence of both spin components in a given Bloch state. In the absence of spin–orbit coupling and q_B mixing, the problem reduces to solving a series of 2×2 secular equations resulting in states of predominantly one spin (and $\mathbf{k}$ value) or the other (and $\mathbf{k} + \mathbf{Q}$).

Watson et al.[27] have obtained illustrative results by perturbing the non-magnetic bands of Figure 6–40 for Tm with a helix of the same Q as was used in the calculations for Tm and a $\mathscr{J}(Q) = \mathscr{J}(0)$ for gadolinium.

The band gaps of 0.34 eV match the observed optical anomaly in Ho and represent a large perturbation of the bands. However, Ho has the smallest spin and hence the smallest gap, among the rare earths which order in this spiral. On going from Ho to Tb, the gap increases by 50%. Because of the increase in gap size over that for Tm, a significant number of Fermi energy intersections will be severed and the c-axis resistivity is again appreciably raised. The increasing size of the gaps implies (1) the destruction of large sections of the Fermi surface and (2) that Bloch states, lying increasingly far from the nonmagnetic Fermi surface, take an active role in the magnetic ordering.

6.5.2.3. Temperature Dependence of the Magnetic Ordering Wave Vector

One of the more interesting properties of the rare earth metals is the observed temperature dependence of the wave vector, $\mathbf{Q}$, which describes the magnetic ordering. Experimental results have been discussed earlier in Chapter 3. The different directions of magnetic order favored in the different metals are the result of Fermi surface effects combined with the large aniso-tropy associated with the crystalline electric field. Various attempts have been made to explain the temperature dependence[1, 2] seen in the metals and their alloys following the early work of Cooper,[123] De Gennes and Saint James,[124] Elliott and Wedgwood,[108] and Miwa.[125] De Gennes and Saint James showed that the variation of $\mathbf{q}$ at the Néel temperature as a function of composition may be largely explained by the effect of spin-disorder scattering[109] which modifies the conduction electron polarization and hence the RKKY exchange. This theory, however, is not applicable to the low temperature variation of $\mathbf{q}$ between the metals where there is no spin disorder scattering. Elliott and Wedgwood[108] emphasized the role of the superzone boundaries produced by the magnetic order and considered the distortions in the Fermi surface and rearrangement of the conduction electrons across the induced energy gaps. They calculated the total energy of the system for finite order and minimized with respect to $\mathbf{q}$. For small order this was shown to lead to the usual RKKY result, but as the order increases the value of $\mathbf{q}$ at the minimum decreases. Later Miwa[125] considered both the effects of long range order and the spin disorder scattering on the variation of $\mathbf{q}$ with temperature. Kaplan[126] also considered the distortion of the Fermi surface arising from the new zone boundaries but presented no detailed calculations. Kasuya[2] has stressed the possible variation of the band structure in the different metals.

We have seen above, especially from the work of Watson et al.,[27] that the energy gaps, introduced by the magnetic order are very large (comparable to some of the band widths involved), introduce strong distortions of the energy bands, and destroy large segments of the nonmagnetic Fermi surface. These effects are very severe and recall the earlier discussion (cf. Section

6.3.3) about the possible importance of nonlinear effects on $\chi(\mathbf{q})$ and the RKKY interaction. For the rare earth metals we have precisely the situation discussed by Herring[89] and quoted earlier. In an earlier section, the peculiar shape of the Fermi surfaces was shown to give rise to a sharp peak in $\chi(\mathbf{q})$ at the $\mathbf{Q}$ of the perturbation with the onset of magnetic ordering. As the temperature is lowered the increase in the antiferromagnetic energy gap gets big enough to destroy in large measure those particular features of the Fermi surface which are responsible for the peak in $\chi(\mathbf{q})$ and will modify $\chi(\mathbf{q})$ and the magnetic ordering. These nonlinear effects are much larger for the actual band structures of the rare earth metals than the effects arising in the case of the free-electron bands considered by Elliott and Wedgwood and by Miwa. The nonlinear effects of the actual magnetically distorted bands and Fermi surface on $\chi(\mathbf{q})$ and the variation of the resultant $\mathbf{Q}$ with temperature have not been calculated as yet.

An interesting different approach has been presented recently by Arai.[127] As in the case of the Kondo effect,[68] which arises from certain spin flip mechanisms suppressed in the case of magnetic ordering, the presence of an antiferromagnetic or spiral ordering of localized spins (and hence the presence of a molecular exchange field) also introduces a spin flip mechanism which introduces a logarithmic divergence in the energy spectrum. This yields the result that the Fourier transform of the effective exchange coupling constant between spins, $\mathscr{J}(\mathbf{q})$, is temperature dependent. As described above,[108] the magnetic ordering creates spin dependent superzone boundaries which split the degeneracy of the zone boundary at P for an electron with $\mathbf{k}$ and spin up and a spin down electron at zone boundary P' with $\mathbf{k}+\mathbf{Q}+\boldsymbol{\tau}$. This corresponds to a spin flip process, namely, whenever an electron $(\mathbf{k},\uparrow)$ near the boundary P collides with a localized spin, the electron is transformed into the state $(\mathbf{k}'+\mathbf{Q}+\boldsymbol{\tau},\downarrow)$ by flipping its spin and, of course, at the same time, the localized spin reverses its direction by emitting a spin wave $(-\mathbf{Q},\uparrow)$ with zero energy. The electron $(\mathbf{k}'+\mathbf{Q}+\boldsymbol{\tau},\downarrow)$ is transformed by collision with a localized spin to the final state $(\mathbf{k},\uparrow)$ by absorbing the spin wave $(-\mathbf{Q},\uparrow)$. Here a spin wave has appeared in place of a localized spin since the localized spins form a lattice. Arai calculates the energy of the spin waves including the molecular field contribution as a part of the electronic energy since these two energies are inseparable in the s–d Hamiltonian. The Elliott and Wedgwood[108] interaction is reproduced when $\boldsymbol{\tau}\neq0$ and other terms appear for $\boldsymbol{\tau}=0$. The Kondo type logarithmic temperature dependence of the energy spectrum coming from the third order terms and the temperature dependence of $\mathscr{J}(\mathbf{q})$ which results are discussed qualitatively and related to the magnon dispersion curve of Møller et al.[128] However detailed calculations, which include the band structure and Fermi surface effects discussed above, have not been carried out as yet. What is required is a full theory which combines the Arai approach with the theoretical description discussed above, before the relative importance of the various contributions can be assessed.

6.5.3. Contribution to Resistivity Anomalies

We have examined the effect of the magnetic ordering on the energy bands and Fermi surfaces of the nonmagnetic electronic structure and have indicated how the striking resistivity anomalies observed for the rare earth metals (treated in detail in Chapter 7) have their origin in the magnetic superzone gaps and the accompanying destruction of large segments of the Fermi surface previously available for conduction. Their effect on the electrical conductivity may be calculated using the general expression for the conductivity tensor.[1, 2, 129] This has, however, only been done in the case of free-electron conduction bands.[1, 2, 108] A review of the earlier experimental and theoretical work was given by Kasuya[2] where the conduction electrons were treated in the free (or nearly free) electron approximation (which, we have seen, is a rather poor approximation for the rare earths). A discussion of the inadequacy of the free-electron theory for the problem of resistivity anomalies was presented by Elliott.[1] Detailed predictions based on this earlier work are given in Chapter 7 along with a variety of transport phenomena data. It is rather remarkable how well the free-electron model "works" in reproducing qualitatively many of the features of the experimental results. This confirms the view that the essential physics underlying the occurrence of the resistivity anomalies is the role of superzone energy gaps and the destruction of large segments of the Fermi surface. While no detailed estimates of these anomalies have been made using the actual band structure and Fermi surfaces of the rare earth metals it is expected that such realistic (and difficult) calculations would yield good agreement with experiment. Here we confine ourselves to a qualitative description of the change in conductivity brought about by the onset of magnetic order.

The simple theory assumes that a relaxation time may be defined for the conduction electrons and hence that the conductivity tensor may be written as

$$\sigma_{ij} = \frac{e^2 \tau}{4\pi^2 \hbar} \int v_i dS_j \qquad (6.46)$$

Here τ is the relaxation time, v_i is the group velocity in the ith direction and the integral is taken over the Fermi surface. We have mentioned in our first discussion of the s–d Hamiltonian that the simplest consequence of the s–d interaction is the effect on the scattering cross-section of the conduction electrons. This interaction enters into the relaxation time τ generally calculated from the spin disorder scattering which is considered to be the dominant scattering mechanism for resistance in dilute rare earth alloys and in the paramagnetic region of concentrated alloys.[1, 2] Assuming that $j_{sf}(\mathbf{k}, \mathbf{k}') = j$ is a constant, the relaxation time of the conduction electrons is given (in first Born approximation) by

$$1/\tau = (8\pi/\hbar) j^2 \langle S^2 \rangle N(E_\mathrm{F}) \qquad (6.47)$$

where $N(E_\mathrm{F})$ is the density of states. The use of a single relaxation time for all the Fermi surface electrons is clearly a crude approximation not justified by the highly anisotropic Fermi surfaces found for the metals and the varying character of the wave functions (on the "arms" versus the "trunk"). In any accurate calculation of the conductivity, relaxation of this constraint will undoubtedly be required.

It is apparent from Eqn (6.46) that the conductivity is sensitive to the Fermi surface area available to the conduction electrons. An immediate

consequence of the strong anisotropy of the Fermi surface is the large anisotropy expected (and observed) for the resistivity of the rare earth metals.
The ratio of the projected area in the c direction (cf. Figure 6–9) is about
twice that in a basal plane direction, in qualitative agreement with experiment.
(So far, detailed estimates of this anisotropy ratio using the calculated Fermi
surfaces have not been made.)

The role of magnetic superzone gaps in giving rise to the anomalous
behavior of the resistivity observed below the Neel temperature is also
apparent. As discussed, the superzone gaps destroy Fermi surface area
along the c direction but not in the basal plane. This greatly reduces the
conductivity in the c-direction (and increases it somewhat in the basal plane),
thereby explaining qualitatively the experimental observations (cf. Chapter 7)
including the striking behavior of the effect of an external magnetic field
on the conductivity of holmium metal.[130] A magnetic field applied along the
b-direction of holmium causes the magnetic structure to change from a cone
(with superzone gaps) to a ferromagnetic structure (without superzone gaps).
In the temperature range where the superzone effects are the only important
ones, the applied field produces about a 30% increase in the conductivity in
the c-direction (and a decrease of only about 1% in the basal plane).

As stated earlier, estimates of the anomalous resistivity based on Eqn
(6.46) and free-electron conduction bands generally yield good qualitative
agreement with experiment. (One of the difficulties of such calculations is
that they require very large values of the exchange integral.) This oversimpli-
fied model of the conduction bands must be abandoned in favor of calculations using the actual band structures of the magnetically ordered state, but
these latter calculations are exceedingly complex to perform. From the
calculations of Watson et al.[27] discussed above, we have seen that the
magnetic ordering induces strong mixing of the states both near to and quite
removed from the Fermi energy, introduces large energy gaps and annihilates
large regions of the surface normal to the c direction. (This is also consistent
with the positron annihilation results, cf. Section 6.2.4, which in the helical
phase of holmium[60] shows a large reduction in the structure observed.)

The only direct estimate of the destruction of the Fermi surface in a rare
earth by the magnetic superzone gaps was made by Freeman et al.[4] using
their non-relativistic APW calculations for thulium metal. Unlike many of
the other rare earth metals which have a period of the wave vector which is
incommensurate with the lattice and varies with temperature, the period of
the longitudinal wave in Tm is constant (between 50 and 32 K) at seven
hexagonal layers.[130a] Freeman et al. emphasize that the usual first order perturbation treatment (band gaps and superzone boundaries at those $\mathbf{k}$ for which
$E(\mathbf{k}) - E(\mathbf{k} \pm \mathbf{q}) = 0$ when the system is subjected to a perturbation of the
wave-vector, $\mathbf{q}$) is only applicable in the vicinity of the ordering temperature
where the magnetic moment is small. However, at lower temperatures, if the
periodic perturbation is large, compared to the separation between energy
bands, a higher order treatment must be considered. In Tm at low temperature, the energy gaps introduced at the superzone boundaries should be
about 0.17 eV wide (cf. Table 6.1). This energy is comparable to the separation between the flat d bands in the vicinity of the Fermi surface which were

obtained for the rare earth metals. So-called higher order terms in perturbation theory will introduce energy gaps of magnitude comparable to those at $k_z = \pm \frac{1}{2}q$. Therefore, because of the fact that these bands are relatively flat and closely spaced the new condition is

$$E(\mathbf{k}) - E(\mathbf{k} \pm n\mathbf{q}) = 0. \qquad (6.48)$$

This, then, leads to superzone boundaries at $k_z = \pm \frac{1}{2}nq$.

The complete hole surface for thulium metal determined from the APW calculations is similar to that shown for Gd in Figure 6–9 in the double zone representation which is valid for the hcp structure in the absence of spin–orbit coupling. Figure 6–42 shows several vertical cross sections of the Fermi surface. [4] The horizontal lines denote superzone boundaries at $k_z = \pm n(2\pi/7c)$ introduced by the periodic magnetic ordering. The expected distortion of the

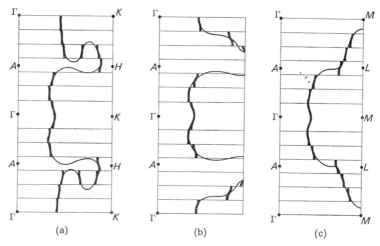

Figure 6–42. Some vertical cross-sections of the Tm Fermi surface containing the c axis. The influence of magnetic ordering is demonstrated by comparing the low-temperature cross-sections, shown as heavy solid curves, with the high-temperature cross-sections, shown as light solid curves. The horizontal lines denote superzone boundaries at $k_z = \pm n(2\pi/7c)$ introduced by the magnetic ordering (after reference 27).

Fermi surface at $T = 0$ is given approximately by the heavy solid curves. The largest portions of the Fermi surface are destroyed by the superzone boundary corresponding to $n = 3$. Notice that this effect is a direct consequence of the presence of several relatively flat d bands in the vicinity of the Fermi energy (cf. Figure 6–41) and would not occur to any extent in the free-electron model. Notice further that the vector $q = 4\pi/7c$ actually spans a section of the calculated Fermi surface as is true of the "webbing" in the RAPW calculations. This is essential to the resistivity anomaly at the transition temperature as discussed above.

Freeman et al.[4] have drawn the perturbed Fermi surface in such a way as to emphasize its relationship to the unperturbed surface (the light solid curves). As discussed earlier in this section, the Bloch states which originally had unique k values within the double Brillouin zone for the hcp structure are now mixtures of k values differing by multiples of q. Therefore, the assignment of a particular one-electron state to a particular point in k space is ambiguous to this degree. However, the essential feature of the result is that, in the magnetic state, the Fermi surface normal to the z axis is largely destroyed while segments parallel to the axis remain, though somewhat perturbed. This implies a resistance anomaly parallel but not perpendicular to the crystal c axis, in agreement with experiment. While some of these features are somewhat different in the RAPW results the general features of the Fermi surface, as shown in Figure 6–9, and of the effects of magnetic ordering, as shown in Figure 6–42, remain even when the relativistic effects are taken into account. It thus appears that the anomalies in the temperature dependence of the resistivity of the heavy rare earth metals can be understood in terms of the calculated energy bands of these materials, and that qualitative agreement, at least, exists between theory and experiment. More quantitative agreement awaits further detailed calculations.

Little has been said so far about the contribution of critical scattering to the resistivity. It is known that the fluctuations of spins, which contribute through electron scattering to the conduction electron resistivity,[131, 132] becomes anomalously large in the vicinity of the magnetic critical point. The possible anomalous electrical resistivity in the antiferromagnetic metals near their Néel temperature has recently been considered by Suezaki and Mori.[133] The anomalous fluctuations do not generate any strong anomaly in the resistivity itself because the lifetime effect of the fluctuations is masked by the energy transfer of conduction electrons during scattering in both ferromagnets and antiferromagnets. They point out that the temperature derivative of the resistivity contains the correlations of fluctuations of four spins. In antiferromagnets the anomalous fluctuations occur near the wave vector $\mathbf{Q}$ and thus contribute to the large angle scattering resulting in a stronger anomaly than for the ferromagnetic case (which is limited by the $(1 - \cos\,\theta)$. geometrical scattering factor).

Suezaki and Mori use the s–d exchange model and write the resistivity in terms of the time correlation function[132] of the localized spins. The calculations predict that the temperature derivative of the electrical resistivity has an anomaly proportional to $\epsilon^{-(\alpha+\gamma-1)}$ where $\epsilon=|\,T-T_N\,|/T_N$ in the antiferromagnetic metals near their Néel temperature. Here γ is the critical index for the susceptibility and under certain conditions for which $\alpha=0$ and $\gamma=\frac{4}{3}$, the anomaly goes as $\epsilon^{-\frac{1}{3}}$.

6.5.4. Magnetic Superzone Band Gaps and Optical Anomalies

Miwa[110] first predicted that optical anomalies are expected as a consequence of the s–f exchange splitting of the conduction bands at low temperatures. The formation of magnetic superzone energy gaps which occur at superzone boundaries created by an antiferromagnetic superlattice (as exists in Tm) or the simple separation into the so called spin up and spin down bands of a ferromagnet (as in Gd) are examples of the exchange splittings which are expected to result in optical absorption or reflectivity anomalies due to transitions between the exchange split bands. The shape and possibly the position of the absorption band might be expected to be somewhat dependent on the magnetic structure. In contrast, as we have seen, the resistivity anomalies, which arise in heavy rare earth metals at the onset of antiferromagnetic ordering[107, 108] are associated primarily with the band gaps introduced by superzone boundaries and are absent in the case of a ferromagnet.

The infra-red reflectivity of a thin film of antiferromagnetic Ho was first measured by Schüler.[111] His data show a temperature dependent anomaly at about 0.35 eV which has been interpreted as due to interband transitions across the energy gaps created by the antiferromagnetic superlattice. Cooper and Redington[112] have observed the infra-red absorption spectra of a thin film of Dy both with (ferromagnetic state) and without (antiferromagnetic state) an external magnetic field. They observed an anomaly at about 0.44 eV quite similar to that seen by Schüler in Ho. The spectra observed in antiferromagnetic and ferromagnetic Dy were essentially identical. Recently, Schüler[134] has observed a similar anomaly in the reflectivity of thin films of ferromagnetic Gd. The frequency of the anomalies in both Dy and Ho are strongly temperature dependent, becoming lower at higher temperatures. In addition, the anomalies vanish altogether above the Néel points. This indicates that they are probably of magnetic origin, due to the exchange splitting of the s–d conduction bands.

This rather remarkable agreement between the antiferromagnetic state (magnetic superzone gaps) and the ferromagnetic state (spin splitting of the up and down bands) is totally unexpected from the theory described above. A rather simple explanation of the optical anomalies was given by Dimmock, Freeman, and Watson[5] using the simple exchange (spin) splitting of the rare earth energy bands and the calculated density of states at the Fermi energy (cf. discussion in Section 6.5.1). At $T=0$ the exchange splitting is given by Eqn (6.41) where the s–f exchange integral has been taken as a parameter (for $q=0$). Dimmock et al.[5] estimated ΔE for Gd by using their calculated density of states at the Fermi energy and fitting Eqn (6.45) to match the observed saturation magnetization (which is 0.55 μ_B greater than expected for an $^8S_{\frac{7}{2}}$ ion alone). This gave a value of 0.61 eV for ΔE in Gd and assuming the same $j_{sf}(0)$ for all the heavy rare earths (and replacing S by $(\lambda-1)J$) they obtained the calculated ΔE's shown in Table 6.1. The numerical agreement between these calculated splittings and the observed optical anomalies for Ho and Dy is quite astonishing; it must be due to a fortuitous cancellation of errors as such agreement is not justified on any grounds.

It should be mentioned that these transitions require a spin-flip (transition between spin up and spin down bands) in the non-relativistic treatment, which is not allowed for in the electric dipole transitions. The inclusion of spin–orbit coupling, however, mixes the band functions and makes these transitions allowed. Further, if the order of the spin orbit energy is comparable, as is the actual case, with the magnetic band gaps then there will be large mixing and strong transitions will occur.

Other anomalies in the optical absorption and reflectivity which may occur are also discussed by Dimmock et al.[5] These are expected whenever the interband transitions between occupied and unoccupied energy levels for those frequencies at which the bands yield a high joint density of states. They discussed the selection rules for allowed optical transitions and their polarizations using their calculated band structures and density of states. Here, they concluded, the relativistic effects are of considerable importance in the interpretation of optical anomalies due to inter-band transitions especially in the visible and infra-red regions.

Interband transitions also expected to occur between the atomic 4f levels and the conduction bands were observed by Blodgett and Spicer[26] in photoemission. They observed emission from 4f levels located about 5.8 eV below the Fermi energy. Since the band calculations yield a high density of states in Gd (and the other hcp metals) at about 1.2 eV above the Fermi energy, transitions between the 4f levels and the conduction bands are expected to yield an anomaly in the ultraviolet at about 7 eV. To date no such anomaly has been observed.

The UV measurements[46, 44] on Eu and the photo-emission experiments on Yb were discussed earlier in Section 6.2.3, and so will not be discussed further here.

Aside from the measurements mentioned above, very little has been done using optical methods. Clearly more optical experiments are needed before the detailed band structure of the rare earth metals will be considered to be well understood. As emphasized by Schüler,[134] in addition to absorption and reflectance experiments, magneto-optic studies should be carried out especially near the anomalies. For this one needs single crystals and the use of polarized radiation to help clarify where in the Brillouin zone the band gap appears.

6.6. CRYSTAL FIELDS

Many of the magnetic properties of the rare earth metals and their compounds are strongly influenced by the crystalline field which acts on the localized 4f electrons (cf. Chapter 1). Its vital role in the phenomenological theory which successfully describes the magnetic ordering of the rare earths and some of their dynamic properties was the subject of Chapter 2. Unfortunately, microscopic theories of the origin of the crystalline field and its role in explaining various observations in metals has been crude (at best) and fairly speculative.[135] What is clearly required is a first principles calculation based on the extensive band structure results currently available and discussed in Section 6.2. A recent attempt[136] at such a calculation offers the hope that this existing situation may soon be turned around and the origin of the crystal field splittings of the 4f electronic levels in the rare earth metals put on a sound footing.

In this section we review the still meagre work which has been done for the case of metals following a discussion of the simple case of the origin of crystal field splittings in paramagnetic salts. For the salts there exists a considerable body of literature and a general semi-empirical understanding has evolved but as yet even for these systems *ab initio* theory has not been successful in reproducing experiment. The techniques of optical absorption and fluorescence cannot be used for the case of metals (as the intensity of the transitions is much weaker than in the salts) but fortunately a number of techniques (particularly inelastic neutron scattering) have been recently employed to determine the crystal field parameters for the ground multiplet in the rare earth metals, their intermetallic compounds and rare earths as dilute impurities in other metals.

6.6.1. Crystal Field Splittings in Paramagnetic Salts

The rare earth salts are generally thought to be the outstanding example of a system for which the electrostatic crystalline field theory is applicable. In this simplest theory one assumes that there is negligible overlap between the 4f electrons and the surrounding ionic charge distributions of the ligands (point charge approximation) so that the static crystalline potential, V_c, which acts solely on the open 4f shell electrons has the (simple) general representation[137]

$$V_c = \sum_{L, M} A_L^M r^L Y_L^M (\theta, \phi) \qquad (6.49)$$

The use of V_c as a perturbation acting on essentially a free rare earth ion described by a total angular momentum $\mathbf{J}$ leads to a set of parameters $V_L^M = A_L^M \langle r^L \rangle$ (where the brackets denote expectation value of the r^L operator with respect to the 4f radial wave functions alone and $L \leq 6$) which when determined empirically have been successful in interpreting their magnetic and optical properties.

For systems of high symmetry, simple expressions are easily obtained for V_c. For example, in the ethylsulfates and trichlorides, the rare earth sites have C_{3h} symmetry (which is a rather high symmetry for rare earth salts), and the effect of V_c is completely represented by the terms $A_2^0 \langle r^2 \rangle$, $A_4^0 \langle r^4 \rangle$, $A_6^0 \langle r^6 \rangle$, and $A_6^6 \langle r^6 \rangle$, while for a crystal field corresponding to a point symmetry of the type C_{3v}, the only nonzero parameters are $A_2^0 \langle r^2 \rangle$, $A_4^0 \langle r^4 \rangle$, $A_4^3 \langle r^4 \rangle$, $A_6^0 \langle r^6 \rangle$, $A_6^3 \langle r^6 \rangle$, and $A_6^6 \langle r^6 \rangle$. [By contrast, in the case of an iron-group ion in a cubic field, the static crystalline field can be represented (for most purposes by a single parameter called $10Dq$). The greater complexity in the case of the rare earths arises from the higher orbital angular momentum of the 4f electrons as well as from the lower crystal-field symmetry which exists for rare earth ions in salts.] The success of this naive (and from the theoretical point of view inadequate) model is thought to lie in the fact that V_c gives the Hamiltonian the correct symmetry and a sufficient number of empirical parameters which somehow absorb all the various environmental effects not included in the simple description given by Eqn (6.49).

Theoretical efforts to determine these parameters from first principles have met with failure. Calculations generally use the non-relativistic Hartree–Fock values of $\langle r^L \rangle$ for the rare earth ions determined by Freeman and Watson[93] (and reproduced for convenience as Table 6.2) and point charges at the ion sites (with and without shielding corrections.) As discussed by Elliott[135] the most detailed treatment has been given to trichlorides RCl_3. For example the parameters determined from experiment[138] in $PrCl_3$ are, in cm^{-1},

$$\tilde{A}_2^0 \langle r^2 \rangle = 48, \quad \tilde{A}_4^0 \langle r^4 \rangle = -40, \quad \tilde{A}_6^0 \langle r^6 \rangle = -40, \quad \tilde{A}_6^6 \langle r^6 \rangle = 400$$

(where the tilde over the A's indicates a different definition was used). By contrast point charge calculations[139] and the $\langle r^L \rangle$ values of Table 6.2 give some large discrepancies:

$$A_2{}^0 \langle r^2 \rangle = 560, \quad A_4{}^0 \langle r^4 \rangle = -35, \quad A_6{}^0 \langle r^6 \rangle = -3, \quad A_6{}^6 \langle r^6 \rangle = 37$$

Similar discrepancies are found for the ethyl sulfates;[140] the garnets[141] (which have a lower symmetry and more parameters) also show that the calculated second order parameters are too large and the sixth order parameters too small. Of some significance is the observation[93] that the $V_L{}^M$ parameters measured for a series of ions when divided by the $\langle r^L \rangle$ of Table 6.II do not result in constant $A_L{}^M$ values as they must if the point charge description is valid and that the sixth order parameters fall off less rapidly than expected from the contraction of the $\langle r^6 \rangle$ values with increasing nuclear charge.

Table 6.2. Calculated Values of $\langle r^2 \rangle$ in $a_0{}^{-2}$ Units, $\langle r^4 \rangle$ in $a_0{}^{-4}$ Units, and $\langle r^6 \rangle$ in $a_0{}^{-6}$ Units, Using the 4f Wave Functions Determined by Freeman and Watson (Reference 93)

	$\langle r^2 \rangle$	$\langle r^4 \rangle$	$\langle r^6 \rangle$
Ce^{3+}	1.200	3.455	21.226
Pr^{3+}	1.086	2.822	15.726
Nd^{3+}	1.001	2.401	12.396
Sm^{3+}	0.883	1.897	8.775
Eu^{2+}	0.938	2.273	11.670
Gd^{3+}	0.785	1.515	6.281
Dy^{3+}	0.726	1.322	5.102
Er^{3+}	0.666	1.126	3.978
Yb^{3+}	0.613	0.960	3.104

These discrepancies point up the inadequacy of the electrostatic field description. One must consider the role of other electrons belonging to the rare earth ion and the various contributions of overlap and covalency with neighboring (ligand) electrons—contributions known in the case of the 3d ions to play an especially significant role.[142] We have seen (cf. Figure 6–2) that the 4f electrons lie well inside the closed ("inner" in energy) 5s and 5p shells of electrons. They are thus able to *shield* the 4f electrons from the external electrostatic potential reducing thereby the effective potential which acts on the open shell electrons.[143, 144] While calculations differ as to the magnitude of these shielding effects, it is generally agreed that the closed 5s and 5p shells are easily distorted so as to shield the second order terms (and to a lesser extent the fourth order terms) thereby accounting in large measure for the discrepancy between the measured and computed $A_L{}^M \langle r^L \rangle$ parameters such as those illustrated above.

The closed 5s–5p shells also make questionable a second postulate of crystal field theory,[144, 145] namely that the ordering of the 4f crystal-field energy levels is determined by the group transformation properties associated with a Hamiltonian which has the symmetry of V_c acting on the 4f shell,

324   A. J. Freeman

i.e. the ordering and relative spacing of the electronic energy levels is determined by the angular operators of V_c acting on the 4f electrons alone. The
validity of this assumption, often cited as one strength of the crystal-field
method, has been used to justify the parametrization procedure. It has been
shown[144, 145] that in addition to producing large shielding effects (called
linear shielding because they simply scale the crystal field strength parameters),
the crystalline potential gives rise to distortions of the ion's charge distributions which, in turn, produce deviations from the predicted 4f crystal-field
level scheme given by V_c directly. Such *nonlinear* deviations will not always
be of significant magnitude but when they are, the standard parametrization
procedure yields parameters which contain crystal-field components of inappropriate symmetry. The application of this procedure to fit rare earth spectra
is most questionable in such a case.

It should be mentioned that the linear and nonlinear p→4f shielding
distortions, and to a lesser extent other terms, have a direct bearing on
observables other than just the crystal field matrix elements. They make small
environment-dependent contributions to the electric and magnetic hyperfine fields in a rare earth ion. They also produce a spin density in the outer
reaches of the ion which can be significant because of the highly local character
of the 4f shell. [This spin density is in addition to the free ion spin polarization, investigated previously[94] in Gd.] The p→4f shielding produces an
anisotropic density in the outer reaches of the ion which is parallel to the spin
of the 4f shell. Since it involves the outer reaches of the ion, this density's
role in ion–ion exchange coupling and in transferred hyperfine effects is
enhanced, and, on occasion, may predominate over direct 4f terms. The
crystal field matrix elements make these p→4f terms environment-dependent.[144, 145]

Much less is known about the role of environmental effects, overlap and
covalency,[146–148] on the crystal field splitting parameters, although some
time ago, Jørgensen et al.[146] suggested that covalency contributions were
dominant in the rare earth ions. Their simple phenomenological model based
on this hypothesis fitted experimental data reasonably well with a single
parameter. The parameter most sensitive to charge penetrations is $\langle r^6 \rangle$
because the outer part of the 4f wave function is heavily weighted. Theoretical
calculations[148–151] are still in a rather crude stage and suffer from approximations made in the Hamiltonian, neglect of many center integrals, insuffici-
ent variational freedom, lack of self-consistency, and treatment of only a
limited number of the interacting electrons. While none of the results can be
taken seriously in a quantitative sense, some qualitative features are correct.
The computed parameters are reduced by the simple charge penetration
mechanism. However, when orthogonalization and covalency are included
the sixth order term is substantially increased[151] (and the results for the
parameters $A_6^0 \langle r^6 \rangle$, $A_6^6 \langle r^6 \rangle$ and $A_4^0 \langle r^4 \rangle$ agree so well with experiment
that Ellis and Newman[151] conclude that the results must be to some extent
fortuitous). We should note that Ellis and Newman[151] conclude that their
calculations do confirm the Jørgensen et al.[146] view that "overlap and
covalency contributions alone give very nearly the correct relative magnitudes
of all the crystal field parameters."

There has been considerable progress in *ab initio* calculations, using a cluster model of metal ion and nearest neighbor ligands, for treating transition metal complexes. [142] In this latest technique full Unrestricted Hartree–Fock molecular orbital calculations are carried out for all electrons in the cluster; matrix elements of the effective Hamiltonian are evaluated directly by means of a Diophantine (numerical) integration procedure, thus overcoming the problem of molecular multi-center integrals. This method has now been applied [152] to the rare earth cluster $(EuF_6)^{4-}$ in order to determine molecular orbital solutions involving all 121 electrons of the system for both the ground and excited states (including the 5d electrons). Preliminary results show the importance of the closed 5s–5p shells, (as had been emphasized earlier [148]) and large hybridization of the 4f levels. Application to the crystal field splitting problem has not yet been made.

6.6.2. Crystal Field Splittings in Metals

We have discussed briefly the complications involved in theoretical estimates of the crystal field parameters in the salts, and have indicated the failure of *ab initio* [9] theory to account for experiments in these simpler systems. In the metals there is the further theoretical complication of the additional large contribution coming from the conduction electrons. The difficulty of doing unambiguous experiments in metals has resulted, until recently, in only very little experimental information on the crystal field. It is therefore not surprising that both the theoretical and experimental picture of crystal fields in metals is in such a poor state.

Some of the experimental methods which have been used to obtain information about the crystal field in the rare earth metals have been discussed in other chapters. A detailed description of the measurement of the crystal field splitting is given in Section 2.3.2. The most detailed information comes from inelastic neutron scattering measurements of the magnon dispersion relations (Chapter 5,) from observed bulk anisotropy of metals, and from anisotropy of the susceptibility (Chapter 5). Hyperfine interaction measurements Chapter 8), such as Mossbauer effect determinations of the nuclear quadrupole splitting and its temperature dependence, may be used to determine the second order (axial field) parameter *times* the ratio of the optical shielding parameter and the Sternheimer quadrupole anti-shielding factor. (The separation of the second order parameter then requires a theoretical estimate of both shielding parameters.) Magnetic resonance measurements [153] of isolated ions in metals (such as Au, Ag, and Mg) together with susceptibility measurements [154] can also give some information in these systems. [The departures from Curie's law found in the susceptibilities and the positions (or absences) of EPR lines can be understood and fitted primarily in terms of crystal field effects. [154]]

As discussed by Elliott, [135] the crystal field parameters determined in these different ways do not completely agree. For example in the heavy metals the overall picture shows a dominant axial anisotropy, a smaller hexagonal component and fairly good (and undoubtedly fortuitous) agreement with the predictions of simple trivalent point charge calculations using the $\langle r^L \rangle$

values given in Table 6.II. The inclusion of conduction electron screening of, and interaction with, the 4f electrons would markedly change the results of the calculations.

Recently the one method which gives direct measurements of the energy level separations—inelastic neutron scattering—has been developed to the point of giving quantitative information about the 4f levels in the metals. [155, 156] Turberfield et al. [156] have reported measurements on a series of rare earth intermetallic compounds—praseodymium monochalco- genides and monopnictides which illustrate in a dramatic way the power of inelastic neutron scattering techniques. The one restriction is that the exchange field is somewhat smaller than the crystal field. The compounds studied all have the rock-salt structure and in a simple picture the chalcogenides may be thought of as essentially ionically bound compounds with the Pr in a $+3$ state and the chalcogen in a -2 state and with the extra electron going into a conduction band which is 5d, 6s-like. The pnictides are more complicated in that while valence arguments would expect them to be semiconductors, the materials behave like metals but with densities of states significantly lower than the corresponding chalcogenides. [156]

These sets of compounds allowed the possibility of doing systematic studies of the crystal fields. The effects thereon of ionic size are found by going down either the Va or VIa columns and those from changing the ligand charge, the covalency, and the carrier concentration are found by comparing equiva- lent compounds in the two columns. The experimental results show a number of unexpected (one could say *startling*) features which must be quoted in detail.

> Firstly, the crystal field is almost entirely fourth-order; there is no apparent enhance- ment of the sixth-order terms as usually occurs in insulators as a result of overlap and covalency. Secondly, there is no appreciable difference between the Va and VIa compounds; indeed, the lattice constant seems to be the only significant parameter. Thirdly, the fourth-order terms are found to follow a simple R^{-5} law with an absolute magnitude which is quite close to that calculated using a nearest-neighbor point- charge model with a charge of -2 at the ligand site. These results may be restated as follows: *All of the crystal-field levels in all seven compounds can be quantitatively accounted for by use of a nearest neighbor point-charge model with an effective charge of* -2. The mean deviation using this model is 10% and, in fact, the majority of the levels are fitted to within the experimental error. This is a remarkable result especially when one considers that the model is of little quantitative value in rare earth insulators such as $PrCl_3$. Indeed, one would have expected that all of the mechanisms operative in insulators, that is, 4f-ligand overlap and covalency, point-charge effects, charge penetration, 5s–5p shielding, and charge redistribution, would be present in the chal- cogenides and pnictides; in addition, one has the complication of possible shielding and 5d virtual bound-state effects arising from the conduction electrons. Furthermore, we, at least, had anticipated that these effects would manifest themselves differently in the chalcogenides and pnictides. In particular, the fact that corresponding chalcogenides and pnictides have virtually identical crystal fields whereas the conduction-electron concentrations differ considerably indicates that the conduction electrons are ineffective both in terms of simple screening and in terms of 5d crystal-field effects. We have not, as yet, come to any satisfactory understanding of why the Stark splittings in these metallic compounds do not seem to reflect the anticipated complications.

We can add little to this discussion of Turberfield et al. [156] except to caution that there may be occurring an incredible cancellation of contribu- tions. [156a] The very small sixth order terms may be a result of the particular crystal structure involved (rock-salt) and it is suggested that experiments on lower symmetry systems should be undertaken. Clearly band calculations such as the one to be described next would be of great interest and value.

The direct contribution of the conduction electrons to the crystalline electric field (and nuclear quadrupole interaction) in dysprosium metal has been calculated by Das and Ray.[136] Using the APW formalism described in Section 6.2.3 they calculated the charge densities of the conduction electrons for different values of l (up to $l = 15$). The predominant charge density was found to arise from s-functions whereas the amount of p and d electrons inside the APW sphere was small. These functions were then used to calculate the contribution of the conduction electrons to the crystalline field.

Das and Ray[157] had earlier pointed out that there is an appreciable contribution by the conduction electrons to all the parameters and that the net effect of the conduction electrons is to enhance the lattice (point charge) contributions to these parameters. Including a shielding effect of the closed 5s, 5p electrons for the second order parameter they[136] use their earlier calculations to estimate the lattice contributions to the parameters and then add to these the direct contributions from the conduction electrons (but do not include any shielding of these conduction electron contributions). The predominant contributions to $A_2^0 \langle r^2 \rangle$ and $A_4^0 \langle r^4 \rangle$ parameters come from the d electrons whereas the f-electronic contribution is dominant for $A_6^0 \langle r^6 \rangle$ as well as $A_6^6 \langle r^6 \rangle$ parameters. The conduction electrons in the central APW sphere make such a large contribution to the $A_2^0 \langle r^2 \rangle$ term as to overpower the lattice contribution and change the *sign* of this term. For the $A_6^0 \langle r^6 \rangle$ term the conduction electron contribution is 1.9 times the lattice value which is close to the 1.8 factor Das and Ray[136] quote as being the experimental factor for the $A_6^6 \langle r^6 \rangle$ term (which they did not calculate). This result (i.e., same anti-shielding factors for both sixth order terms) is consistent with expectations arising from the observation that f-electrons in the central APW sphere make the predominant contribution in both cases.

Das and Ray[136] discuss the origin of the large discrepancy in the second order term in terms of the modification of the radial function of the conduction electrons by the non-spherical part of the muffin-tin potential which was not considered in the APW calculations. In considering the sort of modifications of the radial functions required to produce agreement of the $A_2^0 \langle r^2 \rangle$ parameter they conclude that the indirect effects of the conduction electrons might be important. They promise further detailed calculations which include all possible indirect effects of the conduction electrons.

The Das and Ray calculations must be viewed as preliminary at this stage but indicate a growing interest in applying band structure calculations and results to the problem of crystal field splittings in metals. Further theoretical effort is needed to resolve the many conflicting aspects of our current understanding of this important problem.

ACKNOWLEDGMENT

I am grateful to B. Harmon and Drs. J. V. Mallow and D. D. Koelling for helpful suggestions concerning the contents of this chapter and to Professor

A. R. Mackintosh for comments. I am greatly indebted to Dr. J. V. Mallow
for a careful reading of the manuscript and for discussions, and to Mrs. A.
Mallow for her indispensable help in typing this manuscript. Most of this work
was done at the Hebrew University as a John Simon Guggenheim Memorial
Foundation Fellow and a Fulbright-Hays Fellow (through the U.S.–Israel
Educational Foundation). The support of these institutions and the hospitality
of the Racah Institute of Physics is gratefully acknowledged.

REFERENCES

1. For earlier reviews of this work see Elliott, R. J., in Magnetism (G. T. Rado and
   H. Suhl, eds). Academic Press, New York (1965), Vol. IIA, p. 385; and Kasuya T.
   (reference 2) and references therein.
2. Kasuya, T., in Magnetism, op. cit. (1966), Vol. IIB, p. 215.
3. Dimmock, J. O. and Freeman, A. J., *Phys. Rev. Letters*, **13**, 750 (1964); Dimmock,
   J. O., Freeman, A. J. and Watson, R. E., *J. Appl. Phys.*, **36**, 1142 (1965).
4. Freeman, A. J., Dimmock, J. O. and Watson, R. E., *Phys. Rev. Letters*, **16**, 94 (1966);
   Watson, R. E., Freeman, A. J. and Dimmock, J. O., *Phys. Rev.*, **167**, 497 (1968).
5. Dimmock, J. O., Freeman, A. J. and Watson, R. E., in Proceedings of the International
   Colloquium on Optical Properties and Electronic Structure of Metals and alloys,
   (F. Abeles, ed). North-Holland Publishing Co., Amsterdam (1966), p. 273.
6. Freeman, A. J., Dimmock, J. O. and Watson, R. E., in Quantum Theory of Atoms,
   Molecules and the Solid State, a Tribute to John C. Slater (P. O. Löwdin, ed).
   Academic Press, New York (1966), p. 361.
7. Keeton, S. C. and Loucks, T. L., *Phys. Rev.*, **146**, 429 (1966).
8. Keeton, S. C. and Loucks, T. L., *Phys. Rev.*, **168**, 672 (1968).
9. Fleming, G. S., Liu, S. H. and Loucks, T. L., *Phys. Rev. Letters*, **21**, 1524 (1968);
   J. Appl. Phys. **40**, 1285 (1969).
10. Andersen, O. K. and Loucks, T. L., *Phys. Rev.*, **167**, 551 (1968).
11. Taken from the Hartree–Fock–Slater calculations of F. Herman and S. Skillman, in
    Atomic Structure Calculations. Prentice-Hall, Inc., Englewood Cliffs, New Jersey
    (1963).
12. Slater, J. C., in Quantum Theory of Matter. McGraw Hill Book Co., New York
    (1951), p. 112. Mackintosh, A. R., Proc. Simon Fraser Summer School, Alta Lake, 1970.
13. Slater, J. C., in Quantum Theory of Molecules and Solids, Vol. 2. McGraw-Hill
    Book Co., New York (1965); and Vol. 3 (1967); Advances in Quantum Chemistry
    (P. O. Löwdin, ed), Vol. I, p. 35 (1967); *I. J. of Quantum Chemistry*, **1**, 37 (1967).
14. Slater, J. C., *Phys. Rev.*, **81**, 385 (1951). See also Appendix 22 of Vol. 2 cited in ref. 13.
15. Slater, J. C., *Phys. Rev.*, **51**, 846 (1937); *ibid.*, **92**, 603 (1953); Saffren, M. M. and
    Slater, J. C., *Phys. Rev.*, **92**, 1126 (1953); and descriptions in ref. 13.
16. Kohn, W. and Rostoker, N., *Phys. Rev.*, **94**, 1111 (1954); Korringa, J., *Physica*, **13**,
    992 (1947).
17. Matthiess, L. F., *Phys. Rev.*, **133**, A1399 (1964).
18. Kohn, W. and Sham, L. J., *Phys. Rev.*, **140**, A1133 (1965); Hohenberg, P. and Kohn,
    W., *Phys. Rev.*, **136**, B864 (1964); Sham, L. J. and Kohn, W., *Phys. Rev.*, **145**, 561
    (1966).
19. Gáspár, R., *Alta Phys. Hung.*, **3**, 263 (1954).
20. Loucks, T. L., in Augmented Plane Wave Method. W. A. Benjamin Inc., New York
    (1967).
21. Matthiess, L. F., Wood, J. H. and Switendick, A. C., *Methods in Computational
    Physics*, **8**, 64 (1968).
22. Dimmock, J. O., in Solid State Physics, **26**, 103 (1971) (Seitz, Turnbull and Ehrenreich,
    eds). Academic Press, New York. Most of the unpublished results of Dimmock,
    Freeman, and Watson are contained in this review and we shall make liberal use of
    the material and discussion contained herein.
23. Freeman, A. J. and Watson, R. E., *Phys. Rev.*, **127**, 2058 (1962).

24. Dimmock, J. O., Freeman, A. J. and Furdyna, A. M., *Bull. Am. Phys. Soc.* II **10**, 377 (1965); and unpublished results.
25. Wood, J. H., *Phys. Rev.*, **117**, 714 (1960).
26. Blodgett, A. J., Jr., Spicer, W. E. and Yu, A. Y-C., in Optical Properties and Electronic Structure of Metals and Alloys (F. Abeles, ed). North-Holland Publ. Co., Amsterdam (1966), p. 246.
27. Freeman, A. J., Dimmock, J. O. and Watson, R. E., *Bull. Am. Phys. Soc.* II, **10**, 376 (1965); Watson, R. E., Freeman, A. J. and Dimmock, J. O., *Phys. Rev.*, **167**, 497 (1968).
28. Dimmock, J. O., Freeman, A. J. and Watson, R. E. (unpublished).
29. Loucks, T. L., *Phys. Rev.*, **137**, A1333 (1965).
30. Liberman, D., Waber, J. T. and Cromer, D. T., *Phys. Rev.*, **137**, A27 (1965).
31. The Symmetrized RAPW scheme developed and programmed by Koelling, D. D. [*Phys. Rev.*, **188**, 1049 (1969)] makes full use of symmetry and gives full identification of the relativistic bands. This lack of identification can lead to serious difficulties as was found by Keeton and Loucks in their RAPW calculations on thorium metal (ref. 7).
32. Loucks, T. L., *Phys. Rev.*, **144**, 504 (1966).
32a. Gupta, R. P. and Loucks, T. L., *Phys. Rev.*, **176**, 848 (1968).
33. Williams, R. W., Loucks T. L. and Mackintosh, A. R., *Phys. Rev. Letters*, **16**, 168 (1966).
34. Lomer, W. M., private communication to Loucks, T. L. (ref. 8) and in Proceedings of the International School of Physics (W. Marshall, ed). Academic Press, New York (1967), Vol. 37, p. 19.
35. Mackintosh, A. R., *Phys. Letters*, **28A**, 217 (1968).
36. Jackson, C., *Phys. Rev.*, **178**, 949 (1969).
37. Myron, H. W. and Liu, S. H., *Phys. Rev.*, **B1**, 2414 (1970).
38. Waber, J. T. and Switendick, A. C., in Proc. Fifth Rare Earth Research Conference. Ames, Iowa (1965), Book II, p. 75.
39. See the review articles by Gschneider, K. A. Jr., in Rare Earth Research III (L. Eyring, ed). Gordon & Breach, New York (1965), p. 153 and Rocher, Y. A., *Advances in Physics*, **11**, 233 (1965), and references in Chapters 3 and 7.
40. Mukhopadhyay, G. and Majumdar, C. K., *J. Phys. C* (Solid State Physics), **2**, 924 (1969).
41. Waber, J. T., Liberman, D. and Cromer, D. T., in Proc. 4th Conference on Rare Earth Research (L. Eyring, ed). Phoenix, Arizona 1964. Gordon & Breach, New York, (1965), p. 187.
42. Freeman, A. J. and Dimmock, J. O., *Bull. Am. Phys. Soc.* II, **11**, 216 (1966), and unpublished.
43. Endriz, J. G. and Spicer, W. E., *Phys. Rev.*, **B2**, 1466 (1970).
44. Brodén, G., Hagström, S. B. M. and Norris, C., *Phys. Rev. Letters*, **24**, 1173 (1970).
45. Matthiess, L. F., *Phys. Rev.*, **139**, A1893 (1965).
46. Johansen, G., *Solid State Commun.*, **7**, 731 (1969).
47. Johansen, G. and Mackintosh, A. R., *Solid State Commun.*, **8**, 121 (1970).
48. Dimmock [ref. 22] has suggested that this discrepancy arises from the fact that the energy band scheme is an independent–electron model for the electronic states which ignores correlation energies and the resulting multiplet splittings which dominate the 4f energies in the rare earths.
49. Fadley, C. S., Shirley, D. A., Freeman, A. J., Bagus, P. S. and Mallow, J. V., *Phys. Rev. Letters*, **23**, 1397 (1969).
50. Tanuma and associates made the first de Haas–van Alphen measurement for a rare earth. They first studied f.c.c. Yb metal [Tanuma, S., Ishizawa, Y., Nagasawa, H. and Sugawara, T., *Phys. Letters*, **25A**, 669 (1967)] and later h.c.p. Yb metal (Tanuma, S., Dators, W. B., Doi, H. and Dunsworth, A., *Solid State Commun.*, **8**, 1107 (1970)]. The latter experiment was done at 1.2 K in magnetic fields up to 116 KOe. Recently, Bucher, E., Schmidt, P. H., Jayaramon, A., Andres, K., Maita, J. P., Nassau, K. and Deinier, P. D. [*Phys. Rev.*, **B2**, 3911 (1970)], and independently Kayser F. X. [*Phys. Rev. Letters*, **25**, 662 (1970)] have reported on a first-order magnetic phase transition

(paramagnetic to diamagnetic) associated with an fcc-hcp martensitic transformation in high purity Yb. The fcc phase was obtained from the hcp phase by applying strain at room temperature. They speculate that this transition can be explained by assuming a conversion of a small fraction (about 0.8%) of Yb^{3+} (above the transition) into Yb^{2+} (below the transition). We believe such speculations to be premature in view of the lack of detailed supporting evidence.

50a. Koelling, D. D. and Harmon, B., *Bull. Am. Phys. Soc.*, **17**, 94 (1972).
51. Jepsen, O. and Anderson, O. K. *Solid State Comm.*, **9**, 1763 (1971).
51a. Harmon, B. and Koelling, D. D., private communication.
52. Patrick, L., *Phys. Rev.*, **93**, 384 (1954); McWhan, D. B. and Stevens, A. L., *Phys. Rev.*, **139**, A682 (1959); Bloch, D. and Pauthenet, R., in Proceedings of the International Conference on Magnetism, Nottingham, 1964. The Institute of Physics and The Physical Society, London (1965), p. 255.
53. Robinson, L. B., Tan, S. I. and Sterett, K. F., *Phys. Rev.*, **141**, 548 (1966).
54. Austin, I. G. and Misra, P. K., *Phil. Mag.*, **15**, 529 (1967).
55. Umebayashi, H., Shirane, G., Frazer, B. C. and Daniels, W. B., *Phys. Rev.*, **165**, 688 (1968).
56. Fleming, G. S. and Liu, S. H., *Phys. Rev. B*, **2**, 164 (1970).
57. For a review see Gschneider, K. A. and Smolochowski, R., *Less Common Metals*, **5**, 372 (1963).
58. See for example, Gschneider, K. A., ref. 39.
59. Phillips, R. A., to be published.
60. Williams, R. W. and Mackintosh, A. R., *Phys. Rev.*, **168**, 679 (1968).
61. Ruderman, M. A. and Kittel, C., *Phys. Rev.*, **96**, 99 (1954).
62. Kasuya, T., *Progr. Theoret. Phys. (Japan)*, **16**, 45 (1956); Mitchell, A. H., *Phys. Rev.*, **105**, 1439 (1957) had apparently given an independent derivation of s–d exchange.
63. Yosida, K., *Phys. Rev.*, **106**, 893 (1957).
64. Kondo, J., *Prog. Theoret. Phys. (Kyoto)*, **32**, 37 (1964).
65. van den Berg, G. J., in *Progress in Low Temperature Physics*, IV, 194 (1964); Bailyn, M., *Adv. Phys.*, **15**, 179 (1966); Rocher, Y., *Adv. Phys.*, **11**, 233 (1962) and references therein.
66. Anderson, P. W., *Phys. Rev.*, **124**, 41 (1961). The relationship between the Anderson and s–d models was described by Schrieffer, J. R., *J. Appl. Phys.*, **38**, 1143 (1967).
67. Friedel, J., *Adv. Phys.*, **3**, 446 (1954); *Nuovo Cimento Suppl.*, **7**, 287 (1958).
68. See the recent review by Kondo, J., in Solid State Physics, **23**, 183 (F. Seitz, D. Turnbull, H. Ehrenreich, eds). Academic Press, New York (1969), and references therein.
69. Heeger, A. J., Solid State Physics, **23**, 283 (F. Seitz, D. Turnbull, H. Ehrenreich, eds). Academic Press, New York (1969), gives an extensive review of the experimental situation in such systems.
70. de Gennes, P. G., *Compt. Rend.*, **247**, 1836 (1958).
71. Van Vleck, J. H., *Rev. Mod. Phys.*, **34**, 681 (1962).
72. Liu, S. H., *Phys. Rev.*, **121**, 451 (1961).
73. Watson, R. E., and Freeman, A. J., *Phys. Rev.*, **152**, 566 (1966); and *Phys. Rev. Letters*, **14**, 695 (1965).
74. Watson, R. E., and Freeman, A. J., *Phys. Rev.*, **178**, 725 (1969).
75. Herring, C., in Magnetism (G. T. Rado and H. Suhl, eds). Academic Press, New York (1966), Vol. IV, gives a massive account of susceptibilities in metals.
76. Overhauser, A. W., *J. Appl. Phys.*, **34**, 1019 (1963) and references therein.
77. Kittel, C., in Solid State Physics, **22**, 1 (F. Seitz, D. Turnbull and H. Ehrenreich, eds). Academic Press, New York (1968).
78. Nagamiya, T., *Solid State Physics*, **20**, 306 (1967), and references therein.
78a. Fedro, A. J. and Arai, T., discuss the relative stability of magnetic states [*Phys. Rev.*, **170**, 583 (1968)].
78b. Falicov, L. M. and da Silva, C. E. T. [*Phys. Rev. Lett.*, **26**, 715 (1971)] give a simple theory which stresses the strong correlations in the atomic 4f states.
79. Villain, J., *J. Phys. Chem. Solids*, **11**, 303 (1959).
80. Blandin, A., *J. Phys. Chem. Solids*, **22**, 507 (1961).
81. Gauthier, F., *J. Phys. Chem. Solids*, **24**, 387 (1963).

82. Roth, L., Zeiger H. and Kaplan, T., *Phys. Rev.*, **149**, 519 (1966).
83. Kohn, W., *Phys. Rev. Letters*, **2**, 393 (1959).
84. Bambakidis, G., *J. Phys. Chem. Solids*, **31**, 503 (1970).
85. Wolff, P. A., *Phys. Rev.*, **120**, 814 (1960); **129**, 84 (1963).
86. Mueller, F. M.,Freeman, A. J., Dimmock, J. O. and Furdyna, A. M., *Phys. Rev.*, **B1**, 4617 (1970).
87. Giovanni, B., Peter, M. and Schrieffer, J. R., *Phys. Rev.*, **152**, 566 (1966).
88. Yosida, K., *Progr. Theoret. Phys.* (*Kyoto*), **28**, 759 (1962).
89. Herring, C., ref. 75, p. 337–338.
90. Kim, D. J., *Phys. Rev.*, **149**, 434 (1966); **167**, 545 (1968).
91. This form of interaction was criticized by Kaplan [Kaplan, T. Λ., *Phys. Rev. Letters*, **14**, 499 (1969)] who asserted the importance of treating the k-dependence of the periodic part of the Bloch wave.
92. For a non-spherical moment there will be contributions to $j(\mathbf{k}, \mathbf{k}')$ from $\psi_\mathbf{k}$ and $\psi_{\mathbf{k}'}$ components having different orbital angular moment values, l. While these are expected to be relatively unimportant, their contributions will cause $j(\mathbf{k}, \mathbf{k}')$ to deviate further from $j(\mathbf{q})$.
93. Freeman, A. J. and Watson, R. E., *Phys. Rev.*, **127**, 2058 (1962).
93. Watson, R. E. and Freeman, A. J., *Phys. Rev. Letters*, **6**, 277, 388E (1961).
95. The earliest evidence for this was obtained by nuclear magnetic resonance and electron paramagnetic resonance measurements for rare earths in several host metals. See Jaccarino, V., Matthias, B. T., Peter, M., Suhl, H., and Wernick, J. H., *Phys. Rev. Letters*, **5**, 251 (1960); Peter, M., *J. Appl. Phys.*, **32**, 338S (1961); Peter M. *et al.*, *Phys. Rev.*, **126**, 1395 (1962); Shaltiel, D., Wernick, J. H., Williams, H. J. and Peter, M., *ibid.*, **135**, A1346 (1964).
96. Anderson, P. W. and Clogston, A. M., *Bull. Am. Phys. Soc.*, **2**, 124 (1961); Anderson, P. W., *Phys. Rev.*, **124**, 41 (1961); Kondo, J., *Progr. Theoret. Phys.*, **28**, 846 (1962); Schrieffer, J. R. and Wolff, P. A., *Phys. Rev.*, **149**, 491 (1966).
97. Watson, R. E., Koide, S., Peter, M. and Freeman, A. J., *Phys. Rev.* **139**, A167 (1965).
98. deWiin, H. W., Buschow, K. N. J. and van Diepen, A. M., *Phys. Status Solidi*, **20**, 759 (1968).
99. Watson, R. E., Freeman, A. J. and Koide, S., *Phys. Rev.*, **186**, 625 (1969).
100. Sugawara, T., *J. Phys. Soc. Japan*, **20**, 2252 (1965); Sugawara, T. and Eguchi, H., *J. Phys. Soc. Japan*, **21**, 725 (1966); Sugawara, T., Yamase, I. and Soga, R., *J. Phys. Soc. Japan*, **20**, 618 (1965); Sugawara, T. and Yoshida, S., *J. Phys. Soc. Japan*, **24**, 1399 (1968); Nagasawa, H., Yoshida, S. and Sugawara, T., *Phys. Lett.*, **26A**, 561 (1968); Smith, T. F., *Phys. Rev. Letters*, **17**, 386 (1966); Coqblin, B. and Ratto, C. F., *Phys. Rev. Letters*, **21**, 1065 (1968).
101. See Coqblin, B., Thesis, Orsay, 1967 (unpublished); Coqblin, B. and Blandin, A., *Advan. Phys.*, **17**, 281 (1968); Coqblin, B., *Proceedings of the Seventh Rare Earth Conference*. Coronado, Calif. (1968) and references therein.
102. A recent review of the continuous magnetic–non magnetic transition of dilute alloys with Ce impurities may be found in Coqblin, B., Maples, N. B., and Toulouse, G. *Int. J. of Magnetism*, **1**, 333 (1971).
103. Coqblin, B. and Schrieffer, J. R., *Phys. Rev.*, **185**, 847 (1969).
104. Schrieffer, J. R. and Wolff, P. A., ref. 96.
105. Lomer, W. M., *Proc. Phys. Soc.* (London), **80**, 489 (1962).
106. Yosida, K. and Watabe, A., *Progr. Theoret. Phys.* (*Kyoto*), **28**, 361 (1962).
107. Mackintosh, A. R., *Phys. Rev. Letters*, **9**, 90 (1962).
108. Elliott, R. J. and Wedgwood, F. A., *Proc. Phys. Soc.* (London), **84**, 63 (1964).
109. de Gennes, P. G., *J. Phys. Radium*, **23**, 630 (1962); Kaplan, T. A., *J. Appl. Phys.*, **34**, 1339 (1963).
110. Miwa, H., *Progr. Theoret. Phys.* (*Kyoto*), **29**, 477 (1963).
111. Schüler, C. C., *Phys. Letters*, **12**, 84 (1964).
112. Cooper, B. R. and Redington, R. W., *Phys. Rev. Letters*, **14**, 1066 (1965).
113. Evenson, W. E. and Liu, S. H., *Phys. Rev. Letters*, **21**, 432 (1968).
114. Evenson, W. E. and Liu, S. H., *Phys. Rev.*, **178**, 783 (1969).

114a. Liu, S. H., Gupta, R. P. and Sinha, S. K., *Phys. Rev.*, **B4**, 1100 (1971) have recently recalulcated $\chi(\mathbf{q})$ using a finer mesh and correcting some results of reference 9. They find a maximum unrelated to any Fermi surface feature and stress the importance of the matrix elements.

115. Moon, R. M., Cable, J. W. and Koehler, W. C., *J. Appl. Phys. Suppl.*, **35**, 401 (1964).

116. Cable, J. W., Moon, R. M., Koehler, W. C. and Wollan, E. O., *Phys. Rev. Letters*, **12**, 553 (1964).

117. Koehler, W. C., *J. Appl. Phys.*, **36**, 1078 (1965).

118. Evenson, W. E., Fleming, G. S. and Liu, S. H., *Phys. Rev.*, **178**, 930 (1969).

119. Gupta, R. P. and Sinha, S. K., *J. Appl. Phys.*, **41**, 915 (1970); *Phys. Rev.*, **3B**, 2401 (1971).

120. Gilat, G. and Raubenheimer, L. J., *Phys. Rev.*, **144**, 390 (1966). See also Gilat, G. and Herman, F., *Annals of Physics* (to appear).

121. Mueller, F. M., Garland, J. W., Cohen, M. H. and Bennemann, K. H., *Annals of Physics*, **67**, 19 (1971) and references therein.

122. See discussions in Computational Methods in Band Theory (P. M. Marcus, J. F. Januk and A. R. Williams, eds). Plenum Press, New York, London, (1971) by J. F. Januk (p. 323), R. L. Jacobs and D. Lipton (p. 340), and J. B. Diamond (p. 347).

123. Cooper, B. R., *Phys. Letters*, **6**, 19 (1963).

124. de Gennes, P. G. and Saint James, D., *Solid State Commun.*, **1**, 62 (1963).

125. Miwa, H., *Proc. Phys. Soc.*, **85**, 1197 (1965).

126. Kaplan, T. A., *J. Appl. Phys.*, **34**, 1339 (1963).

127. Arai, T., *Phys. Rev. Letters*, **25**, 1761 (1970). Fedro, A. J. and Arai, T., *Phys. Rev.* (in press).

128. Møller, H. B., Houmann, J. C. G. and Mackintosh, A. R., *Phys. Rev. Letters*, **19**, 312 (1967).

129. Ziman, J. M., in Electrons and Phonons. Clarendon Press, Oxford (1960) and references therein.

130. Mackintosh, A. R. and Spanel, L. E., *Solid State Commun.*, **2**, 388 (1964). See also the discussion on the field effects in Tb metal given by Mackintosh in ref. 107.

130a. Brun, T. O. *et al.*, *Phys. Rev.*, **B1**, 1251 (1970). Koehler, W. C. *et al.*, *Phys. Rev.*, **126**, 1672 (1962).

131. Van Hove, L., *Phys. Rev.*, **95**, 249 (1954).

132. de Gennes, P. G., in Magnetism, op. cit., III.

133. Suezaki, Y., and Mori, H., *Phys. Letters*, **28A**, 70 (1968).

134. Schüler, C. Chr., op. cit. in ref. 5, p. 221 and references therein to earlier work.

135. See the review given by Elliott, R. J., in *Comments Solid State Physics*, **1**, 85 (1968).

136. Das, K. C. and Ray, D. K., *Solid State Commun.*, **8**, 2025 (1970).

137. Crystal-field theory has a long history dating back to Bethe, H., *Ann. Physik*, **3**, 133 (1929). For the rare earths see the more recent work of Stevens, K. W. H., *Proc. Phys. Soc.* (London), **A65**, 209 (1952); Elliott, R. J. and Stevens, K. W. H·, *Proc. Roy. Soc.* (London), **A215**, 437 (1952); *ibid.*, **A218**, 553 (1953); *ibid.*, **A219**, 387 (1953); and Judd, B. R., *ibid.*, **A227**, 552 (1955). For a recent review see Newman, D. J., in *Advan. Phys.*, **20**, 197 (1971).

138. Judd, B. R., *Proc. Roy. Soc.* (London), **A241**, 414 (1957); Margolis, J. S., *J. Chem. Phys.*, **35**, 1367 (1961).

139. Hutchings, M. T., and Ray, D. K., *Proc. Phys. Soc.* (London), **81**, 663 (1963).

140. Elliott, R. J. and Stevens, K. W. H., *Proc. Roy. Soc.* (London), **219**, 387 (1953); Powell, M. J. and Orbach, R., *Proc. Phys. Soc.* (London), **78**, 753 (1961).

141. Hutchings, M. T. and Wolf, W. P., *J. Chem. Phys.*, **41**, 617 (1964).

142. See Freeman, A. J. and Ellis, D. E., *Phys. Rev. Letters*, **24**, 516 (1970) and references to some of the literature cited therein.

143. Burns, G., *Phys. Rev.*, **128**, 2121 (1962); Lenander, C. J. and Wong, E. Y., *J. Phys. Chem.*, **38**, 2750 (1963); Ghatikar, M. N., Raychaudhuri, A. K. and Ray, D. K., *Proc. Phys. Soc.* (London), **84**, 297 (1964); Sternheimer, R. M., *Phys. Rev.*, **146**, 140 (1966); Sternheimer, R. M., Blume, M. and Peierls, R. F., *Phys. Rev.*, **173**, 376 (1968).

144. Watson, R. E. and Freeman, A. J., *Phys. Rev.*, **133**, A1571 (1964).

145. Freeman, A. J. and Watson, R. E., *Phys. Rev.*, **139**, A1606 (1965).

146. Jørgensen, C. K., Pappalado, R. and Schmidtke, H. H., *J. Chem. Phys.*, **39**, 1422 (1963).
147. Axe, J. D. and Burns, G., *Phys. Rev.*, **152**, 331 (1966).
148. Watson, R. E. and Freeman, A. J., *Phys. Rev.*, **156**, 251 (1967).
149. Ellis, M. M. and Newman, D. J., *J. Chem. Phys.*, **47**, 1986 (1967).
150. Raychaudhuri, A. K. and Ray, D. K., *Proc. Phys. Soc.* (London), **90**, 839 (1967).
151. Ellis, M. M. and Newman, D. J., *J. Chem. Phys.*, **49**, 4037 (1968).
152. Ellis, D. E. and Freeman, A. J., *Journal de Physique* (Proceedings of the Seventh International Conference on Magnetism, Grenoble 1970), **32**, 1192 (1971).
153. Coles, B. R. and Griffiths, D., *Phys. Rev. Letters*, **16**, 1093 (1966); Orbach, R. and Burr, C. R., *Phys. Rev. Letters*, **19**, 1133 (1967).
154. Hirst, L. L., Williams, G., Griffiths, D. and Coles, B. R., *J. Appl. Phys.*, **39**, 844 (1968); Williams, G. and Hirst, L. L., *Phys. Rev.*, **185**, 407 (1970).
155. Rainford, B. D., Turberfield, K. C., Busch, G. and Vogt, O., *J. Phys. C: Proc. Phys. Soc.*, London, **1**, 679 (1968), and references to earlier work in insulators cited herein; Rainford, B. and Houmann, J. C. G., *Phys. Rev. Letters*, **26**, 1254 (1971).
156. Turberfield, K. C., Passell, L., Birgeneau, R. and Bucher, E., *Phys. Rev. Letters*, **25**, 752 (1970); *Phys., Rev.*, **B4**, 718 (1971).
157. Das, K. C. and Ray, D. K., *Phys. Rev.*, **187**, 777 (1969).

Chapter 7

Transport Properties†

Sam Legvold

Professor of Physics, Senior Physicist, Ames Laboratory of the A.E.C.
Iowa State University, Ames, Iowa 50010, U.S.A.

7.1. INTRODUCTION

The rare earth metals are notoriously poor conductors and are thus typical transition metals in this respect. As a simple illustration we note that at room temperature the electrical resistivity of gadolinium is one hundred times that of copper. This behavior in the rare earths arises in part because these metals have three conduction electrons per atom (Ce, Eu, and Yb are exceptions) in s–d like bands, in part from crystal structure (phonon) effects, and in part because the conduction electrons are exchange coupled to the magnetic moments of the

† Work was performed in the Ames Laboratory of the U.S. Atomic Energy Commission. Contribution No. 2901.

4f electrons, so there is an additional scattering mechanism present when the 4f moments are incompletely ordered. A consequence of this s–f exchange is the indirect exchange interaction between the moments of the "deeply sequestered" 4f electrons.[1] In view of this it is not surprising that sharp excursions from ordinary transport behavior, particularly near magnetic transition temperatures, are observed in the rare earth metals.

Many papers describing the transport properties of these metals have appeared in the literature. Only the salient features of these papers will be covered here and, for the sake of brevity, references will be listed numerically so the names of individuals whose work is cited will be found at the appropriate number in the list at the end of the chapter. There are some review articles in the literature[2–4] with which we shall be concerned as well as some basic solid state physics treatises[5–10] to which reference will be made from time to time.

Because of the indirect exchange interaction it is expected that those elements having the high moments should show more pronounced effects. By Hund's rules† the light rare earths have low magnetic moments ($J = L - S$) and only minor departures from ordinary transport property behavior while the heavy ones exhibit high magnetic moments ($J = L + S$) with large departures from ordinary transport property behavior.

Significant differences in crystal structure[4] between the light and the heavy elements lead to marked differences in the ease with which single crystals may be grown. The double hexagonal close packed (dhcp) form for La, Ce, Nd, and Pr occurs at relatively low temperatures and this makes it difficult to grow and keep single crystals. Hence, nearly all the transport measurements on these elements described in this chapter have been made on polycrystalline samples. Only recently have a variety of methods for crystal growing[11–13] yielded single crystals of some of the light rare earths and results of transport measurements should be forthcoming soon.

In contrast to this situation with the light elements, the heavy ones (all hcp except Yb) have responded favorably to a variety of crystal growing techniques[13, 14] such as strain anneal, zone melting, electrotransport, and levitation melting. Transport measurements have been made on single crystals for these elements and are described in detail here.

One of the intriguing mysteries of the trivalent nonmagnetic metals Sc, Y, La, and Lu is the fact that of the four only La exhibits superconductivity. This would seem to demonstrate the now well known[15] sensitive role of crystal structure (phonons) and conduction bands in this phenomenon. A second interesting case is that of cerium for which the 4f energy level falls close to the Fermi energy so there is a tendency for the metal to transform at low temperatures to a more dense, nearly quadrivalent form.[4] Europium and ytterbium are divalent because they have lower energy with the resultant half full and full 4f shell, respectively.

Better, more complete data, and greater magnetic moment effects have prompted us to begin with a presentation in Section 7.2 of the electrical

---

† When electrons are in an incomplete shell and so have the same n, l quantum numbers then the "lowest lying" electronic state has (i) the maximum multiplicity, $2S + 1$, allowed consistent with the Pauli exclusion principle and (ii) the maximum L consistent with this multiplicity. For more details see Ref. 1.

resistivity and thermal conductivity measurements on single crystals of the heavy rare earths and of yttrium. Section 7.3 is devoted to the Seebeck coefficient measurements on the heavy elements plus yttrium. We then turn in Section 7–4 to the light elements plus ytterbium and cover the electrical resistivity thermal conductivity and Seebeck coefficient measurements which have been made on these. The simplified magnetic superzone theory is applied to the electrical resistivity in Section 7.5. Finally, in Section 7.6 we describe Hall effect measurements.

7.2. ELECTRICAL RESISTIVITY AND THERMAL CONDUCTIVITY OF THE HEAVY RARE EARTHS AND YTTRIUM

7.2.1. Brief Review of Theory

The heavy rare earths and yttrium all have the hcp crystal structure. The stability of this structure up to high temperatures has made single crystal growing possible, so for these metals the transport properties have been measured on single crystals. The Fermi surfaces of the heavy rare earths show a marked departure from the spherical free electron shape. This can be seen in the case of Tm shown[16] in Figure 7–1; the surfaces for the other heavy rare

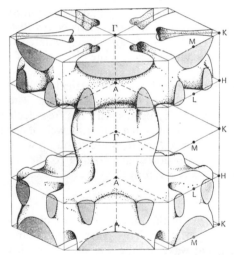

Figure 7–1. The complete Fermi surface for thulium (after Ref. 16). This is the hole surface in the double zone representation.

earths are very much like this. The projected Fermi surface areas in different directions are quite different and vary across the elements. For a detailed description see Chapter 6. The magnetic structures of the heavy elements, described earlier in Chapter 4, are reviewed pictorially in Figure 7–2. It is seen that Yb and Dy have the magnetic moments in the basal plane. Ho has the moment in the basal plane above 19 K and then at lower temperatures the moment moves onto a cone. Sinusoidal, conical, and quasi antiphase domain structures are found in Er and Tm. Gd is not shown because it is a simple

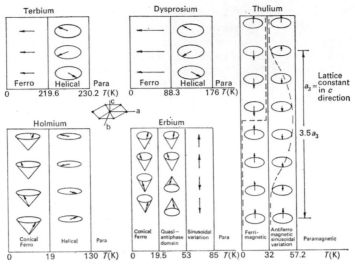

Figure 7–2. Ordered spin structures observed by neutron diffraction.

ferromagnet below 292.7 K with the easy magnetization axis along the c axis above 240 K and along a cone around the c axis below this temperature. Since thermal and electrical conductivities are closely related in metals these two properties will be discussed together with the discussion of Seebeck coefficients to follow in Section 7.3.

We begin with a brief review of some theoretical concepts[5–7] which should aid in the discussion of the experimental results. The usual theoretical approach to electrical resistivity begins with the free electron theory (particle in a box) where the electron moves in a potential well of constant depth ϕ_0 and the quantum mechanical wave functions are standing waves of wave vector **k**. Next the periodic Coulomb potential, ϕ_l, due to the positive ion lattice is introduced so that now ϕ is given by

$$\phi = \phi_0 + \phi_l \tag{7.1}$$

and the wave functions are of the Bloch form, i.e., plane waves modulated by the periodicity of the ion lattice. This periodic Coulomb potential leads to forbidden energy gaps at wave numbers in one dimension of $k = n\pi/a$ where $n = 1, 2, \ldots$ and a is the lattice parameter. For the rare earth metals there is an additional (magnetic) term in the potential, ϕ_m, due to the magnetic interaction between the spin of a conduction electron and the $4f$ magnetic moments of the ions on the lattice sites. The final potential for the conduction electrons then has three terms and is

$$\phi = \phi_0 + \phi_l + \phi_m. \tag{7.2}$$

In analogy with the Coulomb term, the new ϕ_m term leads to additional energy discontinuities[17–19] determined by the magnetic lattice periodicity which can be different from the ionic lattice periodicity. The term magnetic superzones is frequently used to describe the magnetic lattice effect.

Normal transport procedure[5-7] and the relaxation time approach then yield

$$(\rho^{-1})_{ij} = \frac{e^2}{4\pi^3\hbar} \int_{E_F} \tau v_i dS_j \qquad (7.3)$$

where ρ is the resistivity tensor, e is the electronic charge, $\hbar$ is Planck's constant divided by 2π, τ is the relaxation time, v_i is the component of the conduction electron velocity in the i direction and dS_j is the projection of the Fermi surface element in the j direction. The main observable experimental results are usually described by assuming that v_i may be replaced by its average value, v_{av}, over the Fermi surface and that Matthiessen's rule holds[5-7] so that

$$\frac{1}{\tau} = \frac{1}{\tau_0} + \frac{1}{\tau_p} + \frac{1}{\tau_m} \qquad (7.4)$$

where τ_0, τ_p, and τ_m are the relaxation times for electrons scattered by impurities, by phonons (lattice vibrations) and by magnons (magnetic moments or spin waves), respectively.

We call ρ_b the electrical resistivity in a principal crystallographic basal plane direction and let ρ_{0b}, ρ_{pb}, and ρ_{mb} be the corresponding residual, lattice, and magnetic resistivities in a principal basal plane direction;† we also have similar expressions for the resistivity in the principal c direction. We may combine Eqns (7.3) and (7.4), use v_{av}, assume an isotropic τ, and get

$$\rho_b = \frac{A}{\tau_0 \int_{E_F} dS_b} + \frac{A}{\tau_p \int_{E_F} dS_b} + \frac{A}{\tau_m \int_{E_F} dS_b} \qquad (7.5)$$

where A is constant. This expression and a similar one for the c direction have many implications among which is the common additive resistivity form[5-7]

$$\rho_b = \rho_{0b} + \rho_{pb} + \rho_{mb}. \qquad (7.6)$$

At helium temperatures only residual resistivity appears in most instances and the ratio of the residual resistivity in direction b to that in direction c becomes

$$\frac{\rho_{0b}}{\rho_{0c}} = \frac{\int_{E_F} dS_c}{\int_{E_F} dS_b} \qquad (7.7)$$

It is understood that samples must come from the same melt when Eqn (7.7) is applied because residual resistivities vary with the purity of the melt. Also, it is noted that several Fermi surface relationships such as Eqn (7.7) are obtainable in this approximation regime, i.e., if τ is assumed isotropic (some Hall coefficient results and total spin disorder resistivities indicate a slight departure from this). Among the useful relationships we note that for the nonmagnetic

---

† Basal plane transport data are essentially the same along a (line of atoms) and along b; in the text we use the subscript b for either axis.

metals at the same temperature

$$\frac{\rho_b - \rho_{0b}}{\rho_c - \rho_{0c}} = \frac{\int\limits_{E_F} dS_c}{\int\limits_{E_F} dS_b}. \tag{7.8}$$

At higher temperatures the lattice (phonon) terms ρ_{pb} and ρ_{pc} have linear temperature dependence and the magnetic terms[20–24] are constant at their saturation value (paramagnetic range) so we have

$$\frac{d\rho_b/dT}{d\rho_c/dT} = \frac{\int\limits_{E_F} dS_c}{\int\limits_{E_F} dS_b}. \tag{7.9}$$

Another relationship of interest at higher temperatures describes the saturation value of the magnetic term which we will call ρ_{mbsat}. We have

$$\rho_{mbsat} = \rho_b - \rho_{0b} - BT \tag{7.10}$$

where B is the constant high-temperature slope associated with phonon scattering. This expression suggests the following procedure to obtain the saturation magnon resistivity contribution. The linear high temperature resistivity curve should be extrapolated back to zero temperature and then the resistivity intercept minus the residual resistivity will be the desired ρ_{msat}. In our relaxation time approximation then

$$\frac{\rho_{mbsat}}{\rho_{mcsat}} = \int\limits_{E_F} dS_c \Big/ \int\limits_{E_F} dS_b \tag{7.11}$$

It has been shown[20–24] that spin orbit coupling will make the saturation resistivity proportional to

$$(\lambda - 1)^2 J(J+1) \Big/ \int\limits_{E_F} dS$$

for the magnetic rare earths, where λ is the Lande factor appropriate to the $4f$ state. Values of this factor are listed in Table 1.1.

We now add to the above a discussion of thermal conductivity. Several general expressions from formal transport theory[10, 25, 26] are of significance for this discussion. The first is the generally valid statement that in magnetic metals heat is transported by electrons, phonons, and magnons. The thermal conductivity tensor, K_{ij}, is defined by the expression

$$H_i = -K_{ij} \frac{\partial T}{\partial x_j} \tag{7.12}$$

where H_i is the energy flow across unit cross sectional surface area of normal in the i direction per unit time, T is the temperature, and x_j is a Cartesian coordinate along the j direction in space. We will use principal axes a, b in the basal plane and c perpendicular to the basal plane with corresponding principal thermal conductivities K_a, K_b, and K_c each of which is the sum of independent carrier conductivities, K_e for electrons, K_p for phonons, and K_m for magnons, giving equations of the form

$$K_b = K_{eb} + K_{pb} + K_{mb} \tag{7.13}$$

where the subscript interpretation is obvious. For each of the terms on the right there are scattering processes which, to a first approximation, we assume contribute additively to the total thermal resistivity for a given heat carrier. Then we may write the following equations[10, 25, 26]

$$\frac{1}{K_{eb}} = W_{eb} = W_{eb}{}^B + W_{eb}{}^i + W_{eb}{}^e + W_{eb}{}^p + W_{eb}{}^m \tag{7.14}$$

$$\frac{1}{K_{pb}} = W_{pb} = W_{pb}{}^B + W_{pb}{}^i + W_{pb}{}^e + W_{pb}{}^p + W_{pb}{}^m \tag{7.15}$$

$$\frac{1}{K_{mb}} = W_{mb} = W_{mb}{}^B + W_{mb}{}^i + W_{mb}{}^e + W_{mb}{}^p + W_{mb}{}^m \tag{7.16}$$

where the superscripts indicate the scattering agent: B for boundaries of a sample, i for impurities and imperfections, e for electrons, p for phonons, and m for magnons. It is clear from this description that a separation of an experimental thermal conductivity measurement into contributing parts must be approached carefully.

The rare earths are poor electrical conductors and this means that phonons and magnons may contribute significantly to the overall thermal conductivity. This is usually described in terms of the Lorenz function, L,

$$L = \frac{K\rho}{T}. \tag{7.17}$$

When L is larger than L_0, the Wiedemann–Franz ratio, where

$$L_0 = \frac{K_e\rho}{T} = \frac{\pi^2}{3}\left(\frac{\ell}{e}\right)^2 = 2.45 \times 10^{-8}\ \mathrm{W\Omega\ K^{-2}} \tag{7.18}$$

the total thermal conductivity, K, is larger than the electronic thermal conductivity, K_e, making K_p or K_m significant. In Eqn (7.18) ℓ is Boltzmann's constant. Above the highest magnetic ordering temperature the magnetic contribution to the electrical resistivity is constant at the saturation value so one might expect that changes in the magnetic term, K_{bm}, in Eqn (7.13) will be negligible in the paramagnetic range. This leaves the conduction electrons and phonons to dominate the behavior of this regime with the electrons the primary agent in general. That is, the mean free path for magnons in the paramagnetic range (if they exist there) is too short to let the magnons contribute to the thermal conductivity.

7.2.2. Yttrium and Lutecium

The experimental electrical resistivity data[27, 28] for Y and Lu are shown in Figures 7–3 and 7–4, respectively, and are discussed together because both

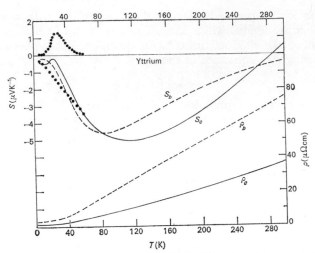

Figure 7–3. Seebeck coefficients and electrical resistivity of yttrium versus temperature
(after Refs. 27, 48).

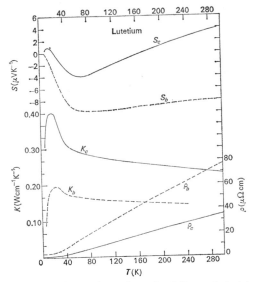

Figure 7–4. Seebeck coefficients, thermal conductivity, and electrical resistivity of
lutecium versus temperature (after Refs. 28, 50).

metals are nonmagnetic, trivalent, hcp metals having similar Fermi surfaces.
It may be seen that the electrical resistivities of both metals are well behaved
and that corresponding curves are almost identical for the two metals. This is
in excellent agreement with Fermi surface calculations which were discussed in
Chapter 6. Because of the importance of the surfaces we again refer to the one
for Tm in Figure 7–1 and use it as typical of all of them. The surface is highly

anisotropic with a considerably greater projected area in the c direction than in a basal plane direction, i.e., along an a or b axis direction. An estimate[29] of the ratio of these projected areas is

$$\int_{E_F} dS_c \bigg/ \int_{E_F} dS_b \sim 2.$$

We may then apply Eqns (7.7)–(7.9) to the electrical resistivity data to make the appropriate comparisons. At a glance we see that the resistivities do indeed show high anisotropy in both metals. The observed ratios found from the data are shown in Table 7.1 and may be described as follows: For Y both samples came from the same single crystal and the residual resistivity ratio was $\rho_{0b}/\rho_{0c}=1.7$. The Lu samples did not come from the same melt but the results for the two different metals certainly do not contradict this anisotropy criterion. The 300 K resistivities give $(\rho_b-\rho_{0b})/(\rho_c-\rho_{0c})=2.0$ for Y and 2.2 for Lu. Next we examine the resistivity slopes at 300 K (these should be the same in the Grüneisen theory) and find

$$\frac{d\rho_b}{dT}\bigg/\frac{d\rho_c}{dT}=1.8 \text{ for Y and 2.0 for Lu.}$$

It is of interest here that the high anisotropy of the resistivity in Y was useful in showing that the proper relationship between single crystal and polycrystal electrical resistivities for hcp metals[30] is

$$\rho_{\text{poly}}=\tfrac{1}{3}\rho_c+\tfrac{2}{3}\rho_a. \tag{7.19}$$

Finally, nonmagnetic metals such as Y and Lu might be expected[5-7] to follow a T^5 power law at low temperature. From a $\log(\rho-\rho_0)$ versus $\log T$ plot the data for Lu gave a $T^{3.9}$ dependence along c and $T^{3.3}$ along b while those for Y gave a $T^{4.4}$ dependence along c and $T^{4.8}$ along b (see Table 7.1).

We turn next to the thermal conductivity observations[28] of Figure 7.4 made on Lu (it is expected that the results for Y will be almost identical to those for Lu). The high peak in K_c of 0.4 W/cm K at 18 K indicates that the sample was of high quality; the residual electrical resistivity of the Lu a-axis sample was 0.76 $\mu\Omega$ cm which at the time was one of the best rare earth metal samples made.

Now, for metals, the general behavior of K at low temperature can be deduced by noting that phonons are not yet numerous enough to conduct heat and that $K\rho/T$ should be nearly constant. This means that since ρ approaches a constant (residual) value at helium temperatures K must be linear in T. The rapid rise in K is, of course, cut off because phonons are generated and these together with the impurities give rise to scattering processes which limit the conductivity, the principal one being electron–phonon scattering in good metals.

The ratio of the thermal conductivities for Lu at 300 K are $K_c/K_b=1.8$. This is to be compared with the 2.2 number for the room temperature electrical resistivity ratio. One is led to the conclusion that the phonons play a part in the conduction of heat at this temperature. The value of $(K_c\rho_c)/T$ is nearly equal to the ideal Lorenz number L_0, so along c nearly all the heat appears to be

12

Table 7.1. Debye θ, ordering temperatures, electrical resistivity, spin disorder resistivity, and thermal conductivity data for the heavy rare earths and yttrium. The resistivities are given in μΩ cm, temperatures in kelvins and room temperature thermal conductivities in W/cm K. See Eqn (7.21) for information on $K_{b\infty}$, $K_{c\infty}$, t_b, and t_c.

	Gd	Tb	Dy	Ho	Er	Tm	Lu	Y
$\theta_D{}^a$	152	158	158	161	163	167	165	218
T_N	—	230.2	176	130	85	57.2	—	—
T_c	292.7	219.6	88.3	19	19.5	(32)	—	—
$[d\rho_b/dT]\,T>300$ K	0.095	0.13	0.15	0.185	0.20	0.22	0.25	0.25
$[d\rho_c/dT]\,T>300$ K	0.08	0.085	0.09	0.11	0.11	0.12	0.12	0.15
Slope ratio	1.2	1.5	1.7	1.7	1.8	1.8	2.1	1.7
ρ_{mb} sat	107	82	62	41	21	12		
ρ_{mc} sat	96	66	44	24.1	13.6	9		
ρ_m ratio	1.1	1.3	1.4	1.6	1.5	1.4		
K_b	0.10	0.10	0.10	0.14	0.13	0.14	0.14	
K_c	0.11	0.15	0.12	0.22	0.18	0.24	0.23	
K_{be}	0.05	0.06	0.07	0.07	0.08	0.08	0.10	
K_{ce}	0.06	0.07	0.10	0.12	0.15	0.15	0.21	
K_{bp}	0.05	0.03	0.04	0.07	0.05	0.06	0.04	
K_{cp}	0.05	0.08	0.02	0.10	0.03	0.08	0.02	
$K_{b\infty}$	0.31	0.20	0.18	0.13	0.12	0.12		
$K_{c\infty}$	0.37	0.30	0.30	0.21	0.23	0.21		
t_b	1200	650	500	240	140	120		
t_c	1240	800	440	220	140	90		

a Debye θ values come from Ref. 4.

carried by electrons. On the other hand, $(K_b\rho_b)/T$ is considerably higher than L_0 and analysis yields an electron part K_{be} of about $\frac{2}{3}K_b$ and a phonon part K_{bp} of about $\frac{1}{3}K_b$. This should be a reasonably valid evaluation since the Debye temperature is 168 K making electronic scattering at 300 K elastic and this would validate the Wiedemann–Franz law.

7.2.3. Gadolinium

Gadolinium, the only rare earth which is a simple ferromagnet, has a half full 4f shell and by Hund's rules the trivalent ion is spherically symmetric and has the spectroscopic designation $^8S_{\frac{7}{2}}$. The transport properties for single crystal samples[31, 32] are shown in Figure 7–5. Here the electrical resistivity due to magnons (spin wave scattering) is the highest for all the metals and leads to the highest resistivity at room temperature for all the metals, the value being 140 $\mu\Omega$ cm in the a axis direction. There is a knee in the curve for the a axis resistivity at the Curie temperature, 293 K, and this a axis resistivity behavior in the vicinity of the highest magnetic ordering temperature is characteristic of all of the heavy rare earths. It is discussed further in Section 7.5.

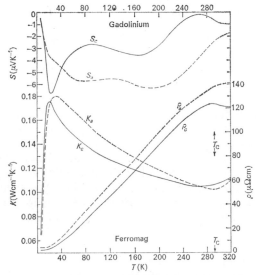

Figure 7–5. Seebeck coefficients, thermal conductivity, and electrical resistivity of Gd versus temperature (after Refs. 32, 48).

Next we use the extrapolation process described earlier to get the saturation magnon resistivity contributions for Gd. The high temperature slopes used and resultant ρ_{mbsat} and ρ_{mcsat} values are shown in Table 7.1. We find there $d\rho_b/dT \sim 0.095$ $\mu\Omega$ cm/K and $d\rho_c/dT \sim 0.08$ $\mu\Omega$ cm/K, the latter being questionable because data have not been taken at sufficiently high temperatures. The ratio is

$$\frac{d\rho_b}{dT} \bigg/ \frac{d\rho_c}{dT} \sim 1.2.$$

The estimated saturation resistivities are

$$\rho_{mbsat} \sim 107 \ \mu\Omega \ \text{cm} \quad \text{and} \quad \rho_{mcsat} \sim 96 \ \mu\Omega \ \text{cm}$$

giving the ratio $\rho_{mbsat}/\rho_{mcsat} \sim 1.1$. We may also compare the residual resistivities in this case since both samples came from the same melt. We use the values from Figure 7–5 and obtain $\rho_{0b}/\rho_{0c} = 1.7$.

By sighting nearly parallel to the page along the a axis resistivity curve of Figure 7–5 it is possible to see a slight bend in the curve at 243 K and this bend correlates nicely with the temperature at which the direction of easy magnetization shifts gradually away from the c axis direction onto a cone around the c axis when the temperature is lowered; the easy direction is along the c axis from 243 to 293 K. In an enlarged plot of the a axis resistivity[33] it was also possible to see a small change in the slope at 210 K which is the temperature at which a cone angle of some 70° is reached. Here is a clear demonstration of the sensitivity of the conduction electrons to the ion spins. We notice though that the shift of the easy direction of magnetization away from the c axis with decreasing temperature is not readily observable in the resistivity along the c axis. It is likely that the abnormally high scattering of electrons near and just below the Curie temperature along the c axis obscures the effect of the change in easy direction. Other ferromagnetic metals do not show as large a hump in the resistivity in the vicinity of the ordering temperature. One realizes that critical scattering arising from short range ordering and from fluctuations accounts for some of the scattering in Gd along the c axis.[34] Alternatively, it is tempting to account for the c axis hump by conjecturing that it is a combined Fermi surface plus fluctuation effect which occurs near the Curie temperature. We assume the Fermi surface suffers fluctuations (e.g. spin up-spin down effects) and exhibits the characteristic flat regions normal to the c direction which could promote some electron scattering. This behavior for Gd then dovetails with the resistivity behavior seen in succeeding members of the $4f$ group and gives smooth variations across the heavy elements. Lattice parameter effects near T_c are related to the hump.[35]

In the case of simple ferromagnets and for temperatures under $T_c/10$ it is expected[36] that $\rho - \rho_0$ will be proportional to T^2. In magnetically anisotropic materials an exponential term is expected.[37] A plot of log $(\rho - \rho_0)$ versus log T was used to obtain the appropriate exponent for T. It was found that the resistivity along the c axis of Gd had a $T^{2.2}$ dependence and that the resistivity along a basal plane direction had a $T^{3.7}$ dependence.

The thermal conductivities[31] of Gd along the a axis and along the c axis are shown in the middle of Figure 7–5. There is the expected sharp rise at low temperatures with rather sharp maxima at 20 and 30 K for the c and a axis samples, respectively. The peaks are high enough to suggest that the sample purity was reasonable. One might expect phonons and magnons to contribute most significantly at or near these temperatures if the theory for conduction in insulators is assumed valid. The expectation is that at temperatures near $\theta_D/10$ to $\theta_D/5$ one should expect the maximum phonon conduction with magnon conduction peaking at or near $T_{mag}/10$ to $T_{mag}/5$ where T_{mag} is the magnetic ordering temperature. In Gd then these conductivity peaks are both in the expected range for heat transport contributions from phonons and magnons. When the $K\rho/T$ values are compared with the ideal electron ratio L_0 it is found that for the basal plane the thermal conductivity runs nearly twice as high as that expected from electrons with the conductivity in the a direction lower than this but still well above the value attributable to electrons alone.

Both samples show marked decreases in conductivity with increasing temperature just above 30 K. This is attributable to the scattering caused by both phonons and magnons in this case. The Curie point is more easily observable for the a axis sample which shows a sharp minimum at this temperature (293 K). The minimum in the c axis data is so broad it would be difficult to pin down the Curie point.

Above the Curie point the thermal conductivity along either direction increases in what appears to be a monotonic fashion. We may understand[31] this by recalling that above the magnetic ordering temperature the ρ_m term in the resistivity expression, Eqn (7.6), is constant along with the constant ρ_0. This gives for the electrical resistivity the form

$$\rho = \rho_0 + \rho_m + \rho_p = c + \alpha T \tag{7.20}$$

where c is the constant $\rho_0 + \rho_m$ and α is the slope in the linear high T region. Since the temperature is high one should have elastic scattering and a separation of the thermal conductivity into the electronic and other terms should be valid.[5,7,10] For K_e we may write

$$K_e = \frac{L_0 T}{\rho} = \frac{L_0 T}{c + \alpha T} = \frac{L_0}{\alpha(1 + (c/\alpha T))} = K_\infty \left(1 - \frac{t}{T}\right) \tag{7.21}$$

where K_∞ is L_0/α and t is $(\rho_0 + \rho_m)/\alpha$, a "characteristic temperature". For Gd the characteristic temperatures are $t_c \sim 1240$ K and $t_b \sim 1200$ K (see Table 7.1). Now K_e makes up a substantial part of K and since K_e increases with increasing T toward an asymptotic value K_∞, one might expect K to increase as the experiments show.

If we assume similar electron scattering contributions from phonons and magnons then at low temperatures it is expected that the electronic thermal conductivity should obey[5]

$$W_e = \frac{1}{K_e} = A/T + BT^2 \tag{7.22}$$

where A and B are constants. To check on this a plot of T/K_e versus T^3 is usually made. When this was done for the data of Figure 7–5 the points were close to a straight line in the 6 to 18 K region. The ordinate intercepts at $T=0$ were considerably below the theoretical ρ_0/L_0 values. This might be expected because of conduction by phonons and magnons.

7.2.4. Terbium

Terbium has eight electrons in the $4f$ shell, one beyond the half filled shell of Gd. The trivalent ion is in the $^7F_6$ state so there is in this case a contribution to the ion magnetic moment from the orbital angular momentum. In terbium, then, one of the eight $4f$ electrons in the $4f$ shell gives rise to a charge distribution which is toroidal in character; this is added to the spherically symmetric half shell of the Gd ion and gives rise to a number of interesting effects related to the magnetic anisotropy and the direction of easy magnetization. The most important feature observed in Tb is the helical magnetic ordering from 220 to 230 K and the effect this has on the electrical resistivity as shown[31, 38] in Figure 7–6. At low temperatures there is just the anisotropy associated with the Fermi surface area projections in the two crystallographic directions. From Figure 7–6 it may be seen that $\rho_{0b}/\rho_{0c} \sim 1.3$. The temperature dependence of the resistivity below 20 K is complicated by the effects of

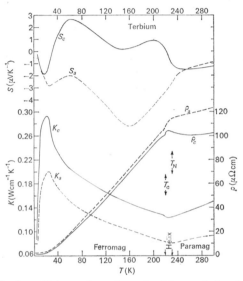

Figure 7–6. Seebeck coefficients, thermal conductivity, and electrical resistivity of terbium versus temperature (after Refs. 31, 38, 48).

magnetic anisotropy and spin wave energy gaps. Analysis of resistivity data[39] from 9–35 K for Tb gives a gap of about 39 K. This appears to be high so this approach may be marginal. Below 20 K the resistivity along b follows $T^{4\cdot3}$ and along c follows $T^{4\cdot3}$. Here, a phonon-like T^5 might be expected.

As the temperature is increased terbium shows very little anisotropy in the resistivity in the ferromagnetic range below 220 K. At the onset of the helical magnetic moment phase there is a sharp jump in the c axis resistivity. This is a manifestation of the magnetic superzones. The magnitude of the jump in resistivity here is not great because the additional scattering is small compared to the spin disorder scattering which is present at this temperature. There is also some chance that part of this jump in resistivity may come from a change in magnon dispersion in going from the ferromagnetic phase to the helical phase. The a axis resistivity shows a knee at the Néel point of 230 K much like that in Gd. The resistivity in the c axis direction shows a well developed hump in the helical moment region from 220–230 K. Above the 230 K ordering temperature the well developed minimum at 260 K is reminiscent of the behavior of Gd above its Curie point. Again in Tb anisotropy develops at temperatures above the magnetic ordering temperature. High temperature resistivity slopes are found in Table 7.1 of $d\rho_b/dT \sim 0.13$ and $d\rho_c/dT \sim 0.09$ and the ratio of the two is 1.4. Linear extrapolations back to $T=0$ from the high temperature region then give $\rho_{mb(sat)} = 82$ $\mu\Omega$ cm and $\rho_{mc(sat)} = 74$ $\mu\Omega$ cm, the ratio of the two being 1.2.

The effect on the Tb c axis resistivity of an external transverse magnetic field applied along the easy b axis direction is shown[38] in Figure 7–7. As the field is increased the anomalous resistivity hump between 200 and 230 K decreases.

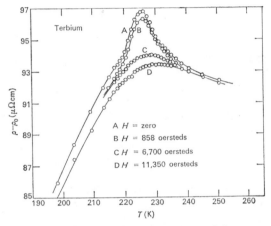

Figure 7–7. Electrical resistivity of terbium along the c axis in a transverse (along b) external magnetic field (after Ref. 38).

At an applied field of 11 kOe the sample is ferromagnetic, the magnetic super-zones are eliminated, and the behavior of the sample is like that for Gd near its Curie point. It appears that short range ordering effects, critical phenomena, and the combined Fermi surface plus fluctuation effects described earlier are manifested in Tb like they were in Gd.

The thermal conductivities[31] of Tb do not rise as smoothly to their low temperature peaks as in the case of Gd and the peaks in the two directions are now at very near the same temperature (~ 22 K) as can be seen in Figure 7–6. This most likely arises because of magnon effects but is really not understood. The rapid fall-off in the K values at temperatures in the vicinity of and above 30 K (especially for K_c) is a significantly different feature as compared with the Gd case. One conjectures that a rapid rise in the scattering of electrons by both phonons and magnons takes place to bring this about.

The onset of the helical structure at 220 K is readily observed on either thermal conductivity curve but is more fully developed along the c axis curve in this case as contrasted with the Gd thermal conductivity data. This is as it should be here because of the strong effect of magnetic superzones on the Fermi surface area projection in the c axis direction as compared with the basal plane direction. It was possible from carefully spaced points in temperature to see both ends of the helical phase in the thermal conductivity along the c axis. The steady rise in the thermal conductivities above the ordering temperature

is in line with the behavior expected when the magnetic scattering has saturated; this was discussed in some detail in connection with Gd. For Tb the
characteristic temperature is 650 K in the basal plane direction and 800 K in
the c direction as seen in Table 7.1. From the same table the thermal conductivity ratio at 300 K is $K_c/K_b \sim 1.5$. We assume that K_c can be separated at
this temperature and find that along c equal amounts of heat are carried by
electrons as compared with phonons. In the basal plane direction the electronic
term is twice as large as the phonon term.

7.2.5. Dysprosium

Dysprosium has nine electrons in the $4f$ shell so the ion is designated
spectroscopically by the symbol $^6H_{15/2}$. Two magnetic transitions are nicely
spaced in temperature at 88 and 176 K making the spin influence on the
transport properties[28] stand out clearly and distinctly as can be seen in
Figure 7–8. The electrical resistivity in the c direction shown in this figure was
among the clues which led to the proposal of the magnetic superzones. The
ferro to helix magnetic transformation at 88 K causes a sharp rise in the
electrical resistivity in the c direction whereas along a basal plane direction the
magnetic transformation causes only a slight irregularity which does not show
on the scale of the figure. The transition from the helical state to the disordered
state shows up as a sharp change in slope for the resistivity along the a axis and
the very unusual hump and resistivity minimum for the resistivity along the c
axis. Both of these curves are consistent with those of terbium when we make
allowances for the closeness of the two ordering temperatures in terbium. The
extension of temperature-dependent spin scattering effects well into the

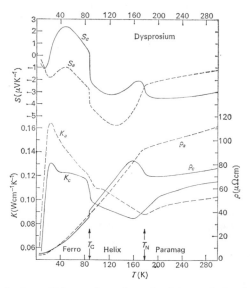

Figure 7–8. Seebeck coefficients, thermal conductivity, and electrical resistivity of
dysprosium versus temperature (after Refs. 28, 48).

paramagnetic region in the c direction is much like that described for Gd and Tb.

Fermi surface effects in Dy may be described as follows: The high temperature resistivity curves yield slopes as shown in Table 7.1 and these have the ratio $(d\rho_b/dT)$ to $(d\rho_c/dT) \sim 1.5$. The ratio of ρ_{mbsat} to ρ_{mcsat} is also found to be 1.5. The ratio of the residual resistivities for two samples from the same melt[40] gives 1.2 (these were not the same samples as those for which the resistivities are shown in Table 7.1).

The observed jump in resistivity along the c axis at T_c amounts to 5 $\mu\Omega$ cm out of a total of 35 $\mu\Omega$ cm. This represents a jump of about 15% and is higher than an estimate of the Fermi surface superzone effect.[29]

In the temperature range below 20 K the electrical resistivity along the a axis follows a T^5 law while the resistivity along the c axis direction obeys a $T^{5.3}$ law. The energy gap in the magnon dispersion law would prevent low temperature magnon scattering and the general T^5 law should describe the behavior here (an analysis of the resistivity data between 11 and 25 K gave a gap[39] of 40 K which seems too high just as in the case of Tb).

The thermal conductivity results[28] for Dy show the magnetic transitions very well indeed in both crystallographic directions as seen in Figure 7–8. There is a sharp drop from 0.12 to 0.10 W/cm K, or 20% in the K_c coefficient at the 88 K ferro to helical magnetic transformation and this change is greater than the change in ρ_c. There is an anomalous drop and leveling off in K_a at this transition where essentially no effect is observable in ρ_a and this indicates that thermal conduction by magnons and phonons must change in an abrupt manner in the a direction to account for the data.

The basal plane (a axis) thermal conductivity of Dy shows a rather sharp minimum at the Néel point (176 K) and also shows the expected monotonic increase with increasing temperature above this temperature as described earlier in this section. The K_c behavior near 176 K is more complex but might be expected because of the electrical resistivity hump near and below this temperature. The peak in the electrical resistivity occurs at 160 K and this coincides with the sharp minimum in K_c. There is a change in the slope of K_c near 200 K which is close to the temperature at which there is a broad minimum in ρ_c. It is believed that critical scattering is partially responsible for this behavior.

The low temperature thermal conductivity maximum in the a direction is higher than that in the c direction; part of this difference is believed to be a consequence of a difference in sample purities. Since the samples did not come from the same melt their purities are different; the residual electrical resistivities indicate that the c axis sample had about twice as many impurities as the a axis sample. The two thermal conductivities hit their low temperature maxima at the same temperature, 25 K. The shapes of the curves below this temperature are much alike. They differ from terbium in this respect and it is difficult to understand this difference since both elements have magnon dispersion energy gaps[39] of ~ 40 K and should behave similarly at low temperatures. The ratio of the high temperature thermal conductivities (see Table 7.1) for Dy is $K_c/K_b \sim 1.2$. A separation of the coefficients shows that electron conduction in the basal plane is nearly twice as large as phonon conduction and that in the c direction the electron term is five times as large as the phonon term.

7.2.6. Holmium

There are 10 electrons in the $4f$ shell in holmium and this leaves just 4 electrons with unpaired spin so the ionic spectroscopic state is $^5I_8$. The ionic anisotropy effect is considerably different from that for Tb because three m_l levels beyond the half shell are occupied and the $4f$ charge cloud is no longer shaped like a torus. This causes the low temperature ($T \sim 19$ K) ferromagnetic state to have the magnetic moments canted from the basal plane. The helical magnetic phase extended from 19 to 130 K in the case of Ho. The lower temperature transition is not detectable in the resistivity measurements along

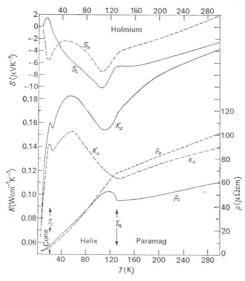

Figure 7–9. Seebeck coefficients, thermal conductivity, and electrical resistivity of holmium versus temperature (after Refs. 31, 41, 48).

a basal plane direction; it is barely detectable in an expanded plot of the resistivity measurements along the c axis and so is not prominent enough to be seen in the resistivity plots[31, 41] of Figure 7–9. The Néel point, on the other hand, is readily observed on the plots with the a axis curve showing the usual knee at 130 K and the c axis curve showing the hump just below the ordering temperature. The latter is a manifestation of the magnetic superzones.

The slopes of the resistivity curves at high temperature give the ratio

$$\frac{d\rho_a}{dT} \bigg/ \frac{d\rho_c}{dT} \sim 1.7$$

and the magnetic disorder anisotropy found for this case is $\rho_{mbsat}/\rho_{mcsat} \sim 1.7$. These are very consistent results and indicate that the ratio of the projected Fermi surface areas in the two crystallographic directions should be nearly this number. Since the samples used here were not from the same melt a comparison of the residual resistivities would be meaningless. According to the residual resistivity the c axis sample had about twice as many impurities as the basal plane sample.

The resistance of Ho along c in the presence of a transverse magnetic field in the easy b direction is shown[42] in Figure 7–10. At 42.2 K $\Delta\rho$ is positive at low fields as a complex fan-like magnetic moment arrangement is induced. Then above an applied field of 20 kOe the helical and fan-like magnetic structures start to break down into a ferromagnetic arrangement and this reduces the resistivity. At 4.2 K there is a large negative $\Delta\rho$ in the presence of a magnetic field until the moments are pulled out of the conical magnetic configuration

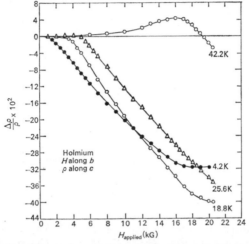

Figure 7–10. Change in electrical resistivity of holmium along the c axis in a transverse (along b) external magnetic field (after Ref. 42).

into the ferromagnetic state along H. The effect is large because magnetic superzones are eliminated. There is a slight positive trend for $\Delta\rho$ beyond magnetic saturation and this is most likely an electron trajectory effect. The high field requirement for saturation was the consequence of the geometry and demagnetization factor. A longitudinal magnetic field applied on a b axis sample gave the results shown[42] in Figure 7–11. At 47.3 K $\Delta\rho$ is positive as a fan structure grows and then goes rapidly negative at 13 kOe when the sample becomes ferromagnetic. As the temperature is lowered below 20 K $\Delta\rho$ becomes less negative because spin disorder effects decrease with temperature. At 4.2 K electron trajectory effects make the resistance increase with increasing fields after the sample becomes ferromagnetic.

A log–log plot of the resistivity data below 15 K led to a temperature dependence of $T^{3.3}$ along the basal plane and T^4 along the c axis. There is a rapidly changing resistivity at low temperatures in this case which comes from magnons.

The thermal conductivities[31] of Ho, as shown in Figure 7–9, have the normal sharp rise at low temperature and then show sharply developed peaks right at the magnetic transition temperature 19 K where the magnetic moments move from a conical ferromagnetic low temperature phase to the basal plane helical phase. After a sharp drop off to a shallow minimum at 24 K both K_a and K_c rise to broad maxima at 55 K. Apparently the magnetic phase transition has suppressed the high conductivity peaks normally encountered at low temperatures. This explains the 55 K "recovery" from the abnormal electron scattering near the phase transition.

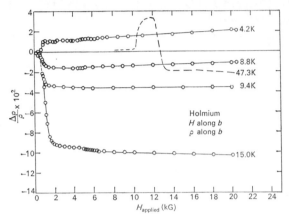

Figure 7–11. Change in electrical resistivity of holmium along the b axis in a longitudinal external magnetic field (after Ref. 42).

The behavior of K_b and K_c near the 130 K Néel point is very much like that for Dy. A broad minimum in K_c occurs at 112 K and this is closely tied to the maximum in the resistivity hump which appears at 115 K. Then K_c shows a well developed slope change at 130 K where K_b shows a sharp minimum. As the temperature rises both K_c and K_b increase in the fashion described by Eqn (7.21) with characteristic temperatures of $t_b \sim 241$ K and $t_c \sim 221$ K.

At room temperature the ratio K_c/K_b is 1.6 which is close to the ratio found for the electrical resistivity slopes. Estimates of the phonon conductivities may be made (see Table 7.I), $K_{bp} \sim 0.07$ W/cm K which amounts to about half the total and $K_{cp} \sim 0.10$ W/cm K which is less than half the total.

7.2.7. Erbium

To display the transport properties of erbium[28, 43] more effectively the temperature axis has been expanded by a factor of two in Figure 7–12. The magnetic ordering temperatures are low (19.5 and 85 K) and the interesting observations are therefore all below 150 K. The spectroscopic designation for the Er ion is $^4I_{15/2}$ since it has four electrons beyond the half filled shell of Gd and this leaves three unpaired spins in the $4f$ shell. The exciting feature in Er is a giant disturbance in the electrical resistivity in the c direction which is caused by the periodic magnetic moment along this direction. We begin the discussion with a look at the low temperature electrical resistivity behavior; here the magnetic moments are on a cone around the c axis giving ferromagnetism with the c axis the easy direction of magnetization (see Chapter 3)

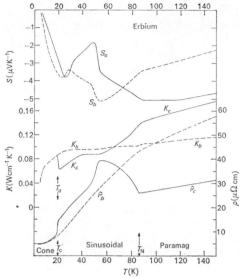

Figure 7–12. Seebeck coefficients, thermal conductivity, and electrical resistivity of erbium versus temperature (after Refs. 28, 48).

and with a helical arrangement (axis of the helix is along the c direction) of the basal plane component of the magnetic moments. The low temperature dependence of the resistivity up to 15 K is roughly estimated as $T^{3.2}$ for the b axis direction and as T^5 for the c axis direction. At 19.5 K Er has a magnetic transformation out of the ferromagnetic phase into the quasi antiphase domain form which has a nearly square wave modulation along the c axis direction with the moments directed alternatedly up and down along the c axis. At this 19.5 K transition the c axis resistivity jumps 5 $\mu\Omega$ cm precipitously because of the magnetic superzones above this temperature. There is no marked effect at 19.5 K in the b direction. As the temperature rises above 19.5 K both samples show the normal resistivity increases which come from magnon and phonon scattering. This continues up to the 85 K Néel point for the b axis sample and then the characteristic change in slope (knee) is seen. On the other hand, ρ_c shows a relatively rapid increase with increasing T starting at 45 K and peaking out at 55 K where the basal plane component of the magnetic moment becomes disordered and where the square wave form gives way and goes into the sinusoidally varying form. As the temperature continues upward the amplitude of the sinusoidal magnetic moment wave decreases and the superzone effect along with the spin scattering decrease until the 85 K order–disorder temperature is reached. This case is discussed more fully in Section 7.6 where a fit of the simplified theory to the data is given.

The two Er samples came from different melts so the residual resistivity ratio is not germane. The high temperature resistivity slope ratio as given in Table 7.1 is

$$\frac{d\rho_b}{dT} \bigg/ \frac{d\rho_c}{dT} \sim 1.8$$

and the spin disorder resistivity ratio is $\rho_{mbsat}/\rho_{mcsat} \sim 1.5$. These should be compared with the estimated[29] Fermi surface area projection ratio

$$\int_{E_F} dS_c \Big/ \int_{E_F} dS_b \sim 2.1$$

which is considerably higher than the experimental values. The precipitous change in ρ_c at 19.5 K of 5 $\mu\Omega$ cm is considerably greater than the 6 % effect estimated for the magnetic superzone effect.

The thermal conductivities[28] for Er are shown across the middle of Figure 7–12. At low temperatures K_b and K_c are identical and increase with temperature up to 19 K where K_c shows a sharp decrease. The latter correlates perfectly with the sharp rise in the electrical resistivity along the c axis at this magnetic transition temperature where the magnetic superzone effect occurs. As the temperature is increased there are only slight increases in K_b at the 55 K square wave to sine wave transition and at the 85 K Néel point. On the other hand, K_c shows sizable changes in slope at both 55 and 85 K with behavior consistent with the c axis electrical resistivity along the way. The characteristic temperatures associated with the increasing thermal conductivity coefficients above 85 K are $t_c \sim 140$ K and $t_b \sim 140$ K as shown in Table 7.1. The ratio of the thermal conductivities at room temperature is $K_c/K_b \sim 1.4$ and it may be seen in Table 7.1 that the estimated phonon conductivity is about 20 % of the total in the c direction as compared with 40 % in the b direction.

7.2.8. Thulium

Thulium has 12 electrons in the $4f$ shell with 2 electrons having unpaired spins and the spectroscopic ion designation is $^3H_6$. Some of the transport

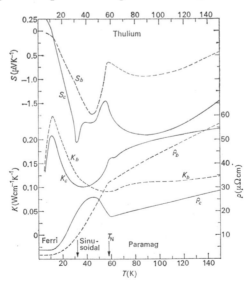

Figure 7–13. Seebeck coefficients, thermal conductivity, and electrical resistivity of thulium versus temperature (after Ref. 44).

properties[44] of Tm are shown in Figure 7–13. The electrical resistivity in the c direction has a beautifully developed anomaly below the 57.2 K Néel point. In this instance the magnetic superzones make the c axis resistivity at the top of the 45 K hump almost twice as large as the phonon-only contribution to the resistivity at this temperature.

Turning to the low temperature region we find that for temperatures below 20 K a fit of the resistivity data to a power law gives $(\rho_c - \rho_{0c}) \sim T^5$ and $(\rho_b - \rho_{0b}) \sim T^{4.6}$ although the resistivity here is expected to have a more complex temperature dependence because of magnon effects. We note next that at helium temperatures the magnetic lattice (see Chapter 3) consists of four layers with spins directed up the c axis followed by three layers with spins directed down the c axis. The incommensurate magnetic lattice superzone effect extends to the residual resistivities in this case because $\rho_{0c} > \rho_{0b}$ and this is opposite to the relationship for the ferromagnetic (commensurate) lattice we have described before when the samples came from the same melt as is the case for the Tm samples here. By measuring the residual resistivities in a strong magnetic field directed along the c axis it would be possible to get a ratio to compare with the projected Fermi surface area ratio obtained from the paramagnetic Fermi surface.

At higher temperatures the resistivity in the b direction shows a knee at the 57.2 K Néel point. The ratio of the high temperature resistivity slopes gives

$$\frac{d\rho_b}{dT} \bigg/ \frac{d\rho_c}{dT} \sim 1.8$$

as seen in Table 7.1. In order to find the saturated c axis spin disorder resistivity for Tm it is necessary to subtract a residual resistivity calculated from the basal plane residual and the high temperature resistivity slope ratio. This must be done because the superzone effect extends to ρ_{0c}. The method yields an estimated ρ_{0c} of about 1 $\mu\Omega$ cm so $\rho_{mcsat} = 9$ $\mu\Omega$ cm as shown in Table 7.1 which also shows $\rho_{mbsat}/\rho_{mcsat} = 1.4$. The estimated Fermi surface projection ratio for Tm is

$$\int_{E_F} dS_c \bigg/ \int_{E_F} dS_b \sim 2$$

and this agrees only tolerably with the experimental high temperature resistivity slope ratio.

Thermal conductivity results[44] for Tm appear in the middle of Figure 7–13. At low temperatures both K_b and K_c rise sharply to maxima at 12 K with K_b reaching a higher maximum than K_c. The higher value for K_b was expected because the c axis electrical resistivity was higher than that in the basal plane as a result of the magnetic superzone and the latter should also have a similar strong influence on the c axis thermal conductivity by way of the electronic term, for certain, and possibly indirectly as well. It would be interesting to see the results of thermal conductivity measurements carried out in a strong magnetic field (50 kOe) directed along the easy c direction for then the samples would be ferromagnetic and the superzone effect would be missing.

The low temperature maxima for both K_b and K_c are typical of relatively pure metals. We apply Eqn (7.22) and obtain plots T/K versus T^3 for both K_b

and K_c which are linear up to 20 K. One can conclude that electronic conduction dominates in this range and that whatever effect magnons have must be like the effect of phonons.

As the temperature is increased above 12 K the coefficient K_c falls to a minimum at 37 K while the coefficient K_b drops off in a somewhat similar fashion to a minimum at the 57.2 K Néel point. The c axis minimum is reasonably close to the c axis resistivity maximum as it should be because of the strong role played by electronic conduction. Near the Néel point K_c shows a leveling off over a narrow temperature interval and this is followed by the usual rise found above the order–disorder temperature. K_a also shows a small rise as the temperature goes into the paramagnetic region. The characteristic temperatures for the paramagnetic region (see Table 7.2) are $t_b \sim 120$ K and $t_c \sim 90$ K. The ratio of the room temperature thermal conductivities is $K_c/K_b \sim 1.7$ and the proportion of the heat carried by phonons is estimated to be 40% along the b axis and 30% along the c axis.

7.2.9. Summary of Electrical Resistivity Results

With the aid of several figures we now summarize the resistivity results and show how some features of these results vary as we go across the heavy series of elements. In Figure 7–14 the slopes of the electrical resistivity curves at high temperatures (taken from Table 7.1) are shown. The basal plane slopes are at the top of the figure and fall close to a straight line except for Gd which seems to be lower than an orderly variation would suggest. The fact that the slope increases in going from Gd to Lu indicates a general decrease in the projected area of the Fermi surface in the basal plane direction. One visualizes that the tree trunk (or torso) of the surface decreases in diameter across the series; de Haas–van Alphen confirmation of this on high purity samples is anticipated. The high temperature resistivity slopes for the c axis samples appear along the lower line of Figure 7–14. The point for Gd is just an estimate based on measurements[32] to 375 K which is not quite high enough for reliability. The general slight upward trend of the slope going toward Lu indicates that the c

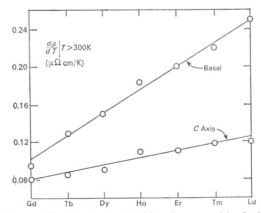

Figure 7–14. Slope of the electrical resistivities along c and in the basal plane for heavy rare earths at $T > 300$ K.

direction projected Fermi surface area changes very slowly across the heavy rare earths.

A plot of the electrical resistivity (residual subtracted) in the c direction versus reduced temperature, T/T_c or T/T_N, for all or the heavy magnetic rare earths appears in Figure 7–15. Among the features which catch the eye is the orderly fashion in which the curves fall one below the other demonstrating the decreasing total spin disorder resistivity in going from Gd to Tm. There is also a very consistent rise in the resistivity at and just below the highest ordering temperature even for gadolinium which is the only one of the lot which is ferromagnetically ordered at all temperatures. Dy shows the highest actual rise

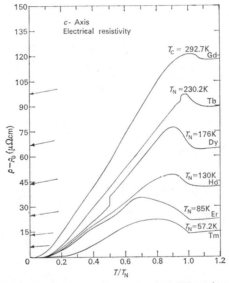

Figure 7–15. Electrical resistivity along the c axis versus T/T_N for the heavy rare earths.

of resistivity (nearly 15 $\mu\Omega$ cm) whereas, relative to the resistivity at the ordering temperatures, Er and Tm show the greatest rises and these are close to 50 %. The curves at the highest temperatures shown are really not quite linear but we may use the slopes of Table 7.1 to estrapolate to zero T and get the actual total spin disorder resistivities, ρ_{mcsat}. The basal plane resistivities with residuals subtracted are shown in Figure 7–16. Again a steady decrease in saturation spin disorder spaces the curves in an orderly fashion and only very small departures from smooth behavior show up except, of course, for the sharp changes in slope of the curves at the highest ordering temperature. The change in slope of the resistivity of Gd when the magnetic moment direction moves away from the c axis is detectable at the reduced temperature 0.8. Also a small precipitous change at the reduced temperature of 0.94 is observable for Tb. The linear high temperature curves are extrapolated to zero temperature to get the total spin disorder resistivities, ρ_{mbsat}, of Table 7.1 and these are discussed next in a fashion which follows presentations in the journals.

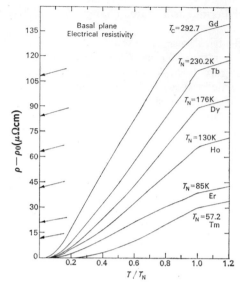

Figure 7–16. Basal plane electrical resistivity versus T/T_N for the heavy rare earths.

Theory indicates[20–24] that the saturated spin disorder resistivity should be proportional to

$$(\lambda-1)^2 J(J+1)\Big/ \int_{E_F} dS$$

where spin-orbit coupling is present and to

$$S(S+1)\Big/ \int_{E_F} dS$$

if spin-orbit effects are neglected. The Fermi surface effect is neglected and the spin disorder values of Table 7.1 are plotted against $(\lambda-1)^2 J(J+1)$ and $S(S+1)$ for the heavy magnetic rare earths[45] at the top of Figure 7–17. The b axis results fall remarkably close to a straight line in the $S(S+1)$ plot; only the point for erbium is off the line an appreciable amount. The c axis results at the middle of the figure show a much better fit to the $(\lambda-1)^2 J(J+1)$ than to the $S(S+1)$ abscissa.

We now recall that the projected Fermi surface area along the c axis changes very little across the series of heavy elements ($d\rho_c/dT$ at high temperatures is nearly constant) whereas the projected Fermi surface area in a basal plane direction changes a great deal ($d\rho_b/dT$ changes by a factor of two). This means that the projected Fermi surface area must be included in examining the overall spin scattering. If we divide the ρ_{mbsat} values by the Fermi surface anisotropy ratios,

$$\left(\frac{d\rho_b}{dT}\Big/\frac{d\rho_c}{dT}\right)$$

from Table 7.1, we obtain corrected values proportional to $(\lambda-1)^2 J(J+1)$ as shown at the lower right of Figure 7–17, indicating that the spin-orbit coupling effect is valid for both crystallographic directions.

This new finding explains why attempts to show the spin-orbit coupling effect on resistivity by use of polycrystalline samples were so obfuscatory except for the results of a dilute alloy study.[46]

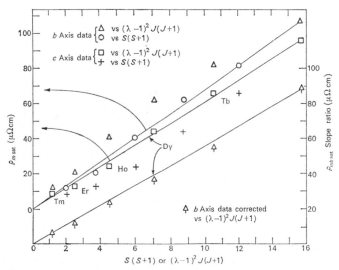

Figure 7–17. Total spin disorder resistivity versus $S(S+1)$ and $(\lambda-1)^2 J(J+1)$ is shown at the top left. Basal plane data divided by the resistivity slope ratio of Table 7.1 versus $(\lambda-1)^2 J(J+1)$ is at the lower right.

7.3. SEEBECK COEFFICIENTS OF THE HEAVY RARE EARTHS AND OF Y

7.3.1. Theory of Seebeck Coefficients

The Seebeck coefficient for a metal depends sensitively on the electronic structure of the metal (the Fermi surface) and on the manner in which the conduction electrons are scattered in their tortuous trajectories when temperature gradients are present. The interpretation of Seebeck measurements is much more obtuse and tenuous than that of the thermal and electrical conductivities. This is particularly true in the case of the heavy rare earth elements because of their complex magnetic phases which can affect the Seebeck coefficient profoundly by way of the magnons and by way of concomitant magnetic superzone energy gaps.

When a magnetic metal is subjected to a temperature gradient the voltage generated or Seebeck tensor coefficient, S_{ij}, arises from three main processes: conduction electron thermal diffusion with which we associate the tensor term S_{ij}^d, phonon drag which implies that in their streaming motion the phonons can carry conduction electrons along and with which we associate the tensor term S_{ij}^p, magnon drag which corresponds to phonon drag and with which we associate the tensor term S_{ij}^m; superscripts identify the origin of the term.

As a first approximation[5-8] the contributions are treated as independent of each other and are added together

$$S_{ij} = S_{ij}{}^d + S_{ij}{}^p + S_{ij}{}^m \qquad (7.23)$$

The diffusion part is treated first. The relaxation time approximation of formal transport theory[5] gives

$$S_{ij}{}^d = \left[\frac{\pi^2 k^2 T}{3e} \right] \sum_l \left[\rho_{il}(E) \frac{d}{dE} \rho_{lj}{}^{-1}(E) \right]_{E=E_F} \qquad (7.24)$$

where ρ_{il} is the electrical resistivity tensor. It is seen here that the Seebeck coefficient does indeed depend on the Fermi surface in a sensitive way. We nevertheless, proceed with the crude simplifying assumptions applied earlier to Eqn (7.3) and assume cubic symmetry so an isotropic Seebeck coefficient expression may be examined. This is

$$S^d = \left[\frac{\pi^2 k^2 T}{3e} \right] \left[\frac{d}{dE}(\tau v) + \frac{dA(E)}{dE} \right]_{E=E_F} \qquad (7.25)$$

where

$$A(E) = \int_E dS$$

is the Fermi surface for energy E, τ is the conduction electron relaxation time and v the conduction electron velocity. The second term of Eqn (7.25) is positive for electron surfaces and negative for hole surfaces, so if the first term is small (τ and v are slowly varying functions of E) we may get some information from the sign of the Seebeck coefficient. For a spherical Fermi surface and a dominant second term S^d will rise linearly with T at low temperature;

$$S^d = \frac{\pi^2 k^2 T}{3eE_F}. \qquad (7.26)$$

There is always the complication of impurity effects which generally show up at low temperatures especially when the impurities carry localized magnetic moments and are imbedded in a nonmagnetic metal.[8]

The contribution to the Seebeck coefficient which comes from phonon drag has been studied theoretically in some detail. For temperatures below the Debye temperature, θ_D, S^p can be important. The theory for an istropic metal predicts the following temperature dependences

$$S^p \sim T^3, \qquad T < \theta_D \qquad (7.27)$$

$$S^p \sim T, \qquad T \sim \theta_D \qquad (7.28)$$

$$S^p \sim \text{constant}, \qquad T > \theta_D. \qquad (7.29)$$

Theoretical investigations of the magnon[47] term indicate that temperature dependences just like those for phonons should be found with the magnetic ordering temperature approximately replacing the Debye temperature. Hence, one looks for a special peak in S in the temperature range

$$\frac{\theta_D}{10} < T < \frac{\theta_D}{5} \quad \text{or} \quad \frac{T_N}{10} < T < \frac{T_N}{5}.$$

Since T_N and θ_D are not far apart in the metals under study there is some difficulty in separating phonon and magnon drag effects. Alloy studies which suppress the phonon term might also affect the magnon term by perturbing the magnetic lattice. On the other hand the magnon contribution may hopefully be isolated by use of the Seebeck coefficients for Y which has the same hcp lattice and in which magnon effects are of course, absent.

7.3.2. Experimental Results

The Seebeck coefficients for Y in the 4 to 300 K temperature range are shown[48] as S_c and S_b at the top of Figure 7–3. Here S_c represents the Seebeck tensor element for the principal c axis and S_b the tensor element for a basal plane principal direction. There appears to be a small impurity or possibly a phonon drag effect in S_c at 24 K (a positive peak superimposed on a negative trend) and then both coefficients change almost linearly with temperature up to about 70 K. In the figure the dotted S_c curves are estimated phonon drag (upper segment) and electron diffusion (lower segment) parts. Both coefficients are negative in sign and this agrees with the principal part of the Fermi surface which describes holes in the third Brillouin zone. We can surmise that the second term in Eqn (7.25) is then possibly larger than the first term. There is some anisotropy in the Seebeck coefficients of yttrium because S_c shows a minimum at 117 K while S_b has a minimum at 80 K. Only very small differences between S_a and S_b have been observed so in the main the metals are isotropic in the basal plane; this is mentioned because there are slight departures from the hcp structure in some cases.[49]

After passing through minima both Seebeck coefficients for Y rise with increasing temperature toward zero and S_c changes sign and becomes positive at 284 K. It appears that S_b will change sign and go positive at about 330 K if its slope at 300 K continues unchanged. This is not too surprising inasmuch as both holes and electrons participate in the conduction process and the total Seebeck coefficient is really a weighted sum[8] of the form

$$S = \sum_i \left[\frac{S_i}{\rho_i}\right] \Big/ \sum_i \left[\frac{1}{\rho_i}\right] \tag{7.30}$$

where ρ_i is the resistivity of carrier i and S_i the Seebeck coefficient of carrier i. The dotted positive peak at 24 K for S_c falls near $\theta_D/10 \sim 22$ K and could well be a phonon drag effect especially since a similar effect appears in S_c for Lu which is discussed next.

At the top of Figure 7–4 the Seebeck coefficients for Lu are shown.[50] The dotted S_c curves are estimated phonon drag (upper segment) and electron diffusion (lower segment) parts. The peak which is at 15 K is very close to the expected temperature for phonon drag in Lu, viz, $\theta_D/10 \sim 17$ K and correlates satisfactorily with the well developed maximum in the thermal conductivity K shown in the middle of Figure 7–4. The sample was of high purity which makes this interpretation plausible and it also agrees with a similar peak in Y. The coefficient S_b for Lu is quite different from that for Y because it remains negative all the way to room temperature and shows no inclination to go positive. Also the coefficient S_b for Lu has twice the magnitude of S_b for Y. This behavior of S_b leads to considerable anisotropy in the Seebeck coefficients of Lu at 300 K whereas in yttrium the anisotropy was small. In Lu the coefficients reach their minima at 70 K along c and 85 K along b and both are lower than the corresponding yttrium values. This follows their Debye temperature, Lu having a much lower value than Y as shown in Table 7.1.

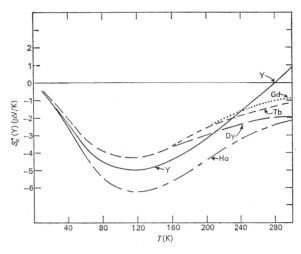

Figure 7–18. Seebeck coefficients of yttrium along c axis, S_c^* (Y), adjusted to make $S_c - S_c^*$ (Y) for gadolinium, terbium, dysprosium, and holmium go to zero above the ordering temperature.

We turn next to the Seebeck coefficients[48] of Gd shown at the top of Figure 7–5. Here S_c and S_a are both negative over the temperature range from 10 to 300 K. They show quite different behavior between 20 and 60 K with S_c having a sharp peak and S_a behaving linearly as though two competing effects were present. Relatively uniform phonon contributions across the heavy rare earths are anticipated if phonon–magnon interactions are not too important. We therefore tried to estimate the magnon contributions to the Seebeck coefficients by subtracting the yttrium coefficients from those measured for the magnetic elements. We obtained the requisite near zero value for the difference above the magnetic ordering temperatures in the basal plane direction (S_b values). However, in the c direction it was necessary to use slightly adjusted S_c values for yttrium to get zero differences above the magnetic ordering temperatures. In Figure 7–18 we show the adjusted yttrium coefficients S_c^*(Y) which yield zero differences just above the magnetic ordering temperatures. When this approach was used the results shown in Figures 7–19 and 7–20 were obtained. For Gd in the c direction there is a strong magnon drag peak at 24 K very close to the 20 K thermal conductivity peak. In the basal plane there is also a magnon drag peak of half the magnitude of the S_c peak and it is best seen in the difference curve of Figure 7–19.

Since Gd is ferromagnetic below 293 K there is an internal magnetic field below this temperature which splits the conduction band into spin up and spin

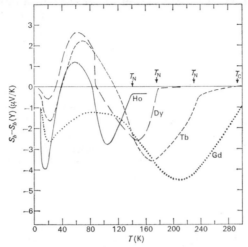

Figure 7–19. Basal plane Seebeck coefficients of gadolinium, terbium, dysprosium, and holmium after S_b for yttrium has been subtracted.

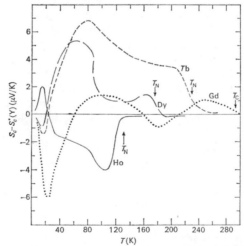

Figure 7–20. Seebeck coefficients along the c axis for gadolinium, terbium, dysprosium, and holmium after an adjusted S_c^* for yttrium (see Figure 7–18) has been subtracted.

down bands. This means that there will be Fermi surface effects which should show up in the Seebeck coefficients according to Eqn (7.25). These effects are over and above the magnon drag effects which were described above and which gave the peaks (negative) at 24 K. It seems reasonable to attribute the two shallow broad maxima in $S_c - S_c^*(Y)$ at 98 and 255 K of Figure 7–20 and the broad substantial minimum in $S_b-S_b(Y)$ at 210 K of Figure 7–19 to such internal magnetic field Fermi surface effects. The fact that an external magnetic field of 1000 Oe lowered or flattened out the 98 K maximum and had no influence on the 24 K magnon drag peak[48] is strong evidence that this is a proper interpretation. We recall that in the absence of an external magnetic field the magnetic moments of Gd are along the c axis at and just below the Curie point and move to a 30° cone around the c axis as the temperature goes down to 300 K. At temperatures above 320 K it appears that for Gd both S_c and S_a of Figure 7–5 will go positive just as they did for yttrium. Both S_a and S_c show only small changes in slope at the 293 K Curie point.

Below 220 K terbium is ferromagnetic and in this sense it is like Gd. On the other hand, orbital angular momentum contributes to the magnetic moment of Tb and gives it strong magnetic anisotropy. Seebeck coefficients[48] at different for the two as shown in Figures 7–6, 7–19, and 7–20 where one notices first that the small Tb magnon drag peaks (both negative) are at different temperatures; for S_c the peak is -1.8 μV/K at 18 K and for S_a the peak is -2.8 μV/K at 26 K. These temperatures are close to the $T_N/10 \sim 23$ K at which the peaks might be expected. The fact that magnon excitation and propagation is different in the different crystallographic directions (see Chapter 5) accounts for the different peak temperatures. Secondly, one notices in Figure 7–6 that S_a has a positive sign (Gd was negative) from 30 to 220 K. To examine this more effectively we turn to Figure 7–20 which shows $S_c-S_c^*(Y)$ rises to a positive maximum of 6 μV/K at 70 K and then levels off at 4 μV/K before dropping to the forced zero at the ordering temperature. This is believed to come from the internal magnetic field effect on the Fermi surface. In the basal plane (see Figure 7–19) $S_b-S_b(Y)$ is positive in sign from 30 to 115 K and shows a broad $+2.5$ μV/K maximum at 70 K. As the temperature goes higher a broad minimum of -3.5 μV/K occurs at 175 K. The latter is very much like that for Gd and appears to scale in temperature as the ordering temperature. At the 230 K Néel point S_c shows sharp curvature while S_a shows a change in slope. By extrapolating the high temperature data of Figure 7–6 it is found that S_c should become positive again at 460 K while S_a goes positive at 380 K.

The Seebeck coefficients[48] for Dy are shown at the top of Figure 7–8. At the 87 K ferromagnetic to helical ordering temperature there is a precipitous drop in both S_c and S_a which shows the sensitivity of the Seebeck coefficients to the magnetic superzones. There also appears to be a small magnon drag peak at 14 K for S_c and a slightly larger one at 23 K for S_a. These are satisfactorily close to the $T_N/10 \sim 18$ K. One is tempted here as in the case of Tb to associate the 14 K peak of S_c with the more easily energized magnon propagated in the c direction and the higher temperature S_a peak with more energetic magnons propagated in the basal plane (see Chapter 4). The Dy data of Figure 7–20 show $S_c-S_c^*(Y)$ is positive from 15 to 176 K with a small (1 μV/K) positive value in the helical range from 88 to 176 K. This contrasts with the nearly $+5$ μV/K at 60 K in the ferromagnetic range. For $S_b-S_b(Y)$ there is also a broad maximum of 3 μV/K at 60 K as shown in Figure 7–19. Then in the helical range a broad minimum of -2 μV/K is reached at 140 K and again this behavior just below the ordering temperature is consistent with that for Tb and Gd. Returning to Figure 7–8 we see there is a sharp change in slope for S_a at the 176 K Néel point while S_c shows an inflection point there. The upward slopes of the curves at 300 K indicate that S_c will become positive again at 530 K and S_a will go positive at 420 K.

Since holmium has a conical ferromagnetic to helical phase transition at 19 K it is reasonable to expect that magnon dispersion and the Seebeck coefficients will differ from those of the elements discussed previously. The experimental results[48] are shown at the top of Figure 7–9 but the comparison with the others can be seen more readily in Figure 7–20 where $S_c - S_c^*(Y)$ for Ho is quite different from the others. There is in Ho a small magnon (or phonon) drag peak of $+1.7$ $\mu V/K$ at 16 K for S_c whereas all others show magnon drag of negative sign. Then at temperatures above 30 K $S_c - S_c^*(Y)$ is negative while all the others shown in Figure 7–20 are positive. In the basal plane the $S_b - S_b(Y)$ curve for Ho in Figure 7–19 exhibits behavior similar to that for the other heavy rare earths shown there. The magnon (or phonon) drag peak of -5.4 $\mu V/K$ at 16 K is the greatest basal plane drag effect seen and is nearly as great as the magnon drag peak along c in Gd. Both magnon drag peaks in Ho occur close in temperature to the estimated $T_N/10 \sim 13$ K temperature and since the peaks occur below 19 K they are related to magnon dispersion in the conical ferromagnetic phase. The Seebeck coefficients do not exhibit sharp changes at the 19 K magnetic phase transition as do the thermal conductivities shown in the middle of Figure 7–9. Both S_a and S_c in Figure 7–9 show sharp slope changes at the 130 K Néel point. In the paramagnetic range coefficients have positive slopes and S_a goes positive at 250 K. By extrapolation it appears that S_c will go positive at 380 K.

The Seebeck coefficients[48] for Er are shown at the top of Figure 7–12. Both S_c and S_b have considerable structure and show high sensitivity to magnetic changes. They may best be compared with the electrical resistivity, particularly ρ_c which appears at the bottom of the figure. S_c has a minimum of -3.8 $\mu V/K$ at 25 K and S_b has an almost identical minimum at 23 K and these are both just above the 19.5 K temperature for the conical ferromagnetic to square wave magnetic phase transition. If there were to be magnon drag in erbium one would look for it near 15 K where Tb, Dy, and Ho show it. Thus it is quite certain that both Seebeck coefficient peaks near 24 K are most likely caused by magnetic effects on the Fermi surface. There is a nearly linear variation in both coefficients in the square magnetic wave region from 30 to 50 K and then sharp changes toward more negative values occur in both the square wave to sinusoidal wave transition range from 50 to 55 K. From 55 to 85 K S_c goes more negative linearly. Both coefficients show sharp slope changes at the 85 K Néel point. They also have positive slopes at high temperatures (paramagnetic range) with S_b going positive at 275 K and S_c going positive by extrapolation at 360 K.

For thulium which is ferromagnetic below 32 K the Seebeck coefficients[44] are shown at the top of Figure 7–13. Several features of the results warrant discussion. For S_c there is an initial positive peak estimated to be about 0.3 $\mu V/K$ centered near 7 K which one is tempted to attribute to magnon drag because the temperature is right and because holmium which is a conical ferromagnet shows a similar S_c peak. It may be said further that one probably should not dismiss the initial positive points for S_a as an impurity effect because the thermal conductivity peaks were high indicating that the samples were good. The next anomaly is also associated with S_c and this is the large narrow negative peak at 32 K which is most likely a magnetically induced Fermi surface effect because neutron diffraction results (see Chapter 3) show that here the low temperature square wave ferromagnetic phase begins to lose the exact four-three spin layer arrangement. At a slightly higher temperature of 45 K the square wave rounds off into a sinusoidal wave and both coefficients are strongly affected. The Néel point is characterized by sharp peaks at 57.2 K for both coefficients. At higher temperatures the data show S_b going positive at 180 K and S_e going positive at 215 K.

7.4. ELECTRICAL RESISTIVITY, THERMAL CONDUCTIVITY, AND SEEBECK COEFFICIENTS OF THE LIGHT RARE EARTHS AND YTTERBIUM

The electrical resistivity of La is shown in Figure 7–21. The curve shown[51] is for a polycrystalline sample which was preponderantly cubic. Each time such a sample was cooled to helium temperatures and then warmed to room temperature it had a tendency to show more of the dhcp phase (double hexagonal close packed phase where the stacking along the hexagonal axis is abac). With the growth of the dhcp phase there is an increase in the room temperature resistivity indicating that the dhcp phase has a higher electrical resistivity than the cubic phase. The dhcp phase will transform to the cubic (abc stacking) at 537 K above which the cubic phase is stable.[4]

The downward curvature of the resistivity curve at temperatures above 40 K is also found in Pr and Nd. There is not, at present, a sound understanding of

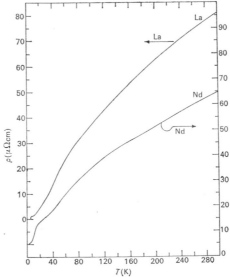

Figure 7–21. Electrical resistivity of lanthanum and neodymium versus temperature (after Ref. 51).

this phenomenon which is different from the behavior of a typical metal, e.g., yttrium which was described earlier. It is not believed that the empty 4f shell is close enough in energy to cause the curvature effect in La but rather it is thought that the phonon spectrum is responsible for the observed behavior.

The electrical resistivity drops to zero at 6 K and this represents the onset of superconductivity for the cubic phase of La. The dhcp phase has a transition temperature of about 5 K and this difference in transition temperature is a clear demonstration that phonons play a strong role in superconductivity as required by the theory for this phenomenon.[15] The most recent transition temperatures[52] for the two phases of La give 6.00 K for the cubic and 4.87 K for dhcp. The other nonmagnetic trivalent elements Sc, Y, Lu are not superconductors but it has been found that yttrium will become a superconductor under a pressure of 110 kbars.[53] A variety of heavy rare earths with high magnetic moments have been dissolved in La to explore the coexistence of magnetic ordering (antiferromagnetic) and superconductivity as well as to explore gapless superconductivity.[54]

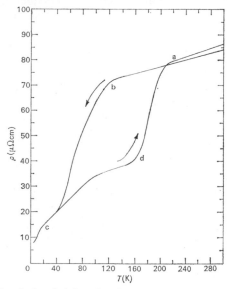

Figure 7–22. Electrical resistivity of cerium versus temperature (after Ref. 55).

Surprises have been frequent in the studies of the transport and other properties of the rare earth metals; cerium was one of the forerunners in this respect and this can be seen in Figure 7–22. Here one of the earliest electrical resistivity curves for cerium[55] is shown. When the temperature of a sample of mixed dhcp and cubic phases is cooled it follows the path from *a* to *b* to *c* in the figure. Then upon warming the path from *c* to *d* to *a* is followed. It was soon learned[56] that the drop in the resistivity along *bc* starting at 120 K was accompanied by a change in crystal structure to a condensed cubic phase triggered by the lone 4f electron's migration to the conduction band. This implied that the 4f energy level was close to the conduction band and when the lattice contracted sufficiently the lower lying energy state for the metal occurred with the 4f electron in the conduction band. (See section 1.2.2.)

Another transition is apparent in Figure 7–22 near 14 K (close to c in the figure). This knee and the resistivity drop with decreasing temperature is more pronounced in those samples which show a higher percentage of the dhcp phase. This transition is magnetic in character and is associated with the onset of magnetic ordering in the dhcp fraction of the sample which retains the 4f electron. This is another case in which there is a manifestation of magnetic superzones. Magnetic effects occur in the sample regions which are dhcp and have an electron in the 4f shell. When the sample of cerium here described was warmed back to room temperature a larger fraction of the sample was dhcp and this explains the higher resistivity at the end of the run at point a in the figure; the dhcp phase is characterized by a higher resistivity than the fcc phase. As indicated earlier in this book (Chapter 3) the magnetically ordered structure below 14 K is of the antiferromagnetic type. A number of studies of cerium and of cerium alloys[24] have been made which give evidence that fractional occupancy of the 4f level sometimes occurs. When small amounts of cerium are introduced into yttrium[57] (dilute alloys) a resistivity minimum occurs because of the proximity of the 4f level to the Fermi energy and it is expected that Yb might also show this effect. The mechanism here is different from the Kondo mechanism.[58]

Praseodymium metal has two electrons in the 4f shell in states which exhibit crystal field effects. The electrical resistivity of a polycrystalline sample is shown[59] in Figure 7–23 where what appears to be a broad magnetic transition is in evidence between 60 and 80 K. However, neutron diffraction and magnetic studies[12] on single crystals have revealed that Pr does not order magnetically and that externally applied magnetic fields induce moments. It is believed that the 4f states are sensitive to crystal defects and that this explains the

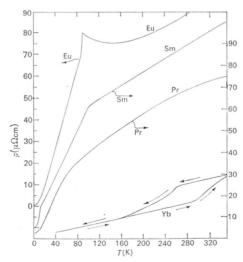

Figure 7–23. Electrical resistivity of europium, samarium, praseodymium, and ytterbium versus temperature (after Refs. 59, 61–63).

divergent observations. The possibility of thermal excitations of electrons in the crystal field controlled 4f levels may give rise to the electrical resistivity behaviour seen in Figure 7–23.

We return to Figure 7–21 for the resistivity of polycrystalline neodymium.[51] Here the outstanding feature is the abnormal behavior in the 5 to 20 K range where a magnetic transition occurs. The jump in resistivity actually arises because of two transitions as revealed by single crystal neutron diffraction studies.[12, 60] These studies show an ordering of the hexagonal sites at 19 K and of the cubic sites at 7.5 K and are described in Chapter 3.

There are several interesting features in the case of samarium for which the resistivity data[61] are shown in Figure 7–23. First of all samarium has a very strange atom stacking arrangement given by ababcbcac (the sites have symmetry chhchhchh). This implies it has hexagonal symmetry and that the range of the interatomic forces and interactions extends across nine atomic layers. Second, it is seen that there are two distinct anomalies in the resistivity, one at 12 K and another at 106 K. Again we point out that the results shown are for a polycrystalline sample. Magnetic susceptibility measurements indicate that antiferromagnetic type transitions occur at both temperatures. (See Chapter 3.)

Europium and ytterbium are divalent metals, the former having a 4f shell which is half full like Gd and the latter a 4f shell which is full as in Lu. Eu is a body centered cubic metal.[4] In Figure 7–23 we see that Eu has a very sharp peak in its resistivity[62] at 90 K. Here is dramatic evidence that magnetic ordering takes place at this temperature. According to neutron diffraction studies (see Chapter 3) the metal has a helical magnetic structure below this temperature.

Recent studies[63, 64] on very pure Yb have revealed that it has a polymorphic transformation from fcc to hcp near room temperature. An earlier study[62] of electrical resistivity showed erratic behavior because of this. The data shown in Figure 7–23 show thermal hysteresis between 250 and 350 K which is the region of the transformation. The resistivity is not as high as for the other rare earths and in this sense Yb is more like the divalent alkaline earths. There are no magnetic transitions inasmuch as the 4f shell is full. Under a pressure[64] of about 15 kbars, Yb apparently becomes an insulator.

This completes the discussion of the electrical resistivity of the light rare earths and Yb. We see that the magnetic structure affects the electrical resistivity and recognize the need for further work on single crystals in many instances.

Thermal conductivity measurements have been made[61] on polycrystalline samarium with results shown at the top left in Figure 7–24. The small maximum at 9 K is followed by a minimum which is caused by the magnetic ordering at 12 K and is reminiscent of the behavior of Ho near its 20 K magnetic transition. Here again one would expect that there should be a normal rapid increase at low temperatures up to the dotted maximum shown in the figure. The magnetic scattering then suppresses this peak and causes the spiked maximum. Above 30 K the temperature dependence is normal with a small increasing value as expected above the 100 K highest ordering temperature. There is no anomaly in K at this transition to paramagnetism. The thermal conductivity

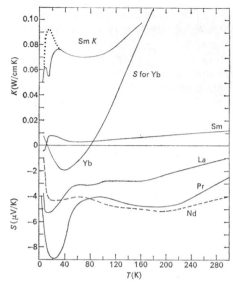

Figure 7-24. Seebeck coefficients of ytterbium, samarium, lanthanum, praseodymium, and
neodymium versus temperature; thermal conductivity of samarium versus temperature
(after Refs. 61, 65). The dotted segment for K of samarium is the estimated curve without
magnetic scattering.

data are consistent with the electrical resistivity for Sm which is shown in
Figure 7-23. Apparently no one has measured the thermal conductivity of the
other light rare earths.

We next turn to the Seebeck measurements which have been carried out on
the light elements. Seebeck coefficients for polycrystalline La, Pr, Nd, Sm, and
Yb are shown[65] at the bottom of Figure 7-24. We shall discuss them in the
order listed: The Seebeck coefficient of La is negative in the temperature range
covered, 7-300 K. Since La goes superconducting at 6 K (fcc form) it would
have a zero Seebeck coefficient below this temperature because a voltage
difference is incompatible with superconductivity. It is seen in the figure that
as we go down in temperature below 20 K there is a sharp turn downward in
the curve near 7 K and this is compatible with the requirement that the
coefficient should be zero below 6 K At higher temperatures there appears
to be just a gradual decrease in the magnitude of the coefficient.

At low temperatures the Seebeck coefficient of praseodymium drops rapidly
with increasing T to a negative peak of -8.7 $\mu V/K$ at 23 K. After passing
through the minimum the magnitude of the coefficient decreases to about
4.5 $\mu V/K$ at 60 K and then remains nearly contsant. When we refer back to the
electrical resistivity of Pr as seen in Figure 7-23 we get no help in explaining
the negative peak but neutron diffraction results (see Chapter 3) show that a
magnetic transition occurs at 25 K, the temperature of the peak in S. This
leads to the suggestion that the cause of the peak is a magnetic effect on the
Fermi surface. It is not likely to be a case of magnon drag. Above 60 K the
Seebeck coefficient changes very little and is much like that of lanthanum
except for its greater magnitude.

There are magnetic transitions in neodymium at 7.5 K (cubic sites) and 19 K (hexagonal sites) but the Seebeck coefficient does not show high response to them as in the case of praseodymium. Rather the coefficient goes sharply negative to -4.3 $\mu V/K$ as the temperature rises from 7 to 20 K and then remains nearly constant between -4 and -5 $\mu V/K$ all the way to room temperature. The general pattern and negative sign are consistent for all three elements.

The crystal structure (nine layer arrangement) of Sm is markedly different from the light elements just discussed so it is not surprising that its Seebeck coefficient (see Figure 7–24) has different characteristics including a difference in sign. Above 20 K the coefficient is flat, positive, and small at about 1 $\mu V/K$. As the temperature is raised past the 12 K magnetic transition the Seebeck coefficient rises rapidly from a small negative value to $+0.6$ μ/VK and this is basically the interesting structure for this element although there is a very slight kink at the 100 K magnetic transition to paramagnetism.

Finally, we come to the divalent nonmagnetic fcc Yb which stands out from all the others because its Seebeck coefficient reaches $+28$ $\mu V/K$ at 300 K (Figure 7–24 only shows the coefficient up to 180 K). The coefficient starts out positive, drops to a minimum of -2 $\mu V/K$ at 40 K, and then rises almost linearly to the large value quoted above.

7.5. SUPERZONE THEORY AND ELECTRICAL RESISTIVITY

We begin by writing out the expression for ϕ_m of Eqn (7.2). This is the magnetic interaction energy[3, 18] of a single conduction electron spin σ at position $\mathbf{r}$ and has the form (see section 6.3.1 for details)

$$\phi_m = \frac{1}{N} \sum_{n=1}^{N} \mathscr{I}_{sf}(\lambda-1)\delta(\mathbf{r}-\mathbf{R}_n)\mathbf{J}_n \cdot \sigma \tag{7.31}$$

where $\mathscr{I}_{sf}$ is the effective exchange interaction between the conduction electron and the ion of total angular momentum $\mathbf{J}_n$ located at the lattice site $\mathbf{R}_n$. The delta function localizes the interaction. It is then assumed that the relaxation time approximation may be used so that the phonon term τ_p is proportional to the temperature T (high temperature assumed) and the magnetic (spin) term is given by

$$1/\rho_m \sim \mathscr{I}_{sf}^2 \left[1 - \frac{\langle \mathbf{J} \rangle^2}{J^2}\right] \tag{7.32}$$

which is essentially just a measure of the spin disorder[20, 66] which exists at temperature T.

The ordering of the spins of the rare earths is described by the following equations:

$$\langle J_n^x \rangle = M'J \cos (\mathbf{q} \cdot \mathbf{R}_n) \tag{7.33}$$

$$\langle J_n^y \rangle = M'J \sin (\mathbf{q} \cdot \mathbf{R}_n) \tag{7.34}$$

$$\langle J_n^z \rangle = MJ \cos (\mathbf{q} \cdot \mathbf{R}_n + \phi) \tag{7.35}$$

where $\mathbf{q}$ is the wave vector associated with the wavelike magnetic moment dependence for translation along the c axis, M' and M are temperature dependent amplitudes which may be found from neutron diffraction data, and ϕ is a phase angle. The helical regions of Tb, Dy, and Ho have $M=0$ and the interlayer turn angle is given by qc where c is the c axis lattice parameter. The high temperature magnetically ordered structures of Er and Tm have $M'=0$.

When the magnetic lattice does not coincide with the ionic lattice, extra planes of energy discontinuity or magnetic superzones[17-19] come into play. A spheroidal Fermi surface is assumed and then first-order perturbation theory and Eqns (7.31)–(7.35) may be used[3, 18] to predict the position **d** of the magnetic superzone planes at

$$\mathbf{d}=\tfrac{1}{2}(\boldsymbol{\eta} \pm \mathbf{q}) \qquad (7.36)$$

where $\boldsymbol{\eta}$ is a reciprocal lattice vector. The energy is

$$2E(\mathbf{k})=E(\mathbf{k})+E(\mathbf{k}+2\boldsymbol{\eta}) \pm \{[E(\mathbf{k})-E(\mathbf{k}+2\boldsymbol{\eta})]^2+\mathscr{J}^2J^2M_\pm^2\}^{\tfrac{1}{2}} \qquad (7.37)$$

with $E(\mathbf{k})$ the unperturbed energy of an electron of wave vector **k** and with

$$M_\pm^2=M^2+2M'^2 \pm 2M'(M^2+M'^2)^{\tfrac{1}{2}}. \qquad (7.38)$$

The new zone energy gap is

$$\Delta=\mathscr{J}_{sf}JM_\pm. \qquad (7.39)$$

Equations (7.3), (7.4), and (7.36) through (7.39) may be used to get the electrical resistivity in the different crystallographic directions:

$$\rho_b=\alpha_1+\beta_1T+\gamma_1(1-\tfrac{1}{2}M^2-M'^2) \qquad (7.40)$$

$$\rho_c=\frac{\alpha_2+\beta_2T+\gamma_2(1-\tfrac{1}{2}M^2-M'^2)}{1-\Gamma(M^2+M'^2)^{\tfrac{1}{2}}}$$

$$\Gamma=(3\pi\mathscr{J}_{sf}J/4E_Fk_F) \sum_i |\mathbf{d}_i|. \qquad (7.41)$$

The sum in the latter is over all superzones which slice the Fermi surface. E_F is the Fermi energy, k_F is the electron wave vector evaluated at the Fermi surface, and α_1, β_1, γ_1, α_2, β_2, and γ_2 are adjustable parameters. Analysis shows ρ_c will have a maximum below the magnetic ordering temperature and ρ_b will have a higher slope below than above the ordering temperature.

When the theory is applied to Dy and Ho for which $M=0$ in Eqn (7.35) the results for the c direction shown at the top of Figure 7–25 are obtained.[18] The general character of the resistivity curve is reproduced, the hump and maximum just below the ordering temperature are faithfully described and the proper behavior at the ferromagnetic transition for Dy is obtained. The latter is greatly exaggerated, however. The fit to the Ho c axis resistivity is better and this may come in part because the complication of the helical to ferromagnetic phase is not a problem here.

For Er and Tm just below their Néel points, Eqns (7.33) and (7.34) have $M'=0$ and only M remains because the magnetic moments point in the c direction while the planar components S_x and S_y are completely random. When the resistivities for these are calculated[18] the results shown in the bottom of Figure 7–25 are obtained. The rapid rise in resistivity of Er with decreasing temperature just below the Néel point is nicely reproduced. In the temperature region where the moment wave squares off, 20 to 55 K, the fit is

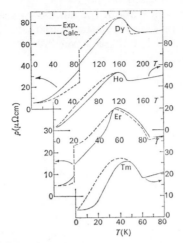

Figure 7–25. Calculated and experimental c axis electrical resistivities for dysprosium, holmium, erbium and thulium (after Refs. 3, 18, 44).

poor and just as in the case of Dy, the theory gives too large a jump in resistivity at the 20 K ferromagnetic to square wave transition.

The fit of the theory to Tm is about as good.[44] Of course Tm does not have a transition to a ferromagnetic phase so the test of the theory is not as severe as in Dy and Er. The energy gaps essentially wipe out some of the Fermi surface area projected in the c direction and leave that projected in a basal plane direction nearly unchanged. To get a better fit of the theory to the data it would be necessary to use Fermi surfaces calculated for each metal together with accurately determined energy gaps. This has been done in some instances.[67]

In a basal plane direction the theory effectively describes the spin wave scattering and provides the change in slope at the Néel point. These results are not shown.

7.6. HALL EFFECT

In nonmagnetic metals the transverse electric field, E_H, observed in a current carrying slab (current density J) when an external magnetic field, B, is applied perpendicular to the slab is called the Hall field. The Hall resistivity, ρ_H, is the Hall electric field per unit current density and is given by

$$\rho_H = \frac{E_H}{J} = R_0 B \qquad (7.42)$$

where R_0 is the Hall coefficient. In the free electron case

$$R_0 = \frac{1}{nec} \qquad (7.43)$$

where c is the velocity of light, e is the electronic charge, and n is the number of conduction electrons per unit volume. Thus, Hall coefficients are of interest

13

because they give information about the current carriers. The rare earths do not fit the free electron model so we assume that conduction is by both holes n_+ and electrons, n_- and that these have different mobilities μ_+ and μ_-, respectively. In this case

$$R_0 = \left(\frac{e}{\sigma^2 c}\right)(n_+\mu_+{}^2 - n_-\mu_-{}^2) \qquad (7.44)$$

where σ is the total electrical conductivity of the metal.

Results of Hall effect measurements[68] on Lu and Y are shown in Figure 7–26. When the magnetic field is applied parallel to the c axis (we call this coefficient R_{0c} and for this case the Hall field is in the basal plane) the Hall coefficients for Y and Lu are both negative and nearly the same in magnitude for temperatures above 80 K. As the temperature decreases below 80 K both coefficients tend toward zero indicating that two carriers are possibly involved

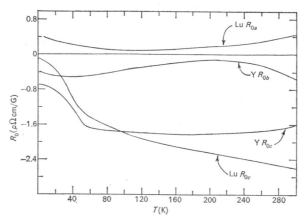

Figure 7–26. Hall coefficients of yttrium and lutetium versus temperature (after Ref. 68).

in conduction. With the magnetic field applied in a basal plane direction (two coefficients characterize the hcp lattice) there is a difference in the sign of the coefficients, R_{0b} for Lu and Y. This is odd because the Fermi surfaces (and so the current carrier behavior) are so much alike. To understand these results we must again use a two-carrier model such as that described in Eqn (7.44). We would then note that the magnitudes of the coefficients are small and only small changes in carrier concentrations, in mobilities, or in Fermi surfaces would be necessary to give the observed data.

The Hall effect is much more complex for magnetic metals. In this case the motion of the conduction electrons is subject to external and internal magnetic fields and it is more appropriate to treat the observations in resistivity terms. We have then the Hall resistivity given by

$$\rho_H = R_0 B + R_s 4\pi M \qquad (7.45)$$

where the first term is the ordinary Hall resistivity described in Eqn (7.43); the second term, which is dependent on the magnetization, M, describes what is called the extraordinary Hall resistivity and R_s is the extraordinary or spin Hall coefficient and is generally much larger than the ordinary coefficient R_0.

It was shown originally[69] that the anomalous term in a ferromagnet results from the coupling of the spin and orbital momenta of the current carriers (conduction electrons). Such a model is clearly inappropriate to explain the giant Hall effect in rare earth metals with their highly localized $4f$ electrons which are distinct from the conduction electrons. More recent work has shown that a Hall field is obtained from the $s-f$ exchange model including either intrinsic spin-orbit coupling of the $4f$ electrons or a crosscoupling[70, 71] between the $4f$ spin and conduction band orbital momentum. The transverse or Hall resistivity term is obtained in the second Born approximation, one order higher in perturbation theory than the $s-d$ exchange result which gives the spin-disorder resistivity.[23]

Experimentally, Hall resistivity data as a function of field greatly resemble magnetization data which suggests the decomposition of terms in Eqn (7.45). The extraordinary coefficient R_s is determined from the saturation Hall resistivity (from data beyond the saturation point for M) extrapolated to zero field. The ordinary coefficient R_0 is determined from the high field (above technical saturation) slope of ρ_H after a correction has been made for the susceptibility $\partial M/\partial H$.

Except for Gd, high magnetic anisotropy in the heavy rare earths makes it impossible to get a complete determination of the Hall coefficients below the Néel points; but it is possible to get coefficients for the field applied in the easy direction of magnetization. It is also possible to get good coefficient data in the paramagnetic range.

To describe the results we let R_{0c} be the ordinary coefficient with the field parallel to the c axis and R_{0b} (or R_{0a}) for the field parallel to a basal plane direction; similarly, R_{sc} and R_{sb} (or R_{0a}) are the spin Hall coefficients. In Figure 7–27 the ordinary coefficients for Gd are shown.[68] At low temperatures R_{0c} is positive and slowly decreases and becomes negative at 150 K; as the temperature rises further R_{0c} decreases steadily and then assumes nearly

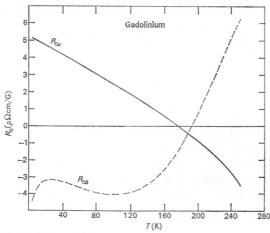

Figure 7–27. Ordinary Hall coefficients of gadolinium versus temperature (after Ref. 68).

infinite slope near 280 K. It can be seen that there is a dependence on spin disorder and this is reminiscent of the spin disorder effect on the electrical resistivity. The other coefficient, R_{0b}, is negative and small (not unlike the Y coefficient of Figure 7–26) at low temperatures, it then reverses sign, going positive at 170 K to reach a broad maximum at 240 K before falling rapidly as the temperature approaches 280 K.

For Gd the spin Hall coefficients, [68, 72] R_{sc} and R_{sa}, are shown in Figure 7–28; they are negative and enormous, being about two orders of magnitude higher than the ordinary coefficients for Gd. The coefficients for Gd are also

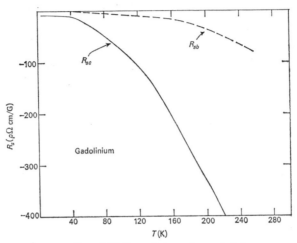

Figure 7–28. Spin (extraordinary) Hall coefficients for gadolinium versus temperature (after Refs. 68, 72).

higher than for the other heavy rare earths by almost an order of magnitude. We see in the figure that both coefficients drop from a zero value at low temperatures to minima of -400 pΩ cm/G at 240 K for R_{sc} and -75 pΩ cm/G at 275 K for R_{sb}.

In Gd it is presumed that the giant Hall effect is derived from the intrinsic $f_{\text{spin}} - s_{\text{orbit}}$ crosscoupling mechanism since the total orbital angular momentum of the $4f$ electrons is zero in this S state ion. The general behavior shows that the spin coefficients follow the rise of the spin disorder electrical resistivity as predicted by theory. [66, 70]

For terbium, the easy direction of magnetization is along the b axis so the coefficients R_{0b} and R_{sb} are measurable. Results [73] are shown in Figure 7–29. Also shown [68, 74] in the figure for purposes of comparison are R_{sa} for Dy and R_{0b} and R_{sa} for Gd. The spin coefficients for Tb and Dy are similar. They have positive maxima near 100 K and then become negative as the temperature approaches their Néel points. The positive maxima are not seen in R_s of spin only Gd so to explain this effect there is an apparent need to include additional matrix elements of the $f_{\text{spin}} - f_{\text{orbit}}$ type in calculations.

In the paramagnetic range the extraordinary coefficients are independent of temperature; there is a variation of Hall resistivity with temperature which is a

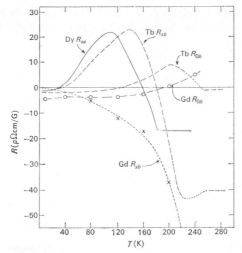

Figure 7–29. Basal plane spin Hall coefficients for gadolinium, terbium, and dysprosium versus temperature; ordinary basal plane Hall coefficients for gadolinium and terbium versus temperature (after Refs. 68, 72–75).

consequence of the temperature dependence of the magnetic susceptibility. The values of the Hall coefficients[75] are given in Table 7.2. There is considerable anisotropy in the Fermi surface and its effect on the thermal and electrical conductivities.

Table 7.2. Values of spin and ordinary Hall coefficients in the paramagnetic region.[75] The second subscript indicates the applied magnetic field direction. The coefficients are in units of $p\Omega$ cm/G

	R_{sc}	R_{sb}	R_{0c}	R_{0b}
Gd	−554.0	−122.0	—	—
Tb	−77.0	−40.0	−3.7	−1.0
Dy	−23.0	−18.0	−3.7	−0.3
Ho	−13.2	+2.2	−3.2	+0.2
Er	−9.4	+4.4	−3.6	+0.3

Hall effect studies have been made on polycrystalline samples of other rare earth metals.[76,77] Above 80 K Pr and Nd have nearly temperature-independent Hall coefficients of +0.7 and +1.0 $p\Omega$ cm/G, respectively. However, Sm showed a changing Hall coefficient with temperature and reversed sign at 180 K showing positive values below and negative values above this temperature; the room temperature value was −0.2 $p\Omega$ cm/G. Ytterbium, which is a divalent metal, had a temperature-independent Hall coefficient of +3.8 $p\Omega$ cm/G.

The author received a great deal of assistance with this manuscript from A. R. Mackintosh, J. J. Rhyne, S. H. Liu, R. J. Elliott, L. R. Edwards, K. A. Gschneidner, Jr., and from his wife Lavona. Thanks also go to C. R.

Andreessen who typed the final draft, to C. M. Brannon for editing the manuscript, to H. Allen for preparing the graphs, and to D. W. Mellon for technical assistance.

REFERENCES

1. Van Vleck, J. H., in Electric and Magnetic Susceptibilities. Oxford University Press, London, England (1932).
2. Kasuya, T., in Magnetism, 11B (G. T. Rado and H. Suhl, eds). Academic Press, New York (1965), p. 252.
3. Elliott, R. J., in Magnetism, 11A. Academic Press, New York (1965), p. 385.
4. Gschneidner, K. A., Jr., in The Rare Earth Alloys. Van Nostrand, Princeton, New Jersey (1961).
5. Ziman, J. M., in Electrons and Phonons. Oxford University Press, London (1960).
6. Ziman, J. M., in Principles of the Theory of Solids. Cambridge University Press, Cambridge, England (1964).
7. Wilson, A. H., in The Theory of Metals, 2nd ed. Cambridge University Press, Cambridge, England (1965).
8. Macdonald, D. K. C., in Thermoelectricity: an Introduction to the Principles. John Wiley and Sons, New York (1962).
9. Freeman, A. J., Dimmock, J. O. and Watson, R. E., in Quantum Theory of Atoms Molecules, and the Solid State (P. O. Löwdin, ed). Academic Press, New York (1966), p. 361.
10. Klemens, P. G., *Handbuck der Physik*, **14**, 198 (1956).
11. Schieber, M., in Crystal Growth, *Suppl. J. Phys. Chem. Solids* (H. S. Peiser, ed). Pergamon Press (1967), p. 271.
12. Johansson, J., Lebech, B., Nielsen, M., Møller, B. and Mackintosh, A. R., *Phys. Rev. Letters*, **25**, 524 (1970).
13. Tonnies, J. J. and Gschneidner, K. A., Jr., "Preparation of Lanthanide Single Crystals", *J. Crystal Growth* (to be published).
14. Nigh, H. E., *J. Appl. Phys.*, **34**, 3323 (1963).
15. Bardeen, J., Cooper, L. N. and Schrieffer, J. R., *Phys. Rev.*, **108**, 1175 (1957).
16. Freeman, A. J., Dimmock, J. O. and Watson, R. E., *Phys. Rev. Letters*, **16**, 94 (1966).
17. Mackintosh, A. R., *Phys. Rev. Letters*, **9**, 90 (1962).
18. Elliott, R. J. and Wedgwood, F. A., *Proc. Phys. Soc.* (London), **81**, 846 (1963).
19. Miwa, H., *Prog. Theoret. Phys.* (Kyoto), **29**, 477 (1963).
20. de Gennes, P. G. and Friedel, J., *J. Phys. Chem. Solids*, **4**, 71 (1958).
21. Dekker, A. J., *J. Appl. Phys.*, **36**, 906 (1965).
22. de Gennes, P. G., *Compt. Rend.*, **247**, 1836 (1958).
23. Kasuya, T., *Prog. Theoret. Phys.* (Kyoto), **22**, 227 (1959).
24. Brout, R. and Suhl, H., *Phys. Rev. Letters*, **2**, 387 (1959).
25. Klemens, P. G., *Solid State Phys.*, **7**, 1 (1958).
26. Mendelssohn, K. and Rosenberg, H. M., *Solid State Phys.*, **12**, 223 (1961).
27. Hall, P. M., Legvold, S. and Spedding, F. H., *Phys. Rev.*, **116**, 1464 (1959).
28. Boys, D. W. and Legvold, S., *Phys. Rev.*, **174**, 377 (1968).
29. Loucks, T. L., private communication, 1967. He estimated the ratio for Er and we have assumed a smooth variation across the heavy rare earths.
30. Alstad, J. K., Colvin, R. V. and Legvold, S., *Phys. Rev.*, **123**, 418 (1961).
31. Nellis, W. J. and Legvold, S., *Phys. Rev.*, **180**, 581 (1969).
32. Nigh, H. E., Legvold, S. and Spedding, F. H., *Phys. Rev.*, **132**, 1092 (1963).
33. Nellis, W. J. and Legvold, S., *J. Appl. Phys.*, **40**, 2267 (1969).
34. de Gennes, P. G. and Friedel, J., *J. Phys. Chem. Solids*, **4**, 71 (1958).
35. Zumsteg, F. C., Cadieu, F. J., Marcelja, S. and Parks, R. D., *Phys. Rev. Letters*, **25**, 1204 (1970).
36. Mannari, I., *Prog. Theoret. Phys.* (Kyoto), **22**, 235 (1959).
37. Mackintosh, A. R., *Phys. Letters*, **4**, 140 (1963).
38. Hegland, D. E., Legvold, S. and Spedding, F. H., *Phys. Rev.*, **131**, 158 (1963).
39. Cress, W. D., unpublished MS thesis, 1970, Iowa State University Library, Ames, Iowa, U.S.A.

40. Jew, T. T., unpublished MS thesis, 1963, Iowa State University Library, Ames, Iowa, U.S.A.
41. Strandburg, D. L., Legvold, S. and Spedding, F. H., *Phys. Rev.*, **127**, 2046 (1962).
42. Mackintosh, A. R. and Spanel, L. E., *Solid State Comm.*, **2**, 383 (1964).
43. Green, R. W., Legvold, S. and Spedding, F. H., *Phys. Rev.*, **122**, 827 (1961).
44. Edwards, L. R. and Legvold, S., *Phys. Rev.*, **176**, 753 (1968).
45. Legvold, S., *Phys. Rev.*, **B3**, 1640 (1970).
46. Mackintosh, A. R. and Smidt, F. A., Jr., *Phys. Letters*, **2**, 107 (1962).
47. Bailyn, M., *Phys. Rev.*, **126**, 2040 (1962).
48. Sill, L. R. and Legvold, S., *Phys. Rev.*, **137**, A1139 (1965).
49. Darnell, F. J., *Phys. Rev.*, **130**, 1825 (1963).
50. Edwards, L. R., Schaeffer, J. and Legvold, S., *Phys. Rev.*, **188**, 1173 (1969).
51. Alstad, J. K., Colvin, R. V., Legvold, S. and Spedding, F. H., *Phys. Rev.*, **121**, 1637 (1961).
52. Johnson, D. L. and Finnemore, D. K., *Phys. Rev.*, **158**, 376 (1967).
53. Wittig, J., *Phys. Rev. Letters*, **24**, 812 (1970).
54. Finnemore, D. K., Johnson, D. L., Ostenson, J. E., Spedding, F. H. and Beaudry, B. J., *Phys. Rev.*, **137**, A550 (1965).
55. James, N. R., Legvold, S. and Spedding, F. H., *Phys. Rev.*, **88**, 1092 (1952).
56. Lawson, A. W. and Tang, T., *Phys. Rev.*, **76**, 301 (1949).
57. Nagasawa, H., Yoshida, S. and Sugawara, T., *Phys. Letters*, **26A**, 561 (1968).
58. Coqblin, B. and Schrieffer, J. R., *Phys. Rev.*, **185**, 847 (1969).
59. Arajs, S. and Dunmyre, G. R., *J. Less-Common Metals*, **12**, 162 (1967).
60. Moon, R. M., Cable, J. W. and Koehler, W. C., *J. Appl. Phys.*, **35**, 1041 (1964)
61. Arajs, S. and Dunmyre, R. G., *Z. Naturforsch.*, **21**, 1856 (1966).
62. Curry, M. R., Legvold, S. and Spedding, F. H., *Phys. Rev.*, **117**, 953 (1960).
63. Kayser, F. X., *Phys. Rev. Letters*, **25**, 662 (1970).
64. Jerome, D. and Rieux, M., *Solid State Comm.*, **7**, 957 (1969).
65. Born, H. J., Legvold, S. and Spedding, F. H., *J. Appl. Phys.*, **32**, 2543 (1961).
66. Kondo, J., *Prog. Theoret. Phys.* (Kyoto), **27**, 772 (1962).
67. Freeman, A. J., Dimmock, J. O. and Watson, R. E., *Phys. Rev. Letters*, **16**, 94 (1966).
68. Lee, R. S. and Legvold, S., *Phys. Rev.*, **162**, 431 (1967).
69. Karplus, R. and Luttinger, J. M., *Phys. Rev.*, **95**, 1154 (1954).
70. Maranzana, F. E., *Phys. Rev.*, **160**, 421 (1967).
71. Irkhin, Yu. P. and Abelskii, Sh. Sh., *Soviet Phys.-Solid State*, **6**, 1283 (1964).
72. Volkenshtein, N. V., Trigorova, T. K. and Federov, G. V., *Soviet Physics-JETP*, **24**, 519 (1967).
73. Rhyne, J. J., *J. Appl. Phys.*, **40**, 1001 (1969).
74. Rhyne, J. J., *Phys. Rev.*, **172**, 523 (1968).
75. Rhyne, J. J., in Proc. Intern. Colloquium on Rare Earth Elements, Grenoble, France (1969).
76. Kevane, C. J., Legvold, S. and Spedding, F. H., *Phys. Rev.*, **91**, 1372 (1953).
77. Anderson, G. S., Legvold, S. and Spedding, F. H., *Phys. Rev.*, **111**, 1257 (1958).

Chapter 8

Hyperfine Interactions

B. Bleaney

Clarendon Laboratory, Oxford

8.1. MAGNETIC HYPERFINE STRUCTURE (THEORY)

We outline first the simple theory of the magnetic hyperfine structure of an atom or ion in a state $4f^n$, L, S, J, leaving the corrections (which are generally small) until later. For such a state the magnetic hyperfine Hamiltonian is

$$\mathcal{H} = A(\mathbf{I} \cdot \mathbf{J}) \qquad (8.1)$$

This may be written in the form

$$\mathcal{H} = - \boldsymbol{\mu}_n \cdot \mathbf{H}_e \qquad (8.2)$$

where for a point nuclear dipole $\mathbf{H}_e$ is just the magnetic field generated by the $4f$ electrons at the point dipole, but for a nucleus of finite volume $\mathbf{H}_e$ represents an average value of this field, taken over the distribution of nuclear magnetism, which may therefore be different for different isotopes of the same element.

For the lanthanides the main contribution comes from the orbital motion of the electrons, which generates a field $\mathbf{H}_l$ at the nucleus equal to

$$\mathbf{H}_l = -2\beta \sum_i \langle r_l^{-3} \rangle_i \mathbf{l}_i \tag{8.3}$$

where the subscript i refers to the ith electron with orbital momentum $\mathbf{l}_i$ and a mean inverse cube $\langle r_l^{-3} \rangle_i$ of its distance from the nucleus, averaged over the orbital wave function. In LS-coupling, if we assume that all electrons in the same shell have the same value of $\langle r_l^{-3} \rangle$, summation over the various electrons gives

$$\mathbf{H}_L = -2\beta \langle r_L^{-3} \rangle \mathbf{L}. \tag{8.4}$$

When $\mathbf{L}$, $\mathbf{S}$ are coupled to form a resultant $\mathbf{J}$ we may project $\mathbf{L}$ onto $\mathbf{J}$ giving

$$\mathbf{H}_L = -2\beta \langle r_L^{-3} \rangle \mathbf{J} \frac{(\mathbf{L} \cdot \mathbf{J})}{J(J+1)} \tag{8.5}$$

where

$$(\mathbf{L} \cdot \mathbf{J}) = \tfrac{1}{2}\{J(J+1) + L(L+1) - S(S+1)\}. \tag{8.6}$$

A useful relation is

$$(\mathbf{L} \cdot \mathbf{J})/J(J+1) = 2 - \lambda \tag{8.7}$$

where λ is just the Landé g-factor.

We now consider the contribution from the spin magnetization of the electrons. For f-electrons, whose wave functions vanish at the nucleus, the magnetic field generated at the nucleus is given by the simple dipolar formula

$$\mathbf{H}_{sC} = +g_s\beta \sum_i \langle r_{sC}^{-3} \rangle_i [\mathbf{s}_i - 3(\mathbf{s}_i \cdot \mathbf{r}_0)\mathbf{r}_0] \tag{8.8}$$

where $g_s = 2.0023$ is the g-factor of the electron spin $\mathbf{s}$, $\langle r_{sC}^{-3} \rangle_i$ is the mean inverse cube of the distance of the ith electron from the nucleus, and $\mathbf{r}_0$ is a unit vector drawn from the nucleus towards the position of the ith electron. It is convenient to replace this rather awkward formula by one using equivalent operators. Inside a manifold (L, S) obeying Hund's rule we then have, assuming the value of $\langle r_{sC}^{-3} \rangle$ to be the same for all electrons in the shell,

$$\mathbf{H}_{sC} = +g_s\beta \langle r_{sC}^{-3} \rangle \xi \{L(L+1)\mathbf{S} - \tfrac{3}{2}\mathbf{L}(\mathbf{L} \cdot \mathbf{S}) - \tfrac{3}{2}(\mathbf{L} \cdot \mathbf{S})\mathbf{L}\} \tag{8.9}$$

where

$$\xi = \frac{2l+1-4S}{S(2l-1)(2l+3)(2L-1)}. \tag{8.10}$$

Finally, projection onto $\mathbf{J}$ gives

$$\mathbf{H}_{sC} = +g_s\beta \langle r_{sC}^{-3} \rangle \frac{\xi}{J(J+1)} \{L(L+1)(\mathbf{S} \cdot \mathbf{J}) - 3(\mathbf{L} \cdot \mathbf{J})(\mathbf{L} \cdot \mathbf{S})\}\mathbf{J} \tag{8.11}$$

where

$$(\mathbf{S} \cdot \mathbf{J}) = \tfrac{1}{2}\{J(J+1) + S(S+1) - L(L+1)\} \tag{8.12}$$

$$(\mathbf{L} \cdot \mathbf{S}) = \tfrac{1}{2}\{J(J+1) - L(L+1) - S(S+1)\}. \tag{8.13}$$

Here again a useful relation is

$$(\mathbf{S} \cdot \mathbf{J})/J(J+1) = \lambda - 1 \tag{8.14}$$

These results are sometimes written[1] in the form

$$\mathbf{H}_e = -2\beta \langle r^{-3} \rangle \mathbf{N} \tag{8.15}$$

in which g_s is taken equal to 2 exactly and it is assumed that

$$\langle r_L^{-3} \rangle = \langle r_{sC}^{-3} \rangle = \langle r^{-3} \rangle.$$

In the manifold (L, S, J), $\mathbf{N}$ is a vector collinear with $\mathbf{J}$, such that

$$\frac{(\mathbf{N} \cdot \mathbf{J})}{J(J+1)} \mathbf{J} = \langle J \| N \| J \rangle \mathbf{J}. \tag{8.16}$$

We can separate out the orbital and spin contributions to $\langle J \| N \| J \rangle$, which is a pure number. We write them as $\langle J \| N_L \| J \rangle$, whose value is given by Eqn (8.7), and as $\langle J \| N_S \| J \rangle$, which from Eqn (8.11) is

$$\langle J \| N_S \| J \rangle = -\frac{\xi}{J(J+1)} \{ L(L+1)(\mathbf{S} \cdot \mathbf{J}) - 3(\mathbf{L} \cdot \mathbf{J})(\mathbf{L} \cdot \mathbf{S}) \}. \tag{8.17}$$

Values of these various quantities are listed in Table 8.1, for the ground states of each configuration f^n. For f^6 the ground state has $J=0$, but the same formulae give values for the $\langle J \| N \| J \rangle$ in the excited states $J=1$ to $J=6$.

Table 8.1. Values of L, S, J and various Hyperfine Parameters for the Ground States of the Lanthanide Ions

	L	S	J	ξ	$\langle J \| N_L \| J \rangle$ $=2-\lambda$	$\langle J \| N_S \| J \rangle$	$\langle J \| N \| J \rangle$
f^1	3	1/2	5/2	2/45	8/7	8/35	48/35
f^2	5	1	4	1/135	6/5	26/225	296/225
f^3	6	3/2	9/2	2/1485	14/11	14/363	476/363
f^4	6	2	4	−1/990	7/5	−7/165	224/165
f^5	5	5/2	5/2	−2/675	12/7	−52/315	488/315
f^6	3	3	0	−1/135	—	—	—
f^7	0	7/2	7/2	2/45	0	0	0
f^8	3	3	6	−1/135	1/2	1/18	5/9
f^9	5	5/2	15/2	−2/675	2/3	2/45	32/45
f^{10}	6	2	8	−1/990	3/4	1/60	23/30
f^{11}	6	3/2	15/2	2/1485	4/5	−4/225	176/225
f^{12}	5	1	6	1/135	5/6	−1/18	7/9
f^{13}	3	1/2	7/2	2/45	6/7	−2/21	16/21
f^{14}	0	0	0	—	—	—	—

It will be seen that the contribution from the electron spin is relatively small compared to that from the orbit for all configurations; the spin contributions to $\langle J \| N \| J \rangle$ are largest for f^1 and f^{13} ($+16.7\%$ and -12.5% respectively), but they fall to 2 to 3% for f^3, f^4, f^{10}, and f^{11}. These latter contributions are, however, enhanced by the effects of intermediate coupling (see Section 8.1.1).

We may now make an estimate of the order of magnitude of H_e. Since we shall be expressing the values of $\langle r^{-3} \rangle$ in atomic units (a.u.), we first give the result

$$2\beta a_0^{-3} = 125.2 \text{ kG} = 12.52 \text{ Tesla}$$

where a_0 is the Bohr radius. In a substance where the electronic magnetic moment assumes its maximum value, the numerical value of N in Eqn (8.15) is just $\langle J\|N\|J\rangle J$, a quantity which is roughly 3.4 for f^1, f^8, and f^{13} and rises approximately to 6 for f^3, f^{10}, and f^{11}. For the $4f$ ions the values of $\langle r^{-3} \rangle$ rise from about 4 a.u. for $4f^1$ to 12 a.u. for $4f^{13}$. The values of H_e should therefore all lie in the region of a few megagauss, the largest approaching 8 megagauss (for Ho^{3+}, Er^{3+}; $4f^{10}$ and $4f^{11}$). The significance of the negative sign in Eqn (8.15), since all $\langle J\|N\|J\rangle$ are positive, is that H_e is oppositely directed to the angular momentum vector $\mathbf{J}$, and hence has the same direction as the electronic magnetization within the ion, whose magnetic moment is $-\lambda\beta\mathbf{J}$. In the literature it is customary to regard the hyperfine magnetic field as positive if it has the same direction as the electronic magnetic moment, and to conform with this convention we shall quote numerical values of the hyperfine field which are essentially the values of $(-H_e)$ as defined above.

Finally we mention that for an s-electron we have

$$\mathbf{H}_s = -\frac{8\pi}{3} g_s\beta |\psi_0|^2 \mathbf{s} \qquad (8.18)$$

where $|\psi_0|^2$ is the square of the amplitude of the wave-function of the s-electron at the nucleus. For a lanthanide ion, $-H_s$ is of order 15 MG for a $6s$ electron and 40 MG for a $5s$ electron, with much larger values for the inner s-shells. In the ground states of the ions all the localized s-shells up to $5s$ are filled with pairs of s-electrons, whose net field at the nucleus vanishes unless the values of $|\psi_0|^2$ are not precisely the same for the two members of a pair. Obviously a small imbalance can produce a sizeable net field (see Section 8.2.3).

The $6s$ (and $5d$) electrons form the band of conduction electrons in the metal. H_s is zero if the mean value of $\langle s \rangle = 0$, but a finite value of $\langle s \rangle$ and hence of H_s may be produced by the applicaton of an external field (giving the Knight shift), or by exchange interaction with the magnetic ions (see Section 8.5.1).

8.1.1. Intermediate coupling

An important correction to the formulae given above arises from the effect known as "intermediate coupling". The ground state of a $4f^n$ configuration can only be written as (L, S, J) in a first approximation where the effects of the spin-orbit coupling in admixing states of different L, S (but the same J) are neglected. Such admixtures are fortunately small in the ground states, a typical example[68] being that of the holmium atom, $4f^{11}$, whose ground state is

$$0.9855\,|^4I_{15/2}\rangle - 0.1691\,|^2K_{15/2}\rangle + 0.0158\,|^2L_{15/2}\rangle + \ldots \qquad (8.19)$$

The effect of intermediate coupling on the Landé g-factor is small (see, for example, reference (3)) because the Zeeman interaction has no matrix elements between the various states which are admixed by the spin–orbit coupling.

In the hyperfine structure the same is true for the orbital contribution, and the correction factor to $\langle J\|N_L\|J\rangle$ differs from unity by less than 1 %. This rule does not hold for the spin–dipolar field $\mathbf{H}_{sc}$, and the corrections are quite appreciable to $\langle J\|N_S\|J\rangle$, as can be seen from Table 8.2. Fortunately the spin contribution is small compared with the orbital contribution, so that the overall correction due to intermediate coupling remains at a few per cent, and vanishes for $4f^1$ and $4f^{13}$, where the spin contribution is largest, because the are no excited states with the same values of J belonging to the same $4f^n$ configuration.

Table 8.2. Correction Factors by which the Quantities $\langle J\|N_L\|J\rangle$, $\langle J\|N_S\|J\rangle$ and $\langle J\|N\|J\rangle$ must be multiplied to allow for Intermediate Coupling

	Atom or ion	$\langle J\|N_L\|J\rangle$	$\langle J\|N_S\|J\rangle$	$\langle J\|N\|J\rangle$	Reference
$4f^3$	Pr	0.9967	1.9203	1.0240	Wybourne[2]
$4f^3$	Nd^{3+}			1.027	Judd[67]
$4f^{10}$	Dy	1.0075	−0.6275	0.972	Childs[62]
$4f^{10}$	Ho^{3+}	1.0082	−0.672	0.972	Wybourne,[2] corrected by Childs (private communication)
$4f^{11}$	Ho	1.0048	2.855	0.9627	Wybourne[68]
$4f^{11}$	Er^{3+}			0.956	Judd[67]
$4f^{12}$	Er	1.0013	1.206	0.9867	Smith and Unsworth[69]

For the lanthanide atoms a complete set of eigenvectors allowing for intermediate coupling is given by Conway and Wybourne;[4] they should not be far wrong for ions of the same configuration $4f^n$.

8.2. THE NUCLEAR MOMENTS

In an ideal world it would be possible to calculate all the electronic quantities in the equations given above, and also some corrections which we have yet to consider. In practice calculations of quantities such as $\langle r^{-3}\rangle$ by methods involving different approximations have shown variations by as much as 25 %. An experimental test of such calculations is possible if the hyperfine constants and the nuclear moments can be determined separately. For the free atoms the hyperfine constants can be measured very accurately by the method of magnetic resonance using atomic beams, and a list of results is given in Table 5.4 of Abragam and Bleaney.[1] These results are not, of course, directly relevant to the hyperfine interactions in metals, but we shall use them in a number of cases for the purposes of extrapolation. For a majority of the lanthanides the ground state of the atom Ln is $4f^{n+1}6s^2$ while that of the dipositive ion Ln^{2+} is $4f^{n+1}$, other closed shells being ignored, and that of the tripositive ion Ln^{3+} (with which we shall mostly be concerned) is $4f^n$.

Direct measurement of the nuclear magnetic dipole moment involves a measurement of the nuclear Zeeman interaction with a known external applied field. Such a field is necessarily small compared with the magnetic field generated at the nucleus by the electrons in the unfilled $4f$ shell, and the precision attained is correspondingly smaller. The nuclear magnetic moments given in Table 8.3 have been measured either by the atomic beam triple resonance method (ATR) or by electron–nuclear double resonance (ENDOR) in the solid. Other values of lower precision have been obtained by calculations based on the magnetic hyperfine constant A and estimated values of $\langle r^{-3} \rangle$.

Table 8.3. Table of Nuclear Moments for the Stable Isotopes of the Lanthanide Group. Two radioactive isotopes (marked †) have been included for Ce, Pm where no stable isotopes with $I \neq 0$ exist

Isotope	I	μ_n n.m.	Q barns	Reference
La 138	5	+3.707	±0.8	
139	7/2	+2.778	+0.22 (3)	Childs[70]
Ce 141†	7/2	±1.15		(See §8.3)
Pr 141	5/2	+4.25 (5)	−0.059 (4)	Hin Lew (private communication)
Nd 143	7/2	−1.063 (5)	−0.484 (20)	Smith and Unsworth[69]
145	7/2	−0.654 (4)	−0.253 (10)	
Pm 147†	7/2	+2.58 (7)	+0.74 (20)	Reader[71]
Sm 147	7/2	−0.8070	−0.20 (2)	Woodgate[7]
149	7/2	−0.665	+0.058 (6)	
Eu 151	5/2	+3.4630 (6)	+1.16	Evans, Sanders and Woodgate[6]
153	5/2	+1.5292 (8)	+2.9	
Gd 155	3/2	−0.258 (1)	+1.59 (16)	Unsworth[72]
157	3/2	−0.336 (1)	+1.70 (16)	Baker, Copland and Wanklyn[16]
Tb 159	3/2	+1.994 (4)	+1.34 (11)	Baker, Chadwick, Garton and Hurrell;[73] Childs[63]
Dy 161	5/2	−0.46 (5)	+2.37 (28)	Childs[62]
163	5/2	+0.65 (6)	+2.51 (30)	
Ho 165	7/2	+4.03 (5)	+2.4	Unsworth (private communication)
Er 167	7/2	−0.565 (2)	+2.83 (1)	Smith and Unsworth[69]
Tm 169	1/2	−0.2310 (15)		Giglberger and Penselin[74]
Yb 171	1/2	+0.49188 (2)		Olschewski and Otten[75]
173	5/2	−0.67755 (2)	+2.8 (2)	
Lu 175	7/2	+2.23	+5.6	

The values given for the nuclear electric quadrupole moment Q are much more uncertain, as there is no method of applying a known electric field gradient to the nucleus.

8.2.1. The hyperfine anomaly

As has already been mentioned, for a nucleus of finite volume the value of H_e is essentially an average value of the magnetic field generated by the electrons, taken over the volume occupied by the nuclear magnetism, and weighted by the density of nuclear magnetism at each point within this volume. For different isotopes of the same element, the distribution of nuclear magnetism may vary appreciably, particularly if the isotopes have different values of the nuclear spin. If the magnetic field generated by the electrons also varies appreciably over the volume of the nucleus, the value of H_e defined by its average value weighted by the density of nuclear magnetization may vary with the isotope concerned. This gives rise to the "hyperfine anomaly", Δ, defined by the relation

$$A_1/A_2 = \{(g_n)_1/(g_n)_2\}(1+\Delta) \tag{8.20}$$

where A_1, A_2 are the magnetic hyperfine constants of a pair of isotopes and $(g_n)_1$ and $(g_n)_2$ their nuclear g-factors. The latter are determined by the nuclear Zeeman interaction with an external magnetic field, which is of course uniform over the nuclear volume, so that the ratio of the two nuclear Zeeman interactions is equal to the ratio of the nuclear g-factors whatever the distribution of nuclear magnetization.

Table 8.4. Ratios of Magnetic Hyperfine Constants, Nuclear g-factors and Electric Quadrupole Hyperfine Constants for pairs of stable Lanthanide Isotopes

		A_1/A_2	$(g_n)_1/(g_n)_2$	B_1/B_2	Reference
Nd	143/145	+1.60860 (36)	+1.622 (22)	+1.893 (16)	Spalding[76]
					Smith and Unsworth[69]
Nd^{3+}	143/145	+1.60883 (4)		+1.96 (2)	Halford[77]
		+1.6089 (1)			Erickson[78]
Sm	147/149	+1.21302 (2)		−3.4602 (2)	Woodgate[7]
Eu	151/153	+2.26498 (8)	+2.26505 (42)	+0.393 (3)	Sandars and
					Woodgate[79]
					Evans, Sandars, and
					Woodgate[6]
Eu^{2+}	151/153	+2.25313 (15)	+2.2632 (26)	+0.387 (4)	Baker and Williams[15]
Gd	155/157	+0.76252 (13)		+0.93867 (3)	Unsworth[72]
Gd^{3+}	155/157	+0.7621 (5)	+0.7633 (45)		Baker, Copland, and
					Wanklyn[16]
Dy	161/163	−0.71415 (2)		+0.94684 (5)	Childs[62]
Yb	171/173		−3.6298 (2)		Owlschewski and
					Otten[75]
Yb^{3+}	171/173	−3.6302 (4)			Baker, Blake and
					Copland[17]

The values of these ratios for pairs of isotopes where they have been determined with high precision are listed in Table 8.4. The results for europium are of special significance, in that the ratio of the magnetic dipole interaction constants $^{151}A/^{153}A$ for the atom is equal, within the experimental error, to the ratio of the nuclear g-factors $^{151}g_n/^{153}g_n$. This is not true for the Eu^{2+} ion, where there is a hyperfine anomaly whose size is well outside the experimental error. Both the Eu atom, $4f^7 6s^2$, and the Eu^{2+} ion, $4f^7$, have a half-filled $4f$ shell whose ground term is $^8S_{7/2}$. From Table 8.1 it can be seen that on simple theory no magnetic hyperfine interaction would be expected, because the orbital momentum $L=0$, and the spin magnetization has spherical symmetry with zero density at the nucleus, a geometrical distribution which (like a hollow uniformly magnetized spherical shell) gives zero magnetic field at the centre. In fact the values of A are not zero because of contributions from three sources: (i) intermediate coupling; (ii) relativistic corrections; (iii) core polarization. As pointed out by Sandars and Beck,[5] the first two of these arise from $4f$ electrons and would not be expected to give a hyperfine anomaly, while the third is due to unpaired s-electrons which have a finite (and non-uniform) density at the nucleus, giving rise to the possibility of a hyperfine anomaly. Evans, Sandars, and Woodgate [6] have shown that the value of A for the Eu atom can be accounted for on the basis of (i) and (ii), so that the absence of an anomaly is consistent with their calculations. On the other hand, for the Eu^{2+} ion the dominant contribution to A comes from the third source, core polarization, and the size of the anomaly is consistent with that observed in optical measurements involving s-electrons.

For configurations other than $4f^7$, the contributions to the magnetic hyperfine interaction from the $4f$ electrons are an order of magnitude greater than in the Eu atom. We may therefore assume with reasonable confidence that in the atoms any hyperfine anomaly will be very small, and that the ratio of the magnetic hyperfine constants A for a pair of isotopes will be very close to the ratio of the nuclear g-factors.

On theoretical grounds no anomaly of comparable size is expected for the nuclear electric quadrupole interaction. We therefore give in Table 8.IV the ratio of the values of the constants B for pairs of isotopes, as calculated from measurements on atoms, and anticipate that the same ratio should be found in measurements on ions in the solid state.

8.2.2. Relativistic corrections

In a non-relativistic approximation the values of $\langle r_l^{-3} \rangle$ and $\langle r_{sc}^{-3} \rangle$ for a single electron are the same. When relativistic equations are used, the result is effectively to give slightly different values of $\langle r^{-3} \rangle$ for the single electron states with $j = l + \frac{1}{2}$ and $j = l - \frac{1}{2}$; the difference may be regarded as arising from the different spin–orbit coupling energy in the two states, which results in a small change in the radial wave-function.

When there is more than one electron in the $4f$ shell, the relativistic effects result in slightly different values of $\langle r^{-3} \rangle$ for the individual electrons, and the quantities $\langle r_l^{-3} \rangle_i$ and $\langle r_{sc}^{-3} \rangle_i$ cannot be simply taken outside the summations in Eqns (8.3) and (8.8) as was assumed above. However, it has been shown,[5] that Eqns (8.4) and (8.9) are correct in form, though the values of $\langle r_L^{-3} \rangle$ and $\langle r_{sc}^{-3} \rangle$ are not in general identical; and, furthermore, that a term linear in the electron spin momentum $\mathbf{S}$ may arise, which we write as

$$\mathbf{H}_S = -2\beta \langle r_s^{-3} \rangle \mathbf{S}. \qquad (8.21)$$

Here the quantity $\langle r_s^{-3} \rangle$ bears no relation to the two previous quantities $\langle r_L^{-3} \rangle$ and $\langle r_{sc}^{-3} \rangle$, which are nearly equal, but is expected to be very much smaller, since it would be zero for $4f$ electrons in the absence of such corrections.

8.2.3. Configuration interaction and core polarization

So far we have assumed that the spectroscopic state of an atom or ion can be accurately represented by a single configuration. Such a configuration is obtained as a solution of the wave-equation for a many-electron atom using the central field approximation, in which an electron is assumed to move in a smoothed-out potential due to the nucleus and the averaged repulsion arising from the other electrons. In a more accurate treatment, such electrostatic interactions result in the admixture of small amplitudes of other configurations, an effect known as "configuration interaction". For our purpose the important feature of this interaction is a change in the radial wave functions which is reflected in small changes in the effective values of $\langle r_L^{-3} \rangle$ and $\langle r_{sC}^{-3} \rangle$ in Eqns (8.4) and (8.9), together with a non-zero value of $\langle r_s^{-3} \rangle$ in Eqn (8.21). As pointed out by Sandars and Beck[5], this result is indistinguishable in form from the effects of the relativistic corrections. Thus in LS-coupling it should be possible to represent the magnetic hyperfine interaction in all terms of an (L, S) multiplet by means of the three equations (8.4), (8.9), and (8.21) with suitable values of the parameters $\langle r_L^{-3} \rangle$, $\langle r_{sC}^{-3} \rangle$, and $\langle r_s^{-3} \rangle$. However, corrections must first be made for the effects of intermediate coupling, which are different for states of different J within the same multiplet.

The effect known as "core polarization" represents an approach to the same problem from a different viewpoint, which may be easier to comprehend in physical terms. The electrostatic repulsion between the electrons gives rise to an exchange interaction between them of the same nature as that which results in ferromagnetism, antiferromagnetism, etc. In this case we are concerned with exchange interaction between electrons within the same atom (or ion), and in particular between electrons in the open ($4f$) shell and electrons in closed shells. In the latter, electrons whose spin is parallel to the net resultant spin S in the $4f$ shell will have, through exchange interaction, a somewhat different energy from those with antiparallel spin. This energy difference is reflected in small changes in the radial wave-functions, giving the results mentioned in the previous paragraph.

For our purpose the most important effect of core polarization is that it gives rise to the term (8.21) which would be zero on our simple theory apart from relativistic effects. It is easy to see how such a term arises. A closed s-shell contains just two electrons, one whose spin is parallel to the net spin S in the $4f$ shell, and one whose spin is antiparallel. The exchange interaction with the $4f$ spin changes the radial wave-functions of the two electrons with "up" and "down" spin, so that they no longer have exactly the same density at the nucleus, and the net magnetic field at the nucleus no longer cancels exactly for a closed s-shell. This is particularly important because the finite density of s-electrons at the nucleus means that (i) a small unbalance gives a large contribution to $\langle r_s^{-3} \rangle$ (in the $3d$ group ions this turns out to be the largest contribution to the magnetic hyperfine interaction); and (ii) their net field at the nucleus is a slightly non-uniform field, so that a hyperfine anomaly is possible.

This last point provides a method of distinguishing experimentally between contributions to (8.21) from relativistic and from core polarization. The relativistic contributions arise from $4f$ electrons whose wave-functions are effectively zero at the nucleus; the "relativistic" magnetic field at the nucleus is uniform and no hyperfine anomaly can arise. The core polarization field is due to s-electrons with unbalanced spin densities at the nucleus, giving a non-uniform magnetic field and the possibility of a hyperfine anomaly. The sole

Table 8.5. Comparison of Magnetic Hyperfine Constants for the Europium Atom and Ion

	A (total)	A (intermediate coupling)	A (relativistic)	A (core polarization)
^{151}Eu atom	-20.0523 (2) MHz	$+7.70$ MHz	$(-28$ MHz)	(0)
^{151}Eu^{2+} ion	-102.9069 (13) MHz	$+8$ MHz	$(-28$ MHz)	$(-83$ MHz)
Eu^{2+} ion hyperfine magnetic field $= (-H_e)$	$-98S = -340$ kG	$+8S = +28$ kG	$-27S = -93$ kG	$-79S = -275$ kG

Values in parentheses are those chosen to fit the experimental values for A (total) assuming that the relativistic contribution is the same for atom and ion, while the core polarization contribution is zero in the atom. The combined value of A_S (relativistic+core polarization) for the ion is -111 MHz for ^{151}Eu, equivalent to a hyperfine field of $-105S = -368$ kG.

clear example comes from the experimental results for Eu and Eu^{2+}, each with the $4f^7$, $^8S_{7/2}$ ground state, to which reference has already been made. The estimated contributions to the magnetic hyperfine constant for the isotope ^{151}Eu in atom and ion are listed in Table 8.5.

In this table the absence of a hyperfine anomaly for the europium atom is interpreted as meaning that the contribution from core polarization is zero. Then the difference between the contribution from intermediate coupling (which can be estimated fairly accurately) and the measured value is ascribed to relativistic effects. The latter are much more difficult to estimate, but Evans, Sandars, and Woodgate[6] have shown that the expected contribution is of the right sign and of the right order of magnitude (in fact they obtain about half the value given in Table 8.5). For the europium ion the contributions from intermediate coupling and relativistic corrections are assumed to be the same, leaving a much larger contribution from core polarization. On this basis, the size of the hyperfine anomaly observed for the ion is in good agreement with that found in optical measurements on states with unpaired s-electrons (for a more detailed discussion see Section 4.7 of Abragam and Bleaney[1]).

To conclude this discussion we refer to a detailed set of measurements by Woodgate[7] on the $J=1$, 2, 3, and 4 levels of the samarium atom, $4f^6$, ^7F. These yield the results (in a.u.) $\langle r_L^{-3} \rangle = 6.393(3)$, $\langle r_{sC}^{-3} \rangle = 6.512(1)$, $\langle r_s^{-3} \rangle = -0.208(1)$. The ratio of the first two, $\langle r_{sC}^{-3} \rangle / \langle r_L^{-3} \rangle = 1.015$, is close to that found by Sandars and Beck in their preliminary relativistic calculations, while the value of $\langle r_s^{-3} \rangle$ is close to that of -0.211 a.u. for the europium atom. This makes it plausible (but not certain) that the "contact" term is again of relativistic origin, and we shall assume that this is true for all the atoms with $4f^n$ ground states. Hereafter we shall also assume that $\langle r_{sC}^{-3} \rangle = \langle r_L^{-3} \rangle$ for both atoms and ions. There is no experimental evidence for this in the ions, but since the overwhelming contribution to the magnetic hyperfine interaction comes from the orbit in all but the $4f^7$ ions, this assumption should not introduce any serious error.

8.3. MAGNETIC HYPERFINE CONSTANTS FOR THE 'ISOLATED' IONS

From measurements on lanthanide salts using electron paramagnetic resonance or Endor it is possible, following the method of Bleaney,[8] to derive fairly accurate values of the hyperfine constant A of Eqn (8.1). Such values are listed by Bleaney (and repeated in Table 5.5 of Abragam and Bleaney[1]). They have been supplemented by a value for $^{141}Ce^{3+}$, calculated from Kedzie, Abraham, and Jeffries,[9] and the value $A = -388.8$ MHz for ^{169}Tm of Jones and Schmidt.[10] In Table 8.6 they have been multiplied by J (for the ground state of each ion), so that the quantity JA represents the separation (if no nuclear electric quadrupole effects are present) between successive nuclear levels in the electronic sub-state $J_z = J$ for the "isolated" ion. The values are close to those given by Bleaney,[11] except that for Nd^{3+} a numerical error has been corrected, and for Tm^{3+} the new value is about 1% smaller than the previous (extrapolated) value. The modifications to be expected in a metal

Table 8.6. Magnetic Hyperfine Splittings (JA) for the state $J_z = J$, where J is the Ground State of each Ion, listed in units of MHz and of millidegrees

Ion	J	Isotope	Abundance (%)	I	JA (MHz)	JA (millideg.)
58 Ce^{3+}	5/2	141	Radioactive	7/2	$(\pm)400$ (10)	$(\pm)19.2$ (5)
59 Pr^{3+}	4	141	100	5/2	$+4372$ (40)	$+209.8$ (20)
60 Nd^{3+}	9/2	143	12.3	7/2	-991.4 (9)	-47.58 (4)
		145	8.3	7/2	-616.0 (5)	-29.57 (2)
61 Pm^{3+}	4	147	Radioactive	7/2	$(+)2396$ (24)	$(+)115.0$ (10)
62 Sm^{3+}	5/2	147	15.0	7/2	-600 (8)	-28.8 (4)
		149	13.9	7/2	-485 (8)	-23.3 (4)
63 Eu^{3+}	0				0	0
63 Eu^{2+}	7/2	151	47.8	5/2	-360.17	-17.29
		153	52.2	5/2	-159.86	-7.67
64 Gd^{3+}	7/2	155	14.7	3/2	$+42.17$	$+2.02$
		157	15.7	3/2	$+55.33$	$+2.66$
65 Tb^{3+}	6	159	100	3/2	$+3180$ (30)	$+152.6$ (2)
66 Dy^{3+}	15/2	161	19.0	5/2	-821 (16)	-38.4 (8)
		163	24.9	5/2	$+1143$ (22)	$+54.9$ (11)
67 Ho^{3+}	8	165	100	7/2	$+6497$ (8)	$+311.8$ (4)
68 Er^{3+}	15/2	167	22.9	7/2	-940 (9)	-45.1 (4)
69 Tm^{3+}	6	169	100	1/2	-2333	-112
70 Yb^{3+}	7/2	171	14.4	1/2	$+3105$ (5)	$+149.0$ (2)
		173	16.2	5/2	-852 (2)	-40.9 (1)

The values given are deduced from electron spin resonance measurements on salts, crystal field effects being eliminated as described by Bleaney.[8] The errors quoted allow for inaccuracies in the measurements and in elimination of crystal field effects. The values quoted for Eu^{2+} correspond to those measured in CaF_2; similarly those for Gd^{3+} correspond to measurements in CeO_2. The remaining values are based on the values of A given by Bleaney[8] and are close to those in Bleaney[11] except for Nd^{3+}, where an error has been corrected, and for Tm^{3+} where a recent result of Jones and Schmidt[10] has been used.

where conduction electrons are present, and the close neighbours are magnetic ions, will be discussed in Section 8.5.

The values of JA in Table 8.6 can be used to deduce the hyperfine field at the nucleus in the fully polarized state $J_z = J$ using the relation

$$-\mathbf{H}_e = (JA)/(g_n \beta_n) = \frac{(JA)\,(\text{MHz})}{762.28\,g_n}\ (\text{megagauss}). \tag{8.22}$$

The accuracy of the calculation depends on the accuracy with which the nuclear magnetic moment is known as well as the experimental error in A, and this is reflected in the errors attributed to the values listed in Table 8.7. In the case of Ce^{3+} where no direct measurement of the nuclear moment has been made, the value of $-\mathbf{H}_e$ (and also of μ_n, Table 8.3) is found instead by the use of Eqn (8.15) with an estimated value of $\langle r^{-3} \rangle = 4.17$ a.u., to which is added the contribution from core polarization given in the last column of Table 8.7. In other cases (Pr^{3+}, Pm^{3+}, Dy^{3+}) the value of μ_n given in Table 8.3 has been used, but this itself has been calculated using a value of $\langle r^{-3} \rangle$ together with the accurate measurement of A for the free atom.

Table 8.7. Values of the Total Magnetic Hyperfine Field and of the Estimated Contribution due to Core Polarization, in units of megagauss ($=10^2$ tesla)

		Hyperfine magnetic field (megagauss)	
		Total	Estimated contribution from core polarization
$4f^1$	Ce^{3+}	+1.83 (5)	+0.035
$4f^2$	Pr^{3+}	+3.37 (4)	+0.08
$4f^3$	Nd^{3+}	+4.30 (2)	+0.12
$4f^4$	Pm^{3+}	+4.26 (12)	+0.16
$4f^5$	Sm^{3+}	+3.42 (1)	+0.18
$4f^6$	Eu^{3+}	0	0
$4f^7$	Eu^{2+}	−0.34	−0.35
	Gd^{3+}	−0.32	−0.35
$4f^8$	Tb^{3+}	+3.14 (3)	−0.30
$4f^9$	Dy^{3+}	+5.70 (10)	−0.25
$4f^{10}$	Ho^{3+}	+7.40 (9)	−0.20
$4f^{11}$	Er^{3+}	+7.64 (3)	−0.15
$4f^{12}$	Tm^{3+}	+6.62	−0.10
$4f^{13}$	Yb^{3+}	+4.13 (1)	−0.05

A positive sign means that the field is in the same direction as the electronic magnetic moment; thus the values quoted are those of $(-H_e)$ as defined in the text.

The values given for the core polarization contribution in the last column of Table 8.7 must be regarded as no more than a guess. They are calculated from the relation

$$-H_S = 0.100(\lambda - 1)J \text{(megagauss)} \qquad (8.23)$$

which is equivalent to Eqn (8.21) after allowing for the projection of **S** onto **J**, and assuming that the numerical constant is the same throughout the $4f$ group. There is no experimental evidence for this, but it has been found to be not a bad assumption for the $3d$ group, and the contribution in the $4f$ group is small compared with the total field in all cases except $4f^7$. For the latter the values given in Table 8.7 for the "total hyperfine field" are those appropriate to Eu^{2+} in CaF_2 and Gd^{3+} in CeO_2, as in Table 8.6. In view of the following discussion, no attempt has been made to assign possible limits of error to these values or to those for $-H_S$. The numerical constant in Eqn (8.23) is of course a conveniently rounded value.

8.4. ELECTRIC QUADRUPOLE HYPERFINE INTERACTION

The hyperfine interaction arising from the nuclear electric quadrupole moment may be written in the form

$$\mathcal{H}_Q = \sum_q \mathbb{A}_2{}^q (\mathbb{B}_2{}^q)^* \qquad (8.24)$$

where the $\mathbb{A}_2{}^q$ are related to the nuclear properties and the $\mathbb{B}_2{}^q$ to the electric field gradient at the nucleus, $(\mathcal{B}_2{}^q)^*$ being the complex conjugate of $\mathcal{B}_2{}^q$. The components of $\mathcal{A}_2{}^q$ are

$$\mathcal{A}_2{}^0 = \frac{eQ}{I(2I-1)}\,[\tfrac{1}{2}\{3I_z{}^2 - I(I+1)\}]$$

$$\mathcal{A}_2{}^{\pm 1} = \frac{eQ}{I(2I-1)}\,[\mp\sqrt{\tfrac{3}{8}}\{I_z I_\pm + I_\pm I_z\}]$$

$$\mathcal{A}_2{}^{\pm 2} = \frac{eQ}{I(2I-1)}\,[\sqrt{\tfrac{3}{8}}\{I_\pm{}^2\}] \tag{8.25}$$

where Q is the nuclear electric quadrupole moment as normally defined. The quantities $\mathcal{B}_2{}^q$ are given by

$$\mathcal{B}_2{}^0 = \tfrac{1}{2}V_{zz}$$

$$\mathcal{B}_2{}^{\pm 1} = \sqrt{\tfrac{1}{6}}\{V_{xz} \pm iV_{yz}\}$$

$$\mathcal{B}_2{}^{\pm 2} = \tfrac{1}{2}\sqrt{\tfrac{1}{6}}\{V_{xx} - V_{yy} \pm 2iV_{xy}\}. \tag{8.26}$$

Here the quantity V_{xy} is a shorthand notation for $(\delta^2 V/\delta x \delta y)_{r=0}$ which is a component of a tensor operator, the electric field gradient tensor at the origin, which is traceless:

$$V_{xx} + V_{yy} + V_{zz} = 0. \tag{8.27}$$

For many purposes it is convenient to have the nuclear electric quadrupole interaction written out in a straightforward Cartesian form. This is

$$\mathcal{H}_Q = \frac{eQ}{I(2I-1)} \begin{bmatrix} \tfrac{1}{6}V_{xx}\{3I_x{}^2 - I(I+1)\} \\ +\tfrac{1}{6}V_{yy}\{3I_y{}^2 - I(I+1)\} \\ +\tfrac{1}{6}V_{zz}\{3I_z{}^2 - I(I+1)\} \\ +\tfrac{3}{4}V_{xy}\{I_x I_y + I_y I_x\} \\ +\tfrac{3}{4}V_{yz}\{I_y I_z + I_z I_y\} \\ +\tfrac{3}{4}V_{zx}\{I_z I_x + I_x I_z\} \end{bmatrix} \tag{8.28}$$

Here a component such as $V_{xy} = V_{yx}$ vanishes if x, y, z are the "principal axes" of the electric field gradient tensor. In this case we may write, using Eqn (8.27),

$$\mathcal{H}_Q = \frac{eQ}{4I(2I-1)}\,V_{zz}[\{3I_z{}^2 - (I(I+1)\} + \eta\{I_x{}^2 - I_y{}^2\}] \tag{8.29}$$

$$= P_{\|}[\{I_z{}^2 - \tfrac{1}{3}I(I+1)\} + \frac{\eta}{3}\{I_x{}^2 - I_y{}^2\}] \tag{8.30}$$

where the symbol q is often written for V_{zz}. The quantity

$$\eta = (V_{xx} - V_{yy})/V_{zz} \tag{8.31}$$

vanishes if we have axial symmetry ($V_{xx} = V_{yy}$) about the z-axis, while

$$P_{\parallel} = \frac{3eQ}{4I(2I-1)} V_{zz}. \tag{8.32}$$

For a free atom or ion with a $4f^n$ configuration for which J is a good quantum number the quantities $\mathfrak{B}_2{}^q$ which represent the electric field gradient at the nucleus due to the $4f$ electrons can be written in terms of electron operators as follows.

$$\mathfrak{B}_2{}^0 = -e\langle r_q{}^{-3}\rangle\langle J\|\alpha\|J\rangle[\tfrac{1}{2}\{3J_z{}^2 - J(J+1)\}]$$

$$\mathfrak{B}_2{}^{\pm 1} = -e\langle r_q{}^{-3}\rangle\langle J\|\alpha\|J\rangle[\mp\sqrt{\tfrac{3}{8}}\{J_z J_{\pm} + J_{\pm} J_z\}]$$

$$\mathfrak{B}_2{}^{\pm 2} = -e\langle r_q{}^{-3}\rangle\langle J\|\alpha\|J\rangle[\sqrt{\tfrac{3}{8}}\{J_{\pm}{}^2\}]. \tag{8.33}$$

Here $\langle r_q{}^{-3}\rangle = (1-R)\langle r^{-3}\rangle$ is the mean inverse cube of the distance of the $4f$ electrons from the nucleus, allowing for the Sternheimer (screening) effect of the other electrons, and $\langle J\|\alpha\|J\rangle$ is a number characteristic of each $4f^n$ configuration given by Elliott and Stevens[12] and also listed in Table 1.3 as the parameter α. The complete operator $\mathcal{H}_Q$ for the $4f$ electrons can thus be written down using Eqns (8.24), (8.25), and (8.33); an alternative form is

$$\mathcal{H}_Q = \frac{B}{2I(2I-1)J(2J-1)}\{3(\mathbf{I}\cdot\mathbf{J})^2 + \tfrac{3}{2}(\mathbf{I}\cdot\mathbf{J}) - I(I+1)J(J+1)\} \tag{8.34}$$

where the parameter B (not to be confused with the $B_2{}^q$) is given by

$$B = -e^2 Q\langle r_q{}^{-3}\rangle\langle J\|\alpha\|J\rangle J(2J-1) \tag{8.35}$$

For the state $|J_z| = J$, the value of $\langle 3J_z{}^2 - J(J+1)\rangle$ is $J(2J-1)$, and for the free ion in this state the value of eQV_{zz} is just that of the parameter B in Eqn (8.35), so that

$$P_{\parallel} = \frac{3B}{4I(2I-1)}. \tag{8.36}$$

Unfortunately we do not have a set of experimental values of B for the lanthanide ions, and we must therefore derive them indirectly. Deductions from E.P.R. or Endor measurements in the solid state are much less reliable than in the case of the magnetic hyperfine interaction for two reasons:

(i) the crystal field may give a direct contribution to the electric field gradient at the nucleus;

(ii) the crystal field lifts the $(2J+1)$ degeneracy, and the calculation of $\langle 3J_z{}^2 - J(J+1)\rangle$ for the sub-states involved in the measurement depends in first order on our knowledge of the electronic wave-functions.

On the other hand very accurate measurements of the values of B exist for the free atoms, and to find the value for the ion we use just the ratio

$$\frac{B\text{ (ion)}}{B\text{ (atom)}} = \frac{\{\langle r_q{}^{-3}\rangle\langle J\|\alpha\|J\rangle J(2J-1)\}\text{ (ion)}}{\{\langle r_q{}^{-3}\rangle\langle J\|\alpha\|J\rangle J(2J-1)\}\text{ (atom)}}. \tag{8.37}$$

This conversion, apart from known numbers, requires only the ratio of the values of $\langle r_q{}^{-3}\rangle$ which we take to be the same as the ratio of the values of

$\langle r^{-3} \rangle$ since calculations[13] indicate that the shielding parameter R varies only very slowly through the $4f^n$ series (from $+0.1308$ for Pr^{3+}, $4f^2$ to $+0.1296$ for Tm^{3+}, $4f^{12}$). For the ratios of the $\langle r^{-3} \rangle$ we use those of Lindgren,[14] and obtain the values of $P_{\parallel}$ (in MHz and in millidegrees K) shown in Table 8.8.

Table 8.8. Values of $P_{\parallel}$ (in MHz and millidegrees), and of P_c (in MHz)

Ion	Isotope	I	$P_{\parallel}$ (MHz)	$P_{\parallel}$ (millideg.)	P_c (MHz)
59 Pr^{3+}	141	5/2	-2.62	-0.128	-0.15
60 Nd^{3+}	143	7/2	-5.3	-0.25	-0.6
	145	7/2	-2.8	-0.13	-0.3
61 Pm^{3+}	147	7/2	-6.6	-0.32	$+0.9$
62 Sm^{3+}	147	7/2	-4.83	-0.232	-0.3
	149	7/2	$+1.40$	$+0.0672$	$+0.1$
63 Eu^{3+}			(0)	(0)	
Eu^{2+}	151	5/2	-0.0589	-0.00283	
	153	5/2	-0.1522	-0.00730	
64 Gd^{3+}	155	3/2	-0.161	-0.0077	$+26$
	157	3/2	-0.172	-0.0083	$+28$
65 Tb^{3+}	159	3/2	$+386$	$+18.5$	$+28$
66 Dy^{3+}	161	5/2	$+219$	$+10.5$	$+17$
	163	5/2	$+228$	$+10.9$	$+18$
67 Ho^{3+}	165	7/2	$+62.7$	$+3.01$	$+9$
68 Er^{3+}	167	7/2	-67.6	-3.24	$+10$
69 Tm^{3+}	169	1/2	—	—	—
70 Yb^{3+}	171	1/2	—	—	—
	173	5/2	-316	-15.2	

Except for Eu^{2+}, Gd^{3+}, Yb^{3+}, where Endor results have been used, the values of $P_{\parallel}$ are deduced from atomic beam measurements on atoms using Eqn (8.37); the probable errors are discussed in Section 8.8. The values of P_c, the lattice+conduction electron contribution in the hexagonal metal, are discussed in Section 8.5.2. Note that $P_{\parallel}$, P_c are not simply additive (see Sections 8.4, 8.5.2, and 8.8).

There are three exceptions in this Table to this method of derivation, being found instead from Endor measurements on cubic crystals. These are Eu^{2+} in CaF$_2$;[15] Gd^{3+} in ThO$_2$;[16]; and Yb^{3+} in CaF$_2$.[17] In the case of Gd^{3+} only ^{157}Gd was measured in this way, the value for ^{155}Gd being deducedthere from using the ratio of the values of B given in Table 8.4.

Basically the nuclear electric quadrupole interaction involves for the electrons only orbital operators, which have no matrix elements between the ground state and other states (of different S) admixed by intermediate coupling (cf. Eqn (8.19)). Thus the latter produces only corrections to B in the second order (unlike the situation for A), and such corrections (see the discussion in Section 8.8) have been ignored in Table 8.8. For the same reason it is probably more realistic to use for the ratios of $\langle r^{-3} \rangle$ the values calculated by Lindgren, rather than values derived from comparisons of the magnetic hyperfine constants in atom and ion, as was done by Bleaney.[11] However, the major differences (particularly those for the light lanthanides) between the values of $P_{\parallel}$ in Table 8.8 and those of Bleaney arise from the difficulty, in interpreting E.P.R. and Endor measurements, of estimating the lattice contribution. In particular, the change in sign and magnitude in the values for Nd^{3+} in the two compilations can be ascribed to a lattice contribution in $LaCl_3$ which outweighs that due to the $4f$ electrons.[8]

The lattice contributes to the electric field gradient only if the lanthanide ion occupies a site of less than cubic symmetry. In its origin, this gradient at the nucleus is identical with that which, acting on $4f$ the electrons, produces the crystal field terms of the second degree and, as discussed in Section 1.3.1, whatever model is used to evaluate the field gradient produced by the lattice itself, the result must be modified to allow for 'shielding'. In the case of the nucleus, the field gradient model must be multiplied by a factor $(1 - \gamma)$ to allow for distortion of the closed shells of electrons on the lanthanide ion. This factor is similar to the Sternheimer factor $(1 - R)$ introduced above after Eqn (8.33), but whereas $R \sim +0.1$ for a lanthanide ion, it turns out that $\gamma \sim -80$. The reason for the tremendous difference between R and γ is discussed by Watson and Freeman[18] (or see Abragam and Bleaney[1], Section 17.7). The result is an enormous "anti-shielding" effect which greatly enhances the "external" field gradient at the nucleus, and for Nd^{3+} in $LaCl_3$ (see above) this is accidentally combined with an abnormally small field gradient from the $4f$ electrons in the ground doublet produced by the crystal field splitting.

In a metal the lattice field gradient is further modified (see Section 8.5.2) by a local non-spherically symmetrical charge distribution for the conduction electrons. The resultant field gradient maintains the lattice symmetry, and in the hexagonal lanthanide metals is necessarily along the c-axis. The field gradient of the $4f$ electrons is parallel to the direction of the electronic magnetization, but if this is not along the c-axis we no longer have axial symmetry. This may result[19] in a small asymmetry parameter η in Eqn (8.30), and a full treatment requires summing the individual components of the field gradient tensors arising from the two sources. However, for all the metals except Gd, the magnetic hyperfine splitting is much greater than any nuclear electric quadrupole splittings, and in a good approximation we need retain only the diagonal matrix elements of the latter. This means that the lattice contribution must be multiplied by $\frac{1}{2}(3 \cos^2 \theta - 1)$ and then added to $P_\parallel$, where θ is the angle between the z-axis (the direction of the electronic magnetic moment) and the c-axis of the crystal. In fact no effects ascribable to a finite value of η have so far been detected in a lanthanide metal.

8.5. THE LANTHANIDE ION IN THE METAL

In the metallic state we have in a first approximation a set of relatively free lanthanide ions immersed in a "sea" of conduction electrons. Each ion is part of a lattice of metallic ions, usually with positive charges. In the pure metal the neighbouring ions will be identical lanthanide ions; in alloys they may be other lanthanide ions, randomly sited, while in intermetallic compounds there may be a regular array of lanthanide and other intervening metallic ions.

If sufficient near neighbours consist of ions with magnetic moments, the exchange interaction between them would be expected to give rise to an ordered magnetic state in which, if no other comparable interactions are present, each ion attains a fully polarized state as the temperature approaches 0 K. For a lanthanide ion, this means $\langle J_z \rangle = J$, and this is substantially the result observed in the heavy lanthanide metals.

Irrespective of whether the neighbours have magnetic moments, they are normally ions carrying electric charges and give rise to a "crystal field," which in the absence of any exchange interaction, causes a Stark splitting of the $(2J+1)$ levels of the ground state of the lanthanide ion. In the heavy lanthanide metals crystal field effects are small compared with those due to exchange interaction, and appear as an "anisotropy" energy which gives rise to complicated spin patterns in the ordered state. In the lighter lanthanide metals crystal field effects are generally larger in importance than exchange interactions. The crystal field splittings may exceed 10^2 K, while the ordered state sets in at 10 to 20 K where only a fraction of the $(2J+1)$ sub-levels is populated. The electronic properties in the ordered state may then be greatly modified, and for non-Kramers ions with a singlet ground state left by the crystal field, an ordered state may not be achieved even at 0 K. For Kramers ions an ordered state must set in, but the saturation electronic moment at 0 K is difficult to predict, and bears no simple relation to the relative strengths of the crystal field interaction and the exchange interaction. For example, it is quite possible for a crystal field to produce quite large level splittings, but to leave as the ground state a doublet for which $\pm J_z = J$, in which case the saturation moment would be the same as for the free ion.

8.5.1. Magnetic hyperfine interaction in the metal

Apart from any external field, the field at the nucleus in a magnetic solid includes contributions from the demagnetizing field, which is shape dependent and zero in a Bloch wall, and the Lorentz field $\mathbf{H} = (4\pi/3)\mathbf{M}$. The latter does not exceed 10 kG, so that it is usually a rather small correction which is not included in results quoted below unless specifically mentioned. Apart from this, we should expect the nuclear level splitting $A\langle J_z \rangle$ to reflect exactly the mean electronic moment of the ion, and an accurate determination of this splitting near 0 K should give a rather precise measurement of $\langle J_z \rangle$ and hence of the saturation magnetic moment. For this purpose we need to know the value of A in the substance under investigation. In practice, we may have to be satisfied with a lower accuracy, based on an estimate of the extent to which A may differ from the free ion value; this is the problem which we examine below.

We consider first effects due to the crystal field, which are essentially similar to those in Lanthanide salts.

(i) The crystal field may admix states with different values J' from the same multiplet into the ground state J, modifying the Zeeman interaction and the magnetic hyperfine interaction by different fractional amounts. The relation between the modified magnetic moment and the modified hyperfine interaction is essentially given by equation (8) of reference 8.

(ii) Fluctuations in the crystal field due to phonons (including the zero point fluctuations) may modify the ground state as calculated from the static crystal field.[20, 21] Such effects are small and unlikely to admix states of different J' so that they do not disturb the relationship between the electronic magnetic moment and the magnetic hyperfine interaction.

(iii) The simple crystal potential approximation is known not to be a wholly accurate representation of the ligand interaction in lanthanide salts, and overlap and covalent bonding may change the effective value of $\langle r^{-3} \rangle$ by a few per cent. It seems unlikely that such effects would be substantially larger in metals.

(iv) The core polarization field may be changed by bonding. A noticeable reduction is well established in Mn^{2+} salts, and for Tb^{4+}, $4f^7$ in ThO_2 the hyperfine field appears to be about -250 kG compared with -340 kG for Eu^{2+} in CaF_2 (see Abragam and Bleaney[1], Table 5.18).

We now consider effects arising from the exchange interaction. One of these is to admix excited states with different values J', as in the case of the crystal field. If we use a molecular field approximation to define the interaction with a lanthanide ion as

$$\mathscr{H}_{ex} = \beta(\mathbf{H}_{ex} \cdot \mathbf{S}) \tag{8.38}$$

then in the ground state J we can write $\mathbf{S} = (\lambda - 1)\mathbf{J}$ and similarly in an excited state J' we have $\mathbf{S} = (\lambda' - 1)\mathbf{J}'$. The excited state is admixed by matrix elements of the form $\langle J' | (\mathbf{H}_{ex} \cdot \mathbf{S}) | J \rangle$. From the fact that matrix elements of $\mathbf{L} + \mathbf{S}$ between states of different J are zero it follows that such off-diagonal matrix elements of $\mathbf{S}$ are the same as those of $\mathbf{L} + 2\mathbf{S}$. These are non-zero only for $J' = J \pm 1$, and the formulae of Elliott and Stevens[12] are readily adapted to calculate the admixtures if the magnitude of H_{ex} is known. So far as the relation between the electronic moment and the magnetic hyperfine interaction is concerned the calculation then proceeds as in the case of admixtures due to the crystal field. However for a fully polarized heavy lanthanide ion with $\langle J_z \rangle = J$, the correction arising from the exchange field is zero since the only excited state which could be admixed is the non-existent state $|J-1, J_z = J\rangle$. For the light lanthanide ions it turns out that the correction is always such as to increase the magnitude of the magnetic hyperfine constant.

The effects mentioned above may occur either in an insulator or a metal, but we now consider an effect peculiar to the latter, namely, the contribution to the hyperfine interaction from the conduction electrons. Like the core polarization field, which can be written in the form

$$\mathbf{H}_S = -\frac{8\pi}{3} g_s \beta \{ |\psi_\uparrow|^2 - |\psi_\downarrow|^2 \} \mathbf{s} \tag{8.39}$$

and which results from the contact interaction with the unbalanced spin density at the nucleus $\{ |\psi_\uparrow|^2 - |\psi_\downarrow|^2 \} \mathbf{s}$, the conduction electron field can be written as

$$\mathbf{H}_{ce} = -\frac{8\pi}{3} N_s g_s \beta |\psi_{ce}^0|^2 \langle \mathbf{s} \rangle. \tag{8.40}$$

Here N_s is the number of conduction electrons per unit volume, and $\langle \mathbf{s} \rangle$ the average value of the unbalanced spin density per conduction electron. In a metal with no permanent magnetic moments on the ions, the quantity $\langle \mathbf{s} \rangle$ is proportional to the external field H_0 and manifests itself in the bulk properties as the conduction electron susceptibility χ_{ce}. The field at the nucleus set up by the polarized conduction electrons is responsible for the "Knight shift" in nuclear magnetic resonance, where the frequency in a metal is changed by a fractional amount

$$K_0 = H_{ce}/H_0 \tag{8.41}$$

which, like χ_{ce}, is independent of the temperature.

In a metal containing ions with electronic magnetic dipole moments a further contribution to the polarization of the conduction electrons results from exchange interaction with the spins of the ions. For a lanthanide ion the Hamiltonian for interaction with a conduction electron

$$\mathscr{H}_{ex} = -\mathscr{J}_{sf}(\lambda - 1)(\mathbf{J} \cdot \mathbf{s}) \tag{8.42}$$

may be compared with that for an external field H_0

$$\mathscr{H} = g_s\beta(\mathbf{H}_0 \cdot \mathbf{s}) \tag{8.43}$$

to yield a value for the effective field acting on the conduction electrons of

$$\mathbf{H}_{eff} = -(\lambda - 1)(\mathscr{J}_{sf}/g_s\beta)\mathbf{J}.$$

If the lanthanide ion is itself not fully polarized we must take the mean value

$$H_{eff} = -(\lambda - 1)(\mathscr{J}_{sf}/g_s\beta)\langle J_z \rangle \tag{8.44}$$

where $\langle J_z \rangle$ is the time averaged component of J_z. For the field of the conduction electrons at the nucleus we thus obtain

$$H_{ce} = -K_0(\lambda - 1)(\mathscr{J}_{sf}/g_s\beta)\langle J_z \rangle \tag{8.45}$$

where $\langle J_z \rangle$ and hence also H_{ce} is temperature dependent. In the paramagnetic state it will be proportional to the external field H_0, and in the ferromagnetic state it will be proportional to the spontaneous magnetic moment. We shall be concerned only with the values of H_{ce} at temperatures sufficiently low for the spontaneous moment to have attained its saturation value.

Experimentally it has been found by Itoh, Kobayashi, and Sano[22] and others (see below) that no increase in line width is observed in N.M.R. measurements on alloys of various lanthanide metals, but a shift in the resonance frequency occurs which is linear in the concentration of the alloying element (cf. also Hüfner and Wernick[23]). These results show that the interaction of the conduction electrons with the ionic moments is sufficiently long range to average the contributions from a large number of neighbours. Thus the net polarization of the conduction electrons depends only on the average interaction, and in an alloy we can replace (8.44) by a sum of similar terms, each weighted by the fractional concentration. Whatever the dilution, however, a finite contribution will remain from polarization of the conduction electrons by the $4f$ electrons of the lanthanide ion itself. Furthermore, in intermetallic compounds or alloys with $3d$ ions carrying a permanent moment we may also expect a contribution to the conduction electron polarization through exchange interaction of the form

$$\mathbf{H}_{ex} = -\mathscr{J}_{sd}(\mathbf{S} \cdot \mathbf{s})$$

where S is the spin of the $3d$ ion.

Table 8.9. Estimated Contributions to the Hyperfine Magnetic Field in Europium Metal and in Gadolinium Metal

	Hyperfine magnetic field (megagauss)	
	Eu[23]	Gd[24]
Core polarization	−0.340 (20)	−0.340 (20)
Conduction electron polarization by own 4f electrons	+0.190 (20)	+0.250 (30)
Neighbour effects: conduction electrons + overlap + covalency	−0.115 (20)	−0.260 (25)
Experimental value for metal	−0.265 (5)	−0.350 (13)

An N.M.R. measurement[65] gives −0.371 (2) MG for Gd metal, including the Lorentz field correction, which the other values do not.

It turns out that the hyperfine field due to the conduction electrons is of the same order as that assigned to core polarization, so that both are small compared with the direct field of the 4f electrons except for ions with $4f^7$ shells (cf. Table 8.7). Most experiments which attempt to distinguish between the various contributions in a metal have therefore been made either with Eu^{2+} or Gd^{3+}. Using the Mössbauer effect, Hüfner and Wernick[23] have studied the hyperfine field of Eu metal as a function of dilution with Yb and with Ba, neither of which carry a permanent moment (Yb^{2+}, $4f^{14}$; Ba^{2+}, $4f^0$) but which have the same electronic structure as Eu^{2+} except for the 4f shell. Assuming the core polarization field to have the same value as for Eu^{2+} in CaF_2, they obtain the results listed in Table 8.9.

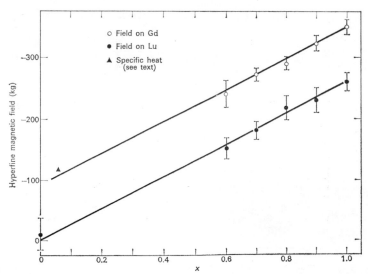

Figure 8–1. Variation of the hyperfine magnetic fields at the Gd and Lu nuclei in Gd_xLu_{1-x} alloys (after Zmora, Blau and Ofer[24]).

Table 8.10. Measured Hyperfine Fields in Various Compounds of Gadolinium

Compound	Hyperfine field (MG)
GdN	$(-)0.35$
GdMn$_2$	$(-)0.27$
GdPt$_2$	$(-)0.17$
GdAl$_2$	-0.16
GdRh$_2$	-0.10
GdFe$_2$	$+0.43$

Signs in parentheses are assumed. After Gegenwarth, Budnick, Skalski, and Wernick,[80] recalculated using $\mu_n = -0.3363$ n.m. for ^{157}Gd. A value of -0.22 (2) MG is obtained for GdAl$_2$ by Dintelmann, Dormann, and Buschow.[81]

In the case of Gd^{3+} the technique of perturbed angular correlation has been used by Zmora, Blau and Ofer[24] to show that for a range of alloys of Gd and Lu the effective field (see Figure 8–1) varies linearly with concentration, and at the same rate, at both Gd and Lu nuclei. Their results for the conduction electron contributions are given in Table 8.9. Extrapolation gives a field of $-0.090(20)$ MG at a Gd^{3+} ion surrounded by diamagnetic ions in a metal, a result which is comparable with those for the intermetallic compounds GdPt$_2$, GdAl$_2$, GdRh$_2$ listed in Table 8.10. This table shows also that in GdFe$_2$ there is a large positive contribution to the hyperfine field, presumably arising from conduction electron polarization by the Fe ions. This is in line with results on other LnFe$_2$ compounds (see Section 8.7).

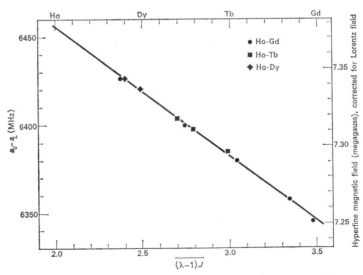

Figure 8–2. NMR measurements of the ^{165}Ho resonance frequency at liquid helium temperatures in various alloys (courtesy of Dr. M. A. H. McCausland).

The change in the nuclear magnetic resonance frequency of [159]Tb and [163]Dy when incorporated into Gd, Tb or Dy metal has been measured by Itoh, Kobayashi, and Sano.[22] A similar and more detailed study of [165]Ho by Mackenzie, McCausland, and Wagg[25] has shown that the shifts are again linear with concentration. One may write (cf. Figure 8–2)

$$a_0 = a_0' + a_0''(\lambda - 1)J + a_0'''\overline{(\lambda - 1)J} + a_L \qquad (8.46)$$

where a_0' is the "free ion" value. The second term is that ascribed to conduction electron polarization by the parent ion (assumed to be proportional to its own value of S projected on J). The third term is that due to neighbours (line width measurements suggest that some 50 neighbours are involved), and $\overline{(\lambda - 1)J}$ is the mean value of the projection of S averaged over the relative concentrations of the various ions. In terms of field, the third contribution amounts to $-0.083\overline{(\lambda - 1)J}$ megagauss for the Ho alloys measured; the second contribution (estimated by assuming $a_0' = JA$) is $+0.060(\lambda - 1)J$ megagauss. These results are quite close to those obtained by Zmora, Blau, and Ofer[24] for Gd (Table 8.IX) and for Lu^{3+} in Gd. The term $a_L \sim 1.1\overline{\lambda J}$ MHz represents the Lorentz field.

It is difficult to distinguish experimentally between the third term in Eqn (8.46) and the effects of direct transfer of unpaired spin density from neighbours due to overlap and covalency. However, the contact field at the nearest F^- nucleus due to lanthanide ions in CaF_2 is of order 1 kG, suggesting that similar effects in the metal are negligible compared with those due to conduction electrons (see, however, the discussion on Ho, Section 8.7).

8.5.2. Electric quadrupole hyperfine interaction in the metal

The contribution V_c of the ionic lattice to the electric field gradient has already been mentioned. For a simple close-packed hexagonal lattice of ions with charge $+Ze$, Das and Pomerantz[26] have derived the result

$$V_c = \left[a^{-3}\left\{0.0065 - 4.3584\left(\frac{c}{a} - 1.633\right)\right\}\right](1 - \gamma)Ze \qquad (8.47)$$

where a, c are the lattice constants, γ the Sternheimer shielding factor. For a hcp lattice with the ideal ratio $c/a = 1.633$, V_c practically vanishes, and estimates of V_c therefore depend sharply on accurate measurements of this ratio for the lanthanide metals. According to Spedding, Beaudry, Croat and Palmer[27] it appears that for the light lanthanide metals (allowing for their more complex crystal structure) the ratio varies only between 1.605 and 1.612, with a mean effective value of

$$(c/a) - 1.633 = -0.024.$$

For the heavier metals (Gd to Lu) which have a simple hcp structure, there is a wider variation, from -0.042 for Gd and -0.049 for Lu up to -0.062 for Ho, Er, and Tm, the mean value being -0.054. The parameter a^{-3} varies from the rather low value of $0.0186 \cdot 10^{24}$ cm^{-3} for La up to 0.0207 for Gd and 0.0232 for Lu. These are room-temperature values, and we need the low-temperature values. The increase in a^{-3} is probably not more than about 1%, but anisotropic thermal expansion and magnetostriction in the ordered phase may change c/a by as much as ± 0.01.[28] In Dy the crystal symmetry is orthorhombic in the ferromagnetic phase.[29] However, neglecting such corrections, we may obtain representative figures for V_c/e by taking $Z = +3$ and $(1 - \gamma) = +80$. For the light rare earths we find V_c/e ranges from 0.47 to 0.64, and for the heavier metals from 0.95 to 1.5 10^{24} cm^{-3}.

In the metal we may expect this electric gradient set up by the neighbouring ions to be modified by the presence of the conduction electrons. We allow for this by introducing an empirical screening factor, since the contribution from the conduction electrons is assumed to be proportional to that from the lattice. We estimate this empirical factor by comparing our calculated values of V_c/e with two experimental values for the actual quadrupole splitting, one for La metal where there are no 4f electrons and the other for Tm metal where the contribution from the 4f electrons has been separated from that due to the lattice + conduction electrons by using the fact that the former is temperature dependent while the latter is not. Both for La[30] and Tm[31] it is found that the total contribution from lattice + conduction electrons is a factor 2 greater than that estimated above, and we therefore scale all our estimates up by such a factor. This leads to the values of P_c given in Table 8.8 but it must be emphasized that these are very approximate, involving not only the assumptions made above but also the values of the nuclear quadrupole moment Q listed in Table 8.3. Three comparisons with experiment are possible, two of them being those we have already used to conjure up our scaling factor of 2. (For a more detailed treatment of this problem see Section 6.6.2.)

(i) ^{139}La. In the double hexagonal close packed lattice there are two sites, the "cubic" and "hexagonal" sites. Poteet, Tipsword, and Williams[32] have resolved two resonances corresponding to $|P_c| = 0.275$ and 0.243 MHz, while our scaled-up estimate gives $P_c = +0.26$ MHz. The other light rare earth metals will also have two values for the two sites in the dhcp structure, but the difference of ca. 10% is small compared with possible errors in our estimates.

(ii) ^{169}Tm. The ground state of ^{169}Tm has $I = 1/2$, but an excited state at 8.42 keV with $I = 3/2$ has been widely used in Mössbauer experiments. The results of Uhrich and Barnes[31] on Tm metal correspond to $P_c = +47$ MHz for the $I = 3/2$ level; our scaled-up estimate is $P_c = +39$ MHz.

(iii) ^{175}Lu. Specific heat data[33] on lutetium metal suggests a possible value of $|P_c| = 15$ MHz, while our scaled-up estimate using $Q = 5.68$ barns comes to $+17$ MHz.

A comparison of the values of $P_{\parallel}$ and P_c in Table 8.8 shows that P_c is usually in the region of 5 to 10% of $P_{\parallel}$ except for Gd^{3+} where it is so large as to rival the magnetic hyperfine interaction. In all other cases the magnetic interaction is dominant, and we must therefore multiply P_c by the factor $\frac{1}{2}(3 \cos^2 \theta - 1)$, where θ is the angle which the electronic magnetic moment makes with the c-axis, before adding it to $P_{\parallel}$. For Tb and Dy we thus get contributions of -18 MHz and -11 MHz respectively, since for these metals the electronic moments are normal to the c-axis.

Itoh, Kobayashi, and Sano[22] have found that the measured value of $P = P_{\parallel} + \frac{1}{2}P_c(3 \cos^2 \theta - 1)$ varies appreciably when Tb or Dy are alloyed in their neighbouring elements, the change being as much as $+16$ MHz ($=$ half their $\Delta\nu_Q$) for Tb in Gd. Though there will be some variation in P_c through the difference in lattice constants, the values of P_c in Table 8.8 suggest that their result could more easily be accounted for by a change in θ, which would permit a maximum change from $-\frac{1}{2}P_c$ to $+P_c$ in the effective contribution from the lattice.

8.6. EXPERIMENTAL METHODS

Three principal methods of determining hyperfine interactions in the lanthanide metals have been used, two of them spectroscopic and the third thermal. We shall outline them only very briefly.

8.6.1. Nuclear Magnetic Resonance

For practical purposes it is generally sufficient to use a Hamiltonian

$$\mathcal{H} = a_0 I_z + P\{I_z^2 - \tfrac{1}{3}I(I+1)\} \qquad (8.48)$$

where the z-axis is the direction of the electronic magnetization, $a_0 = A\langle J_z\rangle$ where $\langle J_z\rangle$ is the mean value of J_z attained as T approaches 0 K, and P includes the contributions to the nuclear electric quadrupole interaction both from the $4f$ electrons and from the lattice and conduction electrons. This equation is valid so long as off-diagonal terms such as that involving η in Eqn (8.30) are small compared with the larger diagonal term in Eqn (8.48), which is usually the first (magnetic) term. In nuclear magnetic resonance the strongly allowed transitions are those for which $\Delta I_z = \pm 1$, and for the transition $I_z = m$ to $(m-1)$ we have

$$h\nu = |a_0 + P(2m-1)|. \qquad (8.49)$$

This gives a set of equally spaced lines from which a_0 and P can be determined; the sign of a_0 can be found by observing the direction in which the resonance moves in frequency when an external magnetic field is applied.

In most ferromagnetic metals the oscillatory magnetic field is greatly enhanced by motion of the domain walls which oscillates the electronic moment and hence also the hyperfine field at the applied radio frequency. At the high frequencies ($\sim$ GHz) needed in some lanthanide metals motion of the domain walls is small. Also the line width may be as much as 10 MHz, so that very high r.f. power is required. However, most of this width is due to inhomogeneous broadening, so that spin–echo methods are particularly advantageous and often the only methods which give sufficient sensitivity to detect the resonance.

Values of a_0 and P determined by nuclear magnetic resonance in the lanthanide metals are typically quoted to about 1 MHz.

8.6.2. Recoilless absorption of γ-rays (Mössbauer effect)

A number of lanthanide nuclei have low-lying nuclear levels which decay to the ground state by emission of γ-rays of energy 10 to 100 keV. If the lifetime

14

of the excited state is in the range 10^{-7} to 10^{-9} s, line widths of order 10 to 10^3 MHz are obtained for resonant absorption in solids.

The resonance lines are split by hyperfine interactions, the spectra being complicated because two nuclear levels are involved, with different values of I, a_0, and P, though in some cases these may be zero in one of the levels (e.g., the ground level of an even–even nucleus). The selection rules giving the number of allowed transitions, and their relative intensities, depend on the nature of the γ-ray transition (M1, E2, etc). When the emitter and absorber are in different states of chemical combination there may also be an "isomer shift" which displaces all the hyperfine lines equally. This arises from changes in the total electric charge density within the nucleus of the electrons, this density being sensitive to valency and chemical binding.

The accuracy quoted for hyperfine parameters derived from Mössbauer experiments ranges from a few MHz up to 50 MHz. It is in general thus less accurate than nuclear magnetic resonance, but measurements can be made up to temperatures of hundreds of Kelvins. In lanthanides, electronic relaxation is usually so rapid that the hyperfine parameters measured in such experiments correspond to thermally averaged values. If only states belonging to the ground manifold J need be considered, the magnetic splitting is proportional to $\langle J_z \rangle$ and hence to the average magnetic moment per ion, while that part of the electric quadrupole splitting arising from the $4f$ electrons is proportional to the thermally averaged value of $\langle 3J_z^2 - J(J+1) \rangle$.

8.6.3. Hyperfine specific heat

The hyperfine splitting of the nuclear levels gives rise to an anomaly of the Schottky type in the specific heat of the metal. This anomaly has its maximum below 1 K where other contributions to the specific heat (e.g., from the lattice, conduction electrons, magnons) become vanishingly small and the Schottky anomaly can therefore be measured quite accurately if measurements can be extended to sufficiently low temperatures. At temperatures well above the maximum, the nuclear specific heat C_N can be expanded in a power series

$$C_N/R = c_2 T^{-2} + c_3 T^{-3} + \dots \tag{8.50}$$

where

$$c_2 = \tfrac{1}{3} a_0^2 I(I+1) + \tfrac{1}{45} P^2 I(I+1)(2I+3)(2I-1) \tag{8.51}$$

$$c_3 = -\tfrac{1}{15} a_0^2 P I(I+1)(2I+3)(2I-1). \tag{8.52}$$

Spin–spin interactions between nuclei, such as the Suhl–Nakamura interaction also contribute to c_2, c_3[11] but can generally be neglected. At higher temperatures the nuclear terms in the specific heat must be separated from other contributions by means of their different temperature dependences. Since in general $a_0 \gg P$, it is difficult to determine P accurately, and obviously both c_2 and c_3 must be measured in order to determine both a_0 and P. The nuclear abundance of each isotope with a spin must be known, and if more than one such isotope is present, only the total combined contributions are measured. If the ratios of the nuclear moments are known as well as spins and abundances the contributions from the two isotopes can be extracted from the combined total.

It is difficult to give a succinct assessment of the accuracy attainable in hyperfine specific heat measurements; a useful review, with particular reference to the lanthanide metals, is given by Lounasmaa.[34]

8.6.4. Other methods

We shall not discuss here, but mention with a reference, some nuclear techniques which may be (or have been) applied to the measurement of hyperfine splittings in magnetic metals. They are:

(i) interaction of polarized neutrons with polarized nuclei (Shore, Reynolds, Sailor, and Brunhart[35]);

(ii) angular correlation of successive γ-rays (Cohen;[36] Matthias;[37] Karlsson[38]);

(iii) nuclear orientation (Stone[39]).

In some cases these techniques may be used to detect transitions induced by nuclear magnetic resonance.

8.7. EXPERIMENTAL RESULTS

In general the most complete and accurate data for the hyperfine constants of the rare earth metals are expected to be those derived from N.M.R. (spin echo) experiments, and we shall therefore summarize these first. A brief discussion of the results, ion by ion, including values obtained by other types of experiment, is given later.

Values of the hyperfine parameters a_0 and P, both in frequency units (MHz), derived from N.M.R. experiments are listed in Table 8.11. The measurements mostly refer to a temperature of about 1.5 K, and may be taken

Table 8.11. Values from N.M.R. Measurements of the Hyperfine Parameters a_0, P (both in MHz) of Lanthanide ions in the Pure Metal, and in a Dilute Alloy with Gadolinium

			In pure metal		Dilute alloy in Gd metal	
		I	a_0	P	a_0	P
Nd	143	7/2			834 (3)	2.3 (5)
Nd	145	7/2			519 (3)	1.8 (5)
Sm	147	7/2			624 (3)	4.8 (5)
Sm	149	7/2			516 (3)	2.5 (5)
Tb	159	3/2	3120	337	3068 (3)	351 (2)
Dy	161	5/2	830 (3)	193 (2)	817 (3)	200 (2)
Dy	163	5/2	1162 (3)	204 (2)	1144 (3)	211 (2)
Ho	165	7/2	6467 (2)	53 (2)	6354 (2)	57 (1)
Er	167	7/2	913 (3)	53 (2)	896 (3)	57 (2)
Tm	169	1/2			2223 (3)	—

The figures for holmium include the Lorentz field correction.

References: dilute alloys (*ca.* 10%), Kobayashi, Sano, and Itoh;[82] except for Ho (extrapolated to zero concentration), Wagg, McCausland, and Mackenzie;[83] pure metals, Tb, Sano and Itoh;[84] Dy, Kobayashi, Sano, and Itoh;[85] Ho (extrapolated, ferromagnetic phase), Wagg, McCausland, and Mackenzie;[83] Er, Sano, Teraoka, Shimizu and Itoh.[86]

14§

as the saturation values as far as the electronic magnetization is concerned. Two sets of measurements are given, one for the pure metal, the other for a low concentration (usually ca. 10 %) in ferromagnetic gadolinium metal. In the former case results exist only for some of the heavier metals, with the simple hcp structure, and we may assume that the dilute alloys in Gd metal also have this structure, including the alloys of Nd and Sm.

A comparison of Tables 8.9 and 8.6 shows that the measured values of a_0 lie within a few per cent of those of JA calculated for the free ion (Table 8.6), except for Nd, where the measured values are about 15% smaller. Similarly, the measured values of P are within some 10% of those of $P_\parallel$ (Table 8.8), except for Nd. The signs of a_0, P are not usually determined in the N.M.R. experiments, but in view of the general overall agreement with the free ion values we may expect that the signs are the same as those given for JA in Table 8.6 and for $P_\parallel$ in Table 8.8. This argument is not conclusive when applied to the dilute alloy of Nd in Gd, where the measured value of P is only about 40% that estimated for $P_\parallel$ in Table 8.8. This may be due to crystal field effects (which may also be responsible for the low value of a_0 compared with that of JA), but the value of P_c in Table 8.8 should be multiplied by a factor of between $2\frac{1}{2}$ and 3 for Nd (and Sm) in Gd if we assume the lattice constants for the dilute alloy to be the same as for pure Gd. This would bring it close to the measured value of P for Nd in Gd. However, without knowing the orientation of the electronic magnetic moment on Nd in Gd we do not know the value of the factor $\frac{1}{2}(3\cos^2\theta - 1)$ by which P_c must be multiplied.

We now consider briefly the experimental results for each of the rare earth metals. For the light rare earth metals data are rather scarce, and for the pure metals the situation is complicated by the fact that at low temperatures the stable phase is usually the double hexagonal close packed (dhcp) phase (though others may be produced), while samarium has yet another structure.

Lanthanum

The metal has the dhcp structure, and the lanthanide ions appear to be La^{3+}, with no unclosed shells. The two slightly different nuclear electric quadrupole interactions (see earlier, Section 8.5.2) are attributed to lattice electric field gradients at the "cubic" and "hexagonal" sites.[32]

Cerium

The stable isotopes all have zero nuclear spin, and no hyperfine measurements are available.

Praseodymium

A ferromagnetic fcc phase with a Curie temperature of 8.7 K has been reported.[40] The high-temperature tail of the hyperfine specific heat corresponds to a (magnetic) splitting of only about 22% of the value for JA given in Table 8.8. This is consistent with a small spontaneous moment induced by exchange interaction in the singlet ground state produced by a cubic crystal field splitting of $J = 4$.

In the dhcp phase (the normal stable phase at low temperatures) the situation has been singularly confused. A simple crystal field model[41] predicts singlet ground states on both "cubic" and "hexagonal" sites, but the exchange interaction is close to that needed to produce a spontaneous ordered moment on the "hexagonal" sites. An ordered state with a moment ($\sim 1\beta$) on one type of site only was found to exist below 25 K in a polycrystalline sample.[42] In a single crystal, on the other hand, Johansson et al. observed no spontaneous magnetic ordering down to 4.2 K[43] They conclude that "pure monocrystalline Pr is not antiferromagnetic but that the exchange is sufficiently great that a small modification of the crystal-field splittings, perhaps due to strains, can lead to spontaneous ordering". From measurements of the induced moments in a high applied field they find that the effective molecular field must be highly anisotropic (see also Lebech and Rainford[44]).

The only hyperfine data available come from specific heat measurements on polycrystalline samples. Measurements from 0.4 to 0.035 K by Holmström, Anderson, and Krusius[45] have been extended down to below 0.02 K by Gregers-Hansen, Krusius and Pickett.[46] In zero magnetic field these are consistent with an electronic moment of about 0.6β on half the sites, the remaining sites having no ordered moment. However, application of an external magnetic field noticeably increases the hyperfine specific heat, and at a field of 19 kG it is approximately doubled. Hence this field is sufficient to induce a moment of similar size on the remaining sites. The lower temperatures in the later work were produced by adiabatic demagnetization of the metal, using the enhanced nuclear cooling effect. As pointed out by Gregers-Hansen, Krusius and Pickett, a single crystal of praseodymium with no ordered moment on any site should provide an excellent substance for enhanced nuclear cooling.

Neodymium

An fcc crystal phase with a ferromagnetic peak in the susceptibility at 29 K has been reported by Bucher, Chu et al.[40] No hyperfine data are available.

In the dhcp phase (the normal stable phase at ordinary temperatures), neutron diffraction measurements of Moon, Cable, and Koehler[47] and Lebech and Rainford[44] show that the hexagonal sites order at 19 K and the cubic sites at 7.5 K. At 4.2 K the moment of each site is close to 2β, and is thus noticeably smaller than the theoretically possible value of 3.27β. This is consistent with the presence of rather large crystal field splittings (see, for example, Lounasmaa and Sundström[48] which remove most of the degeneracy of the $2J+1 = 10$ electronic levels, apart from the Kramers degeneracy.

The hyperfine specific heat has been fitted (knowing the ratios of the nuclear moments of the two odd isotopes) by Anderson, Holmström, Krusius, and Pickett[49] with the values (which we quote only for isotope 143)

$$\text{Site 1} \quad a_0 = (-)763(10) \text{ MHz}, \quad P = -10(8) \text{ MHz}$$

$$\text{Site 2} \quad a_0 = (-)523(8) \text{ MHz}, \quad P = -10(15) \text{ MHz}.$$

Using the value of JA in Table 8.6, and neglecting conduction electron effects, the two values of a_0 correspond to electronic moments of 2.52β and 1.73β.

For a dilute alloy of Nd in Gd we should expect the exchange interaction to be much more important relative to the crystal field. The N.M.R. results (Table 8.11) suggest an electronic moment about 0.85 of the theoretical maximum.

Promethium

No hyperfine data for the metal are available.

Samarium

The metal has a crystal structure repeating every nine layers, but below 14 K it is thought to have a simple antiferromagnetic arrangement with the moments parallel to the c-axis.[50] Assuming the known ratios of the hyperfine constants (Table 8.4) for the two odd isotopes 147, 149, the hyperfine specific heat can be fitted with the following values for all sites[49]

$$\text{Isotope 147} \quad a_0 = (-)590(17) \text{ MHz}, \quad P = -4(6) \text{ MHz}$$
$$\text{Isotope 149} \quad a_0 = (-)486(15) \text{ MHz}, \quad P = +1(2) \text{ MHz}.$$

The values of a_0 are very close to those of JA in Table 8.6.

For Sm in Gd metal the values of a_0 (Table 8.11) are a little higher. This is probably due to admixtures of the $J=7/2$ state arising from the exchange interaction (see Ofer, Segal, Nowik, Bauminger, Grodzins, Freeman, and Schieber[51] for a discussion of the Sm^{3+} ion). Conduction electron polarization by the neighbours would give a smaller splitting for Sm in Gd than for pure Sm, if the signs of the contributions are as in Table 8.9.

Europium

The normal crystal structure of the metal is bcc, and it is regarded as consisting of Eu^{2+} ions in the $^8S_{7/2}$ state. The value of JA in Table 8.6 is mostly due to core polarization, but the Mössbauer measurements of Hüfner and Wernick[23] discussed in Section 8.5.1 show that conduction electron contributions are appreciable (see Table 8.9). The nuclear electric quadrupole coupling is expected to be very small (Table 8.8) with no lattice contribution because of the cubic structure.

Gadolinium

Starting with this element, the normal crystal structures of the heavy metals are all hcp. For Gd^{3+}, $^8S_{7/2}$ the magnetic hyperfine constants are considerably smaller than for the Eu^{2+} ions because the two odd isotopes 155, 157 have rather small nuclear moments. From changes in the transmission of polarized neutrons between 0.05 and 0.2 K a hyperfine splitting of $a_0 = 58(6)$ MHz for isotope 157 was deduced by Shore, Reynolds, Sailor, and Grunhart.[35] This is close to the value of JA in Table 8.6, but Hüfner and Wernick[23] suggest, and the results of Zmora, Blau, and Ofer[24] confirm, that the net conduction electron contribution is fortuitously small (see Table 8.9). The variation of the hyperfine field in different intermetallic compounds is shown in Table 8.10. The size of the nuclear electric quadrupole interaction is unknown; it is probably dominated by the lattice+conduction electron contribution (see Table 8.8).

Hyperfine specific heat measurements of Dreyfus, Michel, andThoulouze[52] give values of c_2 (Eqn (8.51)) of 4.3 (millidegrees)^2 for pure Gd, and 0.39 (millidegrees)^2 per mole of Gd in a dilute alloy (5.87%) of Gd in yttrium. However, the values of JA (Table 8.6) and P_c (Table 8.8) lead to $c_2 = (2.14 + 0.52) = 2.66$ (millidegrees)^2, the contributions $(5/4)(JA)^2$ and P_c^2 being additive whatever the relative orientations of the hyperfine magnetic field and the electric field gradient. Although the value of a_0 is presumably (see Section 8.5.1) much less in the dilute alloy, it is difficult to account for the discrepancies. In pure Gd they would require a 30% increase in a_0, or a massive increase in P_c. On the other hand, in the dilute alloy the measured value of c_2 is less than that which we should expect from P_c alone, whereas the lattice parameters of yttrium metal would lead us to expect P_c to be bigger, not smaller. Also, we might expect a contribution from hyperfine splitting at the ^{89}Y ($I = 1/2$) nucleus, induced by the conduction electron polarization due to the Gd^{3+} ions. However, in assessing the importance of these discrepancies, it must be remembered that the hyperfine specific heat for Gd is exceedingly small (especially in the dilute alloy), and correspondingly difficult to measure accurately.

Terbium

The N.M.R. measurements of the hyperfine constants are given in Table 8.11. Compared with the value of JA in Table 8.6, the measured value of a_0 is some 2% lower in the pure metal, and $3\frac{1}{2}$% lower for the dilute alloy with Gd. For the quadrupole coupling constant P, which is the largest of any of the lanthanide metals, the experimental values are 10 to 15% lower than the estimate for $P_\parallel$ in Table 8.8. However, in the pure metal the electronic moments are perpendicular to the c-axis ($\theta = 90°$), and the discrepancy is somewhat reduced if a lattice contribution of $-\frac{1}{2}P_c = -14$ MHz (as estimated from Table 8.8) is included.

Krusius, Anderson, and Holmström[53] have shown that the hyperfine specific heat down to 0.03 K is in excellent agreement with the values of a_0, P for the pure metal given in Table 8.11.

Dysprosium

The N.M.R. measurements of a_0 (Table 8.11) agree closely with the value of JA in Table 8.6. The measurements of P give values a little lower than that of $P_\parallel$ (Table 8.8), but the discrepancy is reduced if $-\frac{1}{2}P_c = -9$ MHz (as estimated) is added to P_c. Mössbauer measurements[54] are less accurate, but for $DyFe_2$ indicate a hyperfine field ($-H_e$) larger by about 0.6 MG than in Dy. This result is comparable with that for $GdFe_2$ (see Table 8.10).

Holmium

For this element the large nuclear magnetic moment and nuclear spin $I = 7/2$ result in the largest N.M.R. frequency (~ 6.5 GHz) and hyperfine specific heat of any of the pure lanthanide metals.

In a detailed study of the variation of the N.M.R. frequency of ^{165}Ho at different concentrations in Gd, Tb, and Dy, Mackenzie, McCausland, and

Wagg[25] have shown that it can be represented by the formula (cf. Figure 8–2)

$$a_0 = 6602 - 73\overline{(\lambda - 1)J} + 1.1\overline{\lambda J} \text{ (MHz)}. \tag{8.53}$$

The last term is the contribution a_L from the Lorentz field, while the second term is proportional to the mean value $\overline{(\lambda - 1)J}$ averaged over the relative concentrations. This corresponds to a hyperfine field of $-0.083\overline{(\lambda - 1)J}$ MG due to polarization of the conduction electrons by neighbouring ions, a figure which lies between the values for ^{159}Tb and ^{163}Dy in the corresponding alloys.[22] Comparison of the first term in (8.53) with JA in Table 8.6 leads to a positive value of $+0.060(\lambda - 1)J$ MG for the hyperfine field due to polarization of the conduction electrons by the parent ion. The experimental uncertainty in (8.53) is about 5 MHz, and the results apply to the ferromagnetic phase of the alloys. In pure holmium, where the electronic moments have a conical configuration in the low-temperature phase, Wagg and McCausland (private communication) find $a_0 = 6450(5)$ MHz, $P = 54.5(1.5)$ MHz. The slightly higher value of a_0 in Table 8.11, which is an extrapolation for the ferromagnetic phase, is partly due to a Lorentz field of about 13 kG in this phase. This adds to the hyperfine field, and by applying an external field it has also been verified that the hyperfine field is in the same direction as the electronic magnetization (cf. Section 8.1).

The hyperfine specific heat results of Krusius, Anderson, and Holmström[53] for Ho metal yield the values $a_0 = (+)6510(60)$ MHz, $P = +80(20)$ MHz.

N.M.R. measurements have been made on ^{165}Ho in mixed intermetallic compounds of the form $(Ho_xGd_{1-x})X_2$ with X = Fe, Co, Al.[55] In the Fe_2 compounds the hyperfine field is about 0.65 MG higher than in the Al_2 compounds (cf. Table 8.10), and about 0.5 MG higher than in pure Ho (cf. the results for $GdFe_2$ in Table 8.10, and for $DyFe_2$ above). Measurements on $(Ho_{0.03}Gd_{0.97})(Fe_{0.01}Co_{0.99})_2$ and on $(Ho_{0.03}Gd_{0.97})(Fe_{0.99}Co_{0.01})_2$ show a well resolved satellite structure, suggesting that the transferred hyperfine interaction from the $3d$ ions is essentially a nearest neighbour effect. There is also a small (positive) transferred hyperfine interaction from the lanthanide neighbours, which for X = Fe or Co amounts to $+0.03\overline{(\lambda - 1)J}$ MG.

Erbium

Erbium has just one stable isotope of mass 167, abundance 22.9% and nuclear spin $I = 7/2$, the remaining isotopes having $I = 0$. The value of $P_\parallel$ in Table 8.8 is estimated from atomic beam measurements on the atom, but an independent value can be found from E.P.R. measurements on Er^{3+} in MgO. In this cubic crystal the lattice contribution is zero, and the quadrupole interaction observed by Beloritzky, Ayant, Descamps, and Merle d'Aubigné[56] corresponds to a value of $P_\parallel = -71(5)$ MHz, in agreement with the value in Table 8.8.

The N.M.R. values of a_0 (Table 8.11) for Er are somewhat smaller than JA (Table 8.6), and the measured values of P somewhat smaller than that of $P_\parallel$ (Table 8.8). If the Er moment is oriented about 30° from the c-axis (as in the pure metal), we should expect a positive value $\sim 0.6P_c$ for the lattice contribution, which might explain the second discrepancy.

Mössbauer measurements on an excited (81 keV) state of ^{166}Er with $I = 2$

give a value of H_{eff} some 15% higher (about 0.9 MG) in $ErFe_2$ than in Er metal at 21 K[57]. A measurement for Er metal is also reported by Kienle.[58]

Thulium

Hyperfine specific heat measurements down to 0.03 K of Holmström, Anderson, and Krusius[45] on the pure metal yield a value of

$$|a_0| = 2230(20) \text{ MHz},$$

very close to the N.M.R. value of 2223(3) MHz for Tm in Gd (Table 8.11). These results are rather smaller than that of -2333 MHz (Table 8.6), based on Jones and Schmidt,[10] or the earlier extrapolated value of -2360 MHz.[11]

An excited level of ^{169}Tm at 8.4 keV, with $I = 3/2$, has been widely employed for Mössbauer experiments, using the transition from this level to the ground state $I = 1/2$. In the pure metal at 5 K a magnetic splitting of $a_0 = -2180$ MHz has been found by Kalvius, Kienle, Eicher, Wiedemann, and Schüler[59] for the ground nuclear state, and $+1690$ MHz for the $I = 3/2$ state. The accuracy (not quoted) is probably not better than 2%. In the excited nuclear state they find $P = -493$ MHz, which agrees well with the value of $P = -513$ MHz which can be derived from the work of Uhrich and Barnes.[31] As discussed in Section 8.5.2, the latter have separated the $4f$ and lattice contributions, giving $P_{\parallel} = -560$ MHz and $P_c = +47$ MHz.

In the cubic Laves compound $TmFe_2$ the value of a_0 is 1.09(2) greater than that in pure Tm[60], corresponding to an increase of about $+0.6$ MG in the effective field. The quadrupole splitting is also about 10% larger in $TmFe_2$, which would agree well with the value of $P_{\parallel}$ quoted above, since $P_c = 0$ for the cubic site of Tm in this compound.

Ytterbium

The normal stable phase of this metal has the fcc structure, and its properties are consistent with the presence of Yb^{2+} ions, with the closed shell $4f^{14}$. The hyperfine interactions should be zero, with no lattice contribution to an electric field gradient. (The constants given in Tables 8.6 to 8.8 are for Yb^{3+}.)

Lutetium

This metal has the simple hcp structure at ordinary temperatures, consisting of Lu^{3+} ions with the configuration $4f^{14}$. The lattice contribution to the electric field gradient has been discussed in Section 8.5.2.

8.8. CONCLUSION

There is no doubt that the hyperfine interactions arising from the $4f$ electrons are very closely the same for the ions in the lanthanide metals as for the isolated ions with the same electronic configuration. For the heavier lanthanide metals Tb to Tm inclusive, where one might expect the saturation moment in the ordered phase at very low temperatures to correspond to a mean value $\langle J_z \rangle = J$, the measured magnetic hyperfine constant a_0 is within 2% of JA, while that of P is within 10% to 20% of $P_{\parallel}$, where JA and $P_{\parallel}$ are the free ion values. For the light lanthanide metals Pr and Nd, splittings due to crystal fields play an important role, and the measured hyperfine constants (in particular, a_0) fall considerably short of the free ion values for $J_z=J$. For samarium, however, a_0 is very close to JA. This is most likely due to the predominance of exchange interaction, though the crystal field term $V_2{}^0$ (but not the fourth degree terms) would lead to a ground state with $J_z = \pm 5/2 = \pm J$. The fact that in the antiferromagnetic state the moments are aligned along the c-axis suggests that $V_2{}^0$ is the more important term in the anisotropy energy.

For Sm in Gd a_0 is larger than JA; this can be explained by admixture of the $J=7/2$ state through exchange interaction, and this is consistent with the value of a_0 being larger than in pure Sm.

Apart from the large hyperfine field of the $4f$ electrons, there are contributions of a few hundred kilogauss from (i) core polarization, (ii) polarization of the conduction electrons. Since the values of JA in Table 8.6 are derived from measurements on paramagnetic ions in salts, a core polarization contribution is already included. If we assume that it is the same in the metal, and directly transferred hyperfine interactions from neighbours are negligible, the only difference from a salt arises from the conduction electrons.

Polarization of the conduction electrons arises through exchange interaction with (i) the local moment on the parent ion, and (ii) the moments on the neighbouring ions. The effective fields at the nucleus from these two sources appear to be of opposite sign. From measurements on Eu, Gd, and Ho (Sections 8.5.1 and 8.7), the contribution to the effective field (in megagauss) from (i) is of order $+0.06(\lambda-1)J$, where λ, J apply to the parent ion, while that from (ii) is (apart from Eu) of order $-0.08\overline{(\lambda-1)J}$, where $\overline{(\lambda-1)J}$ is an average taken over a large number (~ 50) of neighbours. Obviously only the second of these changes on alloying; its negative sign leads to the value of a_0 being normally a little smaller for a dilute alloy in Gd (which gives the largest negative contribution) than in the pure metal, as is observed in the heavier metals (cf. Table 8.11).

The fact that the conduction electron hyperfine fields due to polarization by the parent ion and by neighbouring ions appear to be of opposite sign can perhaps be understood from the work of Watson, Freeman, and Koide.[61] The conventional exchange interaction leads to a positive polarization within the ion, but, through interband mixing, they find that the polarization is negative at a distance from the ion. The hyperfine field of $-0.08\overline{(\lambda-1)J}$ MG given above seems roughly to fit the results for Tb and Dy in Gd, as well as the alloys of Ho and of Gd discussed in Section 8.5.1. However, the value for Eu (Table 8.9), with its lower density of conduction electrons, is only about half as great. We may also expect different values for intermetallic compounds (cf. Table 8.10). For a number of cubic Laves compounds LnFe$_2$, whose Curie temperatures are in the region of 600 K, the effective fields at the Ln nucleus are higher by some $+0.5$ to $+0.9$ MG. This may be due to polarization of the conduction electrons by the $3d$ ions, or to overlap and covalency effects (see the discussion on Ho in Section 8.7), or to a mixture of the two.

It is difficult to know what accuracy to assign to an individual value of $P_\parallel$ in Table 8.8 without detailed and reliable calculations of the effects of intermediate coupling and relativistic contributions. For the Dy atom

$(4f^{10}6s^2)$, Childs[62] has obtained an excellent fit to his precise atomic beam data for both the $^5I_8$ and $^5I_7$ states after allowing for such effects. These result in a value of B/Q some 6.4% smaller than that given by Eqn (8.35) for pure LS-coupling. Unfortunately we do not have similar calculations for the Dy^{3+} ion $(4f^9)$, and Childs[63] has found it difficult to get a realistic fit to his data for the $4f^96s^2$, $^6H(J=15/2, 13/2, 11/2)$ states of the Tb atom, probably from lack of accurate intermediate coupling wave functions. The value of $P_\parallel$ for Tb^{3+}, $4f^8$, $^7F_6$ given in Table 8.8 is based on his measured value of

$$B = +1449.33(4) \text{ MHz}$$

for the $J=15/2$ state of the atom without allowance for intermediate coupling and relativistic effects; values of $P_\parallel = +375$ and $+404$ MHz would be obtained for Tb^{3+} by the same treatment from the two different values of

$$b_{4f} = e^2 Q \langle r^{-3} \rangle_{4f}$$

given in the last line of Table VIII of reference 63. Altogether it seems unlikely that the errors in $P_\parallel$ in Table 8.8 exceed about 5%; if the corrections (intermediate coupling, relativistic effects) are nearly the same in atom and ion, they may be less.

In a number of the heavy lanthanide metals the value of P is numerically smaller than the value of $P_\parallel$ estimated to arise from the $4f$ electrons. Since all the values of Q involved are positive, we may for simplicity discuss the signs of P rather than of (P/Q), which is proportional to the electric field gradient Then the sign of $P_\parallel$, due to the $4f$ electrons, is the same as that of $-\langle J\|\alpha\|J\rangle$ (Eqns 8.35, 8.36), which is positive for Tb^{3+}, Dy^{3+}, Ho^{3+} and negative for Er^{3+}, Tm^{3+}, Yb^{3+}. However, the sign of $\langle J\|\alpha\|J\rangle$ also determines the sign of the crystal field anisotropy arising from V_2^0, and if this is the dominant effect, the electronic moments in the ordered state would be aligned normal to the c-axis for the first three, and parallel for the second three. On this basis we would expect $P = P_\parallel - \frac{1}{2}P_c$ for the first three (for which $P_\parallel$ is positive), and $P = P_\parallel + P_c$ for the second three (for which $P_\parallel$ is negative). Since P_c is positive it follows that the lattice contribution is always such as to make P numerically smaller than $P_\parallel$. The experimental results follow this pattern, though the intermediate orientation of the electron moments in Ho and Er reduces the effect of P_c.

In the cubic Laves compounds such as $LnFe_2$ the site symmetry at the Ln site is cubic, so that we expect $P_c = 0$ (apart from any distortion due to magnetostriction) and $P = P_\parallel$. This would explain why the observed values of P in such compounds are greater than in the corresponding pure metals. For example, Mackenzie et al.[55] find $P = 60$ to 62 MHz in some intermetallic HoX_2 compounds, in good agreement with the estimated value of $P_\parallel = 62.7$ MHz in Table 8.8, while rather smaller values of 53 to 55 MHz are observed in the hexagonal metal alloys. A direct measurement of the lattice contribution (modified by the conduction electrons) is afforded by the experiments on ^{169m}Tm $(I=3/2)$ in Tm metal, discussed in Sections 8.5.2 and 8.7. The results for Gd metal are rather puzzling, since here we expect $P_\parallel$ to be very small, leaving the lattice contribution P_c as the major term, comparable with the magnetic hyperfine splitting $a_0 \sim 50$ MHz. Both the Mössbauer measurements of Fink[64] and the N.M.R. spectra of Dintelmann, Dormann and Oppelt[65] have been interpreted, however, in terms of a value of P for the isotopes 155, 157 of less than 6 MHz.

Finally we remark that the few existing measurements of P_c (see Section 8.5.2) are some 2 to 3 times the calculated lattice values, the enhancement being attributed to the effect of a non-uniform conduction electron density.[66]

8.9. ACKNOWLEDGMENTS

The author is indebted to Drs. R. J. Elliott, M. A. H. McCausland, and J. M. Machado da Silva for many helpful suggestions and corrections; and to Drs. W. J. Childs, J. Itoh, M. Krusius, and M. A. H. McCausland for making available data in advance of publication.

REFERENCES

1. Abragam, A. and Bleaney, B., in Electron Paramagnetic Resonance of Transition Ions. Clarendon Press, Oxford (1970).
2. Wybourne, B. G., *J. Chem. Phys.* **37**, 1807 (1962).
3. Judd, B. R. and Lindgren, I., *Phys. Rev.* **122**, 1802 (1961).
4. Conway, J. G. and Wybourne, B. G., *Phys. Rev.* **130**, 2325 (1963).
5. Sandars, P. G. H. and Beck, J., *Proc. Roy. Soc.* **A289**, 97 (1965).
6. Evans, L., Sandars, P. G. H., and Woodgate, G. K., *Proc. Roy. Soc.* **A289**, 114 (1965).
7. Woodgate, G. K., *Proc. Roy. Soc.* **A293**, 117 (1966).
8. Bleaney, B., in Proc. Third Quantum Electronics Conf. Dunod, Paris (1964).
9. Kedzie, R. W., Abraham, M., and Jeffries, C. D., *Phys. Rev.* **108**, 54 (1957).
10. Jones, E. D. and Schmidt, V. H., *J. Appl. Phys.* **40**, 1406 (1969).
11. Bleaney, B., *J. appl. Phys.* **34**, 1024 (1963).
12. Elliott, R. J. and Stevens, K. W. H., *Proc. Roy. Soc.* **A218**, 553 (1953).
13. Sternheimer, R. M., *Phys. Rev.* **146**, 140 (1966).
14. Lindgren, I., *Nucl. Phys.* **32**, 151 (1962).
15. Baker, J. M. and Williams, F. I. B., *Proc. Roy. Soc.* **A267**, 283 (1962).
16. Baker, J. M., Copland, G. M., and Wanklyn, B. M., *J. Phys. C (Proc. Phys. Soc.)* **2**, 862 (1969).
17. Baker, J. M., Blake, W. B. J., and Copland, G. M., *Proc. Roy. Soc.* **A309**, 119 (1969).
18. Watson, R. E. and Freeman, A. J., in Hyperfine Interactions. Academic Press, New York and London (1967), pp. 53–94.
19. Bleaney, B., *J. Phys. Soc. Japan* **17**, Supplement B-1, 435 (1962).
20. Inoue, M., *Phys. Rev. Letters* **11**, 196 (1963).
21. Birgeneau, R. J., *Phys. Rev. Letters* **19**, 160 (1967).
22. Itoh, J., Kobayashi, S., and Sano, N., *J. Appl. Phys.* **39**, 1325 (1968).
23. Hüfner, S. and Wernick, J. H., *Phys. Rev.* **173**, 448 (1968).
24. Zmora, H., Blau, M., and Ofer, S., *Phys. Letters* **28A**, 668 (1969).
25. Mackenzie, I. S., McCausland, M. A. H., and Wagg, A. R., to be published in *J. Phys. F* (1972).
26. Das, T. P. and Pomerantz, M., *Phys. Rev.* **123**, 2070 (1961).
27. Spedding, F. H., Beaudry, B. J., Croat, J. J., and Palmer P. E. in Les Éléments des Terres Rare, Coll. Inter. du C.N.R.S. N° 180, **1**, 25 (1970).
28. Banister, J. R., Levgold, S., and Spedding, F. H., *Phys. Rev.* **94**, 1140 (1954).
29. Darnell, F. J. and Moore, E. P., *J. Appl. Phys.* **34**, 1337 (1963).
30. Narath, A., *Phys. Rev.* **179**, 359 (1969).
31. Uhrich, D. L. and Barnes, R. G., *Phys. Rev.* **164**, 428 (1967).
32. Poteet, W. M., Tipsword, R. F., and Williams, C. D., *Phys. Rev.* **B1**, 1265 (1970).
33. Lounasmaa, O. V., *Phys. Rev.* **133**, A219–224 (1964).
34. Lounasmaa, O. V., in Hyperfine Interactions. Academic Press, New York and London (1970), pp. 467–496.

35. Shore, F. J., Reynolds, C. A., Sailor, V. L., and Grunhart, G., *Phys. Rev.* **138**, B1361 (1965).
36. Cohen, S. G., in Hyperfine Interactions. Academic Press, New York and London (1967), pp. 553–593.
37. Matthias, E., in Hyperfine Interactions. Academic Press, New York and London (1967), pp. 595–625.
38. Karlsson, E., in Hyperfine Interactions. Academic Press, New York and London (1967), pp. 627–636.
39. Stone, N. J., in Hyperfine Interactions. Academic Press, New York and London (1967), pp. 659–672.
40. Bucher, E., Chu, C. W., Maita, J. P., Andres, K., Cooper, A. S., Buehler, E., and Nassau, K., *Phys. Rev. Letters* **22**, 1260 (1969).
41. Bleaney, B., *Proc. Roy. Soc.* **A276**, 39 (1963).
42. Cable, J. W., Moon, R. M., Koehler, W. C., and Wollan, E. O., *Phys. Rev. Letters* **12**, 553 (1964).
43. Johansson, T., Lebech, B., Nielsen, M., Bjerrum Møller, H., and Mackintosh, A. R., *Phys. Rev. Letters* **25**, 524 (1970).
44. Lebech, B. and Rainford, B. D., in Proc. Int. Conf. Magnetism. Grenoble (1970).
45. Holmström, B., Andersen, A. C., and Krusius, M., *Phys. Rev.* **188**, 888 (1969).
46. Gregers-Hansen, P. E., Krusius, M., and Pickett, G. R., in Proc. 12th Int. Conf. on Low Temperature Physics. Kyoto (1970).
47. Moon, R. M., Cable, J. W., and Koehler, W., *J. Appl. Phys.* **35**, 1041 (1964).
48. Lounasmaa, O. V. and Sundström, L. J., *Phys. Rev.* **158**, 591 (1967).
49. Anderson, A. C., Holmström, B., Krusius, M., and Pickett, G. R., *Phys. Rev.* **183**, 546 (1969).
50. Schieber, M., Foner, S., Doclo, R., and McNiff, E. J., *J. Appl. Phys.* **39**, 885 (1968).
51. Ofer, S., Segal, E., Nowik, I., Bauminger, E. R., Grodzins, L., Freeman, A. J., and Schieber, M., *Phys. Rev.* **137**, A627 (1965).
52. Dreyfus, B., Michel, J. C., and Thoulouze, D., *Phys. Letters* **24A**, 457 (1967).
53. Krusius, M., Andersen, A. C., and Holmström, B., *Phys. Rev.* **177**, 910 (1969).
54. Ofer, S., Rakavy, M., Segal, E., and Khurgin, B., *Phys. Rev.* **138**, A241 (1965).
55. Mackenzie, I. S., McCausland, M. A. H., Wagg, A. R., Giumaraes, A. P., Holden, E., and Bailey, S., in Proc. XVI Colloque Ampere. Bucharest (1970).
56. Belorizky, E., Ayant, Y., Descamps, D., and Merle D'Aubigné, Y., *J. de Phys.* **27**, 313 (1966).
57. Cohen, R. L. and Wernick, J. H., *Phys. Rev.* **134**, B503 (1964).
58. Kienle, P., *Rev. Mod. Phys.* **36**, 372 (1964).
59. Kalvius, M., Kienle, P., Eicher, H., Wiedemann, W., and Schüler, C., *Z. Phys.* **172**, 231 (1963).
60. Cohen, R. L., *Phys. Rev.* **134**, A94 (1964).
61. Watson, R. E., Freeman, A. J., and Koide, S., *Phys. Rev.* **186**, 625 (1969).
62. Childs, W. J., *Phys. Rev.* **2**, A1692 (1970).
63. Childs, W. J., *Phys. Rev.* **2**, A316 (1970).
64. Fink, J., *Z. Phys.* **207**, 225 (1967).
65. Dintelmann, F., Dormann, E., and Oppelt, A., *Solid State Comm.* **8**, 1257 (1970).
66. Das, K. C. and Ray, D. K., *Solid State Comm.* **8**, 2025 (1970).
67. Judd, B. R., *Proc. Phys. Soc.* **82**, 874 (1963).
68. Wybourne, B. G., in Spectroscopic Properties of Rare Earths. Wiley, New York, London, and Sydney (1965).
69. Smith, K. F. and Unsworth, P. J., *Proc. Phys. Soc.* **86**, 1249 (1965).
70. Childs, W. J., *Phys. Rev.* **3**, A25 (1971).
71. Reader, J., *Phys. Rev.* **141**, 1123 (1966).
72. Unsworth, P. J., *J. Phys. B* **2**, 122 (1969).
73. Baker, J. M., Chadwick, J. R., Garton, G., and Hurrell, J. P., *Proc. Roy. Soc.* **A286**, 352 (1965).
74. Giglberger, D. and Penselin, S., *Z. Phys.* **199**, 244 (1967).
75. Olschewshi, L. and Otten, E. W., *Z. Phys.* **200**, 224 (1967).
76. Spalding, I. J., *Proc. Phys. Soc.* **81**, 156 (1963).
77. Halford, D., *Phys. Rev.* **127**, 1940 (1962).

78. Erickson, L. E., *Phys. Rev.* **143**, 295 (1966).
79. Sandars, P. G. H. and Woodgate, G. K., *Proc. Roy. Soc.* **A257**, 269 (1960).
80. Gegenwarth, R. E., Budnick, J. I., Skalski, S., and Wernick, J. H., *Phys. Rev. Letters* **18**, 9 (1967).
81. Dintelmann, F., Dormann, E., and Buschow, K. H. J., *Solid State Comm.* **8**, 1911 (1970).
82. Kobayashi, S., Sano, N., and Itoh, J., *J. Phys. Soc. Japan* **23**, 474 (1967).
83. Wagg, A. R., McCausland, M. A. H., and Mackenzie, I. S., in Proc. XVI Colloque Ampere. Bucharest (1970).
84. Sano, N. and Itoh, J., to be published in *J. Phys. Soc. Japan* (1972).
85. Kobayashi, S., Sano, N., and Itoh, J., *J. Phys. Soc. Japan* **21**, 1456 (1966); *Progr. Theor. Phys. Suppl.* **46**, 84 (1970).
86. Sano, N., Teraoka, M., Shimizu, K., and Itoh, J., to be published in *J. Phys. Soc. Japan* (1972).

Index